Inverse and Ill-Posed Problems Series 55

Managing Editor

Sergey I. Kabanikhin, Novosibirsk, Russia/Almaty, Kazakhstan

Sergey I. Kabanikhin

Inverse and Ill-Posed Problems

Theory and Applications

De Gruyter

Mathematics Subject Classification 2010: 35-00, 35Qxx, 45Axx, 45Bxx, 45Cxx, 45Dxx, 45Exx, 45Gxx, 45Kxx, 45Qxx, 46Txx, 65Fxx, 65Nxx, 65Pxx, 65Qxx, 65Rxx.

ISBN 978-3-11-022400-9
e-ISBN 978-3-11-022401-6
ISSN 1381-4524

Library of Congress Cataloging-in-Publication Data

Kabanikhin, S. I.
Inverse and ill-posed problems : theory and applications / by Sergey I. Kabanikhin.
p. cm. – (Inverse and ill-posed problems series ; 55)
Includes bibliographical references and index.
ISBN 978-3-11-022400-9 (alk. paper)
1. Inverse problems (Differential equations) 2. Boundary value problems – Improperly posed problems. I. Title.
QA378.5.K33 2011
515'.357–dc23
2011036569

Bibliographic information published by the Deutsche Nationalbibliothek

The Deutsche Nationalbibliothek lists this publication in the Deutsche Nationalbibliografie; detailed bibliographic data are available in the Internet at http://dnb.d-nb.de.

Typesetting: Da-TeX Gerd Blumenstein, Leipzig, www.da-tex.de
Printing and binding: Hubert & Co. GmbH & Co. KG, Göttingen
∞ Printed on acid-free paper

Printed in Germany

www.degruyter.com

Preface

First publications on inverse and ill-posed problems date back to the first half of the 20th century. Their subjects were related to physics (inverse problems of quantum scattering theory), geophysics (inverse problems of electrical prospecting, seismology, and potential theory), astronomy, and other areas of natural sciences. Since the advent of powerful computers, the area of application for the theory of inverse and ill-posed problems has extended to almost all fields of science that use mathematical methods. In direct problems of mathematical physics, researchers try to find exact or approximate functions that describe various physical phenomena such as the propagation of sound, heat, seismic waves, electromagnetic waves, etc. In these problems, the media properties (expressed by the equation coefficients) and the initial state of the process under study (in the nonstationary case) or its properties on the boundary (in the case of a bounded domain and/or in the stationary case) are assumed to be known. However, it is precisely the media properties that are often unknown. This leads to inverse problems, in which it is required to determine the equation coefficients from the information about the solution of the direct problem. Most of these problems are ill-posed (unstable with respect to measurement errors). At the same time, the unknown equation coefficients usually represent important media properties such as density, electrical conductivity, heat conductivity, etc. Solving inverse problems can also help to determine the location, shape, and structure of intrusions, defects, sources (of heat, waves, potential difference, pollution), and so on. Given such a wide variety of applications, it is no surprise that the theory of inverse and ill-posed problems has become one of the most rapidly developing areas of modern science. Today it is almost impossible to estimate the total number of scientific publications that directly or indirectly deal with inverse and ill-posed problems. However, since the theory is relatively young, there is a shortage of textbooks on the subject. This is understandable, since many terms are still not well-established, many important results are still being discussed and attempts are being made to improve them. New approaches, concepts, and theorems are constantly emerging.

The main purpose and characteristic feature of this monograph is the accessibility of presentation and an attempt to cover the rapidly developing areas of the theory of inverse and ill-posed problems as completely as possible. It is motivated by two considerations. On one hand, there are a lot of books providing complete coverage of specific branches of the theory of inverse and ill-posed problems. On the other hand, there is no book providing a description of all the subjects listed in the table of contents of this book under a single cover. In this situation, a book that summarizes previous results and essential characteristics of the theory of inverse and ill-posed problems and its numerical methods using simple and not so simple examples will be useful for senior undergraduate students, graduate and PhD students, recent graduates getting practical experience, engineers and specialists in various fields dealing with mathematical methods who study inverse and ill-posed problems.

This monograph is based on lectures and special courses taught by the author for over 25 years at the Department of Mathematics and Mechanics and the Department and Geology and Geophysics of Novosibirsk State University, universities in Germany (The University of Karlsruhe), Italy (The University of Milan and The University of Perugia), Kazakhstan (Kazakh-British Technical University, Almaty), China (Hangzhou University), Sweden (Royal Institute of Technology, Stockholm), Japan (The University of Tokyo), and other countries.

The sources used for this book include well-known books and papers by Russian and foreign authors (see the main bibliography and the supplementary references). The content of specific chapters and the book as a whole was discussed on many occasions with academicians M. M. Lavrentiev and S. K. Godunov, M. I. Epov, B. G. Mikhailenko, corresponding members of The Russian Academy of Sciences V. G. Romanov and V. V. Vasin, professors A. G. Yagola, V. S. Belonosov, M. Yu. Kokurin, and V. A. Yurko, with my coauthors Drs. G. B. Bakanov, M. A. Bektemesov, K. T. Iskakov, A. L. Karchevsky, K. S. Abdiev, A. S. Alekseenko, S. V. Martakov, A. T. Nurseitova, D. B. Nurseitov, M. A. Shishlenin, B. S. Boboev, A. V. Penenko and D. V. Nechaev. I am grateful to all of them for useful discussions and for their kind permission to use the results of our discussions and joint publications in this book. I also gratefully acknowledge constant support from all my former and present colleagues in the department headed by Mikhail Mikhailovich Lavrentiev at the Sobolev Institute of Mathematics.

The majority of the results included in this monograph were obtained by the founders and representatives of research groups from Moscow, Novosibirsk, Yekaterinburg (Sverdlovsk), and St. Petersburg (Leningrad). The book also includes the results of joint publications and discussions with my coauthors J. Gottlieb (Karlsruhe), M. Grasselli (Milan), A. Hasanoglu (Izmir), S. He (Hangzhou), B. Hofmann (Chemnitz), M. Klibanov (Charlotte, NC), R. Kowar (Linz), A. Lorenzi (Milan), O. Scherzer (Innsbruck), S. Ström (Stockholm), and M. Yamamoto (Tokyo), whom I wish to thank for their contribution.

D. V. Nechaev provided valuable help in the design and technical editing of the book.

The author is very grateful to I. G. Kabanikhina, who kindly proofread and analyzed the manuscript for clarity and accessibility of presentation, persistently and methodically searching for parts that appeared to be vague or difficult for understanding, correcting and sometimes modifying them.

It is hard to overestimate the contribution of the reviewers, V. V. Vasin, A. G. Yagola, and V. S. Belonosov, whose criticism and advice helped to significantly improve the content and quality of this monograph.

The work on this monograph was supported by the Russian Foundation for Fundamental Research (projects 05-01-00171 and 05-01-00559), a grant of the President of the Russian Federation (project NSh-1172.2003.1), integration projects of the Siberian Branch of the Russian Academy of Sciences (Nos. 16, 26, and 48), the "Universities of Russia" program managed by the Federal Education Agency of the Russian Federation (Rosobazovanie) (project UR.04.01.200), the International Foundation for Inverse Problems (Almaty, Kazakhstan), the Non-profit Foundation for Inverse Problems in Science (Novosibirsk, Russia), the president of Kazakh-British Technical University (Almaty) and the staff of its Department of Mathematics and Cybernetics. The author sincerely appreciates their support.

The author thanks I. G. Kabanikhina, D. D. Dotsenko and O. A. Yudina for their very important help with translation of this book.

Finally, he wishes to express his most profound gratitude to his Mentor, Vladimir Gavrilovich Romanov, who helped him discover the wonderful and fascinating world of inverse and ill-posed problems. This book would not have been conceived and written without his influence.

Novosibirsk, July 2011 Sergey I. Kabanikhin

Denotations

$D(A)$ — domain of definition of operator A, $D(A) \subseteq Q$

$R(A)$ — range of values of operator A, $R(A) \subseteq F$

$A(M) = \{f \in F : \exists q \in M : Aq = f\}$
— image of set $M \subseteq Q$, in particular, $A(Q) = R(A)$

I — unit operator (matrix), $\forall q \in Q \;\; Iq = q$

$P \cap Q$ — set of all x such that $x \in P$ and $x \in Q$

$P \cup Q$ — set of all x, such that $x \in P$ or $x \in Q$

A^* — adjoint operator of A

Q^* — dual space of Q

A^{-1} — inverse operator of A, that is, $AA^{-1} = I$

$\mathrm{Ker}\, A = N(A) = \{q \in D_A : Aq = 0\}$
— kernel of operator A

$\mathcal{L}(Q, F)$ — space of all linear continuous operators mapping normed space Q into normed space F

$N^{\perp}$ — orthogonal complement of subspace $N \subset Q$, $N^{\perp} = \{q \in Q : \forall p \in N \; \langle q, p \rangle = 0\}$

$\langle q, p \rangle$ — inner product in Hilbert space or value of functional $q \in Q^*$ of element $p \in Q$: $\langle q, p \rangle = q(p)$

$\mathbb{N} = \{1, 2, 3, \dots\}$
— set of all natural numbers

$\mathbb{Z} = \{\dots, -1, 0, 1, \dots\}$
— set of all integers

$\mathbb{Q}$ — set of all rational numbers

$\mathbb{R}$ — set of all real numbers

$\mathbb{C}$ — set of all complex numbers

$\mathbb{R}_+ = \{x \in \mathbb{R} : x > 0\}$

$\mathbb{R}^n = \{(x_1, x_2, \ldots, x_n) : x_i \in \mathbb{R}, i = \overline{1,n}\}$,
$i = \overline{m,n}$ means that i takes all integers from m to n

$\mathbb{E}^n$ — Euclidean space with inner product $\langle a, b\rangle = \sum_{i=1}^{n} a_i \bar{b}_i$, $\bar{b}_i$ — complex conjugate of b_i

$\mathcal{H}$ — Hilbert space with inner product $\langle \cdot, \cdot \rangle$

$\mathcal{B}$ — Banach space with norm $\|\cdot\|$

$\mathcal{N}$ — normed space with norm $\|\cdot\|$

$\mathcal{M}$ — metric space with metric $\rho(\cdot, \cdot)$

$\mathcal{L}$ — linear space with operations of vector addition and scalar multiplication

$A\big|_M$ — restriction of operator A on $M \subset D(A)$

q_e — exact solution of problem $Aq = f$

q_n^0 — normal (with respect to q_0) solution of problem $Aq = f$

q_n — normal (with respect to the zero element) solution of problem $Aq = f$

q_p — pseudo-solution

q_p^0 — normal (with respect to q^0) pseudo-solution

q_q^M — quasi-solution on set $M \subset Q$

q_np — normal pseudo-solution (that is, pseudo-solution that is normal with respect to the zero element)

$C(G)$ — space of functions continuous on G

$C^m(G)$ — space of functions continuous with their derivatives up to order $m \in \mathbb{N}$ on set G, $C^0(G) = C(G)$

$L_p(G)$ — space of functions integrable with degree p on set G

$W_k^p(G)$ — Sobolev space, $k, p \in \mathbb{N}$, $H^p(G) = W_2^p(G)$, $H^0(G) = L_2(G)$

$\delta(x)$ — Dirac delta-function

$\theta(x)$ — Heaviside theta function, $\theta(x) = \begin{cases} 1, & x \geq 0, \\ 0, & x < 0 \end{cases}$

$\inf E$ — infimum of set $E \subset M$, $c = \inf E \in M$ if, first, $\forall x \in E \ \ x \geq c$, and, second, $\forall \varepsilon > 0 \ \exists y \in E$: $c < y < c + \varepsilon$ ($\sup E$, supremum of set E, is defined by analogy)

$J : Q \to \mathbb{R}$ — functional from Q into R

J' — gradient of functional

$q_* = \arg\min_P J(q)$
— element on which functional $J(q)$ reaches its minimal value for all q satisfying condition P

$B(q, \varepsilon)$ — open ball of radius ε with center at point q,
$B(q, \varepsilon) = \{p \in Q : \rho(p, q) < \varepsilon\}$

$S(q, \varepsilon)$ — sphere of radius ε with center at point q,
$S(q, \varepsilon) = \{p \in Q : \rho(p, q) = \varepsilon\}$

$m \times n$ matrix
— matrix with m rows and n columns

$\mathrm{diag}(a_1, a_2, \dots, a_p, m, n)$
— diagonal $m \times n$ matrix, $p = \min\{m, n\}$, whose elements a_{ij} satisfy condition $a_{ij} = \begin{cases} a_i, & i=j, \\ 0, & i \neq j \end{cases}$

$\{a_n\}$ — sequence $a_1, a_2, a_3, \dots$

$\mathrm{span}\{a_1, a_2, \dots, a_n, \dots\}$
— linear span of elements $a_1, a_2, \dots,$
$\mathrm{span}\{a_n\} = \{\sum_n \beta_n a_n, \beta_n \in \mathbb{R}\}$ (summation may be done also over a finite set of elements)

$\det A$ — determinant of $n \times n$ matrix A is defined by induction on n:
$\det A = \sum_{s=1}^{n} (-1)^{k+s} a_{ks} A_{ks}$, where A_{ks} is determinant of $(n-1) \times (n-1)$ matrix obtained from A by deleting kth row and sth column

$$\Delta v = \Delta_x v = \sum_{i=1}^{n} \frac{\partial^2 v}{\partial x_i^2}$$
— Laplace operator in $\mathbb{R}^n$

$$\Box v = \frac{\partial^2 v}{\partial t^2} - \Delta_x v$$
— D'Alembert operator

$$\nabla g(x) = \nabla_x g(x) = \mathrm{grad}_x\, g(x) = \Big(\frac{\partial g(x)}{\partial x_1}, \frac{\partial g(x)}{\partial x_2}, \dots, \frac{\partial g(x)}{\partial x_n}\Big)$$
— gradient of function $g(x)$, $x \in \mathbb{R}^n$

$\operatorname{div} u = \operatorname{div}_x u = \dfrac{\partial u_1}{\partial x_1} + \dfrac{\partial u_2}{\partial x_2} + \dfrac{\partial u_3}{\partial x_3}$
— divergence of vector-function $u(x) = (u_1(x), u_2(x), u_3(x))$, $x \in \mathbb{R}^3$

$\operatorname{rot} u = \operatorname{rot}_x u = \left(\dfrac{\partial u_3}{\partial x_2} - \dfrac{\partial u_2}{\partial x_3}, \dfrac{\partial u_1}{\partial x_3} - \dfrac{\partial u_3}{\partial x_1}, \dfrac{\partial u_2}{\partial x_1} - \dfrac{\partial u_1}{\partial x_2}\right)$
— rotor of vector-function $u(x) = (u_1(x), u_2(x), u_3(x))$, $x \in \mathbb{R}^3$

Contents

Chapter 1
Basic concepts and examples

Some think the Earth is spherical, others that it is flat and drum-shaped. For evidence they bring the fact that, as the sun rises and sets, the part concealed by the Earth shows a straight and not a curved edge, whereas if the Earth were spherical the line of section would have to be circular ...

... but in eclipses of the moon the outline is always curved: and, since it is the interposition of the Earth that makes the eclipse, the form of this line will be caused by the form of the Earth's surface, which is therefore spherical.

Aristotle. "On the Heavens". Written 350 B.C.E.

In this chapter we give the main definitions and consider various examples of inverse and ill-posed problems. These problems are found everywhere in mathematics (see the right column of Tables 1 and 2) and in virtually any area of knowledge where mathematical methods are used. Scientists have been dealing with them since ancient times (see the epigraph), but it wasn't until the middle of the 20th century that these problems began to be studied systematically and gradually earned the right to be considered a promising area of modern science.

1.1 On the definition of inverse and ill-posed problems

In our everyday life we are constantly dealing with *inverse and ill-posed problems* and, given good mental and physical health, we are usually quick and effective in solving them. For example, consider our visual perception. It is known that our eyes are able to perceive visual information from only a limited number of points in the world around us at any given moment. Then why do we have an impression that we are able to see everything around? The reason is that our brain, like a personal computer, completes the perceived image by interpolating and extrapolating the data received from the identified points. Clearly, the true image of a scene (generally, a three-dimensional color scene) can be adequately reconstructed from several points only if the image is familiar to us, i.e., if we previously saw and sometimes even touched most of the objects in it. Thus, although the problem of reconstructing the image of an object and its surroundings from several points is *ill-posed* (i.e., there is *no uniqueness or stability of solutions*), our brain is capable of solving it rather quickly. This is due to the brain's ability to use its extensive previous experience (*a priori information*). A quick glance at a person is enough to determine if he or she is a child or a senior, but it is usually not enough to determine the person's age with an error of at most five years.

From the example given in the epigraph, it is clear that considering only the shape of the Earth's shadow on the surface of the moon is not sufficient for the unique solution of the inverse problem of projective geometry (reconstructing the shape of the Earth). Aristotle wrote that some think the Earth is drum-shaped based on the fact that the horizon line at sunset is straight. And he provides two more observations as evidence that the Earth is spherical (uses additional information): objects fall vertically (towards the center of gravity) at any point of the Earth's surface, and the celestial map changes as the observer moves on the Earth's surface.

Attempting to understand a substantially complex phenomenon and solve a problem such that the probability of error is high, we usually arrive at an *unstable* (*ill-posed*) problem. Ill-posed problems are ubiquitous in our daily lives. Indeed, everyone realizes how easy it is to make a mistake when reconstructing the events of the past from a number of facts of the present (for example, reconstructing a crime scene based on the existing direct and indirect evidence, determining the cause of a disease based on the results of a medical examination, and so on). The same is true for tasks that involve predicting the future (predicting a natural disaster or simply producing a one week weather forecast) or "reaching into" inaccessible zones to explore their structure and internal processes (subsurface exploration in geophysics or examining a patient's brain using NMR tomography).

Almost every attempt to expand the boundaries of visual, aural, and other types of perception leads to ill-posed problems.

What are inverse and ill-posed problems? While there is no universal formal definition for inverse problems, an "ill-posed problem" is a problem that either has *no solutions* in the desired class, or has *many* (at least two) solutions, or the solution procedure is *unstable* (arbitrarily small errors in the measurement data may lead to indefinitely large errors in the solutions). Most difficulties in solving ill-posed problems are caused by instability. Therefore, the term "ill-posed problems" is often used for unstable problems.

To define various classes of *inverse problems*, we should first define a *direct problem*. Indeed, something "inverse" must be the opposite of something "direct". For example, consider problems of mathematical physics.

In mathematical physics, a *direct problem* is usually a problem of modeling some physical fields, processes, or phenomena (electromagnetic, acoustic, seismic, heat, etc.). The purpose of solving a direct problem is to find a function that describes a physical field or process at any point of a given domain at any instant of time (if the field is nonstationary). The formulation of a direct problem includes

1. the domain in which the process is studied;
2. the equation that describes the process;
3. the initial conditions (if the process is nonstationary);
4. the conditions on the boundary of the domain.

For example, we can formulate the following direct initial boundary value problem for the acoustic equation: In the domain

$$\Omega \subset \mathbb{R}^n \text{ with boundary } \Gamma = \partial\Omega, \quad \mathbb{R}^n \text{ is a Euclidean space,} \tag{1.1.1}$$

it is required to find a solution $u(x,t)$ to the acoustic equation

$$c^{-2}(x)u_{tt} = \Delta u - \nabla \ln \rho(x) \cdot \nabla u + h(x,t) \tag{1.1.2}$$

that satisfies the initial conditions

$$u(x,0) = \varphi(x), \quad u_t(x,0) = \psi(x) \tag{1.1.3}$$

and the boundary conditions

$$\left.\frac{\partial u}{\partial n}\right|_{\Gamma} = g(x,t). \tag{1.1.4}$$

Here $u(x,t)$ is the acoustic pressure, $c(x)$ is the speed of sound in the medium, $\rho(x)$ is the density of the medium, and $h(x,t)$ is the sources function. Like most direct problems of mathematical physics, this problem is well-posed, which means that it has a unique solution and is stable with respect to small perturbations in the data. The following is given in the direct problem (1.1.1)–(1.1.4): the domain Ω, the coefficients $c(x)$ and $\rho(x)$, the source function $h(x,t)$ in the equation (1.1.2), the initial conditions $\varphi(x)$ and $\psi(x)$ in (1.1.3), and the boundary conditions $g(x,t)$ in (1.1.4).

In the inverse problem, aside from $u(x,t)$, the unknown functions include some of the functions occurring in the formulation of the direct problem. These unknowns are called *the solution to the inverse problem.* In order to find the unknowns, the equations (1.1.2)–(1.1.4) are supplied with some additional information about the solution to the direct problem. This information represents *the data of the inverse problem.* (Sometimes the known coefficients of the direct problem are also taken as data of the inverse problem. Many variants are possible.) For example, let the additional information be represented by the values of the solution to the direct problem (1.1.2)–(1.1.4) on the boundary:

$$u|_{\Gamma} = f(x,t). \tag{1.1.5}$$

In the inverse problem, it is required to determine the unknown functions occurring in the formulation of the direct problem from the data $f(x,t)$. Inverse problems of mathematical physics can be classified into groups depending on which functions are unknown. We will use our example to describe this classification.

Classification based on the unknown functions. The inverse problem (1.1.2)–(1.1.5) is said to be *retrospective* if it is required to determine the initial conditions, i.e., the functions $\varphi(x)$ and $\psi(x)$ in (1.1.3). The inverse problem (1.1.2)–(1.1.5) is called an *inverse boundary value problem*, if it is required to determine the function in the boundary condition (the function $g(x,t)$). The inverse problem (1.1.2)–(1.1.5) is called an *extension problem* if the initial conditions (1.1.3) are unknown and the additional information (1.1.5) and the boundary conditions (1.1.4) are specified only on a certain part $\Gamma_1 \subseteq \Gamma$ of the boundary of the domain Ω, and it is required to find the solution $u(x,t)$ of equation (1.1.2) (extend the solution to the interior of the domain). The inverse problem (1.1.2)–(1.1.5) is called an *inverse source problem* if it is required to determine the source, i.e., the function $h(x,t)$ in equation (1.1.2). The inverse problem (1.1.2)–(1.1.5) is called a *coefficient problem* if it is required to reconstruct the coefficients ($c(x)$ and $\rho(x)$) in the main equation.

It should be noted that this classification is still incomplete. There are cases where both initial and boundary conditions are unknown, and cases where the domain Ω (or a part of its boundary) is unknown.

Classification based on the additional information. Aside from the initial boundary value problem (1.1.2)–(1.1.4), it is possible to give other formulations of the direct problem of acoustics (spectral, scattering, kinematic, etc.) in which it is required to find the corresponding parameters of the acoustic process (eigenfrequencies, reflected waves, wave travel times, and so on). Measuring these parameters in experiments leads to new classes of inverse problems of acoustics.

In practice, the data (1.1.5) on the boundary of the domain under study are the most accessible for measurement, but sometimes measuring devices can be placed inside the object under study:

$$u(x_m, t) = f_m(t), \quad m = 1, 2, \dots, \tag{1.1.6}$$

which leads us to *interior inverse problems*. Much like problems of optimal control, retrospective inverse problems include the so-called “final” observations

$$u(x, T) = \hat{f}(x). \tag{1.1.7}$$

Inverse scattering problems are formulated in the case of harmonic oscillations $u(x,t) = \mathrm{e}^{\mathrm{i}\omega t}\bar{u}(x,\omega)$, the additional information being specified, for example, in the form

$$\bar{u}(x, \omega_\alpha) = \bar{f}(x, \alpha), \quad x \in X_1, \quad \alpha \in \Omega, \tag{1.1.8}$$

where X_1 is the set of observation points and $\{\omega_\alpha\}_{\alpha\in\Omega}$ is the set of observation frequencies. In some cases, the eigenfrequencies of the corresponding differential operator

$$\Delta U - \nabla \ln \rho \cdot \nabla U = \lambda U$$

and various eigenfunction characteristics are known (*inverse spectral problems*). It is sometimes possible to register the arrival time at points $\{x_k\}$ for the waves generated by local sources concentrated at points $\{x^m\}$:

$$\tau(x^m, x_k) = \tilde{f}(x^m, x_k), \quad x_k \in X_1, \quad x^m \in X_2.$$

In this case, the problem of reconstructing the speed $c(x)$ is called an *inverse kinematic problem*.

Classification based on equations. As shown above, the acoustic equation alone yields as many as M_1 different inverse problems depending on the number and the form of the unknown functions. On the other hand, one can obtain M_2 different variants of inverse problems depending on the number and the type of the measured parameters (additional information), i.e., the data of the inverse problem. Let q represent the unknown functions, and let f denote the data of the inverse problem. Then the inverse problem can be written in the form of an operator equation

$$Aq = f, \tag{1.1.9}$$

where A is an operator from the space of unknowns Q to the data space F.

It should be noted in conclusion that, instead of the acoustic equation, we could take the heat conduction equation, the radiative transfer equation, Laplace or Poisson equation, or the systems of Lamé or Maxwell equations, and the like. Suppose this could yield M_3 different variants. Then, for equations of mathematical physics alone, it is possible to formulate about $M_1 M_2 M_3$ different inverse problems. Many of these inverse problems became the subjects of monographs.

Remark 1.1.1. It is certainly possible to make the definition of an inverse problem even more general, since even the law (equation) under study is sometimes unknown. In this case, it is required to determine the law (equation) from the results of experiments (observations).

The discoveries of new laws expressed in the mathematical form (equations) are brought about by a lot of experiments, reasoning, and discussions. This complex process is not exactly a process of solving an inverse problem. However, the term "inverse problems" is gaining popularity in the scientific literature on a wide variety of subjects. For example, there have been attempts to view mathematical statistics as an inverse problem with respect to probability theory. There is a lot in common between the theory of inverse problems and control theory, the theory of pattern recognition, and many other areas. This is easy to see, for example, from the results of a search for "inverse problems" on the web site www.amazon.com. At the time when this was written (May 31, 2007), the search produced 6687 titles containing the words "inverse problems", and each of the books found, in its turn, had a list of references to the relevant literature. This monograph, however, deals with only a few mathematical aspects of the theory of inverse problems.

The structure of the operator A. The direct problem can be written in the operator form

$$A_1(\Gamma, c, \rho, h, \varphi, \psi, g) = u.$$

This means that the operator A_1 maps *the data of the direct problem* to *the solution of the direct problem*, $u(x,t)$. We call A_1 the operator of the direct problem. Some data of the direct problem are unknown in the inverse problem. We denote these unknowns by q, and the restriction of A_1 to q by $\overline{A}_1$. The *measurement operator* A_2 maps the solution $u(x,t)$ of the direct problem *to the additional information* f, for example, $A_2 u = u|_\Gamma$ or $A_2 u = u(x_k, t)$, $k = 1, 2, \ldots$, etc. Then equation (1.1.9) becomes as follows:

$$Aq \equiv A_2 \overline{A}_1 q = f,$$

where A is the result of the consecutive application (composition) of the operators $\overline{A}_1$ and A_2. For example, in the retrospective inverse problem we have $q = (\varphi, \psi)$, $f = u(x,T)$, $\overline{A}_1 q = u(x,t)$, and $A_2 \overline{A}_1 q = u(x,T)$; in the coefficient inverse problem, we have $q = (c, \rho)$, $f = u|_\Gamma$, $\overline{A}_1 q = u(x,t)$, and $A_2 \overline{A}_1 q = u|_\Gamma$, etc. The operator A_1 of the direct problem is usually continuous (the direct problem is well-posed), and so is the measurement operator (normally, one chooses to measure stable parameters of the process under study). The properties of the composition $A_2 \overline{A}_1$ are usually even too good (in ill-posed problems, $A = A_2 \overline{A}_1$ is often a *completely continuous* operator or, in other words, a compact operator). This complicates finding the inverse of A, i.e., solving the inverse problem $Aq = f$. In a simple example where A is a constant, the smaller the number A being multiplied by q, generally speaking, the smaller the error (if q is approximate):

$$A(q + \delta q) = \tilde{f},$$

i.e., the operator of the direct problem has good properties. However, when solving the inverse problem with approximate data $\tilde{f} = f + \delta f$, we have

$$\tilde{q} = q + \delta q = \frac{\tilde{f}}{A} = \frac{f}{A} + \frac{\delta f}{A},$$
$$\delta q = \frac{\delta f}{A},$$

and the error $\delta q = \tilde{q} - q$ can be indefinitely large if the number A is sufficiently small.

Inverse and ill-posed problems began to attract the attention of scientists in the early 20th century. In 1902, J. Hadamard formulated the concept of the well-posedness (properness) of problems for differential equations (see supplementary references for Chapter 2: Hadamard, 1902). A problem is called well-posed in the sense of Hadamard if there exists a unique solution to this problem that continuously depends on its data (see Definition 2.1.1). In the same paper, Hadamard also gave an example of an ill-posed problem (the Cauchy problem for Laplace equation). In 1943, A. N. Tikhonov pointed out the practical importance of such problems and the possibility of finding stable solutions to them. In the 1950's and 1960's there appeared a series of new approaches that became fundamental for the theory of ill-posed problems and attracted the attention of many mathematicians to this theory. With the advent of powerful computers, inverse and ill-posed problems started to gain popularity very rapidly. By the present day, the theory of inverse and ill-posed problems has developed into a new powerful and dynamic field of science that has an impact on almost every area of mathematics, including algebra, calculus, geometry, differential equations, mathematical physics, functional analysis, computational mathematics, etc. Some examples of well-posed and ill-posed problems are presented below in Tables 1 and 2. It should be emphasized that, one way or the other, every ill-posed problem (in the right column) can be formulated as an inverse problem with respect to a well-posed direct problem (the corresponding problems in the left column of Tables 1 and 2).

The theory of inverse and ill-posed problems is also widely used in solving applied problems in almost all fields of science, in particular:

- physics (quantum mechanics, acoustics, electrodynamics, etc.);
- geophysics (seismic exploration, electrical, magnetic and gravimetric prospecting, logging, magnetotelluric sounding, etc.);
- medicine (X-ray and NMR tomography, ultrasound testing, etc.);
- ecology (air and water quality control, space monitoring, etc.);
- economics (optimal control theory, financial mathematics, etc.).

More detailed examples of applications of inverse and ill-posed problems are given in the next section, in the beginning of each chapter, and in the sections devoted to specific applications.

Without going into further details of mathematical definitions, we note that, in most cases, *inverse* and *ill-posed* problems have one important property in common: *instability*. In the majority of cases that are of interest, inverse problems turn out to be ill-posed and, conversely, an ill-posed problem can usually be reduced to a problem that is inverse to some direct (well-posed) problem.

To sum up, it can be said that specialists in inverse and ill-posed problems study the properties of and regularization methods for unstable problems. In other words, they develop and study stable methods for approximating unstable mappings, transformations, or operations. From the point of view of information theory, the theory of inverse and ill-posed problems deals with maps from data tables with very small epsilon-entropy to tables with large epsilon-entropy (Babenko, 2002).

Well-posed problems	Ill-posed problems
Arithmetic	
Multiplication by a small number A $Aq = f$	Division by a small number A $q = A^{-1}f \quad (A \ll 1)$
Algebra	
Multiplication by a matrix $Aq = f$	$q = A^{-1}f$, A is ill-conditioned, A is degenerate, the system is inconsistent
Calculus	
Integration $f(x) = f(0) + \int_0^x q(\xi)\,d\xi$	Differentiation $q(x) = f'(x)$
Differential equations	
The Sturm–Liouville problem $u''(x) - q(x)u(x) = \lambda u(x)$, $u(0) - hu'(0) = 0$, $u(1) - Hu'(1) = 0$	The inverse Sturm–Liouville problem $\{\lambda_n, \|u_n\|^2\} \to q(x)$
Integral geometry	
Find the integral of a function $q(x, y)$ along a curve $\Gamma(\xi, \eta)$	Find $q(x, y)$ from a family of integrals $\int_{\Gamma(\xi,\eta)} q(x, y)\,ds = f(\xi, \eta)$
Integral equations	
Volterra equations and Fredholm equations of the second kind $q(x) + \int_0^x K(x,\xi)q(\xi)\,d\xi = f(x)$ $q(x) + \int_a^b K(x,\xi)q(\xi)\,d\xi = f(x)$	Volterra equations and Fredholm equations of the first kind $\int_0^x K(x,\xi)q(\xi)\,d\xi = f(x)$ $\int_a^b K(x,\xi)q(\xi)\,d\xi = f(x)$

Table 1

Well-posed problems	Ill-posed problems
Elliptic equations	
$\Delta u = 0$ Dirichlet and Neumann problems, Robin problem (mixed)	$\Delta u = 0$ Cauchy problem Initial boundary value problem with data given on a part of the boundary
Parabolic equations	
1) Cauchy problem $\Delta u = u_t,\ t > 0$ $u\big\vert_{t=0} = f(x)$ 2) Initial boundary value problem $\Delta u = u_t,\ t > 0$ $u\big\vert_{t=0} = 0$ $u\big\vert_S = g(x,t)$	Cauchy problem in reversed time $-u_t = \Delta u$ $u\big\vert_{t=0} = f$ Initial boundary value problem with data given on a part of the boundary $u_t = \Delta u$ $u\big\vert_{x=0} = f_1,\ u_x\big\vert_{x=0} = f_2$
Hyperbolic equations	
Cauchy problem Initial boundary value problem	Dirichlet and Neumann problems Cauchy problem with data on a time-like surface
Direct problems	Coefficient inverse problems
$\begin{cases} u_{tt} = u_{xx} - q(x)u \\ u\vert_{t=0} = \varphi(x),\ u_t\vert_{t=0} = \psi(x) \end{cases}$ $\begin{cases} u_t = u_{xx} - q(x)u \\ u\vert_{t=0} = 0 \end{cases}$ $\nabla(q(x)\nabla u) = 0,\ u\vert_S = g$	$\begin{cases} u_{tt} = u_{xx} - q(x)u \\ u\vert_{t=0} = \varphi(x),\ u_t\vert_{t=0} = \psi(x) \\ u(0,t) = f(t) \end{cases}$ $\begin{cases} u_t = u_{xx} - q(x)u \\ u\vert_{t=0} = 0,\ u(0,t) = f(t) \end{cases}$ $\nabla(q(x)\nabla u) = 0,\ u\vert_S = g,\ \dfrac{\partial u}{\partial n}\Big\vert_S = f$

Table 2

1.2 Examples of inverse and ill-posed problems

Example 1.2.1 (algebra, systems of linear algebraic equations). Consider the system of linear algebraic equations

$$Aq = f, \tag{1.2.1}$$

where A is an $m \times n$ matrix, q and f are n- and m-dimensional vectors, respectively. Let the rank of A be equal to $\min(m,n)$. If $m < n$, the system has many solutions.

If $m > n$, there may be no solutions. If $m = n$, the system has a unique solution for any right-hand side. In this case, there exists an inverse operator (matrix) A^{-1}. It is bounded, since it is a linear operator in a finite-dimensional space. Thus, all three conditions of well-posedness in the sense of Hadamard are satisfied.

We now analyze in detail the dependence of the solution on the perturbations of the right-hand side f in the case where the matrix A is nondegenerate.

Subtracting the original equation (1.2.1) from the perturbed equation

$$A(q + \delta q) = f + \delta f, \tag{1.2.2}$$

we obtain $A\delta q = \delta f$, which implies $\delta q = A^{-1}\delta f$ and $\|\delta q\| \le \|A^{-1}\|\|\delta f\|$. We also have $\|A\|\|q\| \ge \|f\|$.

As a result, the best estimate for the relative error of the solution is

$$\frac{\|\delta q\|}{\|q\|} \le \|A\|\|A^{-1}\|\frac{\|\delta f\|}{\|f\|}, \tag{1.2.3}$$

which shows that the error is determined by the constant $\mu(A) = \|A\|\|A^{-1}\|$ called the condition number of the system (matrix). Systems with relatively large condition number are said to be ill-conditioned. For normalized matrices ($\|A\| = 1$), it means that there are relatively large elements in the inverse matrix and, consequently, small variations in the right-hand side may lead to relatively large (although finite) variations in the solution. Therefore, systems with ill-conditioned matrices can be considered practically unstable, although formally the problem is well-posed and the stability condition $\|A^{-1}\| < \infty$ holds.

For example, the matrix

$$\begin{pmatrix} 1 & a & 0 & \dots & 0 \\ 0 & 1 & a & \dots & 0 \\ \vdots & \vdots & \vdots & \ddots & \vdots \\ 0 & 0 & 0 & \dots & a \\ 0 & 0 & 0 & \dots & 1 \end{pmatrix}$$

is ill-conditioned for sufficiently large n and $|a| > 1$ because the inverse matrix has elements of the form a^{n-1}.

In the case of perturbations in the elements of the matrix, the estimate (1.2.3) becomes as follows:

$$\frac{\|\delta q\|}{\|q\|} \le \mu(A)\frac{\|\delta A\|}{\|A\|} \Big/ \Big(1 - \mu(A)\frac{\|\delta A\|}{\|A\|}\Big)$$

(for $\|A^{-1}\|\|\delta A\| < 1$).

If $m = n$ and the determinant of A is zero, then the system (1.2.1) may have either no solutions or more than one solution. It follows that the problem $Aq = f$ is ill-posed for degenerate matrices A ($\det A = 0$).

Example 1.2.2 (calculus; summing a Fourier series). The problem of summing a Fourier series consists in finding a function $q(x)$ from its Fourier coefficients.

We show that the problem of summing a Fourier series is unstable with respect to small variations in the Fourier coefficients in the l_2 metric if the variations of the sum are estimated in the C metric. Let

$$q(x) = \sum_{k=1}^{\infty} f_k \cos kx,$$

and let the Fourier coefficients f_k of the function $q(x)$ have small perturbations: $\tilde{f}_k = f_k + \varepsilon/k$. Set

$$\tilde{q}(x) = \sum_{k=1}^{\infty} \tilde{f}_k \cos kx.$$

The difference between the coefficients of these series in the l_2 metric is

$$\Big\{\sum_{k=1}^{\infty}(f_k - \tilde{f}_k)^2\Big\}^{1/2} = \varepsilon\Big\{\sum_{k=1}^{\infty}\frac{1}{k^2}\Big\}^{1/2} = \varepsilon\sqrt{\frac{\pi^2}{6}},$$

which vanishes as $\varepsilon \to 0$. However, the difference

$$q(x) - \tilde{q}(x) = \varepsilon \sum_{k=1}^{\infty} \frac{1}{k} \cos kx$$

can be as large as desired because the series diverges for $x = 0$.

Thus, if the C metric is used to estimate variations in the sum of the series, then summation of the Fourier series is not stable.

Example 1.2.3 (geometry). Consider a body in space that can be illuminated from various directions. If the shape of the body is known, the problem of determining the shape of its shadow is well-posed. The inverse problem of reconstructing the shape of the body from its projections (shadows) on various planes is well-posed only for convex bodies, since a concave area obviously cannot be detected this way.

As noted before, Aristotle was one of the first to formulate and solve a problem of this type. Observing the shadow cast by the Earth on the moon, he concluded that the Earth is spherical.

Example 1.2.4 (integral geometry on straight lines). One of the problems arising in computerized tomography is to determine a function of two variables $q(x, y)$ from the collection of integrals

$$\int_{L(p,\varphi)} q(x, y)\, dl = f(p, \varphi)$$

along various lines $L(p, \varphi)$ in the plane (x, y), where p and φ are the line parameters. This problem is not well-posed because the condition for the existence of a solution for any right-hand side $f(\rho, \varphi)$ does not hold (see Section 5.1).

Example 1.2.5 (integral geometry on circles). Consider the problem of determining a function of two variables $q(x, y)$ from the integral of this function over a collection of circles whose centers lie on a fixed line.

Assume that $q(x, y)$ is continuous for all $(x, y) \in \mathbb{R}^2$. Consider a collection of circles whose centers lie on a fixed line (for definiteness, let this line be the coordinate axis $y = 0$). Let $L(a, r)$ denote the circle $(x - a)^2 + y^2 = r^2$, which belongs to this collection. It is required to determine $q(x, y)$ from the function $f(x, r)$ such that

$$\int_{L(x,r)} q(\xi, \tau)\, dl = f(x, r), \tag{1.2.4}$$

and $f(x, r)$ is defined for all $x \in (-\infty, \infty)$ and $r > 0$.

The solution of this problem is not unique in the class of continuous functions, since for any continuous function $\tilde{q}(x, y)$ such that $\tilde{q}(x, y) = -\tilde{q}(x, -y)$ the integrals

$$\int_{L(x,r)} \tilde{q}(\xi, \tau)\, dl$$

vanish for all $x \in \mathbb{R}$ and $r > 0$. Indeed, using the change of variables $\xi = x + r\cos\varphi$, $\tau = r\sin\varphi$, we obtain

$$\begin{aligned} \int_{L(x,r)} \tilde{q}(\xi, \tau)\, dl &= \int_0^{2\pi} \tilde{q}(x + r\cos\varphi, r\sin\varphi) r\, d\varphi \\ &= \int_0^{\pi} \tilde{q}(x + r\cos\varphi, r\sin\varphi) r\, d\varphi \\ &\quad + \int_{\pi}^{2\pi} \tilde{q}(x + r\cos\varphi, r\sin\varphi) r\, d\varphi. \end{aligned} \tag{1.2.5}$$

Putting $\bar{\varphi} = 2\pi - \varphi$ and using the condition $\tilde{q}(x, y) = -\tilde{q}(x, -y)$, we transform the last integral in the previous formula:

$$\begin{aligned} \int_{\pi}^{2\pi} \tilde{q}(x + r\cos\varphi, r\sin\varphi) r\, d\varphi &= \int_{\pi}^{0} \tilde{q}(x + r\cos\bar{\varphi}, -r\sin\bar{\varphi}) r\, d\bar{\varphi} \\ &= -\int_0^{\pi} \tilde{q}(x + r\cos\bar{\varphi}, r\sin\bar{\varphi}) r\, d\bar{\varphi}. \end{aligned}$$

Substituting the result into (1.2.5), we have

$$\int_{L(x,r)} \tilde{q}(\xi, \tau)\, dl = 0$$

for $x \in \mathbb{R}$, $r > 0$.

Thus, if $q(x, y)$ is a solution to the problem (1.2.4) and $\tilde{q}(x, y)$ is any continuous function such that $\tilde{q}(x, y) = -\tilde{q}(x, -y)$, then $q(x, y) + \tilde{q}(x, y)$ is also a solution to (1.2.4). For this reason, the problem can be reformulated as the problem of determining the even component of $q(x, y)$ with respect to y.

The first well-posedness condition is not satisfied in the above problem: solutions may not exist for some $f(x, r)$. However, the uniqueness of solutions in the class of even functions with respect to y can be established using the method described in Section 5.2.

Example 1.2.6 (a differential equation; Denisov, 1994). The rate of radioactive decay is proportional to the amount of the radioactive substance. The proportionality constant q_1 is called the decay constant. The process of radioactive decay is described by the solution of the Cauchy problem for an ordinary differential equation

$$\frac{du}{dt} = -q_1 u(t), \quad t \geq 0, \tag{1.2.6}$$

$$u(0) = q_0, \tag{1.2.7}$$

where $u(t)$ is the amount of the substance at a given instant of time and q_0 is the amount of the radioactive substance at the initial instant of time.

The direct problem: Given the constants q_0 and q_1, determine how the amount of the substance $u(t)$ changes with time. This problem is obviously well-posed. Moreover, its solution can be written explicitly as follows:

$$u(t) = q_0 \mathrm{e}^{-q_1 t}, \quad t \geq 0.$$

Assume now that the decay constant q_1 and the initial amount of the radioactive substance q_0 are unknown, but we can measure the amount of the radioactive substance $u(t)$ for certain values of t. *The inverse problem* consists in determining the coefficient q_1 in equation (1.2.6) and the initial condition q_0 from the additional information about the solution to the direct problem $u(t_k) = f_k$, $k = 1, 2, \dots, N$.

Exercise 1.2.1. Use the explicit formula for the solution of the direct problem

$$u(t) = q_0 \mathrm{e}^{-q_1 t}$$

to determine if the inverse problem is well-posed, depending on the number of points t_k, $k = 1, 2, \dots, N$, at which the additional information $f_k = u(t_k)$ is measured.

Example 1.2.7 (a system of differential equations). A chemical kinetic process is described by the solution to the Cauchy problem for the system of linear ordinary differential equations

$$\frac{du_i}{dt} = q_{i1} u_1(t) + q_{i2} u_2(t) + \cdots + q_{in} u_n(t), \tag{1.2.8}$$

$$u_i(0) = \bar{q}_i, \quad i = 1, \dots, N. \tag{1.2.9}$$

Here $u_i(t)$ is the concentration of the ith substance at the instant t. The constant parameters q_{ij} characterize the dependence of the change rate of the concentration of the ith substance on the concentration of the substances involved in the process.

The direct problem: Given the parameters q_{ij} and the concentrations $\bar{q}_i$ at the initial instant of time, determine $u_i(t)$.

The following *inverse problem* can be formulated for the system of differential equations (1.2.8). Given that the concentrations of substances $u_i(t)$, $i = 1, \dots, N$, are measured over a period of time $t \in [t_1, t_2]$, it is required to determine the values of the parameters q_{ij}, i.e., to determine the coefficients of the system (1.2.8) from a solution to this system. Two versions of this inverse problem can be considered. In the first version, the initial conditions (1.2.9) are known, i.e., $\bar{q}_i$ are given and the corresponding solutions $u_i(t)$ are measured. In the second version, $\bar{q}_i$ are unknown and must be determined together with q_{ij} (Denisov, 1994).

Example 1.2.8 (differential equation of the second order). Suppose that a particle of unit mass is moving along a straight line. The motion is caused by a force $q(t)$ that depends on time. If the particle is at the origin $x = 0$ and has zero velocity at the initial instant $t = 0$, then, according to Newton's laws, the motion of the particle is described by a function $u(t)$ satisfying the Cauchy problem

$$\ddot{u}(t) = q(t), \quad t \in [0, T], \tag{1.2.10}$$

$$u(0) = 0, \quad \dot{u}(0) = 0, \tag{1.2.11}$$

where $u(t)$ is the coordinate of the particle at the instant t. Assume now that the force $q(t)$ is unknown, but the coordinate of the particle $u(t)$ can be measured at any instant of time (or at certain points of the interval $[0, T]$). It is required to reconstruct $q(t)$ from $u(t)$. Thus, we have the following *inverse problem*: determine the right-hand side of equation (1.2.10) (the function $q(t)$) from the known solution $u(t)$ to the problem (1.2.10), (1.2.11).

We now prove that the inverse problem is unstable.

Let $u(t)$ be a solution to the direct problem for some $q(t)$. Consider the following perturbations of the solution to the direct problem:

$$u_n(t) = u(t) + \frac{1}{n}\cos(nt).$$

These perturbations correspond to the right-hand sides

$$q_n(t) = q(t) - n\cos(nt).$$

Obviously, $\|u - u_n\|_{C[0,T]} \to 0$ as $n \to \infty$, and $\|q - q_n\|_{c[0,T]} \to \infty$ as $n \to \infty$.

Thus, the problem of determining the right-hand side of the linear differential equation (1.2.10), (1.2.11) from its right-hand side is unstable.

Note that the inverse problem is reduced to double differentiation if the values of $u(t)$ are given for all $t \in [0, T]$.

Exercise 1.2.2. Analyze the differentiation operation for stability (see Example 2.5.1).

Example 1.2.9 (Fredholm integral equation of the first kind). Consider the problem of solving a Fredholm integral equation of the first kind (see Section 4.1)

$$\int_a^b K(x,s)q(s)\,ds = f(x), \quad c \le x \le d, \tag{1.2.12}$$

where the kernel $K(x,s)$ and the function $f(x)$ are given and it is required to find $q(s)$. It is assumed that $f(x) \in C[c,d]$, $q(s) \in C[a,b]$, and $K(x,s)$, $K_x(x,s)$, and $K_s(x,s)$ are continuous in the rectangle $c \le x \le d$, $a \le s \le b$. The problem of solving equation (1.2.12) is ill-posed because solutions may not exist for some functions $f(x) \in C[c,d]$. For example, take a function $f(x)$ that is continuous but not differentiable on $[c,d]$. With such a right-hand side, the equation cannot have a continuous solution $q(s)$ since the conditions for the kernel $K(x,s)$ imply that the integral in the left-hand side of (1.2.12) is differentiable with respect to the parameter x for any continuous function $q(s)$.

The condition of continuous dependence of solutions on the initial data is also not satisfied for equation (1.2.12). Consider the sequence of functions

$$q_n(s) = q(s) + n\sin(n^2 s), \quad n = 0,1,2,\ldots.$$

Substituting $q_n(s)$ into (1.2.12), we obtain

$$\begin{aligned} f_n(x) &= \int_a^b K(x,s)q_n(s)\,ds \\ &= f(x) + \int_a^b K(x,s)n\sin(n^2 s)\,ds, \quad n = 0,1,\ldots. \end{aligned}$$

We now estimate $\|f_n - f\|_{C[c,d]}$. Since

$$\begin{aligned} |f_n(x) - f(x)| &\le \left| \int_a^b K(x,s)n\sin(n^2 s)\,ds \right| \\ &= \left| -\frac{1}{n}\cos(n^2 s)K(x,s)\Big|_a^b + \frac{1}{n}\int_a^b K_s(x,s)\cos(n^2 s)\,ds \right| \le \frac{K_1}{n}, \end{aligned}$$

where the constant K_1 does not depend on n, we have

$$\|f_n - f\|_{C[c,d]} \le \frac{K_1}{n}, \quad n = 0,1,\ldots.$$

On the other hand, from the definition of the sequence $q_n(s)$ it follows that

$$\|q_n - q\|_{C[a,b]} \to \infty \quad \text{as } n \to \infty.$$

Thus, as $n \to \infty$, the initial data $f_n(x)$ are as close to $f(x)$ as desired, but the corresponding solutions $q_n(s)$ do not converge to $q(s)$, which means that the dependence of solutions on the initial data is not continuous.

Example 1.2.10 (Volterra integral equation of the first kind). Consider a Volterra integral equation of the first kind (see Section 4.1)

$$\int_0^x K(x,s)\,q(s)\,ds = f(x), \quad 0 \le x \le 1. \tag{1.2.13}$$

Assume that $K(x,s)$ is continuous, has partial derivatives of the first order for $0 \le s \le x \le 1$, and $K(x,x) = 1$ for $x \in [0,1]$. Suppose also that the unknown function $q(s)$ must belong to $C[0,1]$ and $f(x) \in C_0[0,1]$, where $C_0[0,1]$ is the space of functions $f(x)$ continuous on $[0,1]$ such that $f(0) = 0$, endowed with the uniform metric.

Exercise 1.2.3. Prove that the problem of solving equation (1.2.13) is ill-posed.

Example 1.2.11 (the Cauchy problem for Laplace equation). Let $u = u(x,y)$ be a solution to the following problem (see Chapter 7):

$$\Delta u = 0, \quad x > 0,\ y \in \mathbb{R}, \tag{1.2.14}$$

$$u(0,y) = f(y), \quad y \in \mathbb{R}, \tag{1.2.15}$$

$$u_x(0,y) = 0, \quad y \in \mathbb{R}. \tag{1.2.16}$$

Let the data $f(y)$ be chosen as follows:

$$f(y) = f_n(y) = u(0,y) = \frac{1}{n}\sin(ny).$$

Then the solution to the problem (1.2.14)–(1.2.16) is given by

$$u_n(x,y) = \frac{1}{n}\sin(ny)(e^{nx} + e^{-nx}), \quad n \in \mathbb{N}. \tag{1.2.17}$$

For any fixed $x > 0$, the solution $u_n(x,y)$ tends to infinity as $n \to \infty$, while $f(y)$ tends to zero as $n \to \infty$. Therefore, small variations in the data of the problem in C^l or W_2^l (for any $l < \infty$) may lead to indefinitely large variation in the solution, which means that the problem (1.2.14)–(1.2.16) is ill-posed.

Example 1.2.12 (an inverse problem for a partial differential equation of the first order). Let $q(x)$ be continuous and $\varphi(x)$ be continuously differentiable for all $x \in \mathbb{R}$. Then the following Cauchy problem is well-posed:

$$u_x - u_y + q(x)u = 0, \quad (x,y) \in \mathbb{R}^2, \tag{1.2.18}$$

$$u(x,0) = \varphi(x), \quad x \in \mathbb{R}. \tag{1.2.19}$$

Consider the *inverse problem* of reconstructing $q(x)$ from the additional information about the solution to the problem (1.2.18), (1.2.19)

$$u(0,y) = \psi(y), \quad y \in \mathbb{R}. \tag{1.2.20}$$

The solution to (1.2.18), (1.2.19) is given by the formula

$$u(x,y) = \varphi(x+y)\exp\Big(\int_{x+y}^{x} q(\xi)\,d\xi\Big), \quad (x,y) \in \mathbb{R}^2.$$

The condition (1.2.20) implies

$$\psi(y) = \varphi(y)\exp\Big(\int_{y}^{0} q(\xi)\,d\xi\Big), \quad y \in \mathbb{R}. \tag{1.2.21}$$

Thus, the conditions

1) $\varphi(y)$ and $\psi(y)$ are continuously differentiable for $y \in \mathbb{R}$;
2) $\psi(y)/\varphi(y) > 0$, $y \in \mathbb{R}$; $\psi(0) = \varphi(0)$

are necessary and sufficient for the existence of a solution to the inverse problem, which is determined by the following formula (Romanov, 1973a):

$$q(x) = -\frac{d}{dx}\Big[\ln\frac{\psi(x)}{\varphi(x)}\Big], \quad x \in \mathbb{R}. \tag{1.2.22}$$

If $\varphi(y)$ and $\psi(y)$ are only continuous, then the problem is ill-posed.

Example 1.2.13 (the Cauchy problem for the heat conduction equation in reversed time). The Cauchy problem in reversed time is formulated as follows: Let a function $u(x,t)$ satisfy the equation

$$u_t = u_{xx}, \quad 0 < x < \pi, \quad t > 0 \tag{1.2.23}$$

and the boundary conditions

$$u(0,t) = u(\pi,t) = 0, \quad t > 0. \tag{1.2.24}$$

It is required to determine the values of $u(x,t)$ at the initial instant $t = 0$:

$$u(x,0) = q(x), \quad 0 \le x \le \pi, \tag{1.2.25}$$

given the values of $u(x,t)$ at a fixed instant of time $t = T > 0$

$$u(x,T) = f(x), \quad 0 \le x \le \pi. \tag{1.2.26}$$

This problem is inverse to the problem of finding a function $u(x,t)$ satisfying (1.2.23)–(1.2.25) where the function $q(x)$ is given. The solution to the direct problem (1.2.23)–(1.2.25) is given by the formula

$$u(x,t) = \sum_{n=1}^{\infty} \mathrm{e}^{-n^2 t} q_n \sin nx, \tag{1.2.27}$$

where $\{q_n\}$ are the Fourier coefficients of $q(x)$:

$$q_n(x) = \frac{2}{\pi}\int_0^\pi q(x)\sin(nx)\,dx.$$

Setting $t = T$ in (1.2.27), we get

$$f(x) = \sum_{n=1}^{\infty} e^{-n^2 T} q_n \sin nx, \quad x \in [0, \pi], \tag{1.2.28}$$

which implies

$$q_n = f_n e^{n^2 T}, \quad n = 1, 2, \dots,$$

where $\{f_n\}$ are the Fourier coefficients of $f(x)$.

Since the function $q(x) \in L^2(0, \pi)$ is uniquely determined by its Fourier coefficients $\{q_n\}$, the solution of the inverse problem is unique in $L^2(0, \pi)$. Note that the condition (1.2.25) holds as a limit condition:

$$\lim_{t \to +0} \int_0^\pi [u(x,t) - q(x)]^2\,dx = 0.$$

The inverse problem (1.2.23)–(1.2.26) has a solution if and only if

$$\sum_{n=1}^{\infty} f_n^2 e^{2n^2 T} < \infty,$$

which obviously cannot hold for all functions $f \in L^2(0, \pi)$.

Example 1.2.14 (coefficient inverse problem for the heat conduction equation). The solution $u(x,t)$ to the boundary value problem for the heat conduction equation

$$c\rho u_t = (ku_x)_x - \alpha u + f, \quad 0 < x < l, \ 0 < t < T, \tag{1.2.29}$$

$$u(0,t) - \lambda_1 u_x(0,t) = \mu_1(t), \quad 0 \le t \le T, \tag{1.2.30}$$

$$u(\ell,t) - \lambda_2 u_x(l,t) = \mu_2(t), \quad 0 \le t \le T, \tag{1.2.31}$$

$$u(x,0) = \varphi(x), \quad 0 \le x \le l, \tag{1.2.32}$$

describes many physical processes such as heat distribution in a bar, diffusion in a hollow tube, etc. The equation coefficients and the boundary conditions represent the parameters of the medium under study. If the problem (1.2.29)–(1.2.32) describes the process of heat distribution in a bar, then c and k are the heat capacity coefficient and the heat conduction coefficient, respectively, which characterize the material of the bar. In this case, the direct problem consists in determining the temperature of the bar at a point x at an instant t (i.e., determining the function $u(x,t)$) from the known parameters c, ρ, k, α, f, λ_1, λ_2, μ_1, μ_2, and φ. Now suppose that all coefficients and functions that determine the solution $u(x,t)$ except the heat conduction coefficient $k = k(x)$ are known, and the temperature of the bar can be measured at a certain interior point x_0, i.e., $u(x_0,t) = f(t)$, $0 \le t \le T$. The following inverse problem arises: determine the heat conduction coefficient $k(x)$, provided that the function $f(t)$ and all other functions in (1.2.29)–(1.2.32) are given. Other inverse problems can be formulated in a similar way, including the cases where $C = C(u)$, $k = k(u)$, etc. (Alifanov, Artyukhin, and Rumyantsev, 1988).

Example 1.2.15 (interpretation of measurement data). The operation of many measurement devices that measure nonstationary fields can be described as follows: a signal $q(t)$ arrives at the input of the device, and a function $f(t)$ is registered at the output. In the simplest case, the functions $q(t)$ and $f(t)$ are related by the formula

$$\int_0^t g(t-\tau)q(\tau)\,d\tau = f(t). \tag{1.2.33}$$

In this case, $g(t)$ is called the *impulse response function* of the device. In theory, $g(t)$ is the output of the device in the case where the input is the generalized function $\delta(t)$, i.e., Dirac's delta function (see its definition in Appendix A): $\int_a^t g(t-\tau)\delta(t)d\tau = g(t)$. In practice, in order to obtain $g(t)$, a sufficiently short and powerful impulse is provided as an input. The resulting output function is close to the impulse response function in a certain sense.

Thus, the problem of interpreting measurement data, i.e., determining the form of the input signal $q(t)$ is reduced to solving the integral equation of the first kind (1.2.33) (see Section 4.2).

The relationship between the input signal $q(t)$ and the output function $f(t)$ can be more complicated. For a "linear" device, this relationship has the form

$$\int_0^t K(t,\tau)q(\tau)\,d\tau = f(t).$$

The relationship between $q(t)$ and $f(t)$ can be nonlinear (see Section 4.6):

$$\int_0^t K(t,\tau,q)\,d\tau = f(t).$$

This model describes the operation of devices that register alternate electromagnetic fields, pressure and tension modes in a continuous medium, seismographs, which record vibrations of the Earth's surface, and many other kinds of devices.

Remark 1.2.1. To solve the simple equation (1.2.33), one can use Fourier or Laplace transforms. For example, extend all functions in (1.2.33) by zero to $t < 0$, and let

$\tilde{g}(\lambda)$, $\tilde{q}(\lambda)$, and $\tilde{f}(\lambda)$ be the Fourier transforms of the functions $g(t)$, $q(t)$, and $f(t)$, respectively:

$$\tilde{g}(\lambda) = \int_0^\infty e^{i\lambda t} g(t)\, dt, \quad \tilde{q}(\lambda) = \int_0^\infty e^{i\lambda t} q(t)\, dt, \quad \tilde{f}(\lambda) = \int_0^\infty e^{i\lambda t} f(t)\, dt.$$

Then, by the convolution theorem,

$$\tilde{g}(\lambda)\tilde{q}(\lambda) = \tilde{f}(\lambda),$$

and, consequently, the inversion of the Fourier transform yields the formula for the solution of (1.2.33):

$$q(t) = \frac{1}{2\pi} \int_{-\infty}^\infty e^{-i\lambda t} \tilde{q}(\lambda)\, d\lambda = \frac{1}{2\pi} \int_{-\infty}^\infty e^{-i\lambda t} \frac{\tilde{f}(\lambda)}{\tilde{g}(\lambda)}\, d\lambda. \tag{1.2.34}$$

The calculation method provided by formula (1.2.34) is unstable because the function $\tilde{g}(\lambda)$, which is the Fourier transform of the impulse response function of the device, tends to zero as $\lambda \to \infty$ in real-life devices. This means that arbitrarily small variations in the measured value of $\tilde{f}(\lambda)$ can lead to very large variations in the solution $q(t)$ for sufficiently large λ.

Remark 1.2.2. If $g(t)$ is a constant, then the problem of solving (1.2.33) is the differentiation problem.

Example 1.2.16 (continuation of stationary fields). Some problems of interpreting gravitational and magnetic fields related to mineral exploration lead to ill-posed problems equivalent to the Cauchy problem for Laplace equation. If the Earth were a spherically uniform ball, the gravitational field strength on the surface of the Earth would be constant. The inhomogeneity of the land surface and the density distribution within the Earth causes the gravitational field strength on the Earth's surface to deviate from its mean value. Although these deviations are small in terms of percentage, they are reliably registered by physical devices (gravimeters). Gravimetric data are used in mineral exploration and prospecting.

The purpose of gravimetric exploration is to determine the location and shape of subsurface inhomogeneities based on gravimetric measurement data.

If the distance between geological bodies is greater than the distance between either of them and the surface of the Earth, then their locations correspond to the local maxima of the anomalies. Otherwise, the two bodies may be associated with a single local maximum. Geophysical measurements and the interpretation of their results represent the preliminary stage in prospecting for mineral deposits. The main stage of prospecting consists in drilling the exploratory wells and analyzing the drilling data. If the shape of an anomaly leads us to the conclusion that it represents a single body, then a natural choice would be to drill in the center of the anomaly. However, if the conclusion is wrong, the decision to drill in the center will result in the well being drilled between the actual bodies that we are interested in. This was often the case in the practice of geological exploration. Then it was proposed to calculate the anomalous gravitational field at a certain depth under the surface based on the results of the gravitational measurements performed on the surface of the Earth (i.e., to solve the Cauchy problem for Laplace equation). If it turns out that the anomaly at that depth still has a single local maximum, then it is highly probable that the anomaly is generated by a single body. Otherwise, if two local maxima appear as a result of recalculation, then it is natural to conclude that there are two bodies, and the locations for drilling must be chosen accordingly.

A similar problem formulation arises when interpreting anomalies of a constant magnetic field, since the potential and the components of the field strength outside the magnetic bodies also satisfy Laplace equation.

In electrical sounding using direct-current resistivity methods, direct current is applied to two electrodes inserted into the ground and the potential difference on the surface is measured. It is required to determine the structure of the subsurface area under study based on the measurement results. If the sediment layer is a homogeneous conductive medium and the basement resistivity is much higher, then the electric current lines will run along the relief of the basement surface. Consequently, to determine the relief of the basement surface, it suffices to determine the electric current lines in the sediment layer. In a homogeneous medium, the direct current potential satisfies Laplace equation. The normal derivative of the electric potential on the Earth's surface is equal to zero. The potential is measured. Thus, we again arrive at the Cauchy problem for Laplace equation.

Chapter 2
Ill-posed problems

Happy families are all alike;
every unhappy family is unhappy in its own way.
Leo Tolstoy. "Anna Karenina".

There is no universal method for solving ill-posed problems. In every specific case, the main "trouble"—instability—has to be tackled in its own way.

This chapter is a brief introduction to the foundations of the theory of ill-posed problems. We begin with abstract definitions of well-posedness, ill-posedness, and conditional well-posedness (Section 2.1), since it is assumed that the reader is already familiar with relevant examples from various areas of mathematics and other sciences after having read the first chapter. Paying tribute to the founders of the theory of ill-posed problems, we devote the next three sections (2.2–2.4) to the theorems and methods formulated by A. N. Tikhonov, V. K. Ivanov, and M. M. Lavrentiev. Theorem 2.2.1 on the continuity of the inverse mapping $A^{-1} : A(M) \to M$ for any compact set M and continuous one-to-one operator A provides the basis for the theory of conditionally well-posed problems and for constructing a variety of effective numerical algorithms for solving them (Section 2.2). The concept of a quasi-solution introduced by V. K. Ivanov (Section 2.3) serves three purposes: it generalizes the concept of a solution, restores all the well-posedness conditions, and yields new algorithms for the approximate solution of ill-posed problems. The Lavrentiev method (Section 2.4) is used to construct numerical algorithms for solving a wide class of systems of algebraic equations (Chapter 3), linear and nonlinear integral equations of the first kind (Chapter 4). The concept of a regularizing family of operators (Section 2.5) is fundamental to the theory of ill-posed problems. In brief, a regularizing family $\{R_\alpha\}_{\alpha>0}$ (or a regularizing sequence $\{R_n\}$) consists of operators R_α such that there exists a stable procedure for constructing approximate solutions $q_\alpha = R_\alpha f$ to the equation $Aq = f$ which converge to the exact solution q_e as $\alpha \to +0$. Another important property of regularizing operators consists in providing a way to construct an approximate solution $q_{\alpha\delta} = R_\alpha f_\delta$ from the approximate data f_δ for which the solution to the equation $Aq = f_\delta$ may not exist. In many cases, it can be proved that $q_{\alpha\delta}$ converge to the exact solution q_e if the regularization parameter α and the error δ in the right-hand side f tend to zero at concerted rates (see, for example, Theorem 2.5.2). Iterative regularization based on the minimization of the objective functional $J(q) = \|Aq - f\|^2$ by gradient descent methods (Section 2.5) is one of the most effective and commonly used regularization methods. Gradient methods have their origins in the well-known fact that if the gradient $J'q$ of the functional $J(q)$ does not vanish at a point $q \in Q$, then moving in the antigradient direction to the point $q+\delta q$, we can reduce the value of the functional $J(q + \delta q)$, provided that the descent step is sufficiently small. Indeed, choosing $\delta q = -\alpha J'q$, where α is positive, and recalling the definition of the gradient (for simplicity, Q is assumed to be a Hilbert space), we can write

$$J(q - \alpha J'q) - J(q) = \langle J'q, -\alpha J'q\rangle + o(\alpha\|J'q\|) = -\alpha\|J'q\|^2 + o(\alpha\|J'q\|),$$

which shows that the right-hand side becomes negative as $\alpha \to +0$ and, consequently, $J(q + \delta q) = J(q - \alpha J'q) < J(q)$ for sufficiently small α. Therefore, varying the descent step size α_n and sometimes adjusting the direction of the descent (see, for example, the conjugate gradient method), we can construct a minimizing sequence $q_{n+1} = q_n - \alpha_n J'q_n$ such that $J(q_{n+1}) < J(q_n)$. Furthermore, it turns out that the rate of decrease of the functional $J(q_n)$ can be estimated for many gradient methods even if the problem $Aq = f$ is ill-posed (Section 2.7). The strong convergence $\|q_n - q_e\| \to 0$ as $n \to \infty$ is more problematic. Clearly, it is possible only if the equation $Aq = f$ has at least one exact solution q_e. If the sequence q_n is constructed using the method of simple iteration, the steepest descent method, or the conjugate gradient method, the monotonic convergence $\|q_n - q_e\| \searrow 0$ can be easily established (Section 2.6). However, for the numerical solution of ill-posed problems it is especially important to estimate the rate of convergence and to have a stopping criterion for the iterative process. It has been observed experimentally that in solving an ill-posed problem $Aq = f$ with approximate data f_δ ($\|f - f_\delta\| \le \delta$) by a gradient method, the minimizing sequence $\{q_{n\delta}\}$ may approach the exact solution at an early stage, but the value of $\|q_{n\delta} - q_e\|$ may start to increase at later stages. Then how to formulate the criterion for stopping the iterative process? The most commonly used criterion is the minimal residual principle, which is based on the credible speculation that it is probably not worth continuing the process after the residual $\|Aq - f_\delta\|$ becomes less than the measurement error δ. This also applies to the choice of the regularization parameter α, except that we can, for example, choose a set of regularization parameters $0 < \alpha_1 < \alpha_2 < \cdots < \alpha_k$, calculate the corresponding $q_{\alpha_j\delta}$, and compare the residuals $\|Aq_{\alpha_j\delta} - f_\delta\|$ for $j = 1, 2, \ldots, k$ instead of performing sequential iterations.

If further investigation of the problem $Aq = f$ allows one to obtain a conditional stability estimate (Section 2.8), then it is possible to estimate the rate of strong convergence and obtain a new stopping criterion for the iterative process.

The singular value decomposition of a linear compact operator $A : Q \to F$, where Q and F are separable Hilbert spaces, is presented in Section 2.9. This technique allows one to take a closer look into the properties of A and to estimate to what extent the problem $Aq = f$ is ill-posed.

2.1 Well-posed and ill-posed problems

Let an operator A map a topological space Q into a topological space F ($A : Q \to F$). For any topological space Q, let $\mathcal{O}(q)$ denote a neighbourhood of an element $q \in Q$. Throughout what follows, $D(A)$ is the domain of definition and $R(A)$ is the range of A.

Definition 2.1.1 (well-posedness in the sense of Hadamard). The problem $Aq = f$ is *well-posed* on the pair of topological spaces Q and F if the following three conditions hold:

(1) for any $f \in F$ there exists a solution $q_e \in Q$ to the equation $Aq = f$, i.e., $R(A) = F$ (the existence condition);

(2) the solution q_e to the equation $Aq = f$ is unique in Q (the uniqueness condition);

(3) for any neighbourhood $\mathcal{O}(q_e) \subset Q$ of the solution q_e to the equation $Aq = f$, there is a neighbourhood $\mathcal{O}(f) \subset F$ of the right-hand side f such that for all $f_\delta \in \mathcal{O}(f)$ the element $A^{-1} f_\delta = q_\delta$ belongs to the neighbourhood $\mathcal{O}(q_e)$, i.e., the operator A^{-1} is continuous (the stability condition).

Definition 2.1.1 can be made more specific by replacing the topological spaces Q and F by metric, Banach, Hilbert, or Euclidean spaces. In some cases, it is reasonable to take a topological space for Q and a Euclidean space for F, and so on. Fixed in the definition are only the requirements of the existence, uniqueness, and stability of the solution.

Definition 2.1.2. The problem $Aq = f$ is *ill-posed* on the pair of spaces Q and F if at least one of the three well-posedness conditions does not hold.

M. M. Lavrentiev proposed to distinguish the class of conditionally well-posed problems as a subclass of ill-posed problems. Let Q and F be topological spaces and let $M \subset Q$ be a fixed set. We denote by $A(M)$ the image of M under the map $A : Q \to F$, i.e., $A(M) = \{f \in F\colon \exists q \in M \text{ such that } Aq = f\}$. It is obvious that $A(M) \subset F$.

Definition 2.1.3 (conditional/Tikhonov well-posedness). Let $M \subset Q$ and $f \in A(M)$, i.e., there exists a solution $q_e \in M$ to the problem $Aq = f$. Then the problem $Aq = f$ is said to be *conditionally well-posed* on the set M if the following conditions hold:

(1) for any $f \in A(M)$, the equation $Aq = f$ has a unique solution on M;

(2) for any neighbourhood $\mathcal{O}(q_e)$ of the solution to the equation $Aq = f$, there exists a neighbourhood $\mathcal{O}(f)$ such that for any $f_\delta \in \mathcal{O}(f) \cap A(M)$ the solution to the equation $Aq = f_\delta$ belongs to $\mathcal{O}(q_e)$ (*conditional stability*).

It should be emphasized that the variations f_δ of the data f in the second condition are assumed to lie within the class $A(M)$, which ensures the existence of solutions.

Definition 2.1.4. The set M in Definition 2.1.3 is called *the well-posedness set for the problem* $Aq = f$.

Remark 2.1.1. To prove that the problem $Aq = f$ is well-posed, it is necessary to prove the existence, uniqueness and stability theorems for its solutions. To prove that the problem $Aq = f$ is conditionally well-posed, it is necessary to choose a well-posedness set M, prove that the solution of the problem is unique on M and that $q = A^{-1}(f)$ is conditionally stable with respect to small variations in the data f that keep the solutions within M.

Remark 2.1.2. It should be emphasized that proving the well-posedness of a problem necessarily involves proving the existence of a solution, whereas in the case of conditional well-posedness the existence of a solution is assumed. It certainly does not mean that the existence theorem is of no importance or that one should not aim to prove it. We only draw reader's attention to the fact that in the most interesting and important cases, the conditions on the data f which provide the existence of a solution q_e to a conditionally well-posed problem $Aq = f$ turn out to be too complicated to verify and to apply them in numerical algorithms (see the Picard criterion, the solvability conditions for the inverse Sturm–Liouville problem, etc.). In this sense, the title of V. P. Maslov's paper "The existence of a solution of an ill-posed problem is equivalent to the convergence of a regularization process" (Maslov, 1968) is significant. It reveals one of the main problems in the study of ill-posed problems. For this reason, the introduction of the concept of conditional well-posedness shifts the focus to the search for stable methods for the approximate solution of ill-posed problems. We repeat, however, that this does not make the task of detailed mathematical analysis of the solvability conditions for any specific problem less interesting or important. Note that if a problem has no exact solution, we may construct a regularized family of approximate solutions converging to a quasi-solution of the problem (Ivanov et al., 2002).

In the following section we will show that the uniqueness of a solution implies its conditional stability if the well-posedness set M is compact.

2.2 On stability in different spaces

The trial-and-error method (Tikhonov and Arsenin, 1986) was one of the first methods for the approximate solution of conditionally well-posed problems. We will describe this method assuming that Q and F are metric spaces and M is a compact set. For an approximate solution, we choose an element q_K from the well-posedness set M such that the residual $\rho_F(Aq, f)$ attains its minimum at q_K, i.e.,

$$\rho_F(Aq_K, f) = \inf_{q \in M} \rho_F(Aq, f)$$

(see Definition 2.3.1).

Let $\{q_n\}$ be a sequence of elements in Q such that $\lim_{n\to\infty} \rho_F(Aq_n, f) = 0$. We denote by q_e the exact solution to the problem $Aq = f$. If M is compact, then from $\lim_{n\to\infty} \rho_F(Aq_n, f) = 0$ it follows that $\lim_{n\to\infty} \rho_Q(q_n, q_e) = 0$, which is proved by the next theorem.

Theorem 2.2.1. *Let Q and F be metric spaces, and let $M \subset Q$ be compact. Assume that A is a one-to-one map from M onto $A(M) \subset F$. Then, if A is continuous, then so is the inverse map $A^{-1} : A(M) \to M$.*

Proof. Let $f \in A(M)$ and $Aq_e = f$. We will prove that A^{-1} is continuous at f. Assume the contrary. Then there exists an $\varepsilon_* > 0$ such that for any $\delta > 0$ there is an element $f_\delta \in A(M)$ for which

$$\rho_F(f_\delta, f) < \delta \quad \text{and} \quad \rho_Q(A^{-1}(f_\delta), A^{-1}(f)) \geq \varepsilon_*.$$

Consequently, for any $n \in \mathbb{N}$ there is an element $f_n \in A(M)$ such that

$$\rho_F(f_n, f) < 1/n \quad \text{and} \quad \rho_Q(A^{-1}(f_n), A^{-1}(f)) \geq \varepsilon_*.$$

Evidently,

$$\lim_{n\to\infty} f_n = f.$$

Since $A^{-1}(f_n) \in M$ and M is compact, the sequence $\{A^{-1}(f_n)\}$ has a converging subsequence: $\lim_{k\to\infty} A^{-1}(f_{n_k}) = \bar{q}$.

Since $\rho_Q(\bar{q}, A^{-1}(f)) \geq \varepsilon_*$, we conclude that $\bar{q} \neq A^{-1}(f) = q_e$. On the other hand, from the continuity of the operator A it follows that the subsequence $A(A^{-1}(f_{n_k}))$ converges and $\lim_{k\to\infty} A(A^{-1}(f_{n_k})) = \lim_{k\to\infty} f_{n_k} = f = Aq_e$. Hence, $\bar{q} = q_e$. We have arrived at a contradiction, which proves the theorem. □

Theorem 2.2.1 provides a way of minimizing the residual. Let A be a continuous one-to-one operator from M to $A(M)$, where M is compact. Let $\{\delta_n\}$ be a decreasing sequence of positive numbers such that $\delta_n \searrow 0$ as $n \to \infty$. Instead of the exact right-hand side f we take an element $f_n \in A(M)$ such that $\rho_F(f, f_n) \leq \delta_n$. For any n, we use the trial-and-error method to find an element q_n such that $\rho_F(Aq_n, f_n) \leq \delta_n$. The elements q_n approximate the exact solution q_e to the equation $Aq = f$. Indeed, since A is a continuous operator, the image $A(M)$ of the compact set M is compact. Consequently, by Theorem 2.2.1, the inverse map A^{-1} is continuous on $A(M)$. Since

$$\rho_F(Aq_n, f) \leq \rho_F(Aq_n, f_n) + \rho_F(f_n, f),$$

the following inequality holds:

$$\rho_F(Aq_n, f) \leq \delta_n + \delta_n = 2\delta_n.$$

Hence, in view of the continuity of the inverse map $A^{-1} : A(M) \to M$, we have $\rho_Q(q_n, q_e) \leq \varepsilon(2\delta_n)$, where $\varepsilon(2\delta_n) \to 0$ as $\delta_n \to 0$.

Note that Theorem 2.2.1 implies the existence of a steadily increasing function $\beta(\delta)$ such that

1) $\lim_{\delta \to 0} \beta(\delta) = 0$;
2) for all $q_1, q_2 \in M$ the inequality $\rho_F(Aq_1, Aq_2) \leq \delta$ implies $\rho_Q(q_1, q_2) \leq \beta(\delta)$.

Definition 2.2.1. Assume that Q and F are metric spaces, $M \subset Q$ is a compact set, and $A : Q \to F$ is a continuous one-to-one operator that maps M onto $A(M) \subset F$. The function

$$\omega(A(M), \delta) = \sup_{\substack{f_1, f_2 \in A(M) \\ \rho_F(f_1, f_2) \leq \delta}} \rho_Q(A^{-1} f_1, A^{-1} f_2)$$

is called *the modulus of continuity* of the operator A^{-1} on the set $A(M)$.

Given the modulus of continuity $\omega(A(M), \delta)$ or a function that majorizes it (for example, the conditional stability estimate $\beta(\delta)$), one can evaluate the norm of deviation of the exact solution from the solution corresponding to the approximate data. Indeed, let $q_e \in M$ be an exact solution to the problem $Aq = f$ and $q_\delta \in M$ be a solution to the problem $Aq = f_\delta$, where $\|f - f_\delta\| \leq \delta$. Then $\|q - q_\delta\| \leq \omega(A(M), \delta)$ if M is compact. Therefore, after proving the uniqueness theorem, the most important stage in the study of a conditionally well-posed problem is obtaining a conditional stability estimate (see Section 2.8).

Remark 2.2.1. Stability depends on the choice of topologies on Q and F. Formally, the continuity of the operator A^{-1} can be ensured by endowing F with a stronger topology. For example, if A is a linear one-to-one operator and Q and F are normed spaces, then the following norm can be introduced in F:

$$\|f\|_A = \|A^{-1} f\|.$$

In this case

$$\|A^{-1}\| = \sup_{f \neq 0} \frac{\|A^{-1} f\|}{\|f\|_A} = 1$$

and therefore A^{-1} is continuous. In practice, however, the most commonly used spaces are C^m and H^n (see Subsection A.1.5), where m and n are not very large.

Remark 2.2.2. Theorem 2.2.1 can be extended to the case when Q is a topological space and F is a Hausdorff space (see Definition A.1.36) (Ivanov et al., 2002).

Ill-posed problems can be formulated as problems of determining the value of a (generally, unbounded) operator T at a point f:

$$Tf = q, \quad f \in F, \quad q \in Q. \tag{2.2.1}$$

If the operator A is invertible, then the problem $Aq = f$ is equivalent to problem (2.2.1). The operator A^{-1}, however, may not exist. Moreover, in many applied problems (for example, differentiating a function, summing a series, etc.) the representation $Aq = f$ may be inconvenient or even unachievable, although both problems can theoretically be studied using the same scheme (Ivanov et al., 2002; Morozov, 1987). We now reformulate the conditions for well-posedness in the sense of Hadamard for problem (2.2.1) as follows:

1) the operator T is defined on the entire space F: $D(T) = F$;

2) T is a one-to-one map;

3) T is continuous.

Problem (2.2.1) is said to be ill-posed if at least one of the well-posedness conditions is not fulfilled. The case when the third condition does not hold (the case of instability) is the most important and substantial. In this case, problem (2.2.1) is reduced to the problem of approximating an unbounded operator with a bounded one.

2.3 Quasi-solution. The Ivanov theorems

Other approaches can be applied to ill-posed problems which involve a generalization of the concept of a solution.

Let A be a completely continuous operator (see Definition A.2.11). If q is assumed to belong to a compact set $M \subset Q$ and $f \in A(M) \subset F$, then the formula

$$q = A^{-1} f$$

can be used to construct an approximate solution that is stable with respect to small variations in f.

It should be noted that the condition that f belongs to $A(M)$ is essential for the applicability of the formula $q = A^{-1}f$ for finding an approximate solution, since the expression $A^{-1}f$ may be meaningless if this condition does not hold. However, finding out whether f belongs to $A(M)$ is a complicated problem. Furthermore, even if $f \in A(M)$, measurement errors may cause this condition to fail, i.e., f_δ may not belong to $A(M)$. To avoid the difficulties arising when the equation $Aq = f$ has no solutions, the notion of a quasi-solution to $Aq = f$ is introduced as a generalization of the concept of a solution to this equation (Ivanov, 1963).

Assume that F is a metric space.

Definition 2.3.1. A *quasi-solution* to the equation $Aq = f$ on a set $M \subset Q$ is an element $q_K \in M$ that minimizes the residual:

$$\rho_F(Aq_K, f) = \inf_{q \in M} \rho_F(Aq, f).$$

If M is compact, then there exists a quasi-solution for any $f \in F$. If, in addition, $f \in A(M)$, then q_K coincides with an exact solution.

Note that the quasisolution q_K is not necessarily unique. We will give a sufficient condition for a quasi-solution to be unique and continuously depend on the right-hand side f.

Definition 2.3.2. Let h be an element of a metric space F, and let $G \subset F$. An element $g \in G$ is called a *projection* of h onto the set G if

$$\rho_F(h, g) = \rho_F(h, G) := \inf_{p \in G} \rho_F(h, p).$$

The projection g of h onto G is written as $g = P_G h$.

Theorem 2.3.1. *Assume that the equation $Aq = f$ has at most one solution on a compact set M and for any $f \in F$ its projection onto $A(M)$ is unique. Then a quasi-solution to the equation $Aq = f$ is unique and continuously depends on f.*

The proof of this theorem and Theorem 2.3.2 can be found in (Ivanov et al., 2002) and (Tikhonov and Arsenin, 1986).

Note that all well-posedness conditions are restored when passing to quasi-solutions if the assumptions of Theorem 2.3.1 hold. Consequently, the problem of finding a quasi-solution on a compact set is well-posed.

Remark 2.3.1. If the equation $Aq = f$ has more than one quasi-solution on the set M, they form a subset Q_f^k of the compact set M. In this case the set Q_f^k continuously depends on f on the set $A(M)$ in the sense of continuity of multi-valued maps (Ivanov, 1963).

If the operator A is linear, the theorem takes the following specific form (Ivanov, 1963).

Theorem 2.3.2. *Let $A : Q \to F$ be a linear operator and assume that the homogeneous equation $Aq = \mathbf{0}$ has only one solution $q = \mathbf{0}$. Furthermore, assume that M is a convex compact set and any sphere in F is strictly convex. Then a quasi-solution to the equation $Aq = f$ on M is unique and continuously depends on f.*

We now consider the case where Q and F are separable Hilbert spaces. Let $A : Q \to F$ be a completely continuous operator and

$$M = B(\mathbf{0}, r) := \{q \in Q \colon \|q\| \le r\}.$$

By A^* we denote the adjoint of the operator A (see Definition A.3.13).

It is known that A^*A is a self-adjoint positive completely continuous operator from Q into Q (positivity means that $\langle A^*Aq, q\rangle > 0$ for all $q \neq \mathbf{0}$).

Let $\{\lambda_n\}$ be a sequence of eigenvalues of the operator A^*A (in nonincreasing order), and let $\{\varphi_n\}$ be the corresponding complete orthonormal sequence of eigenfunctions.

The element A^*f can be represented as a series:

$$A^*f = \sum_{n=1}^{\infty} f_n\varphi_n, \quad f_n = \langle A^*f, \varphi_n\rangle. \tag{2.3.1}$$

Under these conditions, the following theorem holds (Ivanov, 1963).

Theorem 2.3.3. *A quasi-solution to the equation $Aq = f$ on the set $B(\mathbf{0}, r)$ is determined by the formulae*

$$q_K = \begin{cases} \sum_{n=1}^{\infty} \frac{f_n}{\lambda_n}\varphi_n & \text{if } \sum_{n=1}^{\infty} \frac{f_n^2}{\lambda_n^2} \le r^2, \\ \sum_{n=1}^{\infty} \frac{f_n}{\lambda_n+\beta}\varphi_n & \text{if } \sum_{n=1}^{\infty} \frac{f_n^2}{\lambda_n^2} > r^2, \end{cases}$$

where β satisfies the equation

$$\sum_{n=1}^{\infty} \frac{f_n^2}{(\lambda_n+\beta)^2} = r^2.$$

Proof. If $\sum_{n=1}^{\infty} f_n^2/\lambda_n^2 \le r^2$, then a quasi-solution q_K that minimizes the functional $\rho_F^2(Aq, f) := J(q) = \langle Aq - f, Aq - f\rangle$ on $B(\mathbf{0}, r)$ can be obtained by solving the Euler–Lagrange equation

$$A^*Aq = A^*f. \tag{2.3.2}$$

We will seek a solution to this equation in the form of a series:

$$q_K = \sum_{n=1}^{\infty} q_n \varphi_n.$$

Substituting this series into equation (2.3.2) and using the expansion (2.3.1) for $A^* f$, we obtain

$$A^* A q_K = \sum_{n=1}^{\infty} q_n A^* A \varphi_n = \sum_{n=1}^{\infty} q_n \lambda_n \varphi_n = \sum_{n=1}^{\infty} f_n \varphi_n.$$

Hence, $q_n = f_n/\lambda_n$. Since $\sum_{n=1}^{\infty} f_n^2/\lambda_n^2 \leq r^2$, we deduce that

$$q_K = \sum_{n=1}^{\infty} (f_n/\lambda_n)\varphi_n \in B(\mathbf{0}, r)$$

minimizes the functional $J(q)$ on $B(\mathbf{0}, r)$.

On the other hand, if $\sum_{n=1}^{\infty} f_n^2/\lambda_n^2 > r^2$, then, taking into account that q_K must belong to $B(0, r)$, it is required to minimize the functional $J(q) = \langle Aq - f, Aq - f\rangle$ on the sphere $\|q\|^2 = r^2$.

Applying the method of Lagrange multipliers, this problem is reduced to finding the global extremum of the functional

$$J_\alpha(q) = \langle Aq - f, Aq - f\rangle + \alpha\langle q, q\rangle.$$

To find the minimum of J_α, we solve the corresponding Euler equation

$$\alpha q + A^* A q = A^* f. \qquad (2.3.3)$$

Substituting $q_K = \sum_{n=1}^{\infty} q_n \varphi_n$ and $A^* f = \sum_{n=1}^{\infty} f_n \varphi_n$ into this equation, we obtain $q_n = f_n/(\alpha + \lambda_n)$. The parameter α is determined from the condition $\|q\|^2 = r^2$, which is equivalent to the condition $w(\alpha) := \sum_{n=1}^{\infty} f_n^2/(\alpha + \lambda_n)^2 = r^2$. It remains to take the root of the equation $\omega(\alpha) = r^2$ for β to complete the proof of the theorem. □

Remark 2.3.2. The equation $w(\alpha) = r^2$ is solvable because $w(0) > r^2$ and $w(\alpha)$ steadily decreases with increasing α and tends to zero as $\alpha \to \infty$.

2.4 The Lavrentiev method

If the right-hand side f_δ of the equation $Aq = f_\delta$ does not belong to $A(M)$, then one can try to replace this equation with a close equation

$$\alpha q + Aq = f_\delta, \quad \alpha > 0 \qquad (2.4.1)$$

for which the problem becomes well-posed. In what follows, we prove that in many cases this equation has a solution $q_{\alpha\delta}$ that tends to the exact solution q_e of the equation $Aq = f$ as α and the error δ in the approximation of f tend to zero at concerted rates (Lavrentiev, 1959).

Let $\mathcal{H}$ be a separable Hilbert space and $A : \mathcal{H} \to \mathcal{H}$ be a linear, completely continuous, positive, and self-adjoint operator.

Assume that for $f \in \mathcal{H}$ there exists such $q_e \in \mathcal{H}$ that $Aq_e = f$. Then we take the solution $q_\alpha = (\alpha E + A)^{-1} f$ to the equation $\alpha q + Aq = f$ as an approximate solution to the equation $Aq = f$ (the existence of q_α will be proved below).

If the data f is approximate, i.e., instead of f we have f_δ such that $\|f - f_\delta\| \le \delta$, then we set

$$q_{\alpha\delta} = (\alpha E + A)^{-1} f_\delta.$$

The family of operators $R_\alpha = (\alpha E + A)^{-1}$, $\alpha > 0$, is regularizing for the problem $Aq = f$ (see Definition 2.5.2). Consider this matter in more detail. Let $\{\varphi_k\}$ be a complete orthonormal sequence of eigenfunctions and $\{\lambda_k\}$ $(0 < \cdots \le \lambda_{k+1} \le \lambda_k \le \cdots \le \lambda_1)$ be the corresponding sequence of eigenvalues of the operator A. Assume that the equation

$$Aq = f \tag{2.4.2}$$

has a solution q_e. Substituting the expansions

$$\begin{aligned} q_e &= \sum_{k=1}^{\infty} q_k \varphi_k, \quad q_k = \langle q_e, \varphi_k \rangle, \\ f &= \sum_{k=1}^{\infty} f_k \varphi_k, \quad f_k = \langle f, \varphi_k \rangle, \end{aligned} \tag{2.4.3}$$

into (2.4.2), we conclude that $q_k = f_k/\lambda_k$ and therefore

$$q_e = \sum_{k=1}^{\infty} \frac{f_k}{\lambda_k} \varphi_k.$$

Note that

$$\sum_{k=1}^{\infty} \Big(\frac{f_k}{\lambda_k}\Big)^2 < \infty. \tag{2.4.4}$$

Consider the auxiliary equation

$$\alpha q + Aq = f. \tag{2.4.5}$$

As before, a solution q_α to equation (2.4.5) can be represented in the form

$$q_\alpha = \sum_{k=1}^{\infty} \frac{f_k}{\alpha + \lambda_k} \varphi_k. \tag{2.4.6}$$

Taking into account that $f_k = \lambda_k q_k$, we estimate the difference

$$\begin{aligned} q_e - q_\alpha &= \sum_{k=1}^{\infty} q_k \varphi_k - \sum_{k=1}^{\infty} \frac{\lambda_k q_k}{\alpha + \lambda_k} \varphi_k \\ &= \sum_{k=1}^{\infty} \Big(q_k - \frac{\lambda_k q_k}{\alpha + \lambda_k} \Big) \varphi_k = \alpha \sum_{k=1}^{\infty} \frac{q_k}{\alpha + \lambda_k} \varphi_k. \end{aligned}$$

Consequently,

$$\|q_e - q_\alpha\|^2 = \alpha^2 \sum_{k=1}^{\infty} \frac{q_k^2}{(\alpha + \lambda_k)^2}. \tag{2.4.7}$$

It is now easy to show that $\lim_{\alpha \to +0} \|q_e - q_\alpha\| = 0$. Indeed, let ε be an arbitrary positive number. The series (2.4.7) is estimated from above as follows:

$$\begin{aligned} \alpha^2 \sum_{k=1}^{\infty} \frac{q_k^2}{(\alpha + \lambda_k)^2} &= \alpha^2 \sum_{k=1}^{n} \frac{q_k^2}{(\alpha + \lambda_k)^2} + \alpha^2 \sum_{k=n+1}^{\infty} \frac{q_k^2}{(\alpha + \lambda_k)^2} \\ &\le \frac{\alpha^2}{\lambda_n^2} \|q\|^2 + \sum_{k=n+1}^{\infty} q_k^2. \end{aligned} \tag{2.4.8}$$

Since the series $\sum_{k=1}^{\infty} q_k^2$ converges, there is a number n such that the second term in the right-hand side of (2.4.8) is less than $\varepsilon/2$. Then we can choose $\alpha > 0$ such that the first term is also less than $\varepsilon/2$. It follows that $\lim_{\alpha \to +0} \|q_e - q_\alpha\| = 0$.

We now return to the problem with approximate data

$$Aq = f_\delta, \tag{2.4.9}$$

where $\|f - f_\delta\| \le \delta$, and the regularized problem (2.4.1):

$$\alpha q + Aq = f_\delta.$$

Put $f_{\delta,k} = \langle f_\delta, \varphi_k \rangle$. Then the solution $q_{\alpha\delta}$ to (2.4.1) can be represented as a series

$$q_{\alpha\delta} = \sum_{k=1}^{\infty} \frac{f_{\delta,k}}{\alpha + \lambda_k} \varphi_k.$$

We now estimate the difference

$$\|q_e - q_{\alpha\delta}\| \leq \|q_e - q_\alpha\| + \|q_\alpha - q_{\alpha\delta}\|. \tag{2.4.10}$$

The first term in the right-hand side of (2.4.10) tends to zero as $\alpha \to 0$. The second term is estimated as follows:

$$\begin{aligned} \|q_\alpha - q_{\alpha\delta}\|^2 &= \Big\| \sum_{k=1}^{\infty} \Big(\frac{f_k}{\alpha + \lambda_k} - \frac{f_{\delta,k}}{\alpha + \lambda_k} \Big) \varphi_k \Big\|^2 = \sum_{k=1}^{\infty} \frac{(f_k - f_{\delta,k})^2}{(\alpha + \lambda_k)^2} \\ &\leq \frac{1}{\alpha^2} \sum_{k=1}^{\infty} (f_k - f_{\delta,k})^2 = \frac{1}{\alpha^2} \|f - f_\delta\|^2 \leq \frac{\delta^2}{\alpha^2}. \end{aligned} \tag{2.4.11}$$

We now prove that $q_{\alpha\delta} \to q_e$ when α and δ tend to zero at such rates that $\delta/\alpha \to 0$ as well. Take an arbitrary positive number ε. Choose α such that $\|q_e - q_\alpha\| < \varepsilon/2$. Then find $\delta > 0$ such that $\delta/\alpha < \varepsilon/2$. Then from (2.4.10) and (2.4.11) it follows that

$$\|q_e - q_{\alpha\delta}\| \leq \|q_e - q_\alpha\| + \|q_\alpha - q_{\alpha\delta}\| \leq \varepsilon/2 + \varepsilon/2 = \varepsilon.$$

We will apply the Lavrentiev method for the regularization of Volterra integral equations of the first kind in both linear case (Section 4.2) and nonlinear case (Section 4.6).

In conclusion, we note that if A is not positive and self-adjoint, then the equation $Aq = f$ can be reduced to an equation with a positive self-adjoint operator by applying the operator A^*:

$$A^*Aq = A^*f.$$

Then the regularizing operator is written as

$$R_\alpha = (\alpha E + A^*A)^{-1}A^*.$$

2.5 The Tikhonov regularization method

In many ill-posed problems $Aq = f$, the class $M \subset Q$ of possible solutions is not a compact set, and approximate data f_δ may not belong to the class $A(M)$ for which solutions exist. The regularization method developed by A. N. Tikhonov (1963b, 1964) can be used to construct approximate solutions to such problems.

First, we give the general definition of a regularizing algorithm for the problem $Aq = f$ (Vasin and Ageev, 1995). Let Q and F be Banach spaces and let $A : Q \to F$ be a bounded linear operator. Suppose that, instead of the operator A and the right-hand side f, we have their approximations A_h and f_δ that satisfy the conditions

$$\|A - A_h\| \leq h, \quad \|f - f_\delta\|_F \leq \delta.$$

Recall that the norm of an operator $A : Q \to F$ is defined as

$$\|A\| = \sup_{\substack{q \in Q \\ q \neq \mathbf{0}}} \frac{\|Aq\|_F}{\|q\|_Q}.$$

For simplicity, when writing the norms we will omit the index specifying the space.

Denote $\overline{B(A,h)} = \{A_h\colon Q \to F \text{ such that } \|A - A_h\| \leq h\}$, $\overline{B(f,\delta)} = \{f_\delta \in F\colon \|f - f_\delta\| \leq \delta\}$, $Q_f = \{q \in Q\colon Aq = f\}$.

Definition 2.5.1. A family of mappings $R_{\delta h}\colon \overline{B(f,\delta)} \times \overline{B(A,h)} \to Q$ is called a *regularizing algorithm* for the problem $Aq = f$ if

$$\sup_{\substack{f_\delta \in \overline{B(f,\delta)} \\ A_h \in \overline{B(A,h)}}} \|R_{\delta h}(f_\delta, A_h) - Q_f\| \to 0$$

as $\delta \to 0$ and $h \to 0$ for all $f \in R(A) = A(Q)$. The set $\{R_{\delta h}(f_\delta, A_h)\}$, $\delta \in (0, \delta_0]$, $h \in (0, h_0]$ is called a *regularized family of approximate solutions* to the problem $Aq = f$.

Here the distance between $R_{\delta h}(f_\delta, A_h)$ and the set Q_f is as usual: $\|R_{\delta h}(f_\delta, A_h) - Q_f\| = \min_{q \in Q_f} \|R_{\delta h}(f_\delta, A_h) - q\|$.

If the equation $Aq = f$ is known a priori to have a solution $q_e \in M \subset Q$, then the set Q_f in the above definition can be replaced with $Q_f \cap M$. In the sequel, in most cases we assume that the representation of the operator A is given exactly.

Let q_e be an exact solution to the ill-posed problem $Aq = f$ for some $f \in F$.

Definition 2.5.2. A family of operators $\{R_\alpha\}_{\alpha>0}$ is said to be *regularizing* for the problem $Aq = f$ if

1) for any $\alpha > 0$ the operator $R_\alpha : F \to Q$ is continuous,

2) for any $\varepsilon > 0$ there exists an $\alpha_* > 0$ such that

$$\|R_\alpha f - q_e\| < \varepsilon$$

for all $\alpha \in (0, \alpha_*)$, in other words,

$$\lim_{\alpha \to +0} R_\alpha f = q_e. \tag{2.5.1}$$

If the right-hand side of the equation $Aq = f$ is approximate and the error $\|f - f_\delta\| \leq \delta$ in the initial data is known, then the regularizing family $\{R_\alpha\}_{\alpha>0}$ allows us not only to construct an approximate solution $q_{\alpha\delta} = R_\alpha f_\delta$, but also to estimate the deviation of the approximate solution $q_{\alpha\delta}$ from the exact solution q_e. Indeed, using the triangle inequality, we have

$$\|q_e - q_{\alpha\delta}\| \le \|q_e - R_\alpha f\| + \|R_\alpha f - R_\alpha f_\delta\|. \tag{2.5.2}$$

The first term in the right-hand side of (2.5.2) tends to zero as $\alpha \to +0$. Since the problem is ill-posed, the estimation of the second term as $\alpha \to +0$ and $\delta \to +0$ is a difficult task which is usually solved with due consideration of the specific character of the problem including an a priori and/or a posteriori information about its exact solution.

For example, consider the case where Q and F are Banach spaces, $A : Q \to F$ is a completely continuous linear operator, and R_α is a linear operator for any $\alpha > 0$. Assume that there exists a unique solution q_e for $f \in F$, and $f_\delta \in F$ is an approximation of f such that

$$\|f - f_\delta\| \le \delta. \tag{2.5.3}$$

We now estimate the difference between the exact solution q_e and the regularized solution $q_{\alpha\delta} = R_\alpha f_\delta$ using (2.5.2).

Set $\|q_e - R_\alpha f\| = \gamma(q_e, \alpha)$. The property (2.5.1) of a regularizing family implies that the first term in the right-hand side of (2.5.2) vanishes as $\alpha \to +0$, i.e., $\lim_{\alpha\to+0} \gamma(q_e, \alpha) = 0$.

Since R_α is linear, from the condition (2.5.3) it follows that

$$\|R_\alpha f - R_\alpha f_\delta\| \le \|R_\alpha\|\,\delta. \tag{2.5.4}$$

The norm $\|R_\alpha\|$ cannot be uniformly bounded because the problem is ill-posed. Indeed, otherwise we would have $\lim_{\alpha\to+0} R_\alpha = A^{-1}$ and the problem $Aq = f$ would be well-posed in the classical sense. However, if α and δ tend to zero at concerted rates, then the right-hand side of the obtained estimate

$$\|q_e - q_{\alpha\delta}\| \le \gamma(q_e, \alpha) + \|R_\alpha\|\delta \tag{2.5.5}$$

tends to zero. Indeed, setting $\omega(q_e, \delta) = \inf_{\alpha>0}\{\gamma(q_e, \alpha) + \|R_\alpha\|\delta\}$, we will show that

$$\lim_{\delta\to 0} \omega(q_e, \delta) = 0.$$

Take an arbitrary $\varepsilon > 0$. Since $\lim_{\alpha\to 0} \gamma(q_e, \alpha) = 0$, there exists an $\alpha_0(\varepsilon)$ such that for all $\alpha \in (0, \alpha_0(\varepsilon))$ we have

$$\gamma(q_e, \alpha) < \varepsilon/2.$$

Put $\mu_0(\varepsilon) = \inf_{\alpha\in(0,\alpha_0(\varepsilon))} \|R_\alpha\|$ and take $\delta_0(\varepsilon) = \varepsilon/(2\mu_0(\varepsilon))$. Then

$$\inf_{\alpha>0}\{\|R_\alpha\|\,\delta\} \le \delta \inf_{\alpha\in(0,\alpha_0(\varepsilon))}\{\|R_\alpha\|\} \le \varepsilon/2.$$

for all $\delta \in (0, \delta_0(\varepsilon))$.

Thus, for any $\varepsilon > 0$ there exist $\alpha_0(\varepsilon)$ and $\delta_0(\varepsilon)$ such that

$$\|q_e - q_{\alpha\delta}\| < \varepsilon$$

for all $\alpha \in (0, \alpha_0(\varepsilon))$ and $\delta \in (0, \delta_0(\varepsilon))$.

For specific operators A and families $\{R_\alpha\}_{\alpha>0}$, an explicit formula relating the regularization parameter α to the data error δ can be derived.

One of the well-known methods for constructing a regularizing family is the minimization of the Tikhonov functional

$$M(q, f_\delta, \alpha) = \|Aq - f_\delta\|^2 + \alpha\Omega(q - q^0),$$

where q^0 is a test solution, α is the regularization parameter, and Ω is a stabilizing functional, which is usually represented by the norm (or a seminorm), for example, $\Omega(q) = \|q\|^2$. The stabilizing functional Ω uses the *a priori* information about the degree of smoothness of the exact solution (or about the solution structure) and determines the type of the convergence of approximate solutions to the exact one for a given relationship between $\alpha(\delta) \to 0$ and $\delta \to 0$. For example, the Tikhonov method is effective for $\Omega(q) = \|q\|^2_{W_2^1}$ in the numerical solution of integral equations of the first kind that have a unique solution smooth enough (see (Vasin and Ageev, 1995) and the bibliography therein).

Consider a regularizing functional $M(q, f_\delta, \alpha)$ in the case of $\Omega(q) = \|q\|^2$:

$$M(q, f_\delta, \alpha) = \|Aq - f_\delta\|^2 + \alpha\|q\|^2, \quad \alpha > 0. \tag{2.5.6}$$

Theorem 2.5.1. *Let Q and F be Hilbert spaces and A be a completely continuous linear operator. Then for any $f \in F$ and $\alpha > 0$, the functional $M(q, f, \alpha)$ attains its greatest lower bound at a unique element q_α.*

Applying Theorem 2.5.1, we can construct an approximate solution $q_{\alpha\delta}$ from the approximate data $f_\delta \in F$ satisfying the condition $\|f - f_\delta\| \le \delta$ and prove that $q_{\alpha\delta}$ converges to the exact solution q_e to the equation $Aq = f$ as the parameters α and δ tend to zero at concerted rates (Denisov, 1999). Indeed, let the functional $M(q, f_\delta, \alpha)$ attain its lower bound at a point $q_{\alpha\delta}$, whose existence and uniqueness is guaranteed by Theorem 2.5.1.

Theorem 2.5.2. *Let the assumptions of Theorem 2.5.1 hold. Assume that there exists a unique solution q_e to the equation $Aq = f$ for some $f \in F$. Let $\{f_\delta\}_{\delta>0}$ be a family of approximate data such that $\|f - f_\delta\| < \delta$ for each of its elements. Then, if the regularization parameter $\alpha = \alpha(\delta)$ is chosen so that $\lim_{\delta\to 0} \alpha(\delta) = 0$ and $\lim_{\delta\to 0} \delta^2/\alpha(\delta) = 0$, then the element $q_{\alpha(\delta),\delta}$ which minimizes the regularizing functional $M(q, f_\delta, \alpha)$ tends to the exact solution q_e to the equation $Aq = f$, i.e., $\lim_{\delta\to 0} \|q_{\alpha(\delta),\delta} - q_e\| = 0$.*

Consider now the case when the equation $Aq = f$ is solvable but its solution is not unique. In this case we will seek a solution which is closest to a given element q^0.

Definition 2.5.3. Let Q_f be a set of all solutions to the equation $Aq = f$ and let an element $q^0 \in Q$ be given. The solution $q_{\mathrm{n}}^0 \in Q_f$ minimizing the functional $\|q - q^0\|$ on the set Q_f is called *normal with respect to* q^0: $q_{\mathrm{n}}^0 = \arg\min_{q \in Q_f} \|q - q^0\|$. If $q^0 = \mathbf{0}$, the solution $q_{\mathrm{n}} = \arg\min_{q \in Q_f} \|q\|$ is called *normal solution* to the equation $Aq = f$.

There holds the following theorem (Vasin, 1989).

Theorem 2.5.3. *Let the operator A_h and the right-hand side f_δ satisfy the approximation conditions*

$$\|A_h - A\| \le h, \quad \|f - f_\delta\| \le \delta, \tag{2.5.7}$$

where $h \in (0, h_0)$, $\delta \in (0, \delta_0)$. Assume that A and A_h are bounded linear operators from Q into F, where Q and F are Hilbert spaces. Let q_n^0 be a solution to the problem $Aq = f$ that is normal with respect to q^0. Then for all $\alpha > 0$, $q_0 \in Q$, $h \in (0, h_0)$, and $\delta \in (0, \delta_0)$ there exists a unique solution $q_{h\delta}^\alpha$ to the problem

$$\min\{\|A_h q - f_\delta\|^2 + \alpha\|q - q^0\|^2, \; q \in Q\}. \tag{2.5.8}$$

Furthermore, if the regularization parameter α satisfies the conditions

$$\lim_{\Delta \to 0} \alpha(\Delta) = 0, \quad \lim_{\Delta \to 0} \frac{(h + \delta)^2}{\alpha(\Delta)} = 0,$$

where $\Delta = \sqrt{h^2 + \delta^2}$, then

$$\lim_{\Delta \to 0} \|q_{h\delta}^\alpha - q_{\mathrm{n}}^0\| = 0.$$

Example 2.5.1 (differentiation problem). Suppose that the function $f(x) \in C^1(0, 1)$ is given approximately:

$$\|f - f_\delta\|_{C(0,1)} < \delta.$$

It is required to determine approximately the derivative $f'(x)$ from the function $f_\delta(x) \in C(0, 1)$.

The ill-posedness of this problem readily follows from the fact that the function $f_\delta(x)$ may not have a derivative at all. However, even if $f_\delta'(x)$ exists, the problem may turn out to be unstable. Indeed, with an error of the form $\delta \sin(x/\delta^2)$, the approximate function

$$f_\delta(x) = f(x) + \delta \sin(x/\delta^2)$$

tends to $f(x)$ as $\delta \to 0$. At the same time, the difference between the derivatives

$$f_\delta'(x) - f'(x) = \frac{1}{\delta}\cos(x/\delta^2)$$

increases indefinitely.

Consider a simple regularizing family $R_\alpha f(x) = (f(x+\alpha) - f(x))/\alpha$, where $x \in (0,1)$ and $\alpha \in (0, 1-x)$. We now estimate the deviation $f'(x) - R_\alpha f_\delta(x)$ at a fixed point $x \in (0,1)$. Since $f(x)$ is differentiable, we have

$$f(x+\alpha) = f(x) + f'(x)\alpha + o(\alpha).$$

Hence

$$\frac{f(x+\alpha) - f(x)}{\alpha} - f'(x) = \frac{o(\alpha)}{\alpha}.$$

Therefore,

$$\begin{aligned} |f'(x) - R_\alpha f_\delta(x)| &\le \left| f'(x) - \frac{f(x+\alpha) - f(x)}{\alpha} \right| \\ &\quad + \left| \frac{f(x+\alpha) - f(x)}{\alpha} - \frac{f_\delta(x+\alpha) - f_\delta(x)}{\alpha} \right| \\ &\le \frac{o(\alpha, x)}{\alpha} + \frac{2\delta}{\alpha}. \end{aligned} \tag{2.5.9}$$

In this case, for the right-hand side of inequality (2.5.9) to tend to zero, it is sufficient to require that α and δ tend to zero in such a way that $\delta/\alpha \to 0$, i.e., $\delta = o(\alpha)$.

The estimate (2.5.9) can be improved by imposing stronger requirements on the smoothness of $f(x)$. For example, suppose that $f(x)$ has the second derivative $f''(x)$ whose absolute value is bounded by a constant c_1 in a neighbourhood $O(x, \varepsilon) = \{y \in \mathbb{R} : |x - y| < \varepsilon\}$ of x:

$$\sup_{y \in O(x,\varepsilon)} |f''(y)| < c_1. \tag{2.5.10}$$

Then take $\alpha \in (0, \min\{\varepsilon, 1-x\})$ and apply the Taylor formula with Lagrange remainder:

$$f(x+\alpha) = f(x) + f'(x)\alpha + \frac{f''(x+\theta\alpha)\alpha^2}{2}, \quad \theta \in (0,1).$$

Hence

$$\left| \frac{f(x+\alpha) - f(x)}{\alpha} - f'(x) \right| \le \frac{c_1\alpha}{2}.$$

Thus, we get the following estimate instead of (2.5.9):

$$|f'(x) - R_\alpha f_\delta(x)| \le \frac{c_1\alpha}{2} + \frac{2\delta}{\alpha}. \tag{2.5.11}$$

The conditions expressing the dependence of α on δ can be refined, for example, by choosing $\alpha > 0$ so as to minimize the right-hand side of (2.5.11). Its minimum with respect to α is attained at the positive solution to the equation

$$\frac{c_1}{2} = \frac{2\delta}{\alpha^2},$$

i.e., at $\alpha = 2\sqrt{\delta/c_1}$. In this case we have

$$|f'(x) - R_\alpha f_\delta(x)| \le 2\sqrt{c_1\delta}.$$

Remark 2.5.1. Other families of regularizing operators can be constructed. For example, $R_\alpha f(x) = (f(x) - f(x-\alpha))/\alpha$ or $R_\alpha f(x) = (f(x+\alpha) - f(x-\alpha))/(2\alpha)$.

Remark 2.5.2. We estimated the deviation $|f'(x) - R_\alpha f_\delta(x)|$ in the neighbourhood of an arbitrary fixed point $x \in (0,1)$. Clearly, the estimate (2.5.11) will hold for all $x \in (0,1)$ if the condition (2.5.10) is replaced with the condition $\|f''\|_{C(0,1)} <$ const. (For an extended discussion of the pointwise and uniform regularization, see (Ivanov et al., 2002).)

Remark 2.5.3. In practice, aside from the fact that the values of $f(x)$ are specified approximately, they are often given only at several points of the interval being studied. In such situations, one can use interpolation polynomials (Newtonian and Lagrange polynomials, splines) and take the derivative of the interpolation polynomial for an approximate derivative of $f(x)$. In this case, estimating the deviation of the approximate derivative from the exact one requires additional assumptions on the smoothness of the derivative and a concordance between the measurement accuracy and the discretization step.

Remark 2.5.4. Another regularization method for evaluating $f'(x)$ consists in taking the convolution of the function $f_\delta(x)$ with a smooth function $\omega_\alpha(x)$ enjoying the following properties:

a) $\operatorname{supp}\{\omega_\alpha(x)\} \subset [-\alpha, \alpha]$;

b) $\int_{-\alpha}^{\alpha} \omega_\alpha(x)\,dx = 1$.

For example, the family $\{\omega_\alpha(x)\}$, $\alpha > 0$, can be chosen as follows:

$$\omega_\alpha(x) = \begin{cases} c_\alpha \exp\{-\alpha^2/(\alpha^2 - x^2)\}, & |x| \le \alpha, \\ 0, & |x| > \alpha, \end{cases}$$

$$c_\alpha = \Big(\int_{-\alpha}^{\alpha} \exp\Big\{-\frac{\alpha^2}{\alpha^2 - x^2}\Big\}dx\Big)^{-1}.$$

It is easy to verify that the operators R_α defined by the formula

$$R_\alpha f_\delta(x) = \int_{-\infty}^{\infty} f_\delta(y)\omega'_\alpha(x-y)\,dy$$

constitute a family of regularizing operators.

Regularizing sequence. In the construction of a regularizing family of operators, the real parameter $\alpha \to 0$ is sometimes replaced by a natural number $n \to \infty$. Consider several examples.

Example 2.5.2 (expansion in eigenfunctions). Assume that $\mathcal{H}$ is a separable Hilbert space. As in Section 2.3, let $A : \mathcal{H} \to \mathcal{H}$ be a completely continuous positive self-adjoint linear operator, and let $\{\varphi_n\}$ and $\{\lambda_n\}$ be the sequences of eigenfunctions and eigenvalues of A, respectively, where $\lambda_{k+1} \le \lambda_k$, $k \in \mathbb{N}$. An exact solution q_e to the equation $Aq = f$ can be represented in the form

$$q_e = \sum_{k=1}^{\infty} \frac{f_k}{\lambda_k}\varphi_k, \qquad f_k = \langle f, \varphi_k\rangle. \tag{2.5.12}$$

Since the solution exists and belongs to the Hilbert space $\mathcal{H}$, the series (2.5.12) and $\sum_{k=1}^{\infty}(f_k/\lambda_k)^2$ converge. However, if the equation $Aq = f_\delta$ has no solutions for some $f_\delta \in \mathcal{H}$ satisfying the condition $\|f - f_\delta\| < \delta$, then the series $\sum_{k=1}^{\infty}(f_{\delta k}/\lambda_k)^2$, $f_{\delta k} = \langle f_\delta, \varphi_k\rangle$, diverges (see the Picard criterion in Corollary 2.9.1). We now construct the sequence of operators $\{R_n\}$ such that

$$R_n f_\delta = \sum_{k=1}^{n} \frac{f_{\delta k}}{\lambda_k}\varphi_k \tag{2.5.13}$$

and show that it has all the properties of a regularizing family as $n \to \infty$. The operators R_n are obviously continuous, $\|R_n\| = 1/\lambda_n$, and for all $q \in Q$

$$\lim_{n\to\infty} R_n A q = \lim_{n\to\infty} \sum_{k=1}^{n} q_k\varphi_k = q, \qquad q_k = \langle q, \varphi_k\rangle.$$

Then the following estimate holds for the exact solution (2.5.12) to the equation $Aq = f$ and the regularized solution $q_{\delta n} = R_n f_\delta$:

$$\|q_e - q_{\delta n}\| \le \|q_e - R_n f\| + \|R_n f - R_n f_\delta\| \le \|q_e - R_n A q_e\| + \|R_n(f - f_\delta)\|$$

$$\le \sum_{k=n+1}^{\infty} q_k^2 + \|R_n\|\delta = \sum_{k=n+1}^{\infty} q_k^2 + \frac{\delta}{\lambda_n}.$$

Consequently, for any $\varepsilon > 0$ we can find a number n_0 such that $\sum_{k=n_0+1}^{\infty} q_k^2 < \varepsilon/2$ and then choose $\delta > 0$ to satisfy the condition $\delta < \lambda_n\varepsilon/2$. Then the inequality $\|q_e - q_{\delta n}\| < \varepsilon$ holds for all $n > n_0$ and $\delta \in (0, \lambda_n\varepsilon/2)$.

Example 2.5.3 (the method of successive approximations). Under the assumptions of the preceding example, suppose that $\lambda_1 < 1$. Take a sequence $\{q_n\}$ defined by the formula

$$q_{n+1} = q_n - Aq_n + f, \quad q_0 = f, \quad n = 0, 1, 2, \dots.$$

We will show that if the equation $Aq = f$ has an exact solution $q_e \in \mathcal{H}$, then

$$\lim_{n\to\infty} q_n = q_e.$$

Indeed,

$$q_n = \sum_{k=0}^{n} (E - A)^k f = \sum_{k=0}^{n} (E - A)^k Aq_e.$$

It is not difficult to prove that

$$q_e - q_n = (E - A)^{n+1} q_e. \tag{2.5.14}$$

We expand (2.5.14) in the eigenfunctions φ_n to get

$$q_e - q_n = \sum_{k=1}^{\infty} (1 - \lambda_k)^{n+1} q_n \varphi_k, \quad q_k = \langle q_e, \varphi_k \rangle.$$

Then

$$\|q_e - q_n\|^2 = \sum_{k=1}^{\infty} (1 - \lambda_k)^{2(n+1)} q_k^2.$$

We leave it to the reader to complete the proof.

Exercise 2.5.1. Prove that $\lim_{n\to\infty} \|q_e - q_n\| = 0$.

Exercise 2.5.2. Prove that the sequence of operators defined by the equality $R_n f = \sum_{k=0}^{n} (E - A)^k f$ is regularizing and $\|R_n\| = n + 1$.

The next section deals with gradient methods, which represent one of the most effective and commonly used class of iterative regularization methods.

2.6 Gradient methods

The idea of replacing the original equation (problem)

$$Aq = f \tag{2.6.1}$$

with the problem of finding the minimum of the objective functional

$$J(q) = \|Aq - f\|^2 \to \min \tag{2.6.2}$$

dates back to the works of A. M. Legendre (1806) and K. Gauss (1809), who proposed the least squares method for solving systems of linear algebraic equations. A. Cauchy (1847) proposed the steepest descent method for solving the problem of finding the minimum of a function of n variables. L. V. Kantorovich proposed to solve linear operator equations $Aq = f$ in Hilbert spaces by minimizing the functional $H(q) = \langle Aq, q\rangle - 2\langle f, q\rangle$ using the steepest descent method. He also studied the convergence of this method in the case where $m\langle q,q\rangle \leq \langle Aq,q\rangle \leq M\langle q,q\rangle$, $m > 0$, and pointed out the guidelines for investigating the case of $m = 0$ (Kantorovich, 1952). In what follows, we will minimize the functional $J(q)$, which is more commonly used in practice than $H(q)$. Note that for $A = A^*$, the gradient $H'q$ of the functional $H(q)$ is equal to $2(Aq - f)$, and its decrease with respect to q means that q approaches the solution to the equation $Aq = f$.

In this section, we briefly describe the general structure of several gradient methods for solving ill-posed problems $Aq = f$ and discuss the convergence of these methods. A more detailed description of the gradient methods and their application to inverse problems can be found in (Alifanov et al., 1988).

The structure of gradient methods. We consider the case where Q and F are Hilbert spaces, A is a Fréchet-differentiable operator from Q into F. To solve the equation $Aq = f$, we minimize the objective functional

$$J(q) = \langle Aq - f, Aq - f\rangle = \|Aq - f\|^2.$$

Lemma 2.6.1. *If A is a Fréchet-differentiable operator, then the functional $J(q)$ is also differentiable and its gradient $J'q$ is expressed in terms of A' by the formula*

$$J'q = 2(A'q)^*(Aq - f). \tag{2.6.3}$$

Proof. Formula (2.6.3) follows from the obvious equalities

$$\begin{aligned} J(q+\delta q) - J(q) &= \langle A(q+\delta q) - f, A(q+\delta q) - f\rangle - \langle A(q+\delta q) - f, Aq - f\rangle \\ &\quad + \langle A(q+\delta q) - f, Aq - f\rangle - \langle Aq - f, Aq - f\rangle \\ &= \langle A(q+\delta q) - f, (A'q)\delta q\rangle + \langle (A'q)\delta q, Aq - f\rangle + o(\|\delta q\|) \\ &= \langle 2(A'q)\delta q, Aq - f\rangle + o(\|\delta q\|) \\ &= \langle \delta q, 2(A'q)^*(Aq - f)\rangle + o(\|\delta q\|). \end{aligned}$$ □

Remark 2.6.1. If A is a linear operator, then $J'q = 2A^*(Aq - f)$.

The simplest gradient methods have the form

$$q_{n+1} = q_n - \alpha_n J'q_n. \tag{2.6.4}$$

This formula represents different methods depending on the way of defining the positive parameter α_n. Throughout this section, we assume that A is a linear operator.

Theorem 2.6.1. *Assume that Q and F are Hilbert spaces, and $A : Q \to F$ is a continuous linear operator such that both A and A^* have zero kernels. Then the functional*

$$J(q) = \langle Aq - f, Aq - f \rangle$$

has no more than one stationary point.

Proof. Suppose that $J(q)$ has two stationary points $q_1, q_2 \in Q$. Then $J'q_1 = 0$ and $J'q_2 = 0$. By Lemma 2.6.1, we have $J'q_j = 2A^*(Aq_j - f) = 0$ for $j = 1, 2$, which means that $Aq_j - f \in \operatorname{Ker} A^*$.

Hence, $Aq_j = f$ and therefore $q_1 = q_2$, since the solution is unique. □

Consider several examples of gradient methods

$$q_{n+1} = q_n - \alpha_n J'q_n, \quad q_n \in Q, \quad \alpha_n > 0, \tag{2.6.5}$$

such that the descent parameter α_n is either fixed or determined from the minimum condition for a quality functional. As before, let q_e denote an exact solution to the problem $Aq = f$ (which is not necessarily unique).

The method of simple iteration (also known as the Landweber iteration in the theory of ill-posed problems)**.** The parameter α_n is fixed:

$$\alpha_n = \alpha \in \Big(0, \frac{1}{\|A\|^2}\Big).$$

The minimum error method. The parameter α_n is determined so as to minimize

$$\begin{aligned}\|q_{n+1} - q_e\|^2 &= \|q_n - \alpha J'q_n - q_e\|^2 \\ &= \|q_n - q_e\|^2 - 2\alpha\langle q_n - q_e, J'q_n\rangle + \alpha^2\|J'q_n\|^2.\end{aligned}$$

The quantity $\|q_{n+1} - q_e\|^2$ attains its minimum at the point

$$\begin{aligned}\alpha_n = \frac{\langle q_n - q_e, J'q_n\rangle}{\|J'q_n\|^2} &= \frac{\langle q_n - q_e, 2A^*(Aq_n - f)\rangle}{\|J'q_n\|^2} \\ &= 2\frac{\langle Aq_n - f, Aq_n - f\rangle}{\|J'q_n\|^2} = \frac{2J(q_n)}{\|J'q_n\|^2}.\end{aligned}$$

Note that q_e appears in the expression $\|q_{n+1} - q_e\|^2$, but the descent step does not depend on q_e.

Remark 2.6.2. The conditions $J(q_n) \neq 0$ and $J'q_n \neq 0$ will be used below in the analysis of gradient methods. The first condition must be verified before each descent step. Even if $J(q_n) \neq 0$, further calculations may become meaningless if $J(q_n)$ is extremely small. When $J(q_n) = 0$, the procedure must be stopped, since we get the required solution: $q_e = q_n$.

The second condition $J'q_n \neq 0$ means that we do not arrive at a minimum of the functional.

The steepest descent method. We choose, as the parameter α_n, the minimizer of the expression

$$\begin{aligned} J(q_{n+1}) &= J(q_n - \alpha J'q_n) = \|Aq_n - \alpha AJ'q_n - f\|^2 \\ &= \|Aq_n - f\|^2 - 2\langle Aq_n - f, \alpha AJ'q_n\rangle + \alpha^2\|AJ'q_n\|^2 \\ &= J(q_n) - \alpha\|J'q_n\|^2 + \alpha^2\|AJ'q_n\|^2. \end{aligned}$$

Exercise 2.6.1. Prove that the minimum of the above expression is attained at the point

$$\alpha_n = \frac{\|J'q_n\|^2}{2\|AJ'q_n\|^2}.$$

If there exists a continuous inverse of the operator A, then the rate of convergence of the method of simple iteration, the minimum error method, and the steepest descent method is linear. The implementation of these methods requires minimal amount of calculations per iteration (compared to other methods, such as the conjugate gradient method).

method	**descent parameter α**
simple iteration	$\alpha = \text{const} \in (0, \|A\|^{-2})$
minimum error	$\alpha_n = 2J(q_n)\|J'q_n\|^{-2}$
steepest descent	$\alpha_n = (1/2)\|J'q_n\|^2\|AJ'q_n\|^{-2}$

The conjugate gradient method. At first, an initial approximation q_0 is specified and $p_0 = J'q_0$ is calculated. Suppose that q_n and p_n have already been determined.

At step $n+1$, we first calculate the auxiliary function

$$p_n = J'q_n + \|J'q_n\|^2\|J'q_{n-1}\|^{-2}\, p_{n-1},$$

then the descent parameter

$$\alpha_n = \frac{\langle J'q_n, p_n\rangle}{\|Ap_n\|^2},$$

and then the approximate solution

$$q_{n+1} = q_n - \alpha_n p_n.$$

Convergence of gradient methods. Let A be a continuous linear operator, and let the equation $Aq = f$ have more than one solution. Then one can introduce additional requirements for the desired solution, for example, the requirement that the solution have minimal norm or be the closest to a certain element $q^0 \in Q$. Denote $Q_f = \{q \in Q : Aq = f\}$. Recall that a normal solution to the equation $Aq = f$ with respect to some $q^0 \in Q$ is a solution $q^0_{\mathrm{n}} \in Q_f$ that has the least deviation from q^0, i.e.,

$$\|q^0_{\mathrm{n}} - q^0\| = \min_{q \in Q_f} \|q - q^0\|.$$

Remark 2.6.3. The element q^0 is usually chosen on the basis of some *a priori* information about the solution.

Consider the problem of finding a normal solution to the equation $Aq = f$ with respect to a given $q^0 \in Q$. We denote the range of A by

$$R(A) = A(Q) = \{f \in F : \exists q \in Q \text{ such that } Aq = f\}. \tag{2.6.6}$$

Note that, the results concerning the convergence and stability of gradient methods as applied to well-posed problems remain valid in the case of finding a normal solution with respect to q^0 if the range $R(A)$ of the operator A is closed.

If $R(A)$ is not closed, then some minimizing sequences may be divergent even when the solution of the equation $Aq = f$ is unique, since A^{-1} is not bounded. It can be proved, however, that the minimizing sequence $\{q_k\}$ constructed using gradient methods converges with respect to the functional ($J(q_n) \searrow 0$ as $n \to \infty$) and in norm ($\|q_k - q^0_{\mathrm{n}}\| \to 0$ as $k \to \infty$) to the normal solution to the equation $Aq = f$ with respect to the initial approximation $q^0 = q_0$ (Alifanov et al., 1988).

Let A be a continuous linear operator that has no bounded inverse. Suppose that the equation

$$Aq = f \tag{2.6.7}$$

has a solution (which is not necessarily unique).

Consider the method of simple iteration for solving equation (2.6.7):

$$\begin{gathered} q_{n+1} = q_n - \alpha J' q_n, \quad q_0 \in Q, \quad n \in \mathbb{N} \cup \{0\}, \\ \alpha = \text{const} \in (0, \|A\|^{-2}). \end{gathered} \tag{2.6.8}$$

The condition imposed on α will be needed in the proof of convergence.

Lemma 2.6.2. *For all $n \in \mathbb{N} \cup \{0\}$*

$$J(q_n) - J(q_{n+1}) = \alpha \|J'q_n\|^2 - \alpha^2 \|AJ'q_n\|^2. \tag{2.6.9}$$

Proof. Develop the difference $J(q_n) - J(q_{n+1})$ taking into account (2.6.3) and (2.6.8):

$$\begin{aligned} J(q_n) - J(q_{n+1}) &= \|Aq_n - f\|^2 - \|Aq_n - A\alpha J'q_n - f\|^2 \\ &= 2\langle Aq_n - f, A\alpha J'q_n\rangle - \alpha^2 \|AJ'q_n\|^2 \\ &= \alpha\langle 2A^*(Aq_n - f), J'q_n\rangle - \alpha^2 \|AJ'q_n\|^2 \\ &= \alpha \|J'q_n\|^2 - \alpha^2 \|AJ'q_n\|^2, \end{aligned}$$

which proves the lemma. □

Lemma 2.6.3. *Let $q_e \in Q_f$ be a solution to* (2.6.7)*. Then for all $n \in \mathbb{N} \cup \{0\}$*

$$\|q_n - q_e\|^2 - \|q_{n+1} - q_e\|^2 = 4\alpha J(q_n) - \alpha^2 \|J'q_n\|^2. \tag{2.6.10}$$

Equality (2.6.10) follows from the trivial equalities

$$\begin{aligned} \|q_n - q_e\|^2 - \|q_n - \alpha J'q_n - q_e\|^2 &= 2\alpha\langle q_n - q_e, J'q_n\rangle - \alpha^2 \|J'q_n\|^2 \\ &= 4\alpha\langle Aq_n - f, Aq_n - f\rangle - \alpha^2 \|J'q_n\|^2. \end{aligned}$$

Since $0 < \alpha < \|A\|^{-2}$, from Lemmas 2.6.2, 2.6.3 it follows that

$$\begin{aligned} 0 < J(q_{n+1}) &< J(q_n), \\ 0 < \|q_{n+1} - q_e\| &< \|q_n - q_e\|, \end{aligned}$$

i.e., the sequences $\{J(q_n)\}$, $\{\|q_n - q_e\|\}$ converge.

We now turn to the steepest descent method:

$$q_{n+1} = q_n - \alpha_n J'q_n, \quad q_0 \in Q, \quad n \in \mathbb{N} \cup \{0\} \tag{2.6.11}$$

$$\alpha_n = \arg\min_{\alpha \ge 0} J(q_n - \alpha J'q_n). \tag{2.6.12}$$

Lemma 2.6.4. *The sequence defined by formulae* (2.6.11), (2.6.12) *has the following properties:*

$$J(q_n) - J(q_{n+1}) = \frac{\alpha_n}{2} \|J'q_n\|^2, \tag{2.6.13}$$

$$\alpha_n = \|J'q_n\|^2 / (2\|AJ'q_n\|^2). \tag{2.6.14}$$

Proof. Proving (2.6.14) was suggested as an exercise to the reader (see Exercise 2.6.1). To prove (2.6.13), we substitute $\alpha = \alpha_n$ into (2.6.9) and perform simple manipulations:

$$J(q_n) - J(q_{n+1}) = \alpha_n(\|J'q_n\|^2 - \alpha_n\|AJ'q_n\|^2)$$
$$= \alpha_n\Big(\|J'q_n\|^2 - \frac{\|J'q_n\|^2}{2\|AJ'q_n\|^2}\|AJ'q_n\|^2\Big) = \frac{\alpha_n}{2}\|J'q_n\|^2. \quad \square$$

Lemma 2.6.5. *Let $q_e \in Q$ be a solution to the problem* (2.6.7). *Then for all $n \in \mathbb{N} \cup \{0\}$ there holds*

$$\|q_n - q_e\|^2 - \|q_{n+1} - q_e\|^2 = 2\alpha_n(J(q_n) + J(q_{n+1})). \tag{2.6.15}$$

Proof. Substituting α_n for α in (2.6.10) and using (2.6.13), we have

$$\|q_n - q_e\|^2 - \|q_{n+1} - q_e\|^2 = 4\alpha_n J(q_n) - \alpha_n^2\frac{2}{\alpha_n}(J(q_n) - J(q_{n+1}))$$
$$= 2\alpha_n(J(q_n) + J(q_{n+1})). \tag{2.6.16}$$

□

Lemma 2.6.6. *If α_n is chosen to satisfy the condition* (2.6.12), *then the following inequality holds for all $n \in \mathbb{N} \cup \{0\}$:*

$$\frac{1}{2\|A\|^2} \le \alpha_n. \tag{2.6.17}$$

Proof. Inequality (2.6.17) follows from (2.6.14) and the definition of the operator norm:

$$\alpha_n = \frac{\|J'q_n\|^2}{2\|AJ'q_n\|^2} \ge \frac{\|J'q_n\|^2}{2\|A\|^2\|J'q_n\|^2} = \frac{1}{2\|A\|^2}. \quad \square$$

Theorem 2.6.2. *Let $A : Q \to F$ be a linear continuous operator, and let Q, F be Hilbert spaces. If there exists a unique solution q_e to the problem $Aq = f$, then the sequence of approximate solutions* (2.6.11), (2.6.12) *obtained using the steepest descent method converges to the solution q_e.*

Proof. From Lemmas 2.6.4 and 2.6.5 it follows that

$$0 < J(q_{n+1}) < J(q_n), \tag{2.6.18}$$
$$0 < \|q_{n+1} - q_e\| < \|q_n - q_e\|, \tag{2.6.19}$$

which means that the sequences $\{J(q_n)\}$ and $\{\|q_n - q_e\|\}$ are positive and monotonically decreasing, and therefore convergent.

Summing (2.6.15) over $n = \overline{0,k}$ yields:

$$\|q_0 - q_e\|^2 - \|q_{k+1} - q_e\|^2 = 2\sum_{n=0}^{k} \alpha_n (J(q_n) + J(q_{n+1})). \tag{2.6.20}$$

Since the sequence $\{\|q_n - q_e\|\}$ has a limit, from (2.6.20) it follows that

$$2\sum_{n=0}^{k} \alpha_n (J(q_n) + J(q_{n+1})) \le C$$

for all $k \in \mathbb{N} \cup \{0\}$, where C is a constant. On the other hand, in view of (2.6.17),

$$2\sum_{n=0}^{k} \alpha_n (J(q_n) + J(q_{n+1})) \ge \frac{1}{\|A\|}\sum_{n=0}^{k} (J(q_n) + J(q_{n+1})) \ge \frac{2}{\|A\|}\sum_{n=0}^{k+1} J(q_n).$$

Hence, $\sum_{n=0}^{k} J(q_n) \le C$ for any $k \in \mathbb{N} \cup \{0\}$, i.e.,

$$\sum_{n=0}^{\infty} J(q_n) < \infty. \tag{2.6.21}$$

Then $\lim_{n\to\infty} J(q_n) = 0$, and $\{q_n\}$ is a minimizing sequence.

In view of (2.6.19), the sequence $\{q_n\}$ is bounded. Then there exists a subsequence $\{q_{n_k}\}_{k\in\mathbb{N}}$ and $q_c \in Q$ such that $\lim_{k\to\infty}\langle q, q_{n_k}\rangle = \langle q, q_c\rangle$ for all $q \in Q$.

Note that the continuity of A implies its weak continuity. Indeed, for any $\varphi \in F$, $A^*\varphi = w \in R(A^*) \subset Q$, which implies that $\langle q_{n_k}, w\rangle = \langle Aq_{n_k}, \varphi\rangle$ and $\langle w, q_c\rangle = \langle \varphi, Aq_c\rangle$. Therefore, the sequence $\{Aq_{n_k}\}$, $k \in \mathbb{N}$, weakly converges to Aq_c. On the other hand, $\{q_{n_k}\}_{k\in\mathbb{N}}$ is also a minimizing sequence, i.e., $\{Aq_{n_k}\}_{k\in\mathbb{N}}$ strongly (and therefore weakly) converges to f. As a result, $Aq_c = f$ and, since the solution of the problem $Aq = f$ is unique, we conclude that $q_c = q_e$, which completes the proof. □

A more detailed analysis of these methods and other gradient methods can be found in (Alifanov et al., 1988).

2.7 An estimate of the convergence rate with respect to the objective functional

In this section we estimate the rate of convergence of the steepest descent method with respect to the objective functional. With the inverse problem

$$Aq = f \tag{2.7.1}$$

we associate the objective functional

$$J(q) = \|Aq - f\|^2. \tag{2.7.2}$$

Lemma 2.7.1. *Let Q and F be Hilbert spaces and $A : Q \to F$ be a bounded linear operator. Then for all $q, \delta q \in Q$ there holds*

$$|J(q+\delta q) - J(q) - \langle \delta q, J'q \rangle| \le \|A\|^2 \|\delta q\|^2. \tag{2.7.3}$$

Proof. Taking into account the definition of the functional $J(q)$ and Remark 2.6.1, we have

$$\begin{aligned}
&J(q+\delta q) - J(q) - \langle \delta q, J'q \rangle \\
&\quad = \|Aq + A\delta q - f\|^2 - \|Aq - f\|^2 - \langle A\delta q, 2(Aq-f)\rangle \\
&\quad = \langle Aq - f + A\delta q, Aq - f + A\delta q\rangle - \|Aq - f\|^2 - \langle A\delta q, 2(Aq-f)\rangle \\
&\quad = \langle Aq - f, Aq - f\rangle + \langle A\delta q, Aq - f\rangle + \langle Aq - f, A\delta q\rangle + \|A\delta q\|^2 \\
&\qquad - \|Aq - f\|^2 - \langle A\delta q, 2(Aq-f)\rangle \\
&\quad = \|A\delta q\|^2,
\end{aligned}$$

which yields the inequality (2.7.3). □

Consider the steepest descent method

$$q_{n+1} = q_n - \alpha_n J'q_n, \quad n \in \mathbb{N} \cup \{0\}, \quad q_0 \in Q, \tag{2.7.4}$$

where the descent parameter is determined by the formula

$$\alpha_n = \arg\min_{\alpha \ge 0} J(q_n - \alpha J'q_n). \tag{2.7.5}$$

Theorem 2.7.1 (on the convergence of the steepest descent method with respect to the objective functional). *Let Q and F be Hilbert spaces and A be a bounded linear operator. Assume that there exists a solution $q_e \in Q$ to equation* (2.7.1) *for some $f \in F$. Then the sequence $\{q_n\}$ defined by* (2.7.4) *and* (2.7.5) *converges to zero with respect to the objective functional (i.e., $\lim_{n\to\infty} J(q_n) = 0$) and the following estimate holds:*

$$J(q_n) \le \frac{\|A\|^2 \|q_0 - q_e\|^2}{n}. \tag{2.7.6}$$

Proof. Substituting $q = q_n$ and $\delta q = -\alpha J'q_n$ ($n \in \mathbb{N} \cup \{0\}, \alpha > 0$) into (2.7.3), we obtain

$$|J(q_n - \alpha J'q_n) - J(q_n) + \alpha \|J'q_n\|^2| \le \|A\|^2 \alpha^2 \|J'q_n\|^2. \tag{2.7.7}$$

From (2.7.7) it follows that

$$J(q_n - \alpha J'q_n) - J(q_n) \le \alpha(\alpha\|A\|^2 - 1)\|J'q_n\|^2. \tag{2.7.8}$$

By the definition of α_n, the following inequality holds for all $\alpha > 0$:

$$J(q_{n+1}) = J(q_n - \alpha_n J'q_n) \leq J(q_n - \alpha J'q_n).$$

Then (2.7.8) can be written in the form

$$J(q_{n+1}) - J(q_n) \leq J(q_n - \alpha J'(q_n)) - J(q_n) \leq \alpha(\alpha\|A\|^2 - 1)\|J'q_n\|^2,$$

which implies

$$\alpha(1 - \alpha\|A\|^2)\|J'q_n\|^2 \leq J(q_n) - J(q_{n+1}), \quad \alpha > 0. \tag{2.7.9}$$

The right-hand side of (2.7.9) does not depend on α. Thus, inequality (2.7.9) holds for all α and, in particular, for $\alpha = 1/(2\|A\|^2)$. Then (2.7.9) implies that

$$J(q_{n+1}) + \frac{1}{4\|A\|^2}\|J'q_n\|^2 \leq J(q_n). \tag{2.7.10}$$

From (2.7.10) we conclude that $0 < J(q_{n+1}) < J(q_n)$. Being decreasing and bounded from below, the sequence $\{J(q_n)\}$ has a limit.

Then from (2.7.10) it follows that the sequence $\{\|J'q_n\|^2\}$ has a limit:

$$\lim_{n\to\infty}\|J'q_n\|^2 = 0. \tag{2.7.11}$$

In view of the obvious equalities

$$\begin{aligned} J(q_n) &= \langle Aq_n - f, Aq_n - f\rangle = \langle A(q_n - q_e), A(q_n - q_e)\rangle \\ &= \frac{1}{2}\langle q_n - q_e, 2A^*A(q_n - q_e)\rangle = \frac{1}{2}\langle J'q_n, q_n - q_e\rangle, \end{aligned}$$

we have

$$J(q_n) \leq \frac{1}{2}\|J'q_n\|\|q_0 - q_e\|, \tag{2.7.12}$$

which yields $\lim_{n\to\infty} J(q_n) = 0$ because of (2.7.11).

From inequality (2.7.12) it follows that

$$\frac{2J(q_n)}{\|q_0 - q_e\|} \leq \|J'q_n\|. \tag{2.7.13}$$

Then (2.7.10) implies

$$\frac{J^2(q_n)}{\|A\|^2\|q_0 - q_e\|^2} \leq J(q_n) - J(q_{n+1}). \tag{2.7.14}$$

To establish the estimate (2.7.6), we formulate and prove the following lemma, which is presented in (Vasil'ev, 1988) in a more general form.

Lemma 2.7.2. *Assume that the elements of a sequence $a_0, a_1, a_2, \dots$ are positive and satisfy the conditions*

$$\frac{a_{n-1}^2}{B} \le a_{n-1} - a_n, \quad B = \text{const} > 0,\ n \in \mathbb{N}. \tag{2.7.15}$$

Then for all $k \in \mathbb{N}$

$$a_k < \frac{B}{k}. \tag{2.7.16}$$

Proof. Note that $\{a_n\}$ is a decreasing sequence. Divide (2.7.15) by $a_n a_{n-1}$:

$$0 < \frac{a_{n-1}}{B a_n} \le \frac{1}{a_n} - \frac{1}{a_{n-1}}.$$

Since the sequence is decreasing, we have

$$\frac{1}{B} \le \frac{1}{a_n} - \frac{1}{a_{n-1}}.$$

Then, summing these inequalities over n from 1 to k, we arrive at the estimate

$$\frac{k}{B} \le \sum_{n=1}^{k} \Big(\frac{1}{a_n} - \frac{1}{a_{n-1}} \Big) = \frac{1}{a_k} - \frac{1}{a_0} < \frac{1}{a_k},$$

which implies (2.7.16). The lemma is proved. □

To verify the estimate (2.7.6), it remains to apply Lemma 2.7.2 to the sequence $\{J(q_n)\}$ using the inequality (2.7.14) for (2.7.15). Thereby Theorem 2.7.1 is proved. □

Theorem 2.7.2 (on the convergence of the method of simple iteration with respect to the objective functional). *Let Q and F be Hilbert spaces and A be a bounded linear operator. Assume that there exists an exact solution $q_e \in Q$ to the equation $Aq = f$ for some $f \in F$. Then, for any $q_0 \in Q$ and $\alpha \in (0, \|A\|^{-2})$, the sequence $\{q_n\}$ defined by the equalities*

$$q_{n+1} = q_n - \alpha J' q_n, \quad n \in \mathbb{N} \cup \{0\}, \tag{2.7.17}$$

converges with respect to the objective functional and the following estimate holds:

$$J(q_n) \le \frac{\|q_0 - q_e\|^2}{4n\alpha(1 - \alpha\|A\|^2)}. \tag{2.7.18}$$

The proof is similar to that of the previous theorem. Indeed, it is not difficult to verify that inequalities (2.7.9) and (2.7.13) are valid not only for the steepest descent method but also for the method of simple iteration, that is, when α is fixed. From these inequalities it follows that

$$\frac{4\alpha(1-\alpha\|A\|^2)}{\|q_0-q_e\|^2} J^2(q_n) \leq J(q_n) - J(q_{n+1}). \tag{2.7.19}$$

Putting

$$a_n = J(q_n), \quad B = \frac{\|q_0-q_e\|^2}{4\alpha(1-\alpha\|A\|^2)}$$

in Lemma 2.7.2, we obtain the desired estimate (2.7.18).

2.8 Conditional stability estimate and strong convergence of gradient methods applied to ill-posed problems

Theorem 2.2.1 guarantees the continuity of the operator $A^{-1} : A(M) \to M$ if $M \subset Q$ is compact and the operator $A : Q \to F$ is a continuous one-to-one map from M onto $A(M)$. M. M. Lavrentiev pointed out that for conditionally well-posed problems, Theorem 2.2.1 implies the existence of the modulus of continuity of the operator A^{-1} on the set $A(M)$ (see Definition 2.2.1), i.e., the function $\omega(A(M), \delta)$.

If the form of the function $\omega(\delta, A(M))$ or of its majorant (for example, a conditional stability estimate) is known, it is possible to estimate in norm the difference between the solution of the equation $Aq = f$ and that of $Aq = f_\delta$.

Definition 2.8.1. Let the problem $Aq = f$ be conditionally well-posed on a set $M \subset Q$. We say that a *conditional stability estimate* is obtained *on the well-posedness set* M if an increasing real-valued function $\beta(\delta)$ is found such that

1) $\lim_{\delta\to+0} \beta(\delta) = 0$;

2) for any $f \in A(M)$ and $f_\delta \in A(M)$ satisfying the estimate $\|f - f_\delta\| \leq \delta$, the inequality

$$\|q_e - q_\delta\| \leq \beta(\delta)$$

holds true, where $q_e \in M$ is a solution to the equation $Aq = f$ and $q_\delta \in M$ is a solution to the equation $Aq = f_\delta$.

As an example, we consider two well-known ill-posed problems. (A detailed analysis of these problems is given in the corresponding sections.)

Example 2.8.1 (see Chapter 8). Consider the Cauchy problem for a parabolic equation in reversed time:

$$u_t = -u_{xx}, \quad x \in (0,1), \quad t \in (0,T), \quad T > 0;$$
$$u(x,0) = f(x), \quad x \in (0,1);$$
$$u(0,t) = u(1,t) = 0, \quad t \in (0,T).$$

Let $u_e(x,t)$ be an exact solution to this problem. We denote by $u_\delta(x,t)$ the solution to the perturbed problem

$$u_t = -u_{xx}, \quad x \in (0,1), \quad t \in (0,T);$$
$$u(x,0) = f_\delta(x), \quad x \in (0,1);$$
$$u(0,t) = u(1,t) = 0, \quad t \in (0,T).$$

Then from the conditions $\|u_e(\cdot,T)\|_{L_2(0,1)} \leq C$, $\|u_\delta(\cdot,T)\|_{L_2(0,1)} \leq C$, and $\|f - f_\delta\|_{L_2(0,1)} \leq \delta$ it follows that

$$\|u_e(\cdot,t) - u_\delta(\cdot,t)\|_{L_2(0,1)} \leq (2C)^{t/T}\delta^{(T-t)/T}, \quad t \in (0,T).$$

Example 2.8.2 (see Chapter 9). The Cauchy problem for Laplace's equation has the form

$$u_{xx} + u_{yy} = 0, \quad x > 0, y \in R; \tag{2.8.1}$$
$$u(0,y) = f(y), \quad y \in R; \tag{2.8.2}$$
$$u_x(0,y) = 0, \quad y \in R. \tag{2.8.3}$$

This problem will be considered in the domain $(x,y) \in \Omega = (0,1) \times (0,1)$ with boundary conditions

$$u(x,0) = u(x,1) = 0, \quad x \in (0,1). \tag{2.8.4}$$

The problem of determining $u(x,y)$ in Ω will be solved if we find the trace of $u(x,y)$ for $x = 1$, i.e.,

$$q(y) = u(1,y), \quad y \in (0,1). \tag{2.8.5}$$

Indeed, knowing $q(y)$, we can solve the well-posed Dirichlet problem for Laplace's equation (2.8.1), (2.8.2), (2.8.4), (2.8.5) and determine the exact solution $u_e(x,y)$ to the original problem in Ω.

Assume that the problem (2.8.1)–(2.8.4) has an exact solution u_e with the trace (2.8.5), and $\|q\|_{L_2(0,1)} < C$. Then, if u_δ is a solution to the perturbed problem (with $\|f - f_\delta\|_{L_2(0,1)} \le \delta$)

$$\begin{aligned} u_{xx} + u_{yy} &= 0, \quad x \in (0,1), \quad y \in (0,1); \\ u(0,y) &= f_\delta(y), \quad u_x(0,y) = 0, \quad y \in (0,1); \\ u(x,0) &= u(x,1) = 0, \quad x \in (0,1); \end{aligned}$$

and $q_\delta(y) = u_\delta(1,y)$ satisfies the estimate $\|q_\delta\|_{L_2(0,1)} \le C$, then

$$\|u_e(x,\cdot) - u_\delta(x,\cdot)\|_{L_2(0,1)} \le (2C)^x \delta^{1-x}, \quad x \in (0,1).$$

Strong convergence of the steepest descent method. If a conditional stability estimate $\beta(\delta)$ is obtained on the well-posedness set

$$M = \{q \in Q : \|q\| \le C\},$$

then an estimate of the strong convergence rate for gradient methods can be derived from the estimate of the rate of convergence with respect to the objective functional. For example, consider the steepest descent method

$$q_{n+1} = q_n - \alpha_n J' q_n, \quad n = 0,1,2,\ldots, \quad q_0 \in Q, \tag{2.8.6}$$

$$\alpha_n = \arg\min_{\alpha \ge 0} J(q_n - \alpha J' q_n). \tag{2.8.7}$$

Theorem 2.8.1 (an estimate of the strong convergence rate for the steepest descent method). *Let Q and F be Hilbert spaces and $A : Q \to F$ be a continuous linear operator. Assume that the problem $Aq = f$ has a solution $q_e \in Q$ for $f \in F$ and a conditional stability estimate $\beta(\delta)$ is obtained on the set*

$$M = \{q \in Q : \|q\| \le 2\|q_e\|\}.$$

Then the following estimate holds if $\|q_e - q_0\| \le \|q_e\|$:

$$\|q_e - q_n\| \le \beta\Big(\frac{2\|A\|\,\|q_e - q_0\|}{\sqrt{n}}\Big). \tag{2.8.8}$$

Proof. From (2.6.19) and the assumptions of the theorem, it follows that

$$\forall n \in \mathbb{N} \quad \|q_e - q_n\| \le \|q_e - q_0\| \le \|q_e\|.$$

Hence

$$\|q_n\| \le 2\|q_e\|, \quad n \in \mathbb{N} \cup \{0\},$$

which means that $q_n \in M$.

Since there exists an exact solution $q_e \in M$ to the equation $Aq = f$,

$$J(q_n) = \|Aq_n - f\|^2 = \|Aq_n - Aq_e\|^2.$$

By Theorem 2.7.1,

$$\|Aq_n - Aq_e\|^2 \le \frac{\|A\|^2 \|q_e - q_0\|^2}{n}.$$

By Definition 2.8.1 (the definition of a conditional stability estimate), this inequality implies (2.8.8). □

Theorem 2.8.2. *Under the assumptions of Theorem* 2.8.1, *the convergence rate of the method of simple iteration* (2.6.8) *is*

$$\|q_e - q_n\| \le \beta\Big(\frac{\|q_e - q_0\|}{2\sqrt{\alpha n}\sqrt{1-\alpha\|A\|^2}}\Big).$$

The proof of this theorem is similar to that of Theorem 2.8.1.

Remark 2.8.1. Note that a conditional stability estimate can be obtained under the assumption that the exact solution is *sourcewise representable*, i.e., it can be represented in the form $q_e = Bv$, where B is an operator with sufficiently good properties and v belongs to a certain ball. We give a simple example as an illustration. Let $\mathcal{H}$ be a separable Hilbert space, $A : \mathcal{H} \to \mathcal{H}$ be a self-adjoint compact linear operator, and let $\{\lambda_n, \varphi_n\}$ be the sequence of eigenvalues and orthonormal eigenfunctions of A. Put

$$M = \{q \in \mathcal{H} : q = Av,\ \|v\| \le c\}.$$

Expand $q \in \mathcal{H}$ and $v \in \mathcal{H}$ in the basis $\{\varphi_k\}$:

$$q = \sum_k q_k \varphi_k, \quad f = Aq = \sum_k q_k \lambda_k \varphi_k,$$
$$v = \sum_k v_k \varphi_k, \quad f = AAv = \sum_k \lambda_k^2 v_k \varphi_k,$$

where $q_k = \langle q, \varphi_k \rangle$ and $v_k = \langle v, \varphi_k \rangle$. Note that if $q = Av$, $\|v\| \le c$, and $\|f\| \le \delta$, then

$$\|q\|^2 = \|Av\|^2 = \sum_k v_k^2 \lambda_k^2 = \sum_k |v_k|\,|v_k|\lambda_k^2$$
$$\le \sqrt{\sum_k v_k^2}\,\sqrt{\sum_k v_k^2 \lambda_k^4} = \|v\|\,\|f\| \le c\delta,$$

which yields

$$\|q\| \le \sqrt{c\delta}.$$

Then the modulus of continuity of the operator A on $A(M)$ can be estimated from above:

$$\omega(A(M),\delta) = \sup_{q\in M,\ \|Aq\|\le\delta} \|q\| < \sqrt{c\delta}.$$

In a more general case (Vainikko and Veretennikov, 1986), the well-posedness set can be chosen as follows:

$$M_{p\rho} = \{q \in \mathcal{H} : q = (A^*A)^{p/2}v,\ v \in \mathcal{H} : \|v\| \le \rho\}, \quad p > 0,\ \rho > 0.$$

In this case a conditional stability estimate takes the form

$$\omega(A(M_{p\rho}),\delta) \le \rho^{1/(p+1)}\delta^{p/(p+1)}.$$

Regularizing properties of gradient methods. Suppose that the assumptions of Theorem 2.8.2 hold, but the function $f \in F$ is given approximately and the approximation $f_\delta \in \overline{B(f,\delta)}$ may not belong to $A(M)$. We will show that if $\delta \to 0$ and $n \to \infty$ at concerted rates, then the exact solution q_e can be approximated as accurately as desired. To solve the problem of minimizing the functional

$$J_\delta(q) = \|Aq - f_\delta\|^2, \tag{2.8.9}$$

we will use the method of simple iteration (with a somewhat stronger condition on the parameter α):

$$q_{\delta\,n+1} = q_{\delta n} - \alpha J' q_{\delta n}, \quad q_{\delta 0} = q_0, \tag{2.8.10}$$

$$\alpha \in \Big(0, \frac{1}{2\|AA^*\|}\Big). \tag{2.8.11}$$

We need to find out how large the difference between $q_{\delta\,n+1}$ and q_{n+1} may become as the iteration index n increases, where q_{n+1} is calculated using the method of simple iteration with exact data ($q_{n+1} = q_n - \alpha J' q_n$) and the same q_0 as an initial approximation. Consider the difference

$$\begin{aligned} q_{n+1} - q_{\delta\,n+1} &= q_n - q_{\delta n} - \alpha J' q_n + \alpha J' q_{\delta n} \\ &= q_n - q_{\delta n} - \alpha(J' q_n - J' q_{\delta n}) \\ &= (I - \alpha 2A^*A)(q_n - q_{\delta n}) + 2\alpha A^*(f - f_\delta). \end{aligned} \tag{2.8.12}$$

Using (2.8.11), from (2.8.12) we obtain

$$\|q_{n+1} - q_{\delta\,n+1}\| \le \|q_n - q_{\delta n}\| + 2\alpha\|A^*\|\delta.$$

Consequently,

$$\|q_{n+1} - q_{\delta\,n+1}\| \le \|q_1 - q_{\delta 1}\| + 2\alpha n\|A^*\|\delta. \tag{2.8.13}$$

Since both processes start with the same initial approximation, we have

$$\|q_1 - q_{\delta 1}\| \leq 2\alpha \|A^*\| \delta. \tag{2.8.14}$$

Then (2.8.13) and (2.8.14) imply

$$\|q_n - q_{\delta n}\| \leq \beta_2(n)\delta, \quad n \in \mathbb{N} \cup \{0\}, \tag{2.8.15}$$

where

$$\beta_2(n) = 2\alpha n \|A^*\|. \tag{2.8.16}$$

The derived estimates are used to prove the following statement.

Theorem 2.8.3 (the regularizing properties of the method of simple iteration). *Let Q and F be Hilbert spaces, and let $A : Q \to F$ be a continuous linear operator. Assume that the problem $Aq = f$ has a solution $q_e \in Q$ for $f \in F$, a conditional stability estimate $\beta(\delta)$ is determined on the set*

$$M = \{q \in Q : \|q\| \leq 2\|q_e\|\},$$

and the following conditions hold:

$$\|q_e - q_0\| \leq \|q_e\|, \tag{2.8.17}$$

$$\|f - f_\delta\| \leq \delta. \tag{2.8.18}$$

Then the sequence $q_{\delta n}$ of the method of simple iteration (2.8.10), (2.8.11) *satisfies the estimate*

$$\|q_e - q_{\delta n}\| \leq \beta\Big(\frac{\|q_e - q_0\|}{2\sqrt{\alpha n}\sqrt{1 - \alpha\|A\|^2}}\Big) + 2\alpha n\|A^*\|\delta. \tag{2.8.19}$$

Proof. Let $\{q_n\}$ be the sequence constructed using the method of simple iteration with exact data and q_0 as an initial approximation. Then

$$\|q_e - q_{\delta n}\| \leq \|q_e - q_n\| + \|q_n - q_{\delta n}\|.$$

The estimate of the first term is provided by Theorem 2.8.2 and the second term by (2.8.15). □

Remark 2.8.2. The estimate (2.8.19) shows that the sequence $\{q_{\delta n}\}$ of the method of simple iteration is a regularizing sequence with n as the regularization parameter. Indeed, write (2.8.19) in the form

$$\|q_e - q_{\delta n}\| \leq \beta_1(n) + \delta\beta_2(n), \tag{2.8.20}$$

where

$$\beta_1(x) = \beta\Big(\frac{\|q_e - q_0\|}{2\sqrt{\alpha x}\sqrt{1-\alpha\|A\|^2}}\Big), \quad \beta_2(x) = 2\alpha\|A^*\|x.$$

Since $\beta_1(n)$ monotonically tends to zero and $\beta_2(n)$ monotonically tends to infinity as $n \to \infty$, the iteration number n_* at which the iterative process should be stopped can be chosen in a neighbourhood of a point x_* such that

$$\beta_1'(x_*) + \delta\beta_2'(x_*) = 0.$$

Remark 2.8.3. The estimate (2.8.20) allows one to see the reason of a "strange" behaviour of the approximations $q_{\delta n}$ in practical implementation of gradient methods. It was noticed that $q_{\delta n}$ approaches the exact solution (i.e., $\|q_{\delta n} - q_e\|$ decreases) at early stages of the iterative process, but $\|q_{\delta n} - q_e\|$ may start to increase at later stages. Indeed, if $Aq = f_\delta$ has a solution for some f_δ, then the estimate (2.8.8) holds. Otherwise, if there is no solution for f_δ, then Theorem 2.8.1 is not applicable and the error starts to increase with n. In this case the function $\beta_2(n)$ increases and for $n > x_* = x_*(\delta)$ the right-hand side of (2.8.20) may become as large as desired.

Exercise 2.8.1. Analyze the function $\beta(x) = \beta_1(x) + \delta\beta_2(x)$ for convexity.

Remark 2.8.4. More detailed analysis of the method can be found in (Engl et al., 1996).

Note that if an estimate from below is available for the operator A, for example, if there is a constant $m > 0$ such that the inequality

$$m\langle q, q\rangle \le \langle Aq, q\rangle$$

holds for all $q \in Q$, then the estimates for both weak and strong convergence of gradient methods can be substantially improved (Vainikko and Veretennikov, 1986; Kantorovich and Akilov, 1982). However, we are interested in analyzing strong convergence in the case where the problem is strongly ill-posed (the rate of convergence with respect to the objective functional was already estimated in Theorems 2.7.1 and 2.7.2, since a compact linear operator satisfies the assumptions of these theorems). Consider an already familiar example (see Example 2.8.2):

$$\Delta u = 0, \quad (x, y) \in \Omega, \tag{2.8.21}$$

$$u_x(0, y) = 0, \quad y \in (0, 1), \tag{2.8.22}$$

$$u(0, y) = f(y), \quad y \in (0, 1), \tag{2.8.23}$$

$$u(x, 0) = u(x, 1) = 0, \quad x \in (0, 1). \tag{2.8.24}$$

It was indicated in Example 2.8.2 that in order to solve this problem it is sufficient to find a function $q(y)$ such that

$$u(1, y) = q(y) \tag{2.8.25}$$

and then to solve the well-posed problem (2.8.21), (2.8.22), (2.8.24), (2.8.25) (whose well-posedness is established in Section 9.3). In other words, it suffices to solve an operator equation

$$Aq = f. \tag{2.8.26}$$

The properties of the operator A are analyzed in detail in Chapter 9. In particular, it is shown that the linear operator A acts from $L_2(0,1)$ to $L_2(0,1)$ and $\|A\| \leq 1$. In the same chapter an estimate will be obtained in the form

$$\int_0^1 u^2(x,y)\,dy \leq \Big(\int_0^1 f^2(y)\,dy\Big)^{1-x} \Big(\int_0^1 q^2(y)\,dy\Big)^{x}, \quad x \in (0,1). \tag{2.8.27}$$

The main difference of this estimate from the conditional stability estimate (see Definition 2.8.1) is that (2.8.27) depends on the parameter x. Nevertheless, it can be used to estimate the rate of strong convergence not of the gradient method itself but of the iterative process of approximate calculation of $u_n(x,y)$ generated by this method. We will explain this in detail.

Suppose that the nth approximation $q_n(y) \in L_2(0,1)$ has been constructed in solving the operator equation (2.8.26) using the method of simple iteration

$$q_{n+1} = q_n - \alpha J' q_n, \quad \alpha \in (0, \|A\|^{-2}). \tag{2.8.28}$$

In view of the choice of α, we have

$$\|q_n - q_e\| \leq \|q_{n-1} - q_e\| \leq \cdots \leq \|q_0 - q_e\|. \tag{2.8.29}$$

By Theorem 2.7.2,

$$J(q_n) = \|Aq_n - f\|^2 \leq \frac{\|q_0 - q_e\|^2}{4n\alpha(1 - \alpha\|A\|^2)}.$$

At the same time, $Aq_n = u_n(0,y)$, where $u_n(x,y)$ is a solution to the problem

$$\begin{aligned} \Delta u &= 0, \quad (x,y) \in \Omega, \\ u_x(0,y) &= 0, \quad y \in (0,1), \\ u(1,y) &= q_n(y), \quad y \in (0,1), \\ u(x,0) &= u(x,1) = 0. \end{aligned} \tag{2.8.30}$$

In Section 9.3 we will prove that there exists a unique solution $u_n(x, y)$ to problem (2.8.30) that satisfies the estimate

$$\|u_n\|^2_{L_2(\Omega)} \leq \sqrt{2}\,\|q_n\|^2_{L_2(0,1)}.$$

Moreover, the solution has a trace $u_n(0, y) = f_n(y)$ such that

$$\|f_n\|_{L_2(0,1)} \leq \|q_n\|_{L_2(0,1)}.$$

Since the problem in question is linear, the estimate (2.8.27) implies that

$$\int_0^1 [u_n(x, y) - u_e(x, y)]^2\,dy \\ \leq \Big(\int_0^1 [Aq_n - Aq_e]^2\,dy\Big)^{1-x}\Big(\int_0^1 [q_n(y) - q_e(y)]^2\,dy\Big)^x, \quad x \in (0, 1),$$

where u_e is an exact solution to the problem (2.8.21)–(2.8.24).

From (2.8.29) it follows that

$$\int_0^1 [u_n(x, y) - u_e(x, y)]^2\,dy \\ \leq \Big(\int_0^1 [Aq_n - f(y)]^2\,dy\Big)^{1-x}\|q_0 - q_e\|^{2x}, \quad x \in (0, 1).$$

It remains to apply Theorem 2.7.2:

$$\int_0^1 [u_n(x, y) - u_e(x, y)]^2\,dy \leq \frac{\|q_0 - q_e\|^{2-2x}}{[4n\alpha(1 - \alpha\|A\|^2)]^{1-x}}\|q_0 - q_e\|^{2x} \\ = \frac{\|q_0 - q_e\|^2}{[4n\alpha(1 - \alpha\|A\|^2)]^{1-x}}, \quad x \in (0, 1). \tag{2.8.31}$$

Thus, we have shown that (2.8.27) allows us to estimate the rate of strong convergence not of the gradient method $q_{n+1} = q_n - \alpha_n J'q_n$ itself, but of the iterative process generated by this method. Note that as x tends to 1 from below ($x \to 1 - 0$), the estimate (2.8.31) becomes as follows:

$$\|q_n - q_e\|^2 \leq \|q_0 - q_e\|^2.$$

This reminds us that the assumption that q_e is bounded in the norm is not sufficient to estimate the rate of convergence of $\|q_n - q_e\|$. To this end, we can either impose a stronger smoothness condition on q_e, or assume that q_e is sourcewise representable (Bakushinskii and Goncharskii, 1989; Bakushinskii and Kokurin, 2004, 2006; Vainikko and Veretennikov, 1986), or assume that the well-posedness set is compact.

2.9 The pseudoinverse and the singular value decomposition of an operator

In this section, we study two highly important and interrelated concepts: the pseudoinverse operator and the singular value decomposition of a linear operator. These concepts, which originate from the related areas of linear algebra (see the works by S. K. Godunov, V. V. Voevodin, M. Z. Nashed, and others in the supplementary references to Chapters 2 and 3), provide a way to study extensively the characteristic features of linear ill-posed problems and numerical methods for solving them. The results presented in this section can be regarded as a generalization of the corresponding results of Chapter 3 to the case of separable Hilbert spaces.

Any continuous linear operator A from a Hilbert space Q to a Hilbert space F generates an orthogonal direct sum decomposition of these spaces:

$$Q = N(A) \oplus N(A)^{\perp} = N(A) \oplus \overline{R(A^*)},$$

$$F = N(A^*) \oplus N(A^*)^{\perp} = N(A^*) \oplus \overline{R(A)},$$

where A^* is the adjoint of A, $N(A)$ and $N(A^*)$ are the kernels of the operators A and A^*, and $R(A)$ and $R(A^*)$ are the ranges of A and A^*, respectively. This decomposition induces a pair of orthogonal projectors $P_{N(A)}$ and $P_{\overline{R(A)}}$ such that

$$P_{N(A)} : Q \to Q, \quad R(P_{N(A)}) = N(A);$$

$$P_{\overline{R(A)}} : F \to F, \quad R(P_{\overline{R(A)}}) = \overline{R(A)}.$$

In its turn, this pair of orthogonal projectors uniquely defines a *pseudoinverse operator* $A^{\dagger}$ (which in this case is also called an *orthogonal generalized inverse* or a *generalized inverse operator*) by the following relations (Nashed and Votruba, 1976; Beutler and Root, 1976):

$$D(A^{\dagger}) = R(A) \oplus N(A^*), \tag{2.9.1}$$

$$AA^{\dagger} f = P_{\overline{R(A)}} f \quad \text{for all } f \in D(A^{\dagger}), \tag{2.9.2}$$

$$A^{\dagger} Aq = (I - P_{N(A)})q \quad \text{for all } q \in Q. \tag{2.9.3}$$

The pseudoinverse operator is densely defined in F and can be represented on its domain of definition as

$$A^{\dagger} = \left(A\big|_{R(A^*)}\right)^{-1} P_{\overline{R(A)}},$$

where $A\big|_{R(A^*)}$ is the restriction of A to the linear manifold $R(A^*)$. Moreover, $A^{\dagger}$ is continuous on the entire F if and only if $\overline{R(A)} = R(A)$.

A distinguishing feature of the pseudoinverse operator as compared to other generalized inverses is its close association with pseudo-solutions to the operator equation

$$Aq = f. \tag{2.9.4}$$

Definition 2.9.1. An element $q_{\mathrm{p}} \in Q$ is called a *pseudo-solution* to the equation $Aq = f$ (or a *least squares solution*) if

$$\|Aq_{\mathrm{p}} - f\|^2 = \inf_{q \in Q} \|Aq - f\|^2.$$

In other words, $q_{\mathrm{p}} = \arg\min_{q \in Q} \|Aq - f\|^2$.

We call attention to the fact that the concepts of a pseudo-solution and a quasi-solution (see Definition 2.3.1) are alike, with the only difference that the quasi-solution is required to belong to a given set M. It should be noted, however, that this distinction may be of considerable importance, for example, if M is assumed to be compact.

Let Q_f^{p} denote the set of all pseudo-solutions to the equation $Aq = f$ for a fixed $f \in F$. Then

$$Q_f^{\mathrm{p}} = \{q_{\mathrm{p}} : \|Aq_{\mathrm{p}} - f\| = \inf_{q \in Q} \|Aq - f\|\} = \{q : Aq = P_{\overline{R(A)}} f\},$$

which implies that Q_f^{p} is nonempty if and only if $f \in R(A) \oplus N(A^*)$. In this case, Q_f^{p} is a convex closed set, and it contains an element of minimal length q_{np}, which is nothing but a pseudo-solution of minimal norm (Nashed and Votruba, 1976), also called a *normal pseudo-solution* to the equation $Aq = f$ (with respect to the zero element). The connection between q_{np} and the pseudoinverse operator $A^\dagger$ is described by the following theorem (Nashed and Votruba, 1976).

Theorem 2.9.1. *The equation $Aq = f$ has a normal pseudo-solution q_{np} if and only if $f \in R(A) \oplus N(A^*)$. Moreover, $q_{np} = A^\dagger f$, where $A^\dagger$ is the pseudoinverse operator defined by* (2.9.1)–(2.9.3).

The problems $Aq = f$ with a compact linear operator A represent one of the most important classes of ill-posed problems. It is worth noting that this class includes integral equations of the first kind with kernels satisfying sufficiently general assumptions. The most natural method for the study (and, in many cases, for numerical solution) of such problems is the singular value decomposition, which is presented below.

First, let A be a compact linear self-adjoint operator ($A = A^*$). Assume that the eigensystem $\{\lambda_n, \varphi_n\}$ of the operator A consists of nonzero eigenvalues $\{\lambda_n\}$ and a complete orthogonal sequence of the corresponding eigenvectors $\{\varphi_n\}$ ($A\varphi_n = \lambda_n \varphi_n$). Then A can be "diagonalized", i.e.,

$$Aq = \sum_n \lambda_n \langle q, \varphi_n \rangle \varphi_n$$

for all $q \in Q$.

If the operator $A : Q \to F$ is linear and compact, but not self-adjoint, then we can construct an analog of the eigensystem $\{\lambda_n, \varphi_n\}$ called the singular system

$\{\sigma_n, u_n, v_n\}$ of A. This construction is based on the connection between the equations $Aq = f$ and $A^*Aq = A^*f$. First, we give a formal definition. Let $A : Q \to F$ be a compact linear operator, where Q and F are separable Hilbert spaces (i.e., Hilbert spaces with countable bases). A system $\{\sigma_n, u_n, v_n\}$ with $n \in \mathbb{N}$, $\sigma_n \geq 0$, $u_n \in F$, and $v_n \in Q$ will be called the *singular system of the operator* A if the following conditions hold:

- the sequence $\{\sigma_n\}$ consists of nonnegative numbers such that $\{\sigma_n^2\}$ is the sequence of eigenvalues of the operator A^*A arranged in the nonincreasing order and repeated according their multiplicity;
- the sequence $\{v_n\}$ consists of the eigenvectors of the operator A^*A corresponding to $\{\sigma_n^2\}$ (this system is orthogonal and complete, $\overline{R(A^*)} = \overline{R(A^*A)}$);
- the sequence $\{u_n\}$ is defined in terms of $\{v_n\}$ as $u_n = Av_n/\|Av_n\|$.

The sequence $\{u_n\}$ is a complete orthonormal system of eigenvectors of the operator AA^*. Moreover,

$$Av_n = \sigma_n u_n, \quad A^*u_n = \sigma_n v_n, \quad n \in \mathbb{N},$$

and the following decompositions hold:

$$Aq = \sum_n \sigma_n \langle q, v_n \rangle u_n, \quad A^*f = \sum_n \sigma_n \langle f, u_n \rangle v_n.$$

The infinite series in these decompositions converge in the norm of the Hilbert spaces Q and F, respectively, and are called the *singular value decomposition* of the operator A (similarly to the singular value decomposition of a matrix; see Chapter 3).

The structure of the range $R(A)$ and the range of the adjoint operator A^* is described by the following theorem (Nashed, 1976; Kato, 1995):

Theorem 2.9.2 (on the singular value decomposition of a compact operator). *If Q and F are separable Hilbert spaces and $A : Q \to F$ is a compact linear operator, then there exist orthonormal sequences of functions $\{v_n\} \subset Q$ (right singular vectors), $\{u_n\} \subset F$ (left singular vectors), and a nonincreasing sequence of nonnegative numbers $\{\sigma_n\}$ (singular values) such that*

$$\begin{aligned} Av_n &= \sigma_n u_n, \\ A^*u_n &= \sigma_n v_n, \\ \overline{\operatorname{span}\{v_n\}} &= \overline{R(A^*)} = N(A)^\perp, \\ \overline{\operatorname{span}\{u_n\}} &= \overline{R(A)} = N(A^*)^\perp, \end{aligned}$$

and the set $\{\sigma_n\}$ has no nonzero limit points.

Another important property of compact operators (Kato, 1966; Kantorovich and Akilov, 1982) is that their nonzero singular values have finite multiplicity.

Corollary 2.9.1. *If A is a compact operator, then*

1) *the Picard criterion for the solvability of the problem $Aq = f$ holds: the problem is solvable, i.e., $f \in R(A) \oplus N(A^*)$, if and only if*

$$\sum_{\sigma_n \neq 0} \frac{\langle f, u_n \rangle^2}{\sigma_n^2} < \infty;$$

2) *for any $f \in R(A) \oplus N(A^*)$, the element*

$$A^\dagger f = \sum_{\sigma_n \neq 0} \frac{\langle f, u_n \rangle^2}{\sigma_n} v_n$$

is a normal pseudo-solution to the equation $Aq = f$.

Remark 2.9.1. As noted before, the operator $A^\dagger$ is bounded on the entire space F if and only if its range $R(A)$ is a closed subspace of F. At the same time, the range of a compact operator is closed if and only if the operator has a finite singular system $\{\sigma_n, u_n, v_n\}$. Therefore, if the singular system of a compact linear operator A is infinite, then $A^\dagger$ is unbounded.

If $f \in D(A^\dagger)$, then the equation $Aq = f$ has a unique normal pseudo-solution $q_{\text{np}} = A^\dagger f$, and pseudo-solutions q_{p} are the only solutions to the equation $A^* Aq = A^* f$, which is called *normal with respect to the equation $Aq = f$*.

The decomposition

$$q_{\text{np}} = A^\dagger f = \sum_n \frac{\langle f, u_n \rangle}{\sigma_n} v_n$$

shows how errors in the data f affect the result q_{np}, namely, errors in the components $f_n = \langle f, u_n \rangle$ corresponding to small singular values σ_n are divided by σ_n and therefore increase indefinitely with n. This makes it possible to construct Hadamard's example illustrating the instability of the problem $Aq = f$ with a linear compact operator A. Indeed, assume that $q_{\text{e}} \in Q$ is an exact solution to the problem $Aq = f$ and the perturbation of the right-hand side f has the form δu_n, $\delta \in \mathbb{R}_+$. Then the difference between the solution $q_{\delta n}$ to the problem $Aq = f + \delta u_n$ and the exact solution is

$$q_{\text{e}} - q_{\delta n} = \frac{\langle \delta u_n, u_n \rangle}{\sigma_n} v_n.$$

Then, on the one hand, the error $\|f - f_{\delta n}\| = \delta$ does not depend on n and can be as small as desired. On the other hand, the corresponding error $\|q_{\mathrm{e}} - q_{\delta n}\| = \delta/\sigma_n$ in the solution tends to infinity as n increases, and the faster the decrease of the sequence of singular values $\{\sigma_n\}$, the faster the increase of the error. Thus, the degree of ill-posedness of the problem $Aq = f$ is directly related to the character of the decrease of σ_n. For example, the problem $Aq = f$ is said to be *weakly ill-posed* if $\sigma_n = O(n^{-\gamma})$ for some $\gamma \in \mathbb{R}_+$, and *strongly ill-posed* otherwise (for instance, if $\sigma_n = O(\mathrm{e}^{-n})$).

The singular value decomposition of the operator A can be used to construct a regularization method for the problem $Aq = f$ based on the projection:

$$q_{\delta n} = \sum_{j=1}^{n} \frac{\langle f_{\delta n}, u_j\rangle}{\sigma_j} v_j.$$

It can be proved that

$$\|q_{\delta n} - q_{\mathrm{np}}\| = O\Big(\sigma_{n+1} + \frac{\delta}{\sigma_n}\Big), \quad n \to \infty.$$

We introduce systems of subspaces and the corresponding orthogonal projectors:

$$\begin{aligned}
Q_r &= \operatorname{span}\{v_1, \dots, v_r\} \subset Q, \\
P_{Q_r} &: Q \to Q, \quad P_{Q_r}(Q) = Q_r, \\
F_r &= \operatorname{span}\{u_1, \dots, u_r\} \subset F, \\
P_{F_r} &: F \to F, \quad P_{F_r}(F) = F_r,
\end{aligned}$$

where the numbers r are chosen so that $\sigma_r > \sigma_{r+1}$. The reason for this requirement is that the orthogonal projectors P_{Q_r} and P_{F_r} are not well-defined if $\sigma_r = \sigma_{r+1}$ in the sense that they depend on the indexing of the singular system.

Definition 2.9.2. The operator $A_r^\dagger = A^\dagger P_{F_r} : F \to Q$ is called the *r-pseudoinverse operator*, and the element

$$q_{(r)} = A_r^\dagger f = \sum_{n=1}^{r} \frac{\langle f, u_n\rangle}{\sigma_n} v_n$$

is called a *generalized normal r-solution* (referred to as an "r-solution" in the sequel) to the operator equation (2.9.4).

This operator is obviously defined and bounded on the entire space F, $A_r^\dagger f = A^\dagger f$ for all $f \in F_r \subset R(A)$, and $A_r^\dagger f = P_{Q_r}(A^\dagger f)$ for all $f \in R(A) \oplus N(A^*)$. Furthermore, since $\bigcup_{r\in\mathbb{N}} Q_r = R(A^*)$, it is clear that for any $f \in D(A^\dagger)$ and $\delta > 0$

there exists $r(\delta, f)$ such that $\|A^\dagger f - A_r^\dagger f\| < \delta$. Thus, the family of operators $A_r^\dagger$ can be viewed as approximating the pseudoinverse operator $A^\dagger$ on its domain of definition.

In the numerical solution of equation (2.9.4), the exact representations of the operator A and the data f are not available. Instead, there are A_h and f_δ such that $\|A - A_h\| \le h$ and $\|f - f_\delta\| \le \delta$ (errors in the numerical approximation of the operator and measurement errors), where f_δ does not necessarily belong to the domain of definition of the pseudoinverse operator $A_h^\dagger$ (even if $f \in D(A^\dagger)$), and therefore it does not make sense to analyze the behavior of $\|A^\dagger f - A_h^\dagger f_\delta\|$ as $\delta, h \to 0$. At the same time, since the operators $A_r^\dagger$ and $A_{hr}^\dagger$ are defined and bounded on the entire space F, the difference $\|A_r^\dagger f - A_{hr}^\dagger f\|$ characterizes the accuracy of calculation of the projection of a solution onto the first r singular vectors.

The concept of an r-solution to a system of algebraic equations $Aq = f$ makes it possible to construct numerically stable algorithms for its solution if the *condition number* of the matrix

$$\mu_r(A) = \frac{\sigma_1(A)}{\sigma_r(A)},$$

the *gap in the singular value spectrum*

$$d_r(A) = \frac{\sigma_1(A)}{\sigma_r(A) - \sigma_{r+1}(A)}$$

and the *system inconsistency parameter*

$$\theta_r(A, f) = \frac{\|Aq_{(r)} - f\|}{\sigma_r(A)\|q_{(r)}\|}$$

are not too large. More specifically, the restriction on the size of admissible perturbations h and δ can be expressed in terms of the quantities $\mu_r(A)$, $d_r(A)$, and $\theta_r(A, f)$ (Godunov et al., 1993).

Chapter 3

Ill-posed problems of linear algebra

This chapter is of particular importance for the study of numerical methods for solving linear inverse and ill-posed problems, since all of them can be reduced to systems of linear algebraic equations in one way or another. For this reason, the main concepts of the theory of ill-posed problems, such as regularization, quasi-solution, normal solution, and others, are represented in this chapter.

The study of ill-posed problems of linear algebra is not only a necessary stage in the numerical solution of linear ill-posed problems. Mathematicians were solving applied problems with overdetermined or underdetermined systems of equations long before the terms "ill-posed problem" and "inverse problem" were coined. Linear algebra was the first area where they started to study such concepts as a normal solution (which provides a way for choosing a unique solution among many possible solutions), a pseudo-solution (a generalized solution that coincides with an exact solution if it exists), ill-conditioned systems, and singular value decomposition, which, like an x-ray, highlights the degree of ill-posedness of the problem and indicates possible ways for its numerical solution.

As we will see later, given the matrix A, the problem of finding a normal pseudo-solution to the system $Aq = f$ is well-posed (the pseudoinverse operator can be determined exactly).

In Section 3.1 we introduce the concept of a pseudo-solution (a generalized solution) to the system of linear algebraic equations $Aq = f$ ($q_{\mathrm{p}} = \arg\min_{q\in\mathbb{R}^n} \|Aq - f\|$), the normal system $A^T Aq = A^T f$, and the normal pseudo-solution $q_{\mathrm{np}} = \arg\min_{q_{\mathrm{p}}\in Q_f^{\mathrm{p}}} \|q_{\mathrm{p}}\|$, where Q_f^{p} is the set of all pseudo-solutions. In Section 3.2, we consider regularizing algorithms for constructing a solution to the system $Aq = f$ from the approximate data $\{A_h, f_\delta\}$ based on using a regularization parameter α. Section 3.3 deals with techniques for choosing the regularization parameter α, which is the most important and complicated problem in regularization. Section 3.4 presents iterative algorithms for solving ill-posed problems of linear algebra where the iteration index plays the role of the regularization parameter. Singular value decomposition, which is briefly described in Sections 3.5 and 3.6, plays a special role in the theory and numerical solution of ill-posed problems of linear algebra. Clearly, the singular value decomposition procedure is complicated and laborious. However, if this task is done, i.e., the $m \times n$ matrix A is represented as $A = U\Sigma V^T$, where the matrices U and V are orthogonal (or unitary in the complex case) and $\Sigma = \mathrm{diag}\,(\sigma_1, \sigma_2, \ldots, \sigma_p)$, $0 \le \sigma_p \le \sigma_{p-1} \le \cdots \le \sigma_1$, $p = \min\{m, n\}$, then the obtained decomposition provides an almost complete answer to the question about the degree of ill-posedness of the problem $Aq = f$. Finally, in Section 3.7 we present a method devised for symmetric matrices and especially for band matrices, which arise, for example, in the numerical implementations of the Gelfand–Levitan methods.

Consider a system of m linear algebraic equations for n unknowns

$$\begin{cases} a_{11}q_1 + a_{12}q_2 + \ldots + a_{1n}q_n = f_1, \\ a_{21}q_1 + a_{22}q_2 + \ldots + a_{2n}q_n = f_2, \\ \qquad\qquad \vdots \\ a_{m1}q_1 + a_{m2}q_2 + \ldots + a_{mn}q_n = f_m. \end{cases}$$

This system can be written in the matrix form

$$Aq = f,$$

where A is the $m \times n$ real matrix of system coefficients, $f = (f_1, f_2, \ldots, f_m)^T \in \mathbb{R}^m$ is the column vector of right-hand sides, and $q = (q_1, q_2, \ldots, q_n)^T \in \mathbb{R}^n$ is the column vector of unknowns. In accordance with the previous notation, throughout this chapter we assume $Q = \mathbb{R}^n$, $F = \mathbb{R}^m$, and $A : \mathbb{R}^n \to \mathbb{R}^m$. We will analyze the methods for constructing approximate solutions to the problem $Aq = f$ in the cases where this problem is ill-posed ($m \neq n$, A is degenerate, A is ill-conditioned, etc.).

As known, if $m = n$ and $\det A \neq 0$, then the system has a unique solution. If A is degenerate ($\det A = 0$) or $m \neq n$, then the system may have no solutions or infinitely many solutions. In this case the concept of a normal pseudo-solution is introduced so that any system $Aq = f$ has a unique solution of this type.

Consider the case of a nondegenerate square matrix A in more detail. As noted before, theoretically this case can be thought of as good in the sense that there exists a unique solution. However, in the theory of numerical methods, nondegenerate matrices are categorized as either "ill-conditioned" or "well-conditioned". A matrix is called *ill-conditioned* if the solution of the corresponding system is unstable. In other words, small errors in the right-hand side of the system or the errors that inevitably arise in the numerical implementation lead to a substantial deviation of the resulting solution from the exact one.

The concept of a condition number is an important characteristic of the solution stability for systems of linear equations. The *condition number* of a square matrix A is defined as

$$\mu(A) = \sup_{\substack{x \neq 0 \\ \xi \neq 0}} \left\{ \frac{\|Ax\| \|\xi\|}{\|A\xi\| \|x\|} \right\}.$$

To demonstrate the meaning of $\mu(A)$, we consider the system with perturbed right-hand side $A(q + \delta q) = f + \delta f$, where δq is the deviation of the solution caused by the perturbation δf in the right-hand side. It is obvious that $A\,\delta q = \delta f$. From the definition of the condition number it follows that

$$\frac{\|\delta q\|}{\|q\|} \le \mu(A)\,\frac{\|A\,\delta q\|}{\|Aq\|} = \mu(A)\,\frac{\|\delta f\|}{\|f\|},$$

and $\mu(A)$ is the least constant that satisfies this inequality. Thus, the condition number $\mu(A)$ makes it possible to estimate the relative error of the solution $\|\delta q\|/\|q\|$ in terms of the relative error of the right-hand side $\|\delta f\|/\|f\|$. Ill-conditioned systems have very large condition number $\mu(A)$.

A system of linear algebraic equations $Aq = f$ is an ill-posed problem for rectangular matrices A $(m \ne n)$ and square matrices that are degenerate or ill-conditioned.

Methods for solving systems of linear equations are classified under two types, direct and iterative. *Direct methods* produce a solution in a predetermined number of operations. These methods are relatively simple and universal. However, they usually require large amounts of computer memory and accumulate errors during the solution process, since the calculation results of each stage are used at the subsequent stage. For this reason, direct methods are appropriate for relatively small systems $(n < 200)$ with a dense matrix whose determinant is not close to zero. Direct methods (sometimes also called *exact*, although this term is not always adequate because of inevitable errors of numerical implementations) include, for example, the Gauss method, the Jordan method, the square root method, the singular value decomposition method, etc.

Iterative methods are methods of successive approximations. In such methods, it is difficult to estimate the amount of computation in advance, but they are more memory-efficient than direct methods. Iterative methods are often used in the regularization of ill-posed systems of linear equations. It should be noted, however, that a combination of an iterative approach and a direct method often provides the most effective way of solving linear systems. In mixed algorithms of this type, iterative methods are used to refine the solutions produced by direct methods.

In this chapter, we consider a system of linear algebraic equations with a rectangular matrix A and describe iterative (regularizing) algorithms and some direct algorithms for solving this system.

3.1 Generalization of the concept of a solution. Pseudo-solutions

Before describing methods for the approximate solution of a system of equations $Aq = f$, we discuss the concept of a solution to this system, which may be overdetermined, underdetermined, or ill-conditioned.

Henceforth we assume that $q \in \mathbb{R}^n$, $f \in \mathbb{R}^m$, A is an $m \times n$ real matrix.

A vector $q_{\mathrm{p}} \in \mathbb{R}^n$ minimizing the norm of the residual

$$J(q) = \|Aq - f\|^2 \to \min \tag{3.1.1}$$

is called a *pseudo-solution* to the system $Aq = f$ (see Definition 2.9.1), i.e., $q_{\mathrm{p}} = \arg\min_{q \in \mathbb{R}^n} \|Aq - f\|^2$.

Since the increment of the functional $J(q)$ can be represented as

$$\delta J(q) = J(q+h) - J(q) = 2(Aq, Ah) - 2(Ah, f) + (Ah, Ah),$$

a necessary condition for $J(q)$ to attain its minimum is

$$A^T Aq - A^T f = 0,$$

where A^T is the transpose of A. Consequently, the vector q_{p} is a solution to the system of equations

$$A^T Aq = A^T f. \tag{3.1.2}$$

Definition 3.1.1. The system of equations (3.1.2) is said to be *normal with respect to the system $Aq = f$.*

It is not difficult to prove the converse, namely, that every solution to the system (3.1.2) minimizes the residual in (3.1.2) (Ivanov et al., 2002). Hence, the problems (3.1.1) and (3.1.2) are equivalent. It is easy to verify that the problem (3.1.1) always has a solution, although the solution is not necessarily unique. Therefore, from the equivalence of the above problems it follows that the system (3.1.2) also has a solution for any matrix A and any vector f. Thus, the set of solutions to the normal system (3.1.2) coincides with the set of pseudo-solutions to the system $Aq = f$. This set will be denoted by Q_f^{p}.

Consider the problem of finding the minimum of the functional $\|q - q^0\|^2$ on the set Q_f^p:

$$\min\{\|q - q^0\|^2 : q \in Q_f^{\mathrm{p}}\}, \tag{3.1.3}$$

where q^0 is a fixed vector. There exists a unique solution q_{np}^0 to the problem (3.1.3) because the strictly convex functional $\|\cdot\|_Q^2$ attains its minimum on the convex closed set Q_f^{p} at a unique point.

Definition 3.1.2. The vector q_{np}^0 is called a *normal pseudo-solution* to the system *$Aq = f$ with respect to q^0*. A normal pseudo-solution to the system $Aq = f$ with respect to the zero vector (a pseudo-solution of minimal norm) is called a *normal pseudo-solution* to this system (or a normal generalized solution) and is denoted by q_{np}: $q_{\mathrm{np}} = \arg\min_{q_{\mathrm{p}} \in Q_f^{\mathrm{p}}} \|q_{\mathrm{p}}\|$.

If the system $Aq = f$ is solvable, then a normal pseudo-solution q^0_{np} with respect to q^0 is a normal solution to this system with respect to q^0, i.e., the solution least deviating from the vector q^0 (in the norm). In particular, if the system $Aq = f$ has a unique solution, then its pseudo-solution is unique and coincides with the exact solution.

A normal pseudo-solution exists, is unique, and depends continuously on the right-hand side f. (Indeed, pseudoinverse operators in finite-dimensional spaces are bounded.)

A normal pseudo-solution q_{np} is unstable with respect to perturbations in the matrix elements. The following example illustrates the instability of the normal pseudo-solution.

Example 3.1.1. Consider an inconsistent system

$$\begin{cases} 1q_1 + 0q_2 = 1, \\ 0q_1 + 0q_2 = 1, \end{cases} \quad A = \begin{bmatrix} 1 & 0 \\ 0 & 0 \end{bmatrix}, \quad f = \begin{bmatrix} 1 \\ 1 \end{bmatrix}.$$

Pass on to the normal system (3.1.2) with

$$A^T A = \begin{bmatrix} 1 & 0 \\ 0 & 0 \end{bmatrix}, \quad A^T f = \begin{bmatrix} 1 \\ 0 \end{bmatrix}$$

and determine the set of pseudo-solutions to the system $Aq = f$:

$$Q_f^{\mathrm{p}} = \{(q_1, q_2) : q_1 = 1,\ q_2 \in \mathbb{R}\}.$$

Then the vector $q_{\mathrm{np}} = (1, 0)^T$ is a normal pseudo-solution to the system $Aq = f$.

Assume that a perturbed system has the form

$$\begin{cases} 1q_1 + 0q_2 = 1, \\ 0q_1 + hq_2 = 1, \end{cases}$$

where h is a small parameter. Since this system has a unique solution $q_{\mathrm{np}h} = (1, 1/h)^T$, it is also a unique (and therefore normal) pseudo-solution to the perturbed system. It is obvious that $q_{\mathrm{np}h} \to \infty$ as $h \to 0$, i.e., $\|q_{\mathrm{np}} - q_{\mathrm{np}h}\| \to \infty$ as $h \to 0$, which proves that the normal pseudo-solution is unstable with respect to perturbations of the matrix elements.

3.2 Regularization method

Consider the concepts of a regularization algorithm and a regularized family of approximate solutions introduced by A. N. Tikhonov (1963) (see Section 2.5). Throughout this section we assume that A is a square matrix.

Suppose that the data $\{A, f\}$ of the problem $Aq = f$ are known approximately with error levels h and δ ($h > 0$, $\delta > 0$), that is, we are given $\{A_h, f_\delta\}$ such that $\|A_h - A\| \le h$, $\|f_\delta - f\| \le \delta$. We now consider the ways to construct families of vectors converging to the normal pseudo-solution q_{np} of the equation $Aq = f$ as $h, \delta \to 0$. Generally speaking, the normal pseudo-solution to the perturbed equation

$$A_h q = f_\delta, \tag{3.2.1}$$

must not be chosen as an approximate solution because the normal pseudo-solution is unstable (Example 3.1).

Let $\{R_\alpha\}_{\alpha>0}$ be a family of algorithms that associate the pair $\{A_h, f_\delta\}$ with a unique vector $R_\alpha(A_h, f_\delta) = q^\alpha_{h\delta} \in Q = \mathbb{R}^n$. If the parameter $\alpha = \alpha(h, \delta)$ depends on the data errors δ and h so that

$$\lim_{h,\delta\to 0} \alpha(h,\delta) = 0, \qquad \lim_{h,\delta\to 0} \|q^\alpha_{h\delta} - q_{\mathrm{np}}\| = 0, \tag{3.2.2}$$

then $\{q^\alpha_{h\delta}\}$ ($h > 0$, $\alpha > 0$, $\delta > 0$) is called a *regularized family* of approximate solutions, and the algorithm R_α is called a *regularization algorithm* for the problem $Aq = f$. Recall that the normal pseudo-solution q_{np} to the system $Aq = f$ coincides with its exact solution if the system is uniquely solvable and with its normal solution if the problem $Aq = f$ has many solutions.

If A^{-1} exists, i.e., if A is a nondegenerate matrix, then for sufficiently small h the inverse matrix A_h^{-1} also exists and the solution to (3.2.1) will theoretically converge to the solution of the equation $Aq = f$ as $h, \delta \to 0$. However, if A is ill-conditioned, which means that its condition number

$$\mu(A) = \sup_{\substack{x\neq 0\\ \xi\neq 0}} \left\{ \frac{\|Ax\|\,\|\xi\|}{\|A\xi\|\,\|x\|} \right\}$$

is large, then the deviation of the solution to the perturbed system (3.2.1) from the normal pseudo-solution to the system $Aq = f$ may turn out to be unacceptably large even for small h and δ, so the problem must be perceived as practically unstable (ill-posed).

We now describe procedures of constructing a regularized family of approximate solutions $q^\alpha_{h\delta}$.

First, consider a case where A is a symmetric positive-semidefinite matrix such that the system $Aq = f$ is solvable for a given vector f. We will apply the regularization method by M. M. Lavrentiev (see Section 2.4).

Take the regularized system

$$(A + \alpha I)q = f + \alpha q^0, \tag{3.2.3}$$

where α is a positive parameter, I is the unit matrix, and q^0 is a test solution, i.e., a certain approximation of the solution (if no information on the solution is available, we can put $q^0 = \mathbf{0}$).

Under the above conditions, the system (3.2.3) has a unique solution q^α that converges to a normal solution q_{n}^0 with respect to q^0 as $\alpha \to 0$ (see Section 2.6). If the data $\{A, f\}$ are given approximately, the following lemma is valid (Ivanov et al., 2002).

Lemma 3.2.1. *Let $\{A_h, f_\delta\}$ with $h > 0$ and $\delta > 0$ be such that $\|A_h - A\| \le h$, $\|f_\delta - f\| \le \delta$, and let A_h be a symmetric positive semidefinite matrix. Then the system*

$$(A_h + \alpha I)q = f_\delta + \alpha q^0 \tag{3.2.4}$$

has a unique solution $q_{h\delta}^\alpha$ and if $\alpha(h, \delta) \to 0$ and $(h + \delta)/\alpha(h, \delta) \to 0$ as $h \to 0$ and $\delta \to 0$, then the solution $q_{h\delta}^\alpha$ converges to a normal solution q_n^0 of the equation $Aq = f$ with respect to q^0, i.e., to the solution least deviating from the vector q^0.

By Definition 2.5.1, the solutions $\{q_{h,\delta}^\alpha\}$ to the systems (3.2.4) form a regularized family of approximate solutions for the system $Aq = f$ if we put, for example, $\alpha = \sqrt[p]{h + \delta}$ $(p > 1)$ because $\alpha = \sqrt[p]{h + \delta} \to 0$ and $(h + \delta)/\sqrt[p]{h + \delta} = (h + \delta)^{1-1/p} \to 0$ as $h \to 0$ and $\delta \to 0$.

Remark 3.2.1. Let A be a positive semidefinite degenerate matrix and $\|A\| = 1$ (which can be achieved by an appropriate scaling of the system $Aq = f$). Then $\mu(A) = \infty$, while $\mu(A + \alpha E) \le (1 + \alpha)/\alpha$. Therefore, with a reasonable choice of the parameter α the systems (3.2.3) and (3.2.4) can be made well-conditioned and the approximation $q^\alpha \simeq q_{\mathrm{n}}^0$ can be satisfactory, although these requirements contradict each other.

The role of the regularization parameter α becomes clear if the solution q^α to the system (3.2.3) (for $q^0 = \mathbf{0}$) is represented in the form

$$q^\alpha = \sum_{i=1}^n \frac{f_i}{\lambda_i + \alpha} \varphi_i,$$

where λ_i are the eigenvalues ($\lambda_i \ge 0$) and φ_i are the orthonormal eigenvectors of the matrix A, and $f_i = \langle f, \varphi_i \rangle$. This representation shows that for small λ_i the addition of a positive parameter α substantially increases the denominator and thus decreases the adverse effect of possible errors in the corresponding components f_i ($\tilde{f}_i = f_i + \delta f_i$). At the same time, for $\lambda_i \gg 0$ the influence of small α is negligible.

We now omit the requirement that the matrix A be symmetric and positive semidefinite, still assuming that the system $Aq = f$ has a solution for a given f. Let B be a matrix such that for some α_0 the matrix $A + \alpha_0 B$ is nondegenerate and therefore invertible. Then the regularization of the form

$$(A + \alpha B)q = f \tag{3.2.5}$$

can be applied, where the parameter α is not necessarily positive and $|\alpha| \leq |\alpha_0|$.

Exercise 3.2.1. Analyze a connection between (3.2.5) and the following lemma (Nazimov, 1986).

Lemma 3.2.2. *Let* $\|(A + \alpha B)^{-1} A\| \leq C < \infty$ *(as* $\alpha \to 0$*),* $\|A_h - A\| \leq h$*,* $\|B_\mu - B\| \leq \mu$*,* $\|f_\delta - f\| \leq \delta \|f\|$*. Then for sufficiently small* h*,* μ*, and* δ *the system*

$$(A_h + \alpha B_\mu)q = f_\delta \tag{3.2.6}$$

has a unique solution $q^\alpha_{h\delta\mu}$ *and there holds the error estimate*

$$\|q^\alpha_{h\delta\mu} - q^B_n\| \leq C\Big(|\alpha| + \mu + \frac{h + \delta}{|\alpha|}\Big), \tag{3.2.7}$$

where q^B_n *is a solution to the system* $Aq = f$ *such that*

$$\|Bq^B_n\| = \min\{\|Bq\| : q \in Q_f\}$$

(Q_f *is the set of solutions to the system* $Aq = f$*).*

From the estimate (3.2.7) it follows that if $\alpha(h, \delta, \mu) \to 0$ and $(h+\delta)/|\alpha(h, \delta, \mu)| \to 0$ as $h, \delta, \mu \to 0$, then

$$\lim_{h,\delta,\mu \to 0} \|q^\alpha_{h\delta\mu} - q^B_{\mathrm{n}}\| = 0.$$

The most important step in the above regularization is finding a matrix B such that $A + \alpha_0 B$ is nondegenerate and $\|(A+\alpha B)^{-1} A\| < \infty$. Some methods for constructing such matrices are described in (Nazimov, 1986).

We conclude by analyzing the general case, where the system $Aq = f$ may have no solution. In this case we find a normal pseudo-solution q^0_{np} with respect to some q^0. Consider the problem of stable approximation of q^0_{np} assuming A and f to be given approximately. The normal pseudo-solution q^0_{np} with respect to q^0 is unstable with respect to perturbations of the matrix elements, which means that a regularization is necessary. We take, as a regularized approximate solution, the vector $q^\alpha_{h\delta}$ that satisfies the system

$$(A_h^T A_h + \alpha I)q = A_h^T f_\delta + \alpha q^0. \tag{3.2.8}$$

Lemma 3.2.3. *Let $\|A_h - A\| \le h$, $\|f_\delta - f\| \le \delta$, and $\alpha > 0$. Then the system* (3.2.8) *has a unique solution and the following estimate holds:*

$$\|q^0_{np} - q^\alpha_{h\delta}\| \le c_1\alpha + \frac{h}{\alpha}\,(\|Aq^0_{np} - f\| + 2c_2^2\alpha^2)^{1/2} + \frac{1}{\alpha^{1/2}}\,(c_3 h + \delta), \tag{3.2.9}$$

where c_i $(i = 1, 2, 3)$ are constants that depend on $\|q^0_{np}\|$, i.e., on the norm of the normal pseudo-solution with respect to q^0.

Corollary 3.2.1. *Assume that h and δ are of the same order of magnitude as ε, where ε is sufficiently small. If there exists an exact solution to the equation $Aq = f$ (i.e., $\|Aq^0_{np} - f\| = 0$), then, as a function of α and ε, the right-hand side of the estimate* (3.2.9) *has the form*

$$\varphi(\alpha, \varepsilon) = \alpha + \varepsilon + \frac{\varepsilon}{\alpha^{1/2}}. \tag{3.2.10}$$

For $\alpha = \varepsilon^{2/3}$, $\varphi(\alpha, \varepsilon)$ is of the same order of magnitude as $\varepsilon^{2/3}$. If the system $Aq = f$ has no solutions ($\|Aq^0_{np} - f\| \neq 0$), then the right-hand side of (3.2.9) *is a function of the form*

$$\psi(\alpha, \varepsilon) = \alpha + \frac{\varepsilon}{\alpha} + \frac{\varepsilon}{\alpha^{1/2}}. \tag{3.2.11}$$

For $\alpha = \varepsilon^{1/2}$, $\psi(\alpha, \varepsilon)$ is of the same order of magnitude as $\varepsilon^{1/2}$. These estimates are obtained by minimizing the functions φ and ψ with respect to α (i.e., solving the equations $\varphi'_\alpha(x) = 0$ and $\psi'_\alpha(x) = 0$).

Thus, if the error in the input data of the system $Aq = f$ has the same order of magnitude as ε, then the normal solution with respect to q^0 can be determined with an accuracy of order of $\varepsilon^{2/3}$ if the equation $Aq = f$ is solvable; otherwise we can construct a normal pseudo-solution with respect to q^0 with an accuracy of order of $\varepsilon^{1/2}$.

Note that a more complicated method for choosing the parameter α can be used to improve the accuracy of approximation to the order of $h + \delta$ (Djumaev, 1982).

Problem (3.2.8) is equivalent to the minimization problem

$$\min\{\|A_h q - f_\delta\|^2 + \alpha\|q - q^0\|^2 : q \in \mathbb{R}^n\}. \tag{3.2.12}$$

This way of constructing $q^\alpha_{h\delta}$ is known as the Tikhonov variational regularization method.

In some cases, it is reasonable to use a stabilizing functional of a more general form, namely,

$$\min\{\|A_h q - f_\delta\|^2 + \alpha\|L(q - q^0)\|^2 : q \in \mathbb{R}^n\},$$

where L is a nondegenerate square matrix which, if chosen appropriately, makes it possible to improve the accuracy of the regularized solution.

3.3 Criteria for choosing the regularization parameter

It is necessary to associate the regularization parameter α with the data errors δ and h: $\alpha = \alpha(\delta, h)$. In applied problems, the error level (δ, h) is usually fixed (δ and h do not tend to zero) and it is required to identify a specific $\alpha(\delta, h)$ which is supposed to be the best in a certain sense. The problem is that the condition number of the matrices of regularized systems becomes higher as α decreases, which may cause computational errors. At the same time, an increase of α leads to worse approximation of the exact solution. Therefore, a reasonable compromise is necessary in this case.

We now describe some criteria for choosing the parameter α.

The residual principle (α_r). Suppose that only the right-hand side of the system $Aq = f$ is approximate, i.e., $A_h \equiv A$ $(h = 0)$, $\|f_\delta - f\| \leq \delta$. By q_δ^α we denote a solution to the system (3.2.8) for $h = 0$ and $q^0 = 0$. The value of the parameter α in the residual principle is chosen so that

$$\|Aq_\delta^\alpha - f_\delta\| = \delta$$

(Morozov, 1987; Tikhonov and Arsenin, 1974).

The generalized residual principle (α_{gr}). This principle is appropriate in the general case of approximate data: $\|A - A_h\| \leq h$, $\|f - f_\delta\| \leq \delta$. The parameter α is determined from the following equation (here and in what follows, $q_{h\delta}^\alpha$ is a solution to equation (3.2.8) with $q^0 = \mathbf{0}$):

$$\|A_h q_{h\delta}^\alpha - f_\delta\| = h\|q_{h\delta}^\alpha\| + \delta.$$

There are numerical methods and computer programs developed for solving this equation (Tikhonov et al., 1983).

For a more detailed study, see also (Tikhonov et al., 1995; Lawson and Hanson, 1974; Golub, 1968).

3.4 Iterative regularization algorithms

Iterative methods can be applied for solving ill-conditioned systems $Aq = f$. In this case a stopping criterion should be formulated for the iterative process depending on the error level in the data. Note that it is not necessary in the case of well-conditioned systems. The iteration index plays the role of a regularization parameter. For example, consider the method of simple iteration

$$q_{n+1} = (I - A^T A)q_n + A^T f, \quad n = 0, 1, 2, \ldots. \tag{3.4.1}$$

In the case of approximate data $\{A_h, f_\delta\}$, we have the following iterative procedure instead of (3.4.1):

$$q_{n+1} = (I - A_h^T A_h)q_n + A_h^T f_\delta, \quad n = 0, 1, 2, \ldots, \tag{3.4.2}$$

where $\|A - A_h\| \le h$ and $\|f - f_\delta\| \le \delta$.

Assume that the system $Aq = f$ is solvable, $\|A\| \le 1$, and $\|A_h\| \le 1$. The condition $\|A\| \le 1$ does not restrict the class of problems under study, since it can always be satisfied by multiplying the equation $Aq = f$ by an appropriate constant. As before, let q_{n}^0 be a normal solution with respect to q_0 (i.e., a solution to the system $Aq = f$ such that the norm $\|q_{\mathrm{n}}^0 - q_0\|$ is minimal) where q_0 is an initial approximation in the iterative procedures (3.4.1) and (3.4.2).

We now define the rules for choosing the iteration index n at which the iterative procedure (3.4.1) should be stopped (Vainikko, 1980):

Rule 1. Take numbers $a_1 > 0$ and $a_2 > 0$, and choose $n(h, \delta)$ to be the smallest index such that $\|q_n - q_{n-1}\| \le a_1 h + a_2\delta$.

Rule 2. Take numbers $a_0 \ge \|q_{\mathrm{n}}^0\|$ and $a_1 > 1$, and choose $n(h, \delta)$ to be the smallest index such that $\|A_h q_n - f_\delta\| \le a_0 h + a_1\delta$. (Naturally, in this case we assume that an estimate of $\|q_{\mathrm{n}}^0\|$ from above is known.)

Rule 3. Take numbers $a_1 > 1$, $a_2 > 1$, and $a > 0$, and choose n to be the smallest index satisfying at least one of the following inequalities:

$$\|A_h q_n - f_\delta\| \le a_1\|q_n\|h + a_2\delta, \quad n \ge a/(a_1\|q_n\|h + a_2\delta)^2.$$

Theorem 3.4.1. *Suppose that one of the three above-mentioned rules is used as a stopping criterion for the successive approximations procedure* (3.4.2). *Then*

$$\lim_{h,\delta\to 0} \|q_{n(h,\delta)} - q_n^0\| = 0. \tag{3.4.3}$$

If Rule 1 *is used, then the number of iterations* $n(h, \delta)$ *satisfies the condition*

$$(h + \delta)n(h, \delta) \to 0 \quad \textit{as } h, \delta \to 0.$$

If Rule 2 *or* 3 *is used, then*

$$(h + \delta)^2 n(h, \delta) \to 0 \quad \textit{as } h, \delta \to 0.$$

Formula (3.4.3) means that rules 1, 2, and 3 determine regularization algorithms for solving the equation $Aq = f$.

Remark 3.4.1. Theorem 3.4.1 also holds for the so-called implicit iterative procedure

$$q_{n+1} = q_n - (A_h^T A_h + B)^{-1}(A_h^T A_h q_n - A_h^T f_\delta), \tag{3.4.4}$$

where B is a positive definite matrix that commutes with $A_h^T A_h$. The number of iterations required to achieve the desired accuracy in the procedure (3.4.4) is usually smaller than in (3.4.2), but the cost of each iteration step in (3.4.4) is higher because we need to invert the matrix $(A_h^T A_h + B)$. Also, note that we can choose $B = \alpha I$, $\alpha > 0$.

Remark 3.4.2. If solutions to the system $Aq = f$ are assumed to belong to a convex set M, then we should use the following nonlinear iterative procedures instead of (3.4.2) and (3.4.4) (Vasin, 1985, 1987):

$$q_{n+1} = \Pr_M[(I - A_h^T A_h)q_n + A_h^T f_\delta],$$
$$q_{n+1} = \Pr_M[(A_h^T A_h + B)^{-1}(Bq_n + A_h^T f_\delta)].$$

Here $\Pr_M$ is a metric projection which to every vector q associates a vector in M that is closest to q. The following are examples of *a priori* given sets M for which $\Pr_M$ can be written explicitly:

$$M_1 = \{q \in \mathbb{R}^n : q_i \geq 0\}, \quad i = 1, 2, \dots, n, \quad \Pr_{M_1} q = q^+ = (q_1^+, q_2^+, \dots, q_n^+),$$

where

$$q_i^+ = \begin{cases} q_i & \text{for } q_i \geq 0, \\ 0 & \text{for } q_i < 0; \end{cases}$$

$$M_2 = \{q \in \mathbb{R}^n : \|q - q^0\| \leq r\}, \quad \Pr_{M_2} q = q^0 + \frac{q - q^0}{\|q - q^0\|} r;$$

$$M_3 = \{q \in \mathbb{R}^n : a_i \leq q_i \leq b_i,\ i = 1, 2, \dots, n\}, \quad \Pr_{M_3} q = z,$$

where

$$z_i = \begin{cases} q_i & \text{for } a_i \leq q_i \leq b_i, \\ a_i & \text{for } q_i < a_i, \\ b_i & \text{for } q_i > b_i. \end{cases}$$

3.5 Singular value decomposition

Singular value decomposition is one of the most common and effective methods for the study and numerical solution of inverse and ill-posed problems. The reason is that the singular value decomposition of a matrix (or, in the general case, of a compact linear operator—see Section 2.9) provides a numerical characteristic for the ill-posedness of the problem, namely, the rate of decrease of singular values expresses the degree of instability of the solution of the problem $Aq = f$ with respect to small variations in the data f.

In this section, we describe the singular value decomposition method using the results presented in (Godunov, 1980, 1998; Godunov et al., 1993).

Consider a system of linear algebraic equations

$$Aq = f,$$

where A is an $m \times n$ real matrix, $q \in \mathbb{R}^n$ and $f \in \mathbb{R}^m$. It is required to solve the system for the unknown vector q. The difficulty of solving this system obviously depends on the structure of the matrix A. In terms of linear algebra, the matrix A is the representation of a linear operator in some coordinate system. It is natural to try to transform this coordinate system so that the linear operator would be represented by a simpler matrix in the new system (for example, a block matrix with a number of zero blocks). In other words, the matrix A can be represented as a product of matrices of special form. A series of algorithms based on the decomposition of the matrix A have been developed recently. One of the advantages of this approach to solving systems of linear equations is its universality, i.e., it is applicable to well-conditioned systems as well as to ill-conditioned or degenerate systems.

Orthogonal transformations, which are norm-preserving by definition, play a special role in matrix decompositions. Recall that any orthogonal transformation is represented by an orthogonal matrix, i.e., by a matrix U such that $U^T U = UU^T = I$, where U^T is the transpose of U and I is the unit matrix. In the problems of finding pseudo-solutions to degenerate systems, the norm-preserving property of orthogonal transformations allows us to replace the original problem of minimizing the residual $\|Aq - f\|$ with the problem of minimizing the functional $\|U^T(Aq - f)\|$ where the matrix $U^T A$ has a simpler structure (for example, it may be a block matrix) due to a choice of the orthogonal matrix U.

The most effective (although resource-intensive) means for solving arbitrary systems of linear equations is provided by the so-called complete orthogonal decompositions which, by definition, have the form $A = UKV^T$. Here U and V are orthogonal matrices of dimensions $m \times m$ and $n \times n$, respectively; K is an $m \times n$ matrix of the form

$$K = \begin{pmatrix} W & 0 \\ 0 & 0 \end{pmatrix},$$

where W is a nondegenerate square matrix (of the same rank as the original matrix A).

The most well-known complete orthogonal decomposition is the singular value decomposition, which has the form

$$A = U\Sigma V^T, \tag{3.5.1}$$

where V is an $n \times n$ orthogonal matrix, U is an $m \times m$ orthogonal matrix, and Σ is an $m \times n$ diagonal matrix with $\sigma_{ii} = \sigma_i \geq 0$. The numbers σ_i are called the singular values of the matrix A. Here and in what follows, σ_i are assumed to be indexed in nonincreasing order ($\sigma_{i+1} \le \sigma_i$).

Recall that an $m \times n$ *diagonal matrix* S is an $m \times n$ matrix with elements s_{ij} such that

$$s_{ij} = \begin{cases} 0, & i \ne j, \\ s_i, & i = j. \end{cases}$$

The $m \times n$ diagonal matrix S has the form

$$S = \begin{pmatrix} s_1 & 0 & 0 & & \cdots & & 0 \\ 0 & s_2 & 0 & & \cdots & & 0 \\ \vdots & \ddots & \ddots & \ddots & & & \vdots \\ 0 & \cdots & 0 & s_m & 0 & \cdots & 0 \end{pmatrix}, \qquad n > m,$$

or

$$S = \begin{pmatrix} s_1 & 0 & \cdots & 0 \\ 0 & s_2 & \ddots & \vdots \\ \vdots & \ddots & \ddots & 0 \\ 0 & \cdots & 0 & s_m \\ 0 & & \cdots & 0 \\ \vdots & & \ddots & \vdots \\ 0 & & \cdots & 0 \end{pmatrix}, \qquad n < m.$$

For brevity, an $m \times n$ diagonal matrix of either form will be denoted by $S = \operatorname{diag}(s_1, s_2, \dots, s_p)$, $p = \min\{m, n\}$.

Theorem 3.5.1 (on the singular value decomposition). *For any $m \times n$ real matrix, there exists an $m \times m$ orthogonal matrix U, an $n \times n$ orthogonal matrix V, and an $m \times n$ diagonal matrix*

$$\Sigma = \operatorname{diag}(\sigma_1, \sigma_2, \dots, \sigma_p), \quad p = \min\{m, n\},$$

such that $0 \le \sigma_p \le \sigma_{p-1} \le \cdots \le \sigma_2 \le \sigma_1$ and

$$A = U \Sigma V^T. \tag{3.5.2}$$

The numbers $\sigma_i = \sigma_i(A)$, $i = \overline{1, p}$, are determined uniquely and are called the singular values of the matrix A.

Remark 3.5.1. Theorem 3.5.1 also holds for $m \times n$ complex matrices. In this case, U and V are complex unitary matrices and V^T is replaced by $V^* = \overline{V^T}$ in the decomposition of A.

Given an $m \times n$ matrix A, we construct the following $(m+n) \times (m+n)$ block matrix $\mathcal{A}$ called an *extended matrix*:

$$\mathcal{A} = \begin{pmatrix} O_{nn} & A^T \\ A & O_{mm} \end{pmatrix}. \tag{3.5.3}$$

Exercise 3.5.1. Prove that all eigenvalues $\lambda_i(\mathcal{A}), i = 1, 2, \dots, n+m$, of the extended matrix are real.

Suppose that the eigenvalues are indexed in nonincreasing order:

$$\lambda_{n+m}(\mathcal{A}) \le \lambda_{n+m-1}(\mathcal{A}) \le \dots \le \lambda_2(\mathcal{A}) \le \lambda_1(\mathcal{A}).$$

Prove that this ordered set of eigenvalues coincides with the ordered set

$$-\sigma_1(A) \le -\sigma_2(A) \le \dots \le -\sigma_p(A) \le \underset{|n-m| \text{ zeros}}{\dots} \le \sigma_p(A) \le \dots \le \sigma_1(A)$$

and therefore $\sigma_k(A) = \lambda_k(\mathcal{A}), k = 1, 2, \dots, p$.

Since the scalar product in $\mathbb{R}^n$ defines the norm $\|x\| = \sqrt{\langle x, x \rangle}$, the *norm of an $m \times n$ matrix A* can be defined as follows:

$$\|A\| := \sup_{x \neq 0} \frac{\|Ax\|}{\|x\|}.$$

Exercise 3.5.2. Verify that the function introduced above has all the properties of a norm.

Exercise 3.5.3. Prove that $\|A\| = \sigma_1(A)$.

Let $\sigma_k(A)$ be a singular value of an $m \times n$ matrix A. Since $\sigma_k(A) = \lambda_k(\mathcal{A})$, for each $\sigma = \sigma_k(A)$ the system of equations

$$\mathcal{A}w = \sigma w \tag{3.5.4}$$

has a solution $w_{(k)} = (w_{(k)1}, w_{(k)2}, \dots, w_{(k)m+n})^T$ (which is not necessarily unique; but in any case the total number of linearly independent $w_{(k)}$ is equal to $m+n$). Let two vectors $v_{(k)} = (v_{(k)1}, v_{(k)2}, \dots, v_{(k)n})^T$ and $u_{(k)} = (u_{(k)1}, u_{(k)2}, \dots, u_{(k)m})^T$ be constructed from the vector $w_{(k)}$ as follows: $v_{(k)j} = w_{(k)j}$, $j = 1, 2, \dots, n$; $u_{(k)j} = w_{(k)n+j}$, $j = 1, 2, \dots, m$. In view of (3.5.4), we have

$$Av_{(k)} = \sigma_k(A)u_{(k)}, \quad A^T u_{(k)} = \sigma_k(A)v_{(k)}. \tag{3.5.5}$$

The vector $v_{(k)}$ is called the *right singular vector of the matrix A* corresponding to the singular value $\sigma_k(A)$, and the vector $u_{(k)}$ is called the *left singular vector*.

Lemma 3.5.1 (mutual orthogonality of singular vectors). *Let $v_{(k)}$ and $u_{(k)}$ be the right and the left singular vectors of the matrix A, respectively. Then $\langle u_{(k)}, u_{(j)}\rangle = 0$ and $\langle v_{(k)}, v_{(j)}\rangle = 0$ for $k \neq j$.*

From (3.5.4) and (3.5.5) it follows that

$$A^T A v_{(k)} = A^T \sigma_{(k)} u_{(k)} = \sigma_{(k)}^2 v_{(k)}, \quad A A^T u_{(k)} = \sigma_{(k)}^2 u_{(k)},$$

which means that the right singular vectors of the $m \times n$ matrix A are the eigenvectors of the $n \times n$ matrix $A^T A$, and the left singular vectors of A are the eigenvectors of the $m \times m$ matrix AA^T.

Theorem 3.5.2. *For any $m \times n$ matrix A, there exists an orthonormal system of n right singular vectors and an orthonormal system of m left singular vectors, which are called the singular bases of the matrix A.*

Exercise 3.5.4. Given an $m \times n$ matrix A, rewrite its singular value decomposition $A = U\Sigma V^T$ in the form $AV = \Sigma U$. Similarly, $A^T = V\Sigma U^T$ is rewritten as $A^T U = V\Sigma$. Prove that the columns of the matrices V and U are the right and the left singular vectors of A, respectively.

Exercise 3.5.5. For a given $m \times n$ matrix A, consider the $n \times n$ matrix $A^T A$ and the $m \times m$ matrix AA^T. Let $e_{(1)}, e_{(2)}, \ldots, e_{(n)}$ be the orthonormal eigenvectors of $A^T A$. Prove that

- the vectors $Ae_{(1)}, Ae_{(2)}, \ldots, Ae_{(n)}$ are orthogonal;
- every nonzero vector $Ae_{(k)}$ is an eigenvector of AA^T.

Exercise 3.5.6. Verify that an $n \times n$ matrix A is nondegenerate if and only if all its singular values $\sigma_i(A)$ are positive. Prove the equality

$$|\det A| = \prod_{i=1}^{n} \sigma_i(A).$$

Lemma 3.5.2. *An $n \times n$ square matrix A is normal ($AA^* = A^*A$) if and only if*

$$|\lambda_i(A)| = \sigma_i(A), \quad i = 1, 2, \ldots, n.$$

Lemma 3.5.3. *Let $A = U\Sigma V^T$ be a singular value decomposition of an $m \times n$ matrix A, and let $A^\dagger$ be the pseudoinverse of A. Then*

$$A^\dagger = V\Lambda^\dagger U^T.$$

Here

$$\Lambda^{\dagger} = \operatorname{diag}(\sigma_1^{-1}, \sigma_2^{-1}, \dots, \sigma_r^{-1}, \underbrace{0, \dots, 0}_{p-r}), \quad r = \operatorname{rank} A, \quad p = \min\{m, n\},$$

i. e., $\Lambda^{\dagger}$ is an $m \times n$ diagonal matrix whose diagonal elements are the inverses of nonzero singular values of A and all other elements are zero.

Given a singular value decomposition (3.5.1) of a matrix A, the system $Aq = f$ can be represented in the form

$$U \Sigma V^T q = f \tag{3.5.6}$$

or in an equivalent form (with $z = V^T q$ and $g = U^T f$)

$$\Sigma z = g, \tag{3.5.7}$$

where $g \in \mathbb{R}^m$ and $z \in \mathbb{R}^n$ is the unknown vector. This system is easy to solve, since Σ is a diagonal matrix.

Assume, for definiteness, that the matrix A is overdetermined, i.e., the number of unknowns in the system $Aq = f$ does not exceed the number of rows ($n \le m$). In this case, the system $\Sigma z = g$ can be written in the form

$$\begin{aligned} \sigma_j z_j &= g_j, \quad \text{if } j \le n,\ \sigma_j \neq 0; \\ 0 z_j &= g_j, \quad \text{if } j \le n,\ \sigma_j = 0; \\ 0 &= g_j, \quad \text{if } n < j \le m; \end{aligned} \tag{3.5.8}$$

where the second subsystem is empty if $r = n$ (the matrix A has full rank) and the third subsystem is empty if $n = m$.

Note that the original equation has a solution if and only if $g_j = 0$ whenever $\sigma_j = 0$ or $j > n$. If $\sigma_j = 0$, then the corresponding unknown z_j can take an arbitrary value (playing the role of a parameter). The solution to the original system is given by the formula $q = Vz$.

Another important property of the singular value decomposition $A = U \Sigma V^T$ of an $m \times n$ matrix A is that it provides an explicit representation of the kernel and the image of the map $A : \mathbb{R}^m \to \mathbb{R}^n$, namely, the right singular vectors $v_{(k)}$ corresponding to the zero singular values span the kernel of A, while the left singular vectors $u_{(k)}$ corresponding to the nonzero singular values span the image $R(A) \subset \mathbb{R}^n$.

Recall that the kernel of the matrix A is the set $N(A)$ of vectors q such that $Aq = \mathbf{0}$, and the image of A is the set $R(A)$ of vectors f for which the system $Aq = f$ has a solution. We will show that the singular value decomposition makes it possible to describe the structure of the sets $N(A)$ and $R(A)$. Indeed, let $u_{(k)}$ and $v_{(k)}$ denote the columns of the matrices U and V, respectively. Due to (3.5.5),

$$A v_{(k)} = \sigma_k u_{(k)}, \quad k = 1, 2, \dots, n. \tag{3.5.9}$$

If $\sigma_k = 0$, then (3.5.9) yields $Av_{(k)} = 0$, i.e., $v_{(k)} \in N(A)$. If $\sigma_k \neq 0$, then $u_{(k)} \in R(A)$.

Let U_1 be a set of the left singular vectors corresponding to nonzero singular values and U_0 be a set of the remaining left singular vectors. Define V_1 and V_0 analogously. Then

a) V_0 is an orthonormal basis of the kernel $N(A)$;

b) U_1 is an orthonormal basis of the image $R(A)$.

The singular value decomposition provides additional insights into the already familiar concepts of a pseudo-solution, stability and conditioning, regularization, etc. In this section, we discuss only a few aspects of this subject. For an extended discussion see (Godunov et al., 1993).

Let an $m \times n$ matrix A have rank $r \leq p = \min\{m, n\}$. Then the matrix Σ in the singular value decomposition $A = U\Sigma V^T$ has the form $\Sigma = \operatorname{diag}(\sigma_1, \sigma_2, \dots, \sigma_r, 0, \dots, 0)$. If $r < m$, it is necessary that $g_{r+1} = g_{r+2} = \cdots = 0$ for a system (3.5.7) to be consistent. If $r < n$, the variables $z_{r+1}, z_{r+2}, \dots, z_n$ do not occur in equation (3.5.7) and therefore the solution components $z_{r+1}, z_{r+2}, \dots, z_n$ can be chosen arbitrarily.

Choose z in such a way as to minimize the norm of the residual

$$\Sigma z - g = \begin{pmatrix} \sigma_1 z_1 - g_1 \\ \sigma_2 z_2 - g_2 \\ \vdots \\ \sigma_r z_r - g_r \\ -g_{r+1} \\ \vdots \\ -g_m \end{pmatrix}.$$

Exercise 3.5.7. Prove that $\min \|\Sigma z - g\| = \sqrt{g_{r+1}^2 + g_{r+2}^2 + \cdots + g_m^2}$ and the equality is attained if and only if

$$z_i = g_i/\sigma_i, \quad i \leq r. \tag{3.5.10}$$

If $n = r$, there exists a unique vector (with components (3.5.10)) at which $\|\Sigma z - g\|$ attains its minimum. If $r < n$, the minimum is attained at any vector $z = (z_1, z_2, \dots, z_n)$ with the first r components defined by (3.5.10) and the remaining components being arbitrary (in particular, if $z_i = 0$ for $r < i \leq n$, then z is a vector of minimal norm at which $\|\Sigma z - g\|$ attains its minimum). Consequently, for any $r \leq p := \min\{n, m\}$ we can find a unique element z that has minimal norm among all the vectors minimizing the functional $\|\Sigma z - g\|$. Since $q = Vz$, in view of the equalities

$$\begin{aligned} Aq - f &= U\Sigma V^T q - Ug = U(\Sigma z - g), \\ \|Aq - f\| &= \|\Sigma z - g\|, \quad \|z\| = \|q\|, \quad q = Vz, \end{aligned}$$

it is easy to verify that there is only one vector with minimal norm among the vectors at which $\|Aq - f\|$ attains its minimum, that is, there exists a unique normal pseudo-solution q_{np} to the system $Aq = f$.

Note that the system $Aq = f$, generally speaking, is unsolvable if $r := \operatorname{rank}(A) < p = \min\{m, n\}$, i.e., the residual $y = Aq - f$ may be nonzero for all q. Therefore, the equality $Aq = f$ should be understood in a more general sense. The generalization of the concept of a solution implies the minimization of the functional $\|Aq - f\|$, which yields the set Q_f^{p} of all pseudo-solutions to the system $Aq = f$, and then another minimization (finding a vector q_{np} of minimal norm in Q_f^{p}). However, since a variational definition is not always convenient for practical calculations, we will consider one of the ways to extend the system $Aq = f$ so that solving the extended system would result in finding both a pseudo-solution q_{p} and the corresponding residual vector $y_{\mathrm{r}} = Aq_{\mathrm{p}} - f$ (Godunov et al., 1993).

Theorem 3.5.3. *A solution to the extended system*

$$\begin{aligned} Aq + y &= f, \\ A^T y &= 0 \end{aligned} \tag{3.5.11}$$

consists of a pseudo-solution q_p to the system $Aq = f$ and the corresponding residual vector y_r. The determination of y_r from (3.5.11) *is unique.*

In order to find the normal pseudo-solution to the system $Aq = f$, we can first solve the extended system (3.5.11) and determine y_{r}. Then the system $Aq = f - y_{\mathrm{r}}$ will be consistent, and therefore finding the normal pseudo-solution q_{np} will amount to finding a solution to the system $Aq = f - y_{\mathrm{r}}$ that has minimal norm. It is clear that $\|y_{\mathrm{r}}\| = \min_q \|Aq - f\|$, which means that $\|y_{\mathrm{r}}\|$ characterizes the degree of inconsistency of the system $Aq = f$. Following (Godunov et al., 1993), the number

$$\theta(A, f) = \frac{\min_{\|q\|=1} \|Aq - f\|}{\min_{\sigma_i \neq 0}\{\sigma_i(A)\}} = \frac{\|y_{\mathrm{r}}\|}{\|q\|\sigma_r} \tag{3.5.12}$$

will be called the *inconsistency number* of the system $Aq = f$.

In conclusion, it should be noted that the singular values and singular vectors of a matrix make it possible to study various aspects of the ill-posedness of problems of linear algebra, including the effect of the system inconsistency, the determination and analysis of angles between subspaces, the theory of perturbations of r-solutions, etc.

In Chapter 2, we mentioned only a simple case to associate the notion of an r-solution with regularization.

Substantial difficulty may be encountered in the implementation of the singular value decomposition method if the calculation produces relatively small singular values. In this case, large errors may appear in z_j determined by the formula $z_j = g_j/\sigma_j$. Therefore, the key to appropriate application of the singular value decomposition is to introduce a thereshold τ representing the accuracy of the initial data and computations. Namely, any $\sigma_j > \tau$ is considered to be acceptable and the formula $z_j = g_j/\sigma_j$ is used for the corresponding z_j, while any $\sigma_j < \tau$ is thought of as zero and the corresponding z_j takes an arbitrary value (we can put $z_j = 0$ to obtain a normal solution). Here τ plays the role of the regularization parameter.

3.6 The singular value decomposition algorithm and the Godunov method

Solving an arbitrary system of equations with a rectangular matrix involves reducing the system to a canonical form. In the case of the singular value decomposition, the system is transformed to a diagonal form (a detailed description of the algorithm can be found in (Godunov et al., 1993)).

Reduction to diagonal form is achieved by a series of orthogonal transformations in the space of the right-hand sides of the system and a series of orthogonal transformations in the space of solutions to the system. The elementary transformations within these series are chosen to be either orthogonal reflections with respect to specific planes or sequences of two-dimensional rotations.

The singular value decomposition of an arbitrary rectangular matrix A is performed in several steps:

1) reducing the matrix A to a bidiagonal form D by means of orthogonal reflections (a rectangular matrix $D = (d_{ij})$ is called *bidiagonal* if $d_{ij} = 0$ for $i \neq j$ and $i \neq j + 1$);

2) determining the singular values of the bidiagonal matrix D:
$0 \leq \sigma_n(D) \leq \sigma_{n-1}(D) \leq \cdots \leq \sigma_1(D)$;

3) reducing the bidiagonal matrix D to a diagonal form Σ by means of the procedure of exhaustion that involves sequences of two-dimensional rotations.

Application to the least squares problem. The singular value decomposition can be used to find a pseudo-solution (or a least squares solution)

$$q_p = \arg\min\{\|Aq - f\|^2 : q \in \mathbb{R}^n\}.$$

Since orthogonal matrices are norm-preserving, we have

$$\|Aq - f\| = \|U^T(U\Sigma V^T q - f)\| = \|\Sigma z - g\|,$$
$$g = U^T f, \quad z = V^T q.$$

The minimizer z is expressed as follows:

$$z_j = g_j/\sigma_j \quad \text{if } \sigma_j \neq 0 \quad (j = 1, 2, \dots, r),$$
$$z_j \text{ is arbitrary} \quad \text{if } \sigma_j = 0 \quad (j = r + 1, \dots, n).$$

Note that $\min \|Aq - f\|^2 = \sum_{j=r+1}^{m} g_j^2$. The set of all pseudo-solutions is determined by the formula $q = Vz$. Putting $z_j = 0$ for $\sigma_j = 0$ yields the normal pseudo-solution (i.e., the pseudo-solution of minimal norm).

Numerical analysis of the least squares method is presented in (Lawson and Hanson, 1974).

As mentioned above, if calculated singular values are very small, it is necessary to include the operation of equating them to zero, which can be interpreted as a special regularization procedure.

The Godunov Method (Godunov et al., 1993). We will give a more detailed description of the method proposed by S. K. Godunov, since it is probably the most promising in the case where very large condition and inconsistency numbers of the system under study seem to leave no chances for finding an acceptable solution.

Application of the least squares method $J(q) = \|Aq - f\|^2 \to \min$ or the Tikhonov regularization method (the simplest version of which is $M_\alpha(q) = J(q) + \alpha\|q\|^2 \to \min$) leads to the systems $A^T Aq = A^T f$ and $\alpha q + A^T Aq = A^T f$, respectively. It should be noted that although these methods have well-known advantages, they also have serious disadvantages which complicate the original ill-posed problem $Aq = f$ in a certain sense. Firstly, the already large condition number of the matrix A is squared when passing to $A^T A$. Secondly, the characteristic features of the behavior of the data f are smoothed out substantially as a result of applying the operator A^T (recall that the matrix A is usually obtained by discretization of a compact operator). Thirdly, although introducing a small parameter α improves the properties of the operator to be inverted, it also changes the original problem and poses an additional problem of choosing an appropriate value of α.

The first two disadvantages can be avoided by using the Lavrentiev method $\alpha q + Aq = f$, which is justified for a symmetric positive semidefinite matrix A (see Lemmas 3.2.1 and 3.2.2). S. K. Godunov proposed a regularization method for solving the system $Aq = f$ with an $m \times n$ rectangular matrix A based on the solution of the following extended system:

$$(1 - \alpha)Aq = (1 - \alpha)f, \tag{3.6.1}$$

$$\alpha Bq = 0. \tag{3.6.2}$$

Here α is the regularization parameter and the system (3.6.2) defines the conditions for a solution q to belong to a given well-posedness set. An algorithm of determining a normal generalized r-solution to the system (3.6.1), (3.6.2) is described in detail in (Godunov et al., 1993). It should be emphasized that the algorithm does not use the normal system $A^T Aq = A^T f$. We suggest considering this algorithm based on the following example.

Take the Fredholm integral equation

$$\int_0^5 \frac{q(t)\,dt}{t+s+1} = f(s), \quad s \in [0,4]. \tag{3.6.3}$$

Exercise 3.6.1. Prove that there exists $f \in C^1[0,4]$ for which the problem (3.6.3) has no solution.

Exercise 3.6.2. Prove that as the solution $q_n(t) = n\mathrm{e}^{-n^4(t-1)^2}$ to equation (3.6.3) increases indefinitely with n, the right-hand side $f(s)$ satisfies the estimate

$$|f_n(s)| = \left| \int_0^5 \frac{q_n(t)\,dt}{t+s+1} \right| \le \int_0^5 q_n(t)\,dt \le n \int_{-\infty}^{\infty} \mathrm{e}^{-n^4(t-1)^2}\,dt$$
$$= \frac{1}{n} \int_{-\infty}^{\infty} \mathrm{e}^{-z^2}\,dz = \frac{\sqrt{\pi}}{n}.$$

Exercise 3.6.3. Given the test function

$$q(t) = \begin{cases} (1-\frac{t}{3})^2, & t \in [0,3], \\ (\frac{t}{3}-1)^3, & t \in (3,5], \end{cases}$$

calculate analytically $f(s)$ in (3.6.3).

Exercise 3.6.4. Divide the interval $[0,5]$ into 50 and the interval $[0,4]$ into 80 subintervals of equal length. Substitute a quadrature formula for the integral in the left-hand side of (3.6.3) and use the analytic solution to calculate the right-hand side f_i of the obtained system

$$\sum_{j=1}^{50} a_{ij} q_j = f_i, \quad i = 1,2,\ldots,80. \tag{3.6.4}$$

Thus, the system (3.6.4) is consistent and we know the exact solution and its second-order derivative

$$q''(t) = \begin{cases} -2/3, & t \in [0,3], \\ 2/3, & t \in (3,5]. \end{cases} \tag{3.6.5}$$

Exercise 3.6.5. Calculate the condition number $\mu(A)$, the inconsistency number $\theta(A, f)$ of the system (3.6.4), and the singular values of the matrix A.

Remark 3.6.1. Calculations (Godunov et al., 1993) demonstrated that $\mu(A) > 10^{15}$, $\theta(A, f) \approx 0.6 \cdot 10^{-15}$, and the 39 smallest singular values do not exceed 10^{-13}, which means that the problem (3.6.4) is strongly ill-posed.

We now construct an extended system of the form (3.6.1), (3.6.2). To this end, we choose $\alpha = 10^{-8}$ and define the elements of the matrix

$$B = \begin{pmatrix} b_{21} & b_{22} & \dots & b_{2,50} \\ b_{31} & b_{32} & \dots & b_{3,50} \\ \vdots & \vdots & \ddots & \vdots \\ b_{50,1} & b_{50,2} & \dots & b_{50,50} \end{pmatrix}$$

by the formulas

$$\begin{aligned} b_{j-1,j} = b_{j+1,j} &= 100, \quad b_{j,j} = -200, \\ b_{ij} &= 0 \quad \text{for } |i-j| \geq 2. \end{aligned}$$

It is clear that the matrix B is chosen to stipulate the condition that the second-order derivative of the solution (3.6.5) be relatively small. In other words, we first multiply the discrete analog $Bq = b$ of equality (3.6.5) by $\alpha = 10^{-8}$ and then put $\alpha b = 0$. As a result, we obtain $\alpha Bq = 0$.

Exercise 3.6.6. Calculate the condition number μ and the inconsistency number θ of the extended system (3.6.1), (3.6.2) for α and B specified above.

Remark 3.6.2. Calculations (Godunov et al., 1993) show that $\mu = 0.29 \cdot 10^6$, $\theta = 0.4 \cdot 10^{-6}$, and the normal solution $\tilde{q}_{\mathrm{n}}$ to the system (3.6.1), (3.6.2) with approximate data $\tilde{f}$ such that $\|f - \tilde{f}\| \leq \|f\| \cdot 0.57 \cdot 10^{-3}$ deviates from the exact solution q_e to the system (3.6.4) by $\|q_e - \tilde{q}_{\mathrm{n}}\| \leq 0.39 \cdot 10^{-1} \|q_e\|$.

Thus, extending a system using additional information about its exact solution reduces the condition number substantially and allows us to calculate an approximate solution.

Exercise 3.6.7. Use Lemma 3.2.2 to analyze the matching conditions for the approximation accuracy δ of the right-hand side f ($\|f - f_\delta\| \leq \delta$) and the regularization parameter α.

It should be emphasized that a normal solution to the system

$$\begin{bmatrix} (1-\alpha)A \\ \alpha B \end{bmatrix} q = \begin{bmatrix} (1-\alpha)f \\ 0 \end{bmatrix}$$

with nondegenerate compound coefficient matrix can be obtained as a solution to the normal system

$$\begin{bmatrix} (1-\alpha)A \\ \alpha B \end{bmatrix}^T \begin{bmatrix} (1-\alpha)A \\ \alpha B \end{bmatrix} q = \begin{bmatrix} (1-\alpha)A \\ \alpha B \end{bmatrix}^T \begin{bmatrix} (1-\alpha)f \\ 0 \end{bmatrix},$$

which is a system with square nondegenerate coefficient matrix:

$$[(1-\alpha)^2 A^T A + \alpha^2 B^T B]q = (1-\alpha)^2 A^T f.$$

However, the condition number of the square matrix

$$(1-\alpha)^2 A^T A + \tau B^T B$$

is equal to the square of the condition number of the compound matrix. It turns out to be $(0.3 \cdot 10^6)^2 = 0.9 \cdot 10^{11}$ in the above example for $\alpha = 10^{-4}$. This is exactly the reason why it is reasonable to try to find a normal solution using the technique described in (Godunov et al., 1993; §§ 5–7) and to avoid passing to normal systems.

3.7 The square root method

In the Lavrentiev–Tikhonov regularization procedures, an approximate solution is determined from the systems (3.2.4), (3.2.8) with an $n \times n$ symmetric positive definite matrix, which is also well-conditioned if the value of α is chosen reasonably. Aside from iterative procedures, traditional direct methods (such as the Gaussian method, the Gauss–Jordan method, etc.) can be used to solve these systems (Voevodin, 1966; Fadeev and Fadeeva, 1963). We will confine ourselves to describing the square root method, which is specially adapted for equations with symmetric matrices and is particularly efficient and stable.

The square root method is based on the Cholesky decomposition of a symmetric matrix

$$A = U^T U, \tag{3.7.1}$$

where U is an upper triangular matrix. Denoting by u_{ij} the coefficients of the matrix U and writing equality (3.7.1) in terms of the matrix elements, we arrive at the system

$$\sum_{k=1}^{n} u_{ki} u_{kj} = a_{ij} \quad (i = 1, \ldots, n;\ j = 1, \ldots, n). \tag{3.7.2}$$

Solving (3.7.2) for u_{ij}, we obtain the calculation formulas for the elements of U:

$$u_{ii} = \sqrt{a_{ii} - \sum_{k=1}^{i-1} u_{ki}^2}, \quad u_{ij} = \frac{1}{u_{ii}}\Big(a_{ij} - \sum_{k=1}^{i-1} u_{ki}u_{kj}\Big)$$
$$(j = i+1, \dots, n;\ i = 1, 2, \dots, n).$$

Now, for solving the original system it suffices to solve two triangular systems

$$U^T z = f, \quad Uq = z. \tag{3.7.3}$$

The square root method is especially convenient when applied to band matrices (Voevodin, 1966), since the method can be easily adapted to their structure so that operations on zero elements are eliminated.

3.8 Exercises

Every statement in the following list is either a definition or a proposition. We suggest that the reader identifies the propositions and proves them.

This section can be viewed as a useful appendix to Chapter 3. A more demanding reader can find the proofs of these statements in the books by V. V. Voevodin et al. and S. K. Godunov et al. (see the bibliography).

Pseudo-solutions. If a system $Aq = f$ is consistent, then it has a unique solution of minimal length. This solution q_{n} is said to be normal.

Of all the solutions to a system $Aq = f$, its normal solution is the only one that is orthogonal to the kernel $N(A)$ of the matrix A and belongs to the image $R(A^T)$ of the matrix A^T.

The vector $y = Aq - f$ is called the residual of the vector q.

A solution to the system $A^T Aq = A^T f$ is called a pseudo-solution q_{p} or a generalized solution to the system $Aq = f$.

Pseudo-solutions q_{p} are the only vectors in the space Q that guarantee that

- the residual $y = Aq - f$ is orthogonal to the image of A;
- the residual $y = Aq - f$ has minimal length.

Given a system $Aq = f$, take the image $\hat{f}$ of the vector f under the projection onto the range $R(A)$ of the matrix A. The system $Aq = \hat{f}$ is always consistent and the set of all solutions to this system coincides with the set Q_f^{p} of all pseudo-solutions to the original system.

The normal solution to the system $A^T Aq = A^T f$ is called the normal pseudo-solution to the system $Aq = f$.

Of all the pseudo-solutions $q_{\mathrm{p}} \in Q_f^{\mathrm{p}}$ to the system $Aq = f$, only the normal pseudo-solution q_{np} satisfies the following conditions:

- the pseudo-solution has minimal length, i.e., $q_{\text{np}} = \arg\min_{q_\text{p} \in Q_f^\text{p}} \|q_\text{p}\|$;
- the pseudo-solution is orthogonal to the kernel $N(A)$ of the matrix A, i.e., $q_{\text{np}} \perp N(A)$;
- the pseudo-solution belongs to the image $R(A^T)$ of the matrix A^T, i.e., $q_{\text{np}} \in R(A^T)$.

The normal pseudo-solution q_{np} of the problem $Aq = f$ is equal to zero if and only if the vector f belongs to the kernel of the matrix A^T, i.e., $q_{\text{np}} = 0 \Leftrightarrow f \in N(A^T)$.

If the existence of a solution to the system of linear algebraic equations $Aq = f$ is not guaranteed, we can always solve the system $A^T Aq = A^T f$ instead. In this case, the norm of the residual $Aq - f$ is guaranteed to be minimal, and the same is true for the norm of the pseudo-solution itself if the normal pseudo-solution is found. Passing from the system $Aq = f$ to the normal system $A^T Aq = A^T f$ is called the first Gaussian transformation or the least squares method. The latter term is explained by the fact that formal expression of the condition for the square of the norm of the residual $\|Aq - f\|^2$ to be minimal leads to the equation $A^T Aq = A^T f$.

Note that all the properties of the normal pseudo-solution mentioned above are equivalent. Therefore, verification of any one of them ensures that others hold as well.

Pseudoinverse. A system $Aq = f$ with nondegenerate matrix is solved using the inverse matrix. The inverse plays an important role in many tasks, but it is defined only for nondegenerate matrices. An analog of the inverse can be constructed for degenerate and rectangular matrices on the basis of normal pseudo-solutions.

Let A be an arbitrary rectangular matrix. With every vector $f \in F$ we associate the normal pseudo-solution $q_{\text{np}} \in Q$ to the system $Aq = f$. As a result, we have a map $q_{\text{np}} = A^\dagger f$ from the space F to the space Q represented by an operator $A^\dagger$, which is called a pseudoinverse or a generalized inverse.

A pseudoinverse operator is linear.

The matrix of a pseudoinverse operator in natural bases is called the pseudoinverse or the generalized inverse of the matrix A and is denoted by $A^\dagger$.

If A is an $m \times n$ matrix, then $A^\dagger$ is an $n \times m$ matrix.

The normal pseudo-solution q_{np} to a system of linear algebraic equations $Aq = f$ satisfies the equality $q_{\text{np}} = A^\dagger f$.

The pseudoinverse of a zero matrix is a zero matrix.

Every column vector of $A^\dagger$ is orthogonal to the kernel of A.

Every column vector of $A^\dagger$ is a linear combination of the column vectors of the adjoint A^* (we recall that $A^* = \overline{A^T}$ and in real spaces $A^* = A^T$).

The matrix $A^\dagger$ can be represented in the form $A^\dagger = A^* V$, where V is a square matrix.

Let a_i^* and $a_i^\dagger$ denote the i-th column vectors of the matrices A^* and $A^\dagger$, respectively. A system of column vectors $a^{\dagger}_{i_1}, \ldots, a^{\dagger}_{i_r}$ is linearly dependent (linearly independent) if and only if the system of column vectors $a^*_{i_1}, \ldots, a^*_{i_r}$ is linearly dependent (linearly independent). This is true for any selection of indices $i_1, \ldots, i_r$.

For any matrix A, rank $A^{\dagger}$ = rank A^*.

Consider the matrix equation $A^*AZ = A^*$, where A is a given $m \times n$ matrix and Z is a desired $n \times m$ matrix. The matrix $A^{\dagger}$ satisfies this equation, i.e., $A^*AA^{\dagger} = A^*$.

Every row vector of the matrix A^* is a linear combination of the row vectors of the matrix $A^{\dagger}$.

Every row vector of the matrix $A^{\dagger}$ is a linear combination of the row vectors of the matrix A^*.

The matrix $A^{\dagger}$ can be represented in the form UA^*, where U is a square matrix.

The kernel and image of the pseudoinverse $A^{\dagger}$ coincide with those of the adjoint A^*, i.e., $R(A^{\dagger}) = R(A^*)$ and $N(A^{\dagger}) = N(A^*)$.

The matrix $A^{\dagger}$ has minimal rank among all solutions to the matrix equation $A^*AZ = A^*$.

Among all solutions to the matrix equation $A^*AZ = A^*$, the matrix $A^{\dagger}$ has minimal sum of squares of the absolute values of elements in each column.

Among all solutions to the matrix equation $A^*AZ = A^*$, the matrix $A^{\dagger}$ has minimal sum of squares of the absolute values of all its elements.

Consider an $m \times n$ matrix A and the $n \times n$ unit matrix I. Take an $n \times n$ matrix $B(Z) = AZ - I$, where Z is an arbitrary $n \times m$ matrix. Among all the matrices Z, the matrix $A^{\dagger}$ minimizes the sum of squares of the absolute values of all elements of $B(Z)$. Furthermore, among all the matrices that minimize this sum, $A^{\dagger}$ minimizes the sum of squares of the absolute values of all elements of Z.

If A is nondegenerate, then $B(A^{-1}) = 0$ and therefore the sum of squares of the absolute values of all elements of $B(Z)$ attains its absolute minimum at the matrix A^{-1}. The matrix $A^{\dagger}$ minimizes this sum in the case of an arbitrary matrix A. This circumstance also emphasizes the "similarity" between the pseudoinverse matrix and the inverse matrix.

So far we have been considering the properties of a pseudoinverse matrix as the properties of the matrix of a pseudoinverse operator in natural bases. It is more convenient to study many properties of a pseudoinverse matrix using an equivalent definition in terms of matrices.

The matrix F is a zero matrix if and only if at least one of the matrices F^*F or FF^* is a zero matrix.

For a given $m \times n$ matrix A, consider the matrix relations

$$A^*AX = A^*, \quad X = UA^* = A^*V,$$

for the unknown $n \times m$ matrix X, where U and V are $n \times n$ and $m \times m$ matrices, respectively. The matrix $X = A^{\dagger}$ is the only matrix satisfying these relations.

The above definitions of a pseudoinverse are equivalent. Note that there are many other equivalent definitions of a pseudoinverse matrix in addition to the ones given above. The proof that these definitions are equivalent is made easier by the fact that the matrix expression for a pseudoinverse is known.

If an $m \times n$ matrix A has full rank, then

$$A^\dagger = \begin{cases} (A^*A)^{-1}A^*, & n \le m, \\ A^*(AA^*)^{-1}, & n \ge m. \end{cases}$$

Let A be an $m \times n$ matrix of rank $r > 0$. There exist an $m \times r$ matrix B and an $r \times n$ matrix C such that $\operatorname{rank} B = r$, $\operatorname{rank} C = r$, and $A = BC$. The decomposition $A = BC$ is called a *skeleton decomposition* of the matrix A.

In the skeleton decomposition, any r basis columns of the matrix A can be chosen as columns of the matrix B. Then the columns of the matrix C consist of the coefficients of the linear combinations that express all the columns of A in terms of its basis columns.

If a matrix A is represented by its skeleton decomposition $A = BC$, then

$$A^\dagger = C^*(CC^*)^{-1}(B^*B)^{-1}B^* = C^*(B^*AC^*)^{-1}B^*,$$

$$A^* = C^*B^*, \qquad A^\dagger = C^\dagger B^\dagger,$$

$$C^\dagger = C^*(CC^*)^{-1}, \qquad B^\dagger = (B^*B)^{-1}B^*$$

are skeleton decompositions of the matrices A^*, $A^\dagger$, $C^\dagger$, and $B^\dagger$, respectively.

The matrices A, $A^\dagger$, and A^* are related by the formulas

$$(A^*)^\dagger = (A^\dagger)^*, \qquad (A^\dagger)^\dagger = A,$$

$$(AA^\dagger)^* = AA^\dagger, \qquad (AA^\dagger)^2 = AA^\dagger,$$

$$(A^\dagger A)^* = A^\dagger A, \qquad (A^\dagger A)^2 = A^\dagger A,$$

$$A^\dagger AA^\dagger = A^\dagger, \qquad AA^\dagger A = A.$$

Given an $m \times n$ matrix A, consider the matrix relations

$$AXA = A, \quad X = UA^* = A^*V,$$

where X is an unknown $n \times m$ matrix, U and V are $n \times n$ and $m \times m$ matrices, respectively. The matrix $X = A^\dagger$ is the only matrix satisfying these relations.

The Penrose equations. For a given $m \times n$ matrix A, consider the matrix relations

$$AXA = A, \quad XAX = X,$$

$$(XA)^* = XA, \quad (AX)^* = AX,$$

for an $n \times m$ matrix X. The matrix $X = A^\dagger$ is the only matrix satisfying these relations.

The pseudo-solutions to the system $Aq = f$ are the only solutions to the system $Aq = AA^\dagger f$.

Singular value decomposition. For any $m \times n$ rectangular matrix A, the matrices A^*A and AA^* are Hermitian and their principal minors are nonnegative.

The nonzero eigenvalues of the matrix A^*A are positive and coincide with those of AA^*.

The arithmetical values of the square roots of the common eigenvalues of A^*A and AA^* are called the singular (principal) values of the matrix A.

Throughout what follows, the nonzero singular values of A will be denoted by $\sigma_1, \dots, \sigma_r$ and they will be assumed to be indexed in nonincreasing order, i.e., $\sigma_1 \geq \sigma_2 \geq \dots \geq \sigma_r > 0$ ($r \leq p = \min(m, n)$). The singular values $\sigma_{r+1}, \dots$ will be considered to be equal to zero (if $r < p$).

Let A be an $m \times n$ rectangular matrix. Let $x_1, \dots, x_n$ denote the orthonormal eigenvectors of the matrix A^*A. Then

- the system of vectors $Ax_1, \dots, Ax_n$ is orthogonal;
- any nonzero vector Ax_k is an eigenvector of AA^* corresponding to the eigenvalue σ_k^2;
- $|Ax_k| = \sigma_k$ for all k.

Let A be an $m \times n$ rectangular matrix. There exist orthonormal systems of vectors $x_1, \dots, x_n$ and $y_1, \dots, y_m$ such that

$$Ax_k = \begin{cases} \sigma_k y_k, & k \leq r, \\ 0, & k > r, \end{cases} \qquad A^* y_k = \begin{cases} \sigma_k x_k, & k \leq r, \\ 0, & k > r. \end{cases}$$

The orthonormal systems $x_1, \dots, x_n$ and $y_1, \dots, y_m$ are called the singular bases of the matrix A.

A square matrix is nondegenerate if and only if all of its singular values are nonzero:

$$|\det A|^2 = \det A^*A = \prod_{i=1}^{m} \sigma_i^2.$$

Let $U \Sigma V^T$ be a singular value decomposition of a matrix A. Then the pseudoinverse $A^\dagger$ can be represented in the form $A^\dagger = V \Sigma^\dagger U^T$.

Suppose that the first diagonal elements $\sigma_1, \sigma_2, \dots, \sigma_r$ of an $m \times n$ diagonal matrix Σ are nonzero and the remaining elements are zero. The matrix $\Sigma^\dagger$ is an $n \times m$ diagonal matrix with the first diagonal elements $\sigma_1^{-1}, \sigma_2^{-1}, \dots, \sigma_r^{-1}$ and the remaining elements being zero.

For any $m \times n$ matrix A with elements a_{ij} and singular values σ_k, there holds

$$\sum_{i=1}^{m} \sum_{j=1}^{n} |a_{ij}|^2 = \sum_{k=1}^{\min(m,n)} \sigma_k^2.$$

For any $m \times m$ matrix A with eigenvalues $\lambda_1, \ldots, \lambda_m$ and singular values σ_1, σ_2, $\ldots, \sigma_m$, there holds

$$\sum_{i=1}^{m} \sigma_k^2 \geq \sum_{i=1}^{m} |\lambda_i|^2,$$

the equality being attained if and only if A is a normal matrix.

Connection between the singular value decomposition and the Tikhonov regularization. Suppose that $A = U\Sigma V^T$, where Σ is an $m \times n$ diagonal matrix, $\Sigma = \operatorname{diag}(\sigma_1, \sigma_2, \ldots, \sigma_p)$, $p = \min\{m, n\}$. Then a solution q_α to the equation

$$\alpha q + A^T A q = A^T f, \quad \alpha > 0,$$

has the form

$$q_\alpha = V\Sigma_\alpha U^T f, \quad \Sigma_\alpha = \operatorname{diag}\Big(\frac{\sigma_1}{\sigma_1^2 + \alpha}, \frac{\sigma_2}{\sigma_2^2 + \alpha}, \ldots, \frac{\sigma_p}{\sigma_p^2 + \alpha}\Big).$$

Chapter 4

Integral equations

Integral equations of the first kind considered in Sections 4.1 and 4.2 are the most suitable subject for the introduction of ill-posed problems in infinite-dimensional spaces. On one hand, there is no simpler example to be found. On the other hand, in the general case, the compactness of integral operators makes their inversion an almost insurmountable problem.

The reader should not be misled by the relative simplicity of the first sections of this chapter. We recommend that the reader finds a practical example for each of the problems to be considered here and, conversely, tries to come up with a generalization and to analyze it.

Sections 4.3–4.6, which are devoted to the Volterra operator equations, are the most difficult for understanding, but the results presented there are applicable to a large class of coefficient inverse problems for hyperbolic equations.

4.1 Fredholm integral equations of the first kind

Consider the Fredholm integral equation of the first kind

$$Aq(x) = \int_a^b K(x,s)q(s)\,ds = f(x), \quad x \in (a,b). \tag{4.1.1}$$

The ill-posedness of equation (4.1.1) can be illustrated by the following simple example. Assume the kernel $K(x,s)$ and the function $f(x)$ to be constant. Then problem (4.1.1) becomes the problem of recovering a curve from the area under the curve. Evidently, this problem has at least continuum of solutions.

Let the kernel $K(x,s)$ be real and symmetric, i.e., $K(s,x) = K(x,s)$. We also assume that $K(x,s) \in L_2((a,b)^2)$, $q \in Q = L_2(a,b)$, and $f(x) \in F = L_2(a,b)$. Then there exists a complete orthonormal sequence of eigenfunctions $\{\varphi_n(x)\}$ and a sequence of the corresponding eigenvalues $\{\lambda_n\}$ of the operator A ($|\lambda_{n+1}| \le |\lambda_n|$, $\lambda_n \in \mathbb{R}$) such that

$$A\varphi_n(x) = \int_a^b K(x,s)\varphi_n(s)\,ds = \lambda_n \varphi_n(x), \quad n \in \mathbb{N}.$$

The completeness of the system $\{\varphi_n(x)\}$ means that any function $f(x) \in F$ such that $\|f\|_{L_2(a,b)} < \infty$ is representable as a Fourier series:

$$f(x) = \sum_{n=1}^{\infty} f_n \varphi_n(x).$$

It will be shown below that the case where all λ_n are nonzero is substantially different from the general case. Let the kernel $K(x, s)$ be represented in the form of a series

$$K(x,s) = \sum_{n=1}^{\infty} \lambda_n \varphi_n(x)\varphi_n(s)$$

that converges in the norm:

$$\|K(x,s)\| = \sqrt{\int_a^b \int_a^b K^2(x,s)\,dx\,ds} = \sum_{n=1}^{\infty} \lambda_n^2 < \infty.$$

In this case $|\lambda_n| \to 0$ as $n \to \infty$.

The kernel $K(x, s)$ of an integral operator A is said to be degenerate if there exists such $n_0 \in \mathbb{N}$ that $\lambda_n \neq 0$ for $1 \le n \le n_0$ and $\lambda_n = 0$ for all $n > n_0$.

Consider the problem (4.1.1) assuming $K(x, s)$ as degenerate:

$$K(x,s) = \sum_{n=1}^{n_0} \lambda_n \varphi_n(x)\varphi_n(s).$$

In this case

$$\begin{aligned}\int_a^b K(x,s)q(s)ds &= \sum_{n=1}^{n_0} \lambda_n \int_a^b \varphi_n(x)\varphi_n(s)q(s)ds \\ &= \sum_{n=1}^{n_0} \lambda_n(\varphi_n, q)\varphi_n(x) = f(x).\end{aligned}$$

Therefore, the problem (4.1.1) has a solution only if $f(x)$ is a linear combination of $\varphi_1(x), \dots, \varphi_{n_0}(x)$, i.e., if it can be written as

$$f(x) = \sum_{n=1}^{n_0} f_n \varphi_n(x).$$

Exercise 4.1.1. Verify that if the problem (4.1.1) with the degenerate kernel $K(x, s)$ is solvable, then any function $q(x)$ of the form

$$q(x) = \sum_{n=1}^{n_0} \frac{f_n}{\lambda_n} \varphi_n(x) + \sum_{n=n_0+1}^{\infty} c_n \varphi_n(x),$$

where $\sum_{n=n_0+1}^{\infty} c_n^2 < \infty$, is a solution to the equation (4.1.1).

Thus, if the kernel is degenerate, the problem (4.1.1) has either no solutions or infinitely many solutions, depending on the right-hand side $f(x)$.

Consider the case where $\lambda_n \neq 0$ for all n. If $\|f\| = \sqrt{\int_a^b f^2(x)\,dx} < \infty$, then $f(x)$ can be represented by a Fourier series converging in the norm of the space L_2:

$$f(x) = \sum_{n=1}^{\infty} f_n \varphi_n(x), \quad f_n = \langle f, \varphi_n \rangle, \quad \sum_{n=1}^{\infty} f_n^2 = \|f\|^2.$$

Put $S = \sum_{n=1}^{\infty} \left(\frac{f_n}{\lambda_n}\right)^2$.

Exercise 4.1.2. Prove that the function $q(x) = \sum_{n=1}^{\infty} \frac{f_n}{\lambda_n} \varphi_n(x)$ is a solution to the equation (4.1.1) if $S < \infty$.

Exercise 4.1.3. Prove that the problem (4.1.1) has no solutions if $S = \infty$.

Exercise 4.1.4. Show that the problem (4.1.1) cannot have two different solutions if $\lambda_n \neq 0$ for all $n \in \mathbb{N}$ (the solutions that differ from each other on a set of zero measure are considered to be the same).

Thus, if $\lambda_n \neq 0$ for all $n \in \mathbb{N}$, then the problem (4.1.1) has no more than one solution, and the solution exists only if the right-hand side satisfies the condition $S < \infty$.

Exercise 4.1.5. Study the case where infinitely many λ_n are nonzero and infinitely many λ_n are equal to zero.

It is clear that the solution to the equation (4.1.1) can be represented by a series only if $\{\lambda_n\}$ and $\{\varphi_n(x)\}$ are given. For the numerical solution of applied problems having the form (4.1.1), the following approaches are commonly used:

- discretization of (4.1.1) combined with the Tikhonov regularization (Vasin and Ageev, 1995);
- iterative regularization (Vainikko, 1982);
- singular value decomposition combined with regularization (Godunov et al., 1993).

Problem with perturbed data. The following situation is frequent in the problems of interpreting observational data, also known as the problems of processing experimental data.

Suppose that a function $q(x)$ is the function of interest in a certain research, but the function measured in the experiment is a different function $f(x)$ that is related to $q(x)$ by the formula $f(x) = \int_a^b K(x,s)q(s)ds$. Furthermore, the values of $f(x)$ have measurement errors.

Thus, the problem

$$\int_a^b K(x,s)q(s)ds = f(x)$$

is known to have a solution, but the solution can be found only approximately from the equation

$$\int_a^b K(x,s)q(s)\,ds = f_\delta(x) \tag{4.1.2}$$

with such function $f_\delta(x)$ that

$$\|f - f_\delta\| \le \varepsilon. \tag{4.1.3}$$

Let $q_e(x)$ be a solution to the problem (4.1.1) and $q_\delta(x)$ be a solution to (4.1.2). Set $\tilde{q} = q - q_\delta$, $\tilde{f} = f - f_\delta$. Then $\tilde{q}(x)$ satisfies the integral equation

$$\int_a^b K(x,s)\tilde{q}(s)\,ds = \tilde{f}(x). \tag{4.1.4}$$

Let $\tilde{f}(x) = \sum_{n=1}^{\infty} \tilde{f}_n\varphi_n(x)$. The condition (4.1.3) means that $\sum_{n=1}^{\infty} |\tilde{f}_n|^2 \le \delta^2$.

First, consider the case where $\lambda_n \ne 0$ for all $n \in \mathbb{N}$. If the series $\sum_{n=1}^{\infty} |\frac{\tilde{f}_n}{\lambda_n}|^2$ diverges, then equation (4.1.4) has no solutions. Even if this series converges, this cannot guarantee that the error $\tilde{q}(x)$ will tend to zero as $\delta \to 0$.

Indeed, among all right-hand sides $\tilde{f}$ such that $\|\tilde{f}\| \le \delta$ there is a right-hand side $\tilde{f} = \delta\varphi_n(x)$ with index n such that $|\lambda_n| \le \delta$. Then $\tilde{q} = (\delta/\lambda_n)\varphi_n(x)$, i.e., $\|\tilde{q}\| = \delta/|\lambda_n| > 1$.

Regularization. Consider the equation

$$\alpha q(x) + \int_a^b K(x,s)q(s)\,ds = f_\delta(x), \quad \alpha > 0, \tag{4.1.5}$$

where α is the regularization parameter.

Theorem 4.1.1. *Assume that all eigenvalues of the operator*

$$Aq = \int_a^b K(x,s)q(s)\,ds, \quad x \in (a,b),$$

are positive. Suppose that there exists a solution $q_e \in L_2(a,b)$ to equation (4.1.1) *for some $f \in L_2(a,b)$. Then there exist $\alpha_*, \delta_* \in R_+$ such that for all $\alpha \in (0,\alpha_*)$, $\delta \in (0,\delta_*)$ and for any $f_\delta \in L_2(a,b)$ satisfying* (4.1.3), *equation* (4.1.5) *has a solution $q_{\alpha\delta}$ and the following estimate holds:*

$$\|q_{\alpha\delta} - q_e\| \le \frac{\delta}{\alpha} + \omega(\alpha), \tag{4.1.6}$$

where the function $\omega(\alpha)$ tends to zero as $\alpha \to 0$.

Proof. Let $q_{\alpha\delta}(x)$ denote a solution to equation (4.1.5). We will seek the solution in the form of a series in terms of the eigenfunctions of the operator A:

$$q_{\alpha\delta}(x) = \sum q_{\alpha\delta n}\varphi_n(x). \tag{4.1.7}$$

Let $\tilde{f}(x) = f_\delta(x) - f(x)$. Then

$$f_\delta(x) = \sum_{n=1}^{\infty} f_{\delta n}\varphi_n(x) = \sum_{n=1}^{\infty}(\tilde{f}_n + f_n)\varphi_n(x), \tag{4.1.8}$$

where $f_{\delta n} = \langle f_\delta, \varphi_n\rangle$, $\tilde{f}_n = (\tilde{f}, \varphi_n)$.

Multiply (4.1.5) by $\varphi_n(x)$ and integrate it with respect to x from a to b. Taking into account the form of the kernel, the orthonormality of the system of functions $\{\varphi_n(x)\}$ and using (4.1.7), (4.1.8), we get

$$\alpha q_{\alpha\delta n} + \lambda_n q_{\alpha\delta n} = \tilde{f}_n + f_n = f_{\delta n}. \tag{4.1.9}$$

Consequently, a regularized solution obtained from the approximate data has the form

$$q_{\alpha\delta}(x) = \sum_{n=1}^{\infty} q_{\alpha\delta n}\varphi_n(x), \tag{4.1.10}$$

$$q_{\alpha\delta n} = \frac{f_{\delta n}}{\alpha + \lambda_n} = \frac{\tilde{f}_n + f_n}{\alpha + \lambda_n}. \tag{4.1.11}$$

Let

$$q_e(x) = \sum_{n=1}^{\infty} q_n\varphi_n(x).$$

Then the error

$$\tilde{q}(x) = q_e(x) - q_{\alpha\delta}(x) = \sum_{n=1}^{\infty}\Big(q_n - \frac{\tilde{f}_n + f_n}{\alpha + \lambda_n}\Big)\varphi_n(x)$$

can be represented as

$$\tilde{q}(x) = \tilde{\beta}_\alpha(x) + \beta_\alpha(x), \tag{4.1.12}$$

where

$$\tilde{\beta}_\alpha(x) = -\sum_{n=1}^{\infty}\frac{\tilde{f}_n}{\alpha + \lambda_n}\varphi_n(x),$$

$$\beta_\alpha(x) = \sum_{n=1}^{\infty}\frac{\alpha q_n}{\alpha + \lambda_n}\varphi_n(x).$$

First, we estimate $\tilde{\beta}_\alpha$ taking into account that $\lambda_n \geq 0$:

$$\|\tilde{\beta}_\alpha\|^2 = \sum_{k=1}^{\infty} \frac{\tilde{f}_n^2}{(\alpha + \lambda_n)^2} \leq \frac{1}{\alpha^2} \sum_{n=1}^{\infty} \tilde{f}_n^2 = \frac{\delta^2}{\alpha^2}. \tag{4.1.13}$$

To estimate the second term $\beta_\alpha(x)$, we recall that, by the assumption of the theorem, there exists an exact solution $q_e(x)$ that belongs to $L_2(a,b)$. Therefore, $\|q_e\|^2 = \sum_{n=1}^{\infty} q_n^2 < \infty$. Denote $\omega(\alpha) = \|\beta_\alpha\|$. Let us prove that

$$\lim_{\alpha \to 0} \omega(\alpha) = 0.$$

Let $\varepsilon > 0$. We will find α_* such that $\omega(\alpha) \leq \varepsilon$ for all $\alpha \in (0, \alpha_*)$.

For $\varepsilon^2/2$ there exists an $n_* \in \mathbb{N}$ such that $\sum_{n=n_*}^{\infty} q_n^2 < \frac{\varepsilon^2}{2}$ and therefore

$$\sum_{n=n_*}^{\infty} \frac{\alpha^2 q_n^2}{(\alpha + \lambda_n)^2} \leq \frac{\varepsilon^2}{2}.$$

On the other hand,

$$\sum_{n=1}^{n_*-1} \frac{\alpha^2 q_n^2}{(\alpha + \lambda_n)^2} \leq \frac{\alpha^2}{\mu_*^2} \|q_e\|^2,$$

where $\mu_* = \min_{n=1,n_*-1}\{\lambda_n\}$. Set

$$\alpha_* = \varepsilon\mu_*/(\sqrt{2}\,\|q_e\|).$$

Then for all $\alpha \in (0, \alpha_*)$ we have

$$\omega^2(\alpha) = \|\beta_\alpha\|^2 \leq \sum_{n=1}^{n_*-1} \frac{\alpha^2 q_n^2}{(\alpha + \lambda_n)^2} + \sum_{n=n_*}^{\infty} \frac{\alpha^2 q_n^2}{(\alpha + \lambda_n)^2} \leq \varepsilon^2,$$

i.e., $\omega(\alpha) \leq \varepsilon$.

Using (4.1.12) and (4.1.13), we obtain

$$\|q_e - q_{\alpha\delta}\| \leq \|\tilde{\beta}_\alpha\| + \|\beta_\alpha\| \leq \delta/2 + \omega(\alpha),$$

which proves the theorem. □

We now show that the regularization method described above is also applicable in the case where some λ_n are equal to zero.

Let N_1 be the set of such n that $\lambda_n \neq 0$, and let N_0 consist of n corresponding to $\lambda_n = 0$. (Each of these two sets may be finite or infinite, but they cannot both be finite at the same time). Let $q_e(x) = \sum_{n=1}^{\infty} q_n\varphi_n(x)$ ($\sum_{n=1}^{\infty} |q_n|^2 < \infty$) be a solution to equation (4.1.1). Then for any $a_n \in \mathbb{R}$ such that $\sum_{n\in N_0} a_n^2 < \infty$ and $\sum_{n\in N_1} a_n^2 = 0$, the function $\sum_{n=1}^{\infty}(q_n + a_n)\varphi_n(x)$ is also a solution to (4.1.1).

Putting $a_n = -q_n$ for $n \in N_0$, we obtain the function $q^0(x) = \sum_{n\in N_1} q_n\varphi_n(x)$ satisfying (4.1.1).

Exercise 4.1.6. Let $q(x) \neq q^0(x)$ be a solution to equation (4.1.1). Prove that

$$\|q\| > \|q^0\|. \tag{4.1.14}$$

We recall that if (4.1.1) has more than one solution, then the solution that has minimal norm is called normal. From (4.1.14) it follows that $q^0(x)$ is a normal solution to the problem.

Theorem 4.1.2. *Let* $q^0(x) = \sum_{n\in N_1} q_n\varphi_n(x)$, $\|q^0\|^2 = \sum_{n\in N_1} q_n^2 < \infty$, *be a normal solution to equation* (4.1.1). *Then the following inequality holds:* $\|q_{\alpha\delta} - q^0\| \leq \delta/\alpha + \omega(\alpha)$, *where* $\omega(\alpha) \to 0$ *as* $\alpha \to 0$.

The proof of this theorem is similar to that of Theorem 4.1.1.

Consider the case where the kernel $K(x,s)$ is not symmetric. We define the operator A^* as follows:

$$A^*p(x) = \int_a^b K(s,x)p(s)\,ds, \quad p \in L_2(a,b).$$

Applying the operator A^* to the equation (4.1.2) yields the integral operator equation with symmetric kernel:

$$A^*Aq = A^*f_\delta.$$

Let $q^*_{\alpha\delta}(x)$ denote a solution to the equation

$$\alpha q(x) + A^*Aq = A^*f_\delta(x).$$

Exercise 4.1.7. Prove that $q^*_{\alpha\delta}$ converges to a quasi-solution of equation (4.1.1) as α and δ tend to zero with concerted rates.

4.2 Regularization of linear Volterra integral equations of the first kind

Volterra integral equations of the first kind

$$\int_0^x K(x,t)q(t)\,dt = f(x), \quad x \in [0,T], \tag{4.2.1}$$

can be viewed as a special case of the Fredholm equation of the first kind

$$\int_0^1 \overline{K}(x,t)q(t)\,dt = f(x), \quad x \in [0,T],$$

whose kernel has the form

$$\overline{K}(x,t) = \begin{cases} K(x,t), & t \in [0,x], \\ 0, & t \in (x,T]. \end{cases}$$

Therefore, all the arguments of the previous section remain valid as to equation (4.2.1). However, the variable limit of integration in (4.2.1) (the distinctive property of a Volterra equation) makes it possible to construct other regularization methods. The simplest one is differentiation. Assume that equation (4.2.1) has a solution $q_e \in C[0,T]$ and the functions $K(x,t)$ and $f(x)$ have continuous derivatives with respect to x. Then, differentiating (4.2.1) with respect to x, we get

$$q(x)K(x,x) + \int_0^x \frac{\partial K}{\partial x}(x,t)q(t)\,dt = f'(x). \tag{4.2.2}$$

If $K(x,x) \neq 0$ for all $x \in [0,T]$, then equation (4.2.2) is a special case of the Volterra integral equation of the second kind.

In what follows we shall also consider a Volterra regularization. Aside from the differentiation and the Volterra regularization, a discrete regularization can be applied to equation (4.2.1), which leads sometimes to the so-called self-regularization where the discretization step plays the role of a regularization parameter.

Volterra integral equations of the second kind in $C[0,T]$ and $L_2(0,T)$. Consider the second-kind Volterra equation

$$q(x) = g(x) + \int_0^x P(x,t)q(t)\,dt, \quad x \in (0,T), \tag{4.2.3}$$

assuming $q(x)$ is to be found.

Let us show that problem (4.2.3) is well-posed.

Theorem 4.2.1. *Let $g \in C[0,T]$ and $P(x,t) \in C(0 \leq t \leq x \leq T)$. Then the problem* (4.2.3) *is well-posed in* $C[0,T]$.

Proof. We define an operator V as follows:

$$(Vq)(x) = g(x) + \int_0^x P(x,t)q(t)\,dt, \quad x \in [0,T].$$

Introduce a norm depending on a parameter $\beta \geq 0$:

$$\|q\|_\beta = \sup_{x\in[0,T]} \{|q(x)|e^{-\beta x}\}.$$

From the definition of the norm it follows that

$$\|q\|_0 = \|q\|_{C[0,T]}, \quad \|q\|_\beta = \|q\|_{C[0,T]} \leq e^{\beta T}\|q\|_\beta. \tag{4.2.4}$$

Set

$$G_0 = \|g\|_{C[0,T]}, \quad P_0 = \sup_{\substack{x\in[0,T] \\ t\in[0,x]}} |P(x,t)|.$$

We will prove that for any fixed $T > 0$, $\gamma > 0$, and any β satisfying the condition

$$\max\{P_0, P_0(G_0+\gamma)/\gamma\} < \beta \tag{4.2.5}$$

the operator V maps the ball

$$B(g,\gamma,\beta) = \{p(x) \in C[0,T] : \|p-g\|_\beta \le \gamma\}$$

into itself and is a contraction on this ball. Let $r \in B(g,\gamma,\beta)$, i.e., $\|g-r\|_\beta \le \gamma$ and therefore $\|r\|_\beta \le \|g\|_\beta + \gamma \le G_0 + \gamma$. We have

$$\begin{aligned}|(Vr)(x) - g(x)| &\le \int_0^x |P(x,t)r(t)|dt \le P_0 \int_0^x |r(t)|e^{-\beta t}e^{\beta t}dt \\ &\le P_0\|r\|_\beta \int_0^x e^{\beta t}dt = P_0\|r\|_\beta \frac{1}{\beta}(e^{\beta t}-1), \quad x\in[0,T].\end{aligned}$$

Hence, for all $x \in [0,T]$,

$$|(Vr)(x) - g(x)|e^{-\beta t} \le P_0\|r\|_\beta \frac{1}{\beta}(1-e^{-\beta x}) \le \frac{P_0}{\beta}\|r\|_\beta,$$

and therefore, in view of (4.2.5),

$$\|Vr - g\|_\beta \le \frac{P_0}{\beta}\|r\|_\beta \le \frac{P_0(G_0+\gamma)}{\beta} < \gamma,$$

which means that $Vr \in B(g,\gamma,\beta)$.

Now let $r_1, r_2 \in B(g,\gamma,\beta)$. By the same argument as before, we get the estimate

$$\begin{aligned}|(Vr_1)(x) - (Vr_2)(x)| &\le \int_0^x |P(x,t)||r_1(t)-r_2(t)|\,dt \\ &\le P_0\|r_1-r_2\|_\beta \frac{1}{\beta}(e^{\beta x}-1),\end{aligned}$$

whence

$$\|Vr_1 - Vr_2\| \le \frac{P_0}{\beta}\|r_1-r_2\|_\beta.$$

Due to (4.2.5), $P_0/\beta < 1$. This means that the operator V maps the ball $B(g, \gamma, \beta)$ into itself and is a contraction on this ball. Therefore, by the Banach fixed point theorem (Theorem A.2.1), the operator V has a unique fixed point in this ball:

$$r(x) = (Vr)(x) = g(x) + \int_0^x K(x,t)r(t)\,dt,$$

which is obviously a solution to equation (4.2.3).

We now turn to the proof of uniqueness and stability of the solution of problem (4.2.3) with respect to small perturbations of $g(x)$. We start by proving the stability. Assume that $q_j(x)$, $j = 1, 2$, are continuous solutions to the equations

$$q_j(x) = g_j(x) + \int_0^x P(x,t)q_j(t)\,dt, \quad j = 1, 2, \tag{4.2.6}$$

respectively. As before, the functions g_1, g_2, and P are assumed to be continuous. Then $\tilde{q} = q_1 - q_2$ and $\tilde{g} = g_1 - g_2$ satisfy the equality

$$\tilde{q}(x) = \tilde{g}(x) + \int_0^x P(x,t)\tilde{q}(t)dt,$$

which implies

$$|\tilde{q}(x)| \le |\tilde{g}(x)| + \left| \int_0^x P(x,t)\tilde{q}(t)\,dt \right|.$$

Therefore,

$$|\tilde{q}(x)| \le |\tilde{g}(x)| + P_0 \|\tilde{q}\|_\beta \frac{1}{\beta} (\mathrm{e}^{\beta x} - 1).$$

Multiplying the obtained inequality by $\mathrm{e}^{-\beta x}$ and taking the supremum with respect to $x \in [0, T]$ (first in the right-hand side and then in the left-hand side), we have

$$\|\tilde{q}\|_\beta \le \|\tilde{g}\|_\beta + \frac{P_0}{\beta} \|\tilde{q}\|_\beta.$$

Then

$$\left(1 - \frac{P_0}{\beta}\right) \|\tilde{q}\|_\beta \le \|\tilde{g}\|_\beta.$$

Since $\beta > P_0$ because of (4.2.5), we have

$$\|\tilde{q}\|_\beta \le \frac{\beta}{\beta - P_0} \|\tilde{g}\|_\beta.$$

Finally, using (4.2.4) we obtain

$$\|q_1 - q_2\|_{C[0,T]} \le \frac{\beta}{\beta - P_0} \mathrm{e}^{\beta T} \|g_1 - g_2\|_{C[0,T]}. \tag{4.2.7}$$

The stability estimate (4.2.7) also implies the uniqueness of the solution. □

Remark 4.2.1. Estimate (4.2.7) can be improved by applying the Gronwall lemma (see Subsection A.6.2) to the inequality

$$|\tilde{q}(x)| \leq |\tilde{g}(x)| + \left| \int_0^x P(x,t)\tilde{q}(t)\,dt \right|$$

obtained when proving the theorem.

Remark 4.2.2. It can be proved that if $g \in L_2(0,T)$ and $P(x,t) \in L_2$ $(0 \leq t \leq x \leq T)$, then problem (4.2.3) is well-posed in $L_2(0,T)$.

Volterra regularization of the first-kind integral equations. The above example of reducing the ill-posed problem (4.2.1) to the well-posed problem (4.2.2) is rarely applicable in practice for two main reasons. Instead of the actual right-hand side $f(x)$ we usually have only its approximation $f_\delta(x)$, and even if the derivative $f'_\delta(x)$ is known, small errors in specifying $f(x)$ may lead to very large errors in the solution. On the other hand, application of the general approach described in Section 4.1, i.e., reducing the problem (4.2.1) to the Fredholm equation of the second kind

$$\alpha q(x) + A^* A q = A^* f_\delta(x)_g$$

results in the loss of the property of being a Volterra equation (A^*A is not a Volterra operator), which considerably complicates the analysis and solution of the regularized equation.

Consider a Volterra regularization of the problem (4.2.1) (Sergeev, 1971; Denisov, 1975), which can also be applied to nonlinear Volterra operator equations of the first kind (Kabanikhin, 1989) that arise in coefficient inverse problems for hyperbolic equations (see Chapter 10). As before, we assume that equation (4.2.1), i. e.,

$$\int_0^x K(x,t)q(t)\,dt = f(x), \quad x \in [0,T],$$

has a solution, but instead of the function $f(x)$ we know only its approximation $f_\delta(x)$ satisfying the condition

$$\|f - f_\delta\|_{C[0,T]} \leq \delta.$$

We will approximate equation (4.2.1) by the following Volterra integral equation of the second kind with a positive parameter α:

$$\alpha q(x) + \int_0^x K(x,t)q(t)\,dt = f_\delta(x), \quad x \in [0,T]. \tag{4.2.8}$$

Theorem 4.2.2. *Assume that for $f, f_\delta \in C[0,T]$ there exists a solution $q_e \in C^1[0,T]$ to equation (4.2.1) such that $q_e(+0) = 0$. Let K, $\frac{\partial K}{\partial x}$ and $\frac{\partial^2 K}{\partial x^2}$ be continuous and $K(x,x) \equiv 1$.*

Then there exist positive numbers α_*, δ_*, and K_* such that, for $\alpha \in (0, \alpha_*)$ and $\delta \in (0, \delta_*)$, equation (4.2.8) has a unique solution $q_{\alpha\delta}$ in $C[0,T]$ that satisfies the estimate

$$\|q_{\alpha\delta} - q_e\|_{C[0,T]} \le K_*\Big(\alpha + \frac{\delta}{\alpha}\Big). \tag{4.2.9}$$

Proof. The existence and uniqueness of a solution $q_{\alpha\delta}(x)$ follow from Theorem 4.2.1. Set

$$\tilde f(x) = f(x) - f_\delta(x), \quad \tilde q(x) = q(x) - q_{\alpha\delta}(x).$$

Then, upon subtracting (4.2.8) termwise from (4.2.1) and rearranging, we obtain

$$\alpha \tilde q(x) + \int_0^x K(x,t)\tilde q(t)\,dt = \alpha q_e(x) + \tilde f(x). \tag{4.2.10}$$

The arguments used in the proof of Theorem 4.2.1 cannot be used to estimate $\tilde q$ because equation (4.2.10) includes a small parameter α. For this reason, we will introduce a more complicated weight function $\mathrm{e}^{-(x-\xi)/\alpha}$ instead of $\mathrm{e}^{\beta t}$ and use additional integration with respect to x. Change from the variable x to ξ in (4.2.10), multiply the resulting equation by the function $\alpha^{-2}\mathrm{e}^{-(x-\xi)/\alpha}$ and integrate it with respect to ξ from 0 to x:

$$\begin{aligned}\int_0^x \mathrm{e}^{-(x-\xi)/\alpha}\Big[\frac{1}{\alpha}\tilde q(\xi) + \frac{1}{\alpha^2}\int_0^\xi K(\xi,t)\tilde q(t)\,dt\Big]\,d\xi \\ = \frac{1}{\alpha}\int_0^x \mathrm{e}^{-(x-\xi)/\alpha} q_e(\xi)\,d\xi + \frac{1}{\alpha^2}\int_0^x \mathrm{e}^{-(x-\xi)/\alpha}\tilde f(\xi)\,d\xi.\end{aligned} \tag{4.2.11}$$

We divide (4.2.10) by α and subtract (4.2.11) from the resulting equality. On rearranging, we get

$$\begin{aligned}&\tilde q(x) + \frac{1}{\alpha}\int_0^x K(x,t)\tilde q(t)\,dt - \frac{1}{\alpha}\int_0^x \mathrm{e}^{-(x-\xi)/\alpha}\tilde q(\xi)\,d\xi \\ &- \frac{1}{\alpha^2}\int_0^x \mathrm{e}^{-(x-\xi)/\alpha}\Big[\int_0^\xi K(\xi,t)\tilde q(t)\,dt\Big]\,d\xi \\ &= q_e(x) + \frac{1}{\alpha}\tilde f(x) + \frac{1}{\alpha}\int_0^x \mathrm{e}^{-(x-\xi)/\alpha} q_e(\xi)\,d\xi - \frac{1}{\alpha^2}\int_0^x \mathrm{e}^{-(x-\xi)/\alpha}\tilde f(\xi)\,d\xi.\end{aligned} \tag{4.2.12}$$

We integrate by parts the fourth term on the left side of (4.2.12) twice using the condition $K(x,x) \equiv 1$:

$$\begin{aligned}
&\frac{1}{\alpha^2}\int_0^x e^{-(x-\xi)/\alpha}\Big[\int_0^\xi K(\xi,t)\tilde{q}(t)\,dt\Big]\,d\xi \\
&\quad= \frac{1}{\alpha}\int_0^x K(x,t)\tilde{q}(t)\,dt - \frac{1}{\alpha}\int_0^x e^{-(x-\xi)/\alpha}\frac{\partial}{\partial\xi}\Big[\int_0^\xi K(\xi,t)\tilde{q}(t)\,dt\Big]\,d\xi \\
&\quad= \frac{1}{\alpha}\int_0^x K(x,t)\tilde{q}(t)\,dt - \frac{1}{\alpha}\int_0^x e^{-(x-\xi)/\alpha}\tilde{q}(\xi)\,d\xi \\
&\qquad - \frac{1}{\alpha}\int_0^x e^{-(x-\xi)/\alpha}\Big[\int_0^\xi D^{(1,0)}K(\xi,t)\tilde{q}(t)\,dt\Big]\,d\xi \\
&\quad= \frac{1}{\alpha}\int_0^x K(x,t)\tilde{q}(t)\,dt - \frac{1}{\alpha}\int_0^x e^{-(x-\xi)/\alpha}\tilde{q}(\xi)\,d\xi \\
&\qquad - \int_0^x D^{(1,0)}K(x,t)\tilde{q}(t)\,dt + \int_0^x e^{-(x-\xi)/\alpha}D^{(1,0)}K(\xi,\xi)\tilde{q}(\xi)\,d\xi \\
&\qquad + \int_0^x e^{-(x-\xi)/\alpha}\Big[\int_0^\xi D^{(2,0)}K(\xi,t)\tilde{q}(t)\,dt\Big]\,d\xi.
\end{aligned}$$

Here $D^{(n_1,n_2)}K(x_1,x_2) = \frac{\partial^{n_1}}{\partial x_1^{n_1}}\frac{\partial^{n_2}}{\partial x_2^{n_2}}K(x_1,x_2)$.

Changing the order of integration in the last term and substituting the resulting expression in (4.2.12), we come to

$$\begin{aligned}
&\tilde{q}(x) + \int_0^x D^{(1,0)}K(x,t)\tilde{q}(t)\,dt - \int_0^x e^{-(x-\xi)/\alpha}D^{(1,0)}K(\xi,\xi)\tilde{q}(\xi)\,d\xi \\
&\quad - \int_0^x\Big[\int_t^x e^{-(x-\xi)/\alpha}D^{(2,0)}K(\xi,t)\,d\xi\Big]\tilde{q}(t)\,dt \\
&\quad = q_e(x) + \frac{1}{\alpha}\tilde{f}(x) - \frac{1}{\alpha}\int_0^x e^{-(x-\xi)/\alpha}q_e(\xi)\,d\xi \\
&\qquad - \frac{1}{\alpha^2}\int_0^x e^{-(x-\xi)/\alpha}\tilde{f}(\xi)\,d\xi.
\end{aligned}$$

Evidently, the derived equality is representable in the form

$$\tilde{q}(x) + \int_0^x B_\alpha(x,t)\tilde{q}(t)dt = \psi_{\alpha\delta}(x), \tag{4.2.13}$$

where

$$\begin{aligned} B_\alpha(x,t) &= \frac{1}{\alpha}\,K(x,t) - \frac{1}{\alpha}\,\mathrm{e}^{-(x-t)/\alpha} - \frac{1}{\alpha^2}\int_t^x K(\xi,t)\mathrm{e}^{-(x-\xi)/\alpha}\,d\xi \\ &= D^{(1,0)}K(x,t) - D^{(1,0)}K(t,t)\mathrm{e}^{-(x-t)/\alpha} \\ &\quad - \int_t^x D^{(2,0)}K(\xi,t)\mathrm{e}^{-(x-\xi)/\alpha}\,d\xi, \\ \psi_{\alpha\delta}(x) &= q_e(x) - \frac{1}{\alpha}\int_0^x \mathrm{e}^{-(x-\xi)/\alpha} q_e(\xi)\,d\xi + \frac{\tilde{f}(x)}{\alpha} \\ &\quad - \frac{1}{\alpha^2}\int_0^x \mathrm{e}^{-(x-\xi)/\alpha}\tilde{f}(\xi)\,d\xi \\ &= \int_0^x \mathrm{e}^{-(x-\xi)/\alpha} q_e'(\xi)\,d\xi + \frac{\tilde{f}(x)}{\alpha} - \frac{1}{\alpha^2}\int_0^x \mathrm{e}^{-(x-\xi)/\alpha}\tilde{f}(\xi)\,d\xi. \end{aligned}$$

From inequality (4.2.13) it follows that

$$|\tilde{q}(x)| \le \|\psi_{\alpha\delta}\|_{C[0,T]} + \Big|\int_0^x \max_{x\in[0,T]} |B_\alpha(x,t)|\tilde{q}(t)\,dt\Big|, \qquad x \in [0,T].$$

Using the Gronwall lemma (see Subsection A.6.2), we have

$$\|\tilde{q}\|_{C[0,T]} = \max_{x\in[0,T]} |\tilde{q}(x)| \le \|\psi_{\alpha\delta}\|_{C[0,T]} \exp\Big\{\int_0^x \max_{x\in[0,T]} |B_\alpha(x,t)|\,dt\Big\}.$$

To complete the proof, it suffices to verify that, for some $\alpha_* > 0$, $\delta_* > 0$, $K_1 > 0$, $K_2 > 0$, the following inequalities hold:

$$\max_{x\in[0,T]} \int_0^x |B_\alpha(x,t)|\,dt \le K_1, \quad 0 < \alpha < \alpha_*, \tag{4.2.14}$$

$$\max_{x\in[0,T]} |\psi_{\alpha\delta}(x)| \le K_2\Big(\alpha + \frac{\delta}{\alpha}\Big), \quad 0 < \alpha < \alpha_*, \quad 0 < \delta < \delta_*. \tag{4.2.15}$$

We leave the proof of these inequalities to the reader as an exercise. □

4.3 Volterra operator equations with boundedly Lipschitz-continuous kernel

This section is concerned with an approach to the study of inverse problems which involves weighted estimates. This technique enables one to estimate the rate of convergence of numerical algorithms such as the method of successive approximations, the Landweber iteration, the steepest descent method, the Newton–Kantorovich method

(Subsection 10.2.2), and the method of inversion of finite-difference scheme (Kabanikhin, 1988).

Many inverse problems for hyperbolic and parabolic equations can be reduced to nonlinear Volterra integral equations. The inverse problems can be written in the general form as a Volterra operator equation of the second kind

$$q(x) = f(x) + \int_0^x (K_x q)(\tau)\, d\tau, \quad x \in S, \tag{4.3.1}$$

where $S = [0, T]$, $T \in \mathbb{R}_+$, $\{K_x\}_{x \in S}$ is a family of nonlinear Volterra operators (see Definition 4.3.1 below), the (vector) function $q(x)$ represents an unknown coefficient (or coefficients), and $f(x)$ is the data of the inverse problem.

As an example we consider the inverse problem

$$u_{tt} = u_{xx} - q(x)u, \quad x \in \mathbb{R}, \quad t \in \mathbb{R}_+, \tag{4.3.2}$$

$$u(x,0) = \varphi_1(x), \quad u_t(x,0) = \varphi_2(x), \tag{4.3.3}$$

$$u(0,t) = f_1(t), \quad u_x(0,t) = f_2(t). \tag{4.3.4}$$

where the coefficient $q(x)$ of equation (4.3.2) is to be found from given functions $\varphi_j(x)$, $f_j(x)$, $j = 1, 2$. We will return to this problem in Chapter 10 (Subsection 10.1.2) assuming $\varphi_1(x) \equiv 0$, $\varphi_2(x) = \delta(x)$.

Let us show that the problem (4.3.2)–(4.3.4) is ill-posed, and consequently, so is (4.3.1).

We simplify the problem by assuming $\varphi_1 \equiv -1$, $\varphi_2 \equiv 0$, $f_2 \equiv 0$, $f(t) = f_1''(t)$ and introducing the function $v(x,t) = u_{tt}(x,t)$. Differentiating (4.3.2) and (4.3.4) twice with respect to t and taking into account that

$$v(x,0) = u_{tt}\big|_{t=0} = u_{xx}\big|_{t=0} - u(x,0)q(x) = q(x),$$
$$v_t(x,0) = u_{ttt}\big|_{t=0} = u_{txx}\big|_{t=0} - u_t(x,0)q(x) = 0,$$

we obtain a new inverse problem

$$v_{tt} = v_{xx} - q(x)v, \quad x \in \mathbb{R}, \quad t > 0, \tag{4.3.5}$$

$$v(x,0) = q(x), \quad v_t(x,0) = 0, \tag{4.3.6}$$

$$v(0,t) = f(t), \quad v_x(0,t) = 0. \tag{4.3.7}$$

Taking the even extensions of $v(x,t)$ and $f(t)$ with respect to t and using d'Alembert's formula, we have

$$v(x,t) = \frac{1}{2}[f(t+x) + f(t-x)] + \frac{1}{2}\int_0^x \int_{t-x+\xi}^{t+x-\xi} q(\xi)v(\xi,\tau)\, d\tau\, d\xi. \tag{4.3.8}$$

Put $t = 0$ in (4.3.8). In view of the first condition in (4.3.6), we have

$$q(x) = f(x) + \int_0^x \int_0^{x-\xi} q(\xi)v(\xi,\tau)\,d\tau\,d\xi. \tag{4.3.9}$$

As a result, we obtained the nonlinear operator equation (4.3.9) for the coefficient $q(x)$. Note that the function $v(\xi,\tau)$ in equation (4.3.9) depends on $q(x)$ and is determined by solving the direct problem (4.3.5), (4.3.6) if $q(x)$ is known.

We will show that the operator equation (4.3.9) may have no solutions on $[0,T]$ even if the data f is infinitely smooth. For example, let $f \equiv c = \text{const} > 0$. Then from (4.3.5), (4.3.7) it follows that v does not depend on t. Consequently, (4.3.6) implies that $v(x,t) = q(x)$. Then (4.3.9) becomes as follows:

$$q(x) = c + \int_0^x \int_0^{x-\xi} q^2(\xi)\,d\tau\,d\xi \tag{4.3.10}$$

or, in the differential form,

$$q''(x) = q^2(x), \tag{4.3.11}$$

$$q(0) = c, \quad q'(0) = 0. \tag{4.3.12}$$

At the same time, the solution to the Cauchy problem (4.3.11), (4.3.12) tends to infinity as $x \to T_* = \sqrt{\frac{3}{2}}\int_c^\infty \frac{dy}{\sqrt{y^3-c^3}}$.

To prove this, we introduce a new variable $p = q'$. Note that $q'' = p'_q q' = p'_q p$. By (4.3.11),

$$p\,\frac{dp}{dq} = q^2.$$

Solving this equation in view of (4.3.12), we obtain

$$q' = p = \sqrt{\frac{2}{3}\,(q^3 - c^3)},$$

which implies

$$x(q) = \sqrt{\frac{3}{2}}\int_c^q \frac{dy}{\sqrt{y^3 - c^3}}.$$

Since the integral in the right-hand side of this equality is uniformly bounded for all $q \in \mathbb{R}_+$, the solution to equation (4.3.15) tends to infinity as $x \to T_*$.

Lavrentiev (1972) was the first to study the Volterra operator equations. One of the main difficulties arising in the study of operator equations is caused by the nonlinearity of the operators K_x. As was proved above, the inverse problem (4.3.1) may have no solutions for some data $f(x)$. Finding conditions on $f(x)$ $(x \in [0, T])$ that are necessary and sufficient for the existence of a solution for an arbitrary $T \in \mathbb{R}_+$ is a complicated problem. It is solved only for the case of one-dimensional inverse problems for hyperbolic equations (Subsection 10.1.3). A large class of Volterra operator equations was examined in (Lavrentiev, 1972) and (Romanov, 1975). In this section, we assume the family of operators $\{K_x\}_{x \in S}$ to be such that the problem (4.3.1) possesses the main properties of such inverse problems for hyperbolic equations.

We will seek a solution to the operator equation (4.3.1) that belongs to the space $C(S; X)$, $S = [0, T]$. This means that $q(x)$ is a continuous function of the parameter $x \in S$ with values in a real Banach space X, whose norm will be denoted by $\|\cdot\|$. The integral in (4.3.1) is understood as the Bochner integral (see Subsection A.5.2).

The following two types of statements can be proved for many one-dimensional and some classes of multi-dimensional inverse problems reducible to the form (4.3.1). First, the inverse problem is locally well-posed, i.e., for a sufficiently small $T \in \mathbb{R}_+$ there exists a solution to equation (4.3.1) which continuously depends on the data $f(x)$. Second, for any $T \in \mathbb{R}_+$, a solution to the inverse problem is unique and conditionally stable, i.e., small variations in $f(x)$ keeping the solvability of (4.3.1) cause small variations in the solutions.

Since the data in applied inverse problems are always approximate, finding out what variations in the data $f(x)$ keep it within the set of functions for which the inverse problem has a solution is very important in practice. Under rather general assumptions about the family of operators $\{K_x\}_{x \in S}$, we will show that the set of data $\{f(x)\}$ for which the operator equation (4.3.1) has a solution is open in $C(S; X)$, i.e., if (4.3.1) is solvable for some $f(x)$, then it admits a solution for all $f_\delta(x)$ close enough to $f(x)$.

We will now specify these properties. Let $S_x = [0, x]$, $x \in S$, and X be a Banach space with the norm $\|\cdot\|$.

Following (Gajewski et al., 1974), we give the below definitions.

Definition 4.3.1. An operator $K_x \in (C(S_x; X) \to C(S_x; X))$ is called a *Volterra operator* if the relation $q(\lambda) = r(\lambda)$, $q, r \in C(S_x; X)$, which holds for all $\lambda \in S_{x_1}$ and $x_1 \in S_x$ implies that for all $\lambda \in S_{x_1}$

$$(K_x q)(\lambda) = (K_x r)(\lambda).$$

Definition 4.3.2. A family of operators $K_x \in (C(S_x; X) \to C(S_x; X))$, $x \in S$, is said to be *boundedly Lipschitz-continuous* if there exists such a real-valued function $\mu(y_1, y_2)$ increasing in y_1 and y_2 that for all $x \in S$ and all $q_1, q_2 \in C(S; X)$ the following estimate holds:

$$\|K_x q_1 - K_x q_2\|_{C(S_x;X)} \le \mu(q_1^*(x), q_2^*(x)) \|q_1 - q_2\|_{C(S_x;X)}. \tag{4.3.13}$$

Here

$$q_j^*(x) = \|q_j\|_{C(S_x;X)}, \quad j = 1, 2;$$
$$\|q\|_{C(S_x;X)} = \sup_{\tau \in S_x} \{\|q(\tau)\|\}, \quad x \in S.$$

Definition 4.3.3. A family $\{K_x\}_{x \in S}$ is said to *belong to the class* $\mathcal{K}(T, \mu)$ if K_x is a Volterra operator such that $K_x \mathbf{0} = \mathbf{0}$ for any $x \in S = [0, T]$ and the family $\{K_x\}_{x \in S}$ is boundedly Lipschitz-continuous.

The following lemma immediately follows from the above definition.

Lemma 4.3.1. *Assume that the family of operators* $\{K_x\}_{x \in S}$ *belongs to the class* $\mathcal{K}(T, \mu)$*. Then the following inequality holds for all* $x \in S$, $x_1 \in S_x$, *and* $q_1, q_2 \in C(S; X)$*:*

$$\|K_x q_1 - K_x q_2\|_{C(S_{x_1};X)} \le \mu(q_1^*(x), q_2^*(x)) \|q_1 - q_2\|_{C(S_{x_1};X)}.$$

Proof. Let $x_1 \in S_x$. For $j = 1, 2$, we set

$$q_j^{x_1}(s) = \begin{cases} q_j(s), & 0 \le s \le x_1, \\ q_j(x_1), & x_1 < s \le x. \end{cases}$$

Obviously, $q_j^{x_1}(s) \in C(S_x; X)$ for $j = 1, 2$. Since K_x is a Volterra operator, we have

$$\begin{aligned} \|K_x q_1 - K_x q_2\|_{C(S_{x_1};X)} &\le \|K_x q_1^{x_1} - K_x q_2^{x_1}\|_{C(S_x;X)} \\ &\le \mu(q_1^*(x), q_2^*(x)) \|q_1^{x_1} - q_2^{x_1}\|_{C(S_x;X)} \\ &= \mu(q_1^*(x), q_2^*(x)) \|q_1 - q_2\|_{C(S_{x_1};X)}. \end{aligned}$$

The lemma is proved. □

Let $\varepsilon \in \mathbb{R}_+$, $k \ge 0$, $q \in C(S; X)$. We define a (C, k)*-norm* and a ball of radius ε as follows:

$$\begin{aligned} \|q\|_{C,k} &= \sup_{x \in S} \{\|q\|_{C(S_x;X)} \exp(-kx)\}, \\ \Phi_k(t, q, \varepsilon) &= \{r \in C(S_t; X) : \|q - r\|_{C,k} < \varepsilon\}. \end{aligned} \tag{4.3.14}$$

In the sequel, we will use some obvious inequalities that follow from the definition of the (C, k)-norm, namely,

$$\|q\|_{C,k} \le \|q\|_{C(S;X)} \le \exp(kT) \|q\|_{C,k}. \tag{4.3.15}$$

4.4 Local well-posedness and uniqueness on the whole

The next theorem was proved by Romanov (1973b) for a one-dimensional inverse problem. The technique used in this paper was later applied to the inverse problems for elasticity equations, the system of Maxwell's equations (Romanov and Kabanikhin, 1994), the transport equation and its $\mathcal{P}_n$-approximation (Romanov et al., 1984), the acoustic equation (Kabanikhin, 1981), and many others.

Theorem 4.4.1. *Assume that $f \in C(S; X)$ and the family of operators $\{K_x\}_{x \in S}$ belongs to the class $\mathcal{K}(T, \mu)$. Then there exists a $T_* \in (0, T)$ such that for all $t \in (0, T_*)$ the operator equation*

$$q(x) = f(x) + \int_0^x (K_x q)(\tau)\, d\tau, \quad x \in S_t, \tag{4.4.1}$$

has a unique solution in the class $C(S_t; X)$ which continuously depends on the data.

Proof. We will construct a solution using the Banach's contraction mapping principle (Theorem A.2.1). For $q \in C(S; X)$ we define the operator

$$(Uq)(x) = f(x) + \int_0^x (K_x q)(\tau)\, d\tau, \quad x \in S.$$

Since the family $\{K_x\}_{x \in S}$ belongs to the class $\mathcal{K}(T, \mu)$, we conclude that $U \in \big(C(S; X) \to C(S; X)\big)$. Using Lemma 4.3.1 and properties of the Bochner integral (Gajewski et al., 1974), we obtain the inequalities

$$\begin{aligned}
\|(Uq)(x) - f(x)\| &\le \left\| \int_0^x (K_x q)(\tau)\, d\tau \right\| \\
&\le \int_0^x \|(K_x q)(\tau)\|\, d\tau \le \int_0^x \|K_x q\|_{C(S_\tau; X)}\, d\tau \\
&\le \mu(q^*(T), 0) \int_0^x \|q\|_{C(S_\tau; X)}\, d\tau \\
&\le \mu(q^*(T), 0) x \|q\|_{C(S_x; X)}, \quad x \in S.
\end{aligned} \tag{4.4.2}$$

Here $q^*(T) = \|q\|_{C(S; X)}$.

We fix an arbitrary $\delta \in \mathbb{R}_+$ and assume that $q \in \Phi_0(t, f, \delta)$ (see (4.3.14)) with $t \in S$. Since (4.4.2) holds for all $x \in S_t \subset S$, we come to

$$\|Uq - f\|_{C(S_t; X)} \le \mu(\|f\|_{C(S_t; X)} + \delta, 0) t (\|f\|_{C(S_t; X)} + \delta) \le \mu(P, 0) t P, \tag{4.4.3}$$

where $P = \|f\|_{C(S;X)} + \delta$. Put

$$T_* = \min\left\{T, \frac{\delta}{\mu(P,0)P}, \frac{1}{\mu(P,P)}\right\}. \tag{4.4.4}$$

Then from (4.4.3) it follows that for all $t \in (0, T_*)$ the operator U maps the ball $\Phi_0(t, f, \delta)$ into itself.

Repeating the arguments used to obtain the estimate (4.4.2), we arrive at the following inequality for any $q_1, q_2 \in \Phi_0(t, f, \delta)$, $t \in S$:

$$\|Uq_1 - Uq_2\|_{C(S_t;X)} \le \mu(P,P)t\|q_1 - q_2\|_{C(S_t;X)}, \quad t \in S. \tag{4.4.5}$$

Since $T_* \le 1/\mu(P,P)$, for all $t \in (0, T_*)$ the operator U maps the ball $\Phi_0(t, f, \delta)$ into itself and is a contraction on $\Phi_0(t, f, \delta)$. By the Banach fixed point theorem, the map U has a fixed point in $\Phi_0(t, f, \delta)$, $t \in (0, T_*)$, i.e., there exists an element $q \in \Phi_0(t, f, \delta)$ such that

$$q(x) = f(x) + \int_0^x (K_x q)(\tau)\, d\tau, \quad x \in S_t.$$

The uniqueness of the solution and its continuous dependence on the data follow from the next statement.

Theorem 4.4.2. *Assume that for $f_j \in C(S;X)$, $j = 1, 2$, there exists a solution $q_j \in C(S;X)$ to the operator equation*

$$q_j(x) = f_j(x) + \int_0^x (K_x q_j)(\tau)\, d\tau, \quad x \in S. \tag{4.4.6}$$

Then the following estimate holds if the family $\{K_x\}_{x \in S}$ belongs to the class $\mathcal{K}(T, \mu)$:

$$\|q_1 - q_2\|_{C(S;X)} \le \exp\{T\mu(q_1^*(T), q_2^*(T))\}\|f_1 - f_2\|_{C(S;X)}. \tag{4.4.7}$$

Proof. Upon subtracting the equality (4.4.6) for $j = 1$ termwise from the same equality for $j = 2$, we obtain the inequalities

$$\begin{aligned}\|q_1(x) - q_2(x)\| &\le \|f_1(x) - f_2(x)\| + \int_0^x \|(K_x q_1)(\tau) - (K_x q_2)(\tau)\|\, d\tau \\ &\le \|f_1 - f_2\|_{C(S;X)} + \mu(q_1^*(T), q_2^*(T)) \int_0^x \|q_1 - q_2\|_{C(S_\tau;X)}\, d\tau, \\ &\qquad x \in S.\end{aligned}$$

Hence

$$\|q_1 - q_2\|_{C(S_x;X)} \le \|f_1 - f_2\|_{C(S;X)} + \mu(q_1^*(T), q_2^*(T)) \int_0^x \|q_1 - q_2\|_{C(S_\tau;X)}\, d\tau, \quad x \in S.$$

Applying the Gronwall inequality, we arrive at the desired inequality

$$\|q_1 - q_2\|_{C(S;X)} \leq \exp\{T\mu(q_1^*(T), q_2^*(T))\}\|f_1 - f_2\|_{C(S;X)}. \qquad \square$$

The estimate (4.4.7) implies that the solution of equation (4.4.1) is unique and continuously depends on the data. □

Returning to the proof of Theorem 4.4.1, we note that the size of the domain S_t was used as the "smallness" parameter (see (4.4.4)). However, the estimates (4.4.3) and (4.4.5) include the factors $\mu(P, 0)$ and $\mu(P, P)$, which depend on $\|f\|_{C(S;X)}$.

The question arises: *is it possible to use* $\|f\|_{C(S;X)}$ *as a "smallness" parameter without assuming that* T_* *is small?*

The answer is positive and in the next section we use this idea to prove that the problem (4.3.1) is well-posed in a neighborhood of the exact solution.

4.5 Well-posedness in a neighborhood of the exact solution

First, we will show that the problem (4.3.1) is well-posed if $\|f\|_{C(S;X)}$ is sufficiently small. From this result we will deduce that the problem (4.3.1) is well-posed in a neighborhood of the exact solution. The technique presented in this section can be used for developing and justifying numerical algorithms for finding an approximate solution to the problem, and in particular, for estimating the rate of convergence of the finite-difference inversion method.

Theorem 4.5.1. *Assume that the family of operators* $\{K_x\}_{x\in S}$ *belongs to the class* $\mathcal{K}(T, \mu)$. *Then there exists a* $\delta \in \mathbb{R}_+$ *such that for all* $f \in \Phi_0(T, \mathbf{0}, \delta)$ *equation* (4.3.1) *has a unique solution in the class* $C(S; X)$ *which continuously depends on the data.*

Proof. As before, we will construct a solution $q(x)$ using the Banach fixed point theorem. For $q \in C(S; X)$, we define the integral operator

$$(Uq)(x) = f(x) + \int_0^x (K_x q)(\tau)\, d\tau, \quad x \in S.$$

Let us prove that if we choose, for example,

$$m > 2\mu(1, 1), \quad \delta = \frac{1}{2}\exp\{-mT\}, \tag{4.5.1}$$

then for any

$$f \in \Phi_0(T, \mathbf{0}, \delta) \subset \Phi_m(T, \mathbf{0}, \delta) \tag{4.5.2}$$

the operator U will map the ball $\Phi_m(T, \mathbf{0}, 2\delta)$ into itself and will be a contraction on $\Phi_m(T, \mathbf{0}, 2\delta)$ in the (C, m)-norm (we recall that the ball $\Phi_m(t, q, \varepsilon)$ and the (C, m)-norm are defined by formulae (4.3.14)).

Indeed, by the assumptions of the theorem and by Lemma 4.3.1,

$$\|(Uq)(x) - f(x)\| \le \mu(q^*(T), 0) \int_0^x \|q\|_{C(S_\tau;X)}\, d\tau, \quad x \in S. \tag{4.5.3}$$

We now estimate the integral in the right-hand side of (4.5.3) using the (C, m)-norm:

$$\begin{aligned}
\int_0^x \|q\|_{C(S_\tau;X)}\, d\tau &= \int_0^x \sup_{\lambda \in S_\tau} \{\|q(\lambda)\| \exp(-m\lambda) \exp(m\lambda)\}\, d\tau \\
&\le \int_0^x \sup_{\lambda \in S_\tau} \{\|q(\lambda)\| \exp(-m\lambda)\} \exp(m\tau)\, d\tau \\
&\le \|q\|_{C,m} \int_0^x \exp(m\tau)\, d\tau = m^{-1} \|q\|_{C,m} [\exp(mx) - 1], \quad x \in S.
\end{aligned} \tag{4.5.4}$$

Substituting (4.5.4) into (4.5.3) and multiplying the resulting inequality by $\exp(-mx)$, we obtain

$$\|(Uq)(x) - f(x)\| \exp(-mx) \le m^{-1} \mu(q^*(T), 0) \|q\|_{C,m}, \quad x \in S.$$

The right-hand side of this inequality does not depend on x, which allows us to take the supremum of the left-hand side over S:

$$\|Uq - f\|_{C,m} \le m^{-1} \mu(q^*(T), 0) \|q\|_{C,m}. \tag{4.5.5}$$

Suppose that $q \in \Phi_m(T, \mathbf{0}, 2\delta)$, i.e., $\|q\|_{C,m} \le 2\delta$. In view of (4.3.15) and (4.5.1), we have

$$q^*(T) = \|q\|_{C(S;X)} \le 2\delta \exp(mT) = 1. \tag{4.5.6}$$

Substituting (4.5.6) into (4.5.5), we get

$$\|Uq - f\|_{C,m} \le 2\delta m^{-1} \mu(1, 1). \tag{4.5.7}$$

Then the conditions (4.5.1) and (4.5.7) imply that

$$\|Uq - f\|_{C,m} \le \delta.$$

Since $\|f\|_{C,m} \le \delta$ for $f \in \Phi_m(T, \mathbf{0}, \delta)$, we have $\|Uq\|_{C,m} \le 2\delta$, which means that the operator U maps the ball $\Phi_m(T, \mathbf{0}, 2\delta)$ into itself.

Now let $q_j \in \Phi_m(T, \mathbf{0}, 2\delta)$, $j = 1, 2$. It is not difficult to verify that

$$\|Uq_1 - Uq_2\|_{C,m} \le m^{-1} \mu(q_1^*(T), q_2^*(T)) \|q_1 - q_2\|_{C,m}. \tag{4.5.8}$$

The conditions $q_j \in \Phi_m(T, \mathbf{0}, 2\delta)$, $j = 1, 2$, and inequality (4.3.15) yield

$$q_j^*(T) \le 2\delta \exp(mT), \quad j = 1, 2. \tag{4.5.9}$$

Substituting (4.5.9) into (4.5.8) and taking into account that μ is an increasing function, we get

$$\|Uq_1 - Uq_2\|_{C,m} \le m^{-1}\mu(2\delta \exp(mT), 2\delta \exp(mT))\|q_1 - q_2\|_{C,m}. \tag{4.5.10}$$

Then, recalling that m and δ were chosen to satisfy the conditions (4.5.1), from (4.5.10) we conclude that the operator U is a contraction on $\Phi_m(T, \mathbf{0}, 2\delta)$. Consequently, the operator U has a unique fixed point q_e in $\Phi_m(T, \mathbf{0}, 2\delta)$. Obviously, this point is a solution to equation (4.3.1). From (4.3.15) it follows that $q_e \in C(S; X)$. The uniqueness and continuous dependence of the solution on the data in $C(S; X)$ can be proved in the same way as in Theorem 4.4.2. □

Thus, we found out that our assumption that either the interval S is small (Theorem 4.4.1) or the initial data f is small (Theorem 4.5.1) makes equation (4.3.1) uniquely solvable in $C(S; X)$ and its solution continuously depending on the data of the problem. On the other hand, the example given in the beginning of Section 4.3 shows that equation (4.3.1) may have no solutions for arbitrary T and $f \in C(S; X)$ (i.e., without assuming that either T or f is small). Nevertheless, the requirement that the family of operators $\{K_x\}_{x \in S}$ belong to the class $\mathcal{K}(T, \mu)$ makes it possible to prove the following statement.

Theorem 4.5.2. *Let the family of operators $\{K_x\}_{x \in S}$ belong to the class $\mathcal{K}(T, \mu)$. Assume that for some $f \in C(S; X)$ there exists a solution $q_e \in C(S; X)$ to equation* (4.3.1). *Then a $\delta \in \mathbb{R}_+$ can be found such that for all $f_\delta \in \Phi_0(T, f, \delta)$ there exists a solution $q_\delta \in C(S; X)$ to the equation*

$$q(x) = f_\delta(x) + \int_0^x (K_x q)(\tau)\, d\tau, \quad x \in S. \tag{4.5.11}$$

This solution is unique in the class $C(S; X)$ and continuously depends on the data.

Proof. For $p \in C(S; X)$, we define the operator

$$(Vp)(x) = f(x) - f_\delta(x) + \int_0^x [(K_x q_e)(\tau) - (K_x(q_e - p))(\tau)]\, d\tau.$$

By assumption, $f, f_\delta, q_e \in C(S; X)$. Consequently, $V \in (C(S; X) \to C(S; X))$. It is easy to verify that since the family $\{K_x\}_{x \in S}$ belongs to the class $\mathcal{K}(T, \mu)$, so does the family of operators $\{H_x\}_{x \in S}$ defined by the equality $(H_x p)(\tau) = (K_x q_e)(\tau) - (K_x(q_e - p))(\tau)$. By Theorem 4.5.1, the equation

$$p(x) = (Vp)(x), \quad x \in S, \tag{4.5.12}$$

has a solution if the difference $f - f_\delta$ is sufficiently small in the norm $\|\cdot\|_{C(S;X)}$. More precisely, such a $\delta > 0$ can be found that for all $f_\delta \in \Phi_0(T, f, \delta)$ there exists a solution $p_\delta \in C(S; X)$ to equation (4.5.12). By the definition of the operator V, we have

$$p_\delta(x) = f(x) - f_\delta(x) + \int_0^x (K_x q_e)(\tau)\,d\tau - \int_0^x [K_x(q_e - p_\delta)](\tau)\,d\tau, \quad x \in S. \tag{4.5.13}$$

Setting $q_\delta(x) = q_e(x) - p_\delta(x)$ and taking into account that $q_e(x)$ is a solution to (4.3.1), from (4.5.13) we conclude that $q_\delta(x)$ is a solution to equation (4.5.11), which completes the proof. □

To estimate the value of the parameter δ in the formulation of the theorem, we shall discuss the conditions under which the operator V has a fixed point in greater detail. Let $\tilde{f} = f - f_\delta$. As before, for $p \in C(S; X)$ we have

$$\begin{aligned}\|(Vp)(x) - \tilde{f}(x)\| &\le \int_0^x \|(K_x q_e)(\tau) - [K_x(q_e - p)](\tau)\|\,d\tau \\ &\le \mu(q_e^*, p^*) \int_0^x \|p\|_{C(S_\tau;X)}\,d\tau, \quad x \in S,\end{aligned}$$

where $p^* = \|q_e - p\|_{C(S;X)}$, $q_e^* = \|q_e\|_{C(S;X)}$. Using the (C, m)-norm again, we get

$$\|Vp - \tilde{f}\|_{C,m} \le m^{-1}\mu(q_e^*, p^*)\|p\|_{C,m}. \tag{4.5.14}$$

Consequently, if $\tilde{f} \in \Phi_m(T, \mathbf{0}, \delta)$ and $p \in \Phi_m(T, \mathbf{0}, 2\delta)$, then

$$\|Vp\|_{C,m} \le \delta + 2\delta m^{-1}\mu(q_e^*, p^*). \tag{4.5.15}$$

Furthermore, if we assume that

$$2\mu(q_e^*, p^*) \le m, \tag{4.5.16}$$

then (4.5.15) implies that $(Vp) \in \Phi_m(T, \mathbf{0}, 2\delta)$.

Now let $p_j \in \Phi_m(T, \mathbf{0}, 2\delta)$ for $j = 1, 2$. Then

$$\|Vp_1 - Vp_2\|_{C,m} \le m^{-1}\mu(p_1^*, p_2^*)\|p_1 - p_2\|_{C,m}, \tag{4.5.17}$$

where $p_j^* = \|q_e - p_j\|_{C(S;X)}$, $j = 1, 2$.

Therefore, the operator V is a contraction on $\Phi_m(T, \mathbf{0}, 2\delta)$ for all m such that

$$\mu(p_1^*, p_2^*) < m. \tag{4.5.18}$$

In view of (4.3.15), from the condition $p_j \in \Phi_m(T, \mathbf{0}, 2\delta)$, $j = 1, 2$, it follows that

$$p_j^* \le \|p_j\|_{C,m} \exp(mT) + q_e^* \le 2\delta \exp(mT) + q_e^*, \quad j = 1, 2. \tag{4.5.19}$$

We choose the parameter m to satisfy the condition

$$2\mu(2 + q_e^*, 2 + q_e^*) < m, \tag{4.5.20}$$

and define δ in terms of m as follows:

$$\delta = \exp(-mT). \tag{4.5.21}$$

With this choice of m and δ (taking into account that μ is an increasing function), it is clear that the inequalities (4.5.16) and (4.5.18) hold for all $\tilde{f} \in \Phi_m(T, \mathbf{0}, \delta)$, $p, p_1, p_2 \in \Phi_m(T, \mathbf{0}, 2\delta)$. It follows that the operator V maps the ball $\Phi_m(T, \mathbf{0}, 2\delta)$ into itself and is a contraction on this ball with the contraction coefficient

$$\omega = m^{-1}\mu(q_e^* + 2, q_e^* + 2). \tag{4.5.22}$$

A simple analysis of the estimates (4.5.20) and (4.5.21) shows that the larger q_e^* and T, the larger the parameter m and, consequently, the smaller should be δ in the condition (4.5.21). In other words, the "deeper" is the level at which we want to solve the inverse problem (4.3.1), the more rigid should be the requirements for the measurement accuracy:

$$\|f - f_\delta\|_{C(S;X)} \le \exp(-mT). \tag{4.5.23}$$

Note that the proof of Theorem 4.5.2 is not constructive, since solving (4.5.12) yields only $p_\delta = q_e - q_\delta$, i.e., the difference between two unknown functions. To construct an approximate solution to equation (4.3.1), one can apply the method of successive approximations

$$q_{n+1} = f(x) + \int_0^x (K_x q_n)(\xi)\, d\xi,$$

and estimate its convergence in the case where the initial approximation $q_0(x)$ is sufficiently close to the exact solution.

4.6 Regularization of nonlinear operator equations of the first kind

In many practical problems, it is possible to find only a first-kind operator equation for the unknown coefficient:

$$0 = f(x) + \int_0^x (K_x q)(\tau)\, d\tau, \quad x \in S. \tag{4.6.1}$$

In many cases, equations of the form (4.6.1) can be reduced to equations of the second kind by differentiation. The operation of differentiation, however, is not correct in the sense that small variations in the data $f(x)$ in the norm of $C(S;X)$ may lead to indefinitely large variations in its derivative. The problem of finding the derivative of the function $f(x)$ can also be viewed as the Volterra operator equation of the first kind

$$f(x) - f(0) = \int_0^x q(\tau)\,d\tau,$$

where $q(\tau) = f'(\tau)$. There are many ways to regularize the operation of differentiation (see for example, (Denisov, 1975)).

A regularization method for the first-kind operator equation of the form

$$Aq = f \tag{4.6.2}$$

was proposed by Lavrentiev (1972) in the case where A is a completely continuous positive linear operator. It was assumed that the problem (4.6.2) is conditionally well-posed and the well-posedness set consists of functions $u = Bv$, $\|v\| \le 1$, where B is a completely continuous positive linear operator that commutes with A. It was shown that, given an approximation of the right-hand side of equation (4.6.2), i.e., a function f_δ satisfying the condition $\|f - f_\delta\| < \delta$, the auxiliary equation

$$\alpha q + Aq = f_\delta \tag{4.6.3}$$

has a solution $q_{\alpha\delta}$ that tends to the exact solution q_e of equation (4.6.2) if $\alpha(\delta)$ tends to zero as $\delta \to 0$.

Based on the Lavrentiev method, A. M. Denisov (1975) developed a regularization of the linear Volterra equation of the first kind (see Theorem 4.2.2)

$$f(x) = \int_0^x K(x,\tau)q(\tau)\,d\tau, \quad x \in S. \tag{4.6.4}$$

(A similar result was obtained by Sergeev (1971).) The class of functions bounded by a certain constant in the norm of $C^1(S)$ was chosen as the well-posedness set. It was shown in (Denisov, 1975) that if $K(x,x) \equiv 1$, the derivatives $\partial^j K(x,\tau)/\partial x^j$, $j = \overline{0,2}$, are continuous, and $q_{\alpha\delta}$ is a solution to the auxiliary equation

$$\alpha q(x) + \int_0^x K(x,\tau)q(\tau)\,d\tau = f_\delta(x), \quad x \in S, \tag{4.6.5}$$

with f_δ such that

$$\|f - f_\delta\| \le \delta,$$

then the following estimate holds:

$$\|q_{\alpha\delta} - q_e\| \le \mathcal{D}_1\Big(\alpha + \frac{\delta}{\alpha}\Big),$$

where $\mathcal{D}_1$ is a positive constant that does not depend on α and δ.

Based on the procedure proposed by Denisov (1975), A. V. Baev (1985) considered a nonlinear version of equation (4.6.4) arising in solving a one-dimensional inverse dynamic problem of seismology and constructed a regularizing algorithm using a scheme analogous to (4.6.3).

We now consider two ways of regularizing equation (4.6.1). The first one is based on regularizing the differentiation operation, while the other follows the Lavrentiev method.

Theorem 4.6.1. *Assume that for some $f \in C^2(S; X)$ there exists a solution $q_e \in C^2(S; X)$ to equation* (4.6.1). *Furthermore, assume that the operators $K_x, x \in S$, are differentiable with respect to x, there holds*

$$(K_x q)(x) \equiv q(x), \quad x \in S, \tag{4.6.6}$$

and the family of operators $\{\partial K_x/\partial x\}_{x\in S}$ belongs to the class $\mathcal{K}(T, \mu)$.

Then there exist constants $\delta_0 \in \mathbb{R}_+$, $\mathcal{D}_0 \in \mathbb{R}_+$ and a function $\alpha = \alpha(\delta)$ ($\alpha(\delta) \to 0$ as $\delta \to 0$) such that for all $\delta \in (0, \delta_0)$ and $f_\delta \in \Phi_0(T, f, \delta)$ the equation

$$-q(x) = \frac{1}{\alpha}\,[f_\delta(x+\alpha) - f_\delta(x)] + \int_0^x \Big(\frac{\partial}{\partial x} K_x q\Big)(\tau)\, d\tau, \quad x \in S_{T-\alpha}, \tag{4.6.7}$$

has a unique solution $q_{\alpha\delta}$ in $C(S_{T-\alpha}; X)$ and the following estimate holds:

$$\|q_e - q_{\alpha\delta}\|_{C(S_{T-\alpha};X)} \le \mathcal{D}_0\Big(\alpha + \frac{\delta}{\alpha}\Big). \tag{4.6.8}$$

Proof. Upon differentiating (4.6.1) with respect to x, we obtain

$$-q(x) = f'(x) + \int_0^x \Big(\frac{\partial}{\partial x} K_x q\Big)(\tau)\, d\tau, \quad x \in S. \tag{4.6.9}$$

Since $f \in C^2(S; X)$, using the well-known results on the regularization of differentiation, we conclude that the difference

$$f'(x) - \frac{1}{\alpha}\,[f_\delta(x+\alpha) - f_\delta(x)]$$

can be made as small as desired if $\delta \in \mathbb{R}_+$ is small enough and the regularization parameter α is chosen to agree with δ. This means that we can apply Theorem 4.5.2 to equation (4.6.7), which implies the existence of a unique solution $q_{\alpha\delta} \in C(S_{T-\alpha}; X)$.

The proof of the estimate (4.6.8) is similar to that of (4.2.9) in Theorem 4.2.2. □

We now construct a Volterra regularization of equation (4.6.1) based on the Lavrentiev method. Consider the following auxiliary Volterra operator equation of the second kind:

$$-\alpha q = f(x) + \int_0^x (K_x q)(\tau)\, d\tau, \quad x \in S, \quad \alpha \in \mathbb{R}_+. \tag{4.6.10}$$

We will assume that the family of operators $\{K_x\}_{x\in S}$ satisfies the conditions

$$\Big\{\frac{\partial^n}{\partial x^n} K_x\Big\}_{x\in S} \in \mathcal{K}(T,\mu), \quad n = 0,1,2; \tag{4.6.11}$$

and the condition (4.6.6).

Without loss of generality, it can be assumed that

$$f'(0) = 0. \tag{4.6.12}$$

The condition (4.6.12) is equivalent to specifying the value of $q(x)$ at $x = 0$.

Two questions concerning the auxiliary equation (4.6.10) need to be resolved, namely: whether equation (4.6.10) is solvable (under the natural assumption that the main equation (4.6.1) is solvable) and, if so, whether its solution q_α converges to the solution of equation (4.6.1) as α tends to zero. The positive answer to these questions will follow from the results of Section 4.5.

First, suppose that (4.6.10) has a solution $q_\alpha(x)$ in $C^1(S;X)$. Setting $\nu = \alpha^{-1}$, differentiating (4.6.10) with respect to x, and rearranging the resulting equation, we obtain

$$q'_\alpha(x) = -\nu[f'(x) + q_\alpha(x)] - \nu \int_0^x \Big(\frac{\partial}{\partial x} K_x q_\alpha\Big)(\tau)\, d\tau, \quad x \in S. \tag{4.6.13}$$

Note that $f(0) = 0$ in view of (4.6.1). Therefore, (4.6.10) implies $q_\alpha(0) = 0$ and from (4.6.13) it follows that $q'_\alpha(0) = 0$.

The following identity is obvious:

$$\int_0^x q(\tau)\, d\tau = \nu \int_0^x \exp\{\nu(\tau - x)\} \int_0^\tau q(\xi)\, d\xi\, d\tau + \int_0^x \exp\{\nu(\tau - x)\} q(\tau)\, d\tau,$$
$$x \in S.$$

Substituting the function $q'_\alpha(\tau)$ for $q_\alpha(\tau)$ in this identity and taking into account that $q'_\alpha(0) = 0$, we have

$$q_\alpha(x) = \nu \int_0^x \exp\{\nu(\tau - x)\} q_\alpha(\tau)\, d\tau + \int_0^x \exp\{\nu(\tau - x)\} q'_\alpha(\tau)\, d\tau, \quad x \in S.$$

Substituting the right-hand side of (4.6.13) for $q'_\alpha(\tau)$ in this equality and integrating by parts, we obtain

$$q_\alpha(x) = R_0[f, q_\alpha](x) + \sum_{j=1}^{2} R_j[f, q_\alpha, \nu](x), \quad x \in S.$$

Here R_j, $j = 0, 1, 2$, are as follows:

$$R_0[f, q](x) = -f'(x) - \int_0^x \Big(\frac{\partial}{\partial x} K_x q\Big)(\tau)\, d\tau,$$

$$R_1[f, q, \nu](x) = \int_0^x \exp\{\nu(\tau - x)\}\Big[f''(\tau) - \Big(\frac{\partial}{\partial \tau} K_\tau q\Big)(\tau)\Big]\, d\tau,$$

$$R_2[f, q, \nu](x) = \int_0^x \exp\{\nu(\tau - x)\} \int_0^\tau \Big(\frac{\partial^2}{\partial^2 \tau} K_\tau q\Big)(\xi)\, d\xi\, d\tau.$$

Consequently, the solution $q_\alpha \in C^1(S; X)$ to equation (4.6.10) also solves the equation

$$q(x) = R_0[f, q](x) + \sum_{j=1}^{2} R_j[f, q, \nu](x), \quad x \in S. \tag{4.6.14}$$

Note that the above notation allows us to write (4.6.9) in the form

$$q(x) = R_0[f, q](x), \quad x \in S. \tag{4.6.15}$$

Theorem 4.6.2. *Assume that* $f \in C^2(S; X)$, $f'(0) = 0$, *and* (4.6.1) *has a solution* $q_e \in C^2(S; X)$. *Furthermore, assume that the family of operators* $\{K_x\}_{x \in S}$ *satisfies the conditions* (4.6.6) *and* (4.6.11). *Then there exists an* $\alpha_* \in \mathbb{R}_+$ *such that for all* $\alpha \in (0, \alpha_*)$ *equation* (4.6.14) *has a unique solution* q_α *in* $C(S; X)$ *and the following estimate holds:*

$$\|q_e - q_\alpha\|_{C(S;X)} \le \mathcal{D}_* \alpha, \tag{4.6.16}$$

where the constant $\mathcal{D}_*$ *depends on* $\|q_e\|_{C(S;X)}$, $\|f\|_{C^2(S;X)}$, T, *and the function* μ.

The proof of this theorem can be found in (Kabanikhin, 1998).

Theorem 4.6.3. *Under the assumptions of Theorem* 4.6.2, *there exists an* $\alpha_* \in \mathbb{R}_+$ *such that for all* $\alpha \in (0, \alpha_*)$ *equation* (4.6.10) *has a unique solution in* $C(S, X)$ *and the estimate* (4.6.16) *holds.*

Proof. Since the assumptions of Theorem 4.6.2 hold, one can find such $\alpha_* > 0$ that for all $\alpha \in (0, \alpha_*)$ there exists a unique solution $q_\alpha(x)$ to equation (4.6.14). Let us prove that $q_\alpha(x)$ satisfies (4.6.10) as well.

Upon integrating by parts in (4.6.14), we have

$$q_\alpha(x) = -\nu \int_0^x \exp\{\nu(\tau - x)\}\Big[f'(\tau) + \int_0^\tau \Big(\frac{\partial}{\partial\tau} K_\tau q_\alpha\Big)(\xi)\, d\xi\Big]\, d\tau$$

or, equivalently,

$$\begin{aligned} q_\alpha(x) &= \nu \int_0^x \exp\{\nu(\tau - x)\} q_\alpha(\tau)\, d\tau \\ &\quad - \nu \int_0^x \exp\{\nu(\tau - x)\}\Big[f'(\tau) + q_\alpha(\tau) + \int_0^\tau \Big(\frac{\partial}{\partial\tau} K_\tau q_\alpha\Big)(\xi)\, d\xi\Big]\, d\tau. \end{aligned} \tag{4.6.17}$$

By the assumptions of the theorem, the function $q_\alpha(x)$ is differentiable, and therefore the following identity holds:

$$\int_0^x \exp\{\nu(\tau - x)\} q'_\alpha(\tau)\, d\tau = q_\alpha(x) - \nu \int_0^x \exp\{\nu(\tau - x)\} q_\alpha(\tau)\, d\tau.$$

Then (4.6.17) implies that

$$\int_0^x \exp\{\nu(\tau - x)\}\omega(\tau)\, d\tau = 0, \quad x \in S,$$

where

$$\omega(x) = q'_\alpha(x) + \nu \frac{\partial}{\partial x}\Big[f(x) + \int_0^x (K_x q_\alpha)(\tau)\, d\tau\Big].$$

Consequently, $\omega(x) \equiv 0$ for $x \in S$ and, since $f(0) = 0$ and $q'_\alpha(0) = 0$, we finally have

$$-q_\alpha(x) = \nu\Big[f(x) + \int_0^x (K_x q_\alpha)(\tau)\, d\tau\Big],$$

i. e., $q_\alpha(x)$ solves equation (4.6.10), which is the required result. □

Now, assume that the function f is given approximately, that is, we know a function f_δ such that

$$\|f - f_\delta\|_{C(S;X)} \le \delta.$$

Consider the auxiliary equation

$$-\alpha q(x) = f_\delta(x) + \int_0^x (K_x q)(\tau)\, d\tau, \quad x \in S. \tag{4.6.18}$$

Theorem 4.6.4. *Under the assumptions of Theorem* 4.6.2, *there exists an* $\alpha_* \in \mathbb{R}_+$ *such that for all* $\alpha \in (0, \alpha_*)$ *and* $\delta \in (0, (\alpha/2)\exp(-T\gamma))$ *equation* (4.6.18) *has a unique solution* $q_{\alpha\delta}$ *in* $C(S, X)$ *and the following estimate holds:*

$$\|q_e - q_{\alpha\delta}\|_{C(S;X)} \le \mathcal{D}_*\alpha + \frac{2\delta}{\alpha}\exp(T\gamma), \tag{4.6.19}$$

where q_e *is a solution to equation* (4.6.1), $\gamma = (2/\alpha)\mu(q_e^* + 2, q_e^* + 2)$, $\mathcal{D}_*$ *is a constant dependent on* $q_e^* = \|q_e\|_{C(S;X)}$, $\|f\|_{C^2(S;X)}$, T *and* μ.

The proof of this theorem is also presented in (Kabanikhin, 1988).

From Theorem 4.6.4 it follows that a solution to the operator equation of the first kind

$$0 = f_\delta(x) + \int_0^x (K_x q)(\tau)\,d\tau, \quad x \in S,$$

can be approximated by a solution to the regularized equation (4.6.18) with accuracy given by the estimate (4.6.19).

Chapter 5
Integral geometry

This chapter deals with problems of integral geometry. They are most generally defined as problems of reconstructing a function on a linear space from its values on a given family of manifolds embedded in this space. We will consider the Radon problem of reconstructing a function from the values of its integrals over different lines (Section 5.1), the problem of reconstructing a function from spherical means (Section 5.2), the problem of moments (Section 5.3), and the direct and inverse kinematic problems of seismology (Section 5.4). The concepts and methods of integral geometry are of essential importance in tomography. The term "tomography" (from Greek *tomos*: slice, section) usually means a method for nondestructive examination of plane sections of internal structures of an object by means of multiple scanning.

Tomography can be most briefly defined as a means of determining the internal structure of opaque bodies. Commercially used *computerized tomography* employs powerful computers performing complex mathematical calculations. The mathematical problems at the heart of computerized tomography (and computerized examination/diagnostics of objects in general) are reduced to operator equations of the first kind $Aq = f$. Consider several specific examples of computerized tomography.

Example 1: X-ray computerized tomography (X-ray CT). While traditional X-ray imaging reveals only general shapes of internal structures of a body distorted by overlying organs and shadows, CT uses mathematical algorithms to produce a digital image whose contrast and brightness can be adjusted within a wide range. The development and application of X-ray CT made it possible to diagnose internal injuries, necrotic tissues, tumors and other pathologies, and detect their form and location with high accuracy.

Example 2: Nondestructive testing. Aside from X-rays, nondestructive testing employs hard gamma rays, various heavy particles and electrons, electromagnetic and acoustic waves, etc.

Example 3: Microscopy. Almost all examinations of biological tissues are performed with electronic microscopes that employ tomographic techniques.

Example 4: Geophysics. Placing the sources of (electromagnetic or seismic) waves in one well and detectors in another well, we obtain the problem of reconstructing the structure of the area between the wells, i.e., the tomography problem.

Example 5: Astrophysics. The principles of computerized tomography were used for creating the map of ultra-high frequency radiation of the sun, reconstructing the X-ray structure of space objects, estimation of the distribution of the density of electrons on a spherical surface around the sun.

Example 6: Ecology. It is hard to overestimate the importance of computerized tomography in space-based atmospheric sounding, especially in measuring air pollutant concentrations.

Example 7: NMR tomography. Diagnostic techniques based on nuclear magnetic resonance are becoming more popular and effective. Unlike X-ray tomography, it helps visualize the internal structure of human body and abnormalities in soft tissues, joints, and other organs without any considerable side effects for patients.

5.1 The Radon problem

Consider the section of an object (body) produced by a plane (x, y). Assume that all X-rays lie within this plane. Upon passing through the body, the rays are registered by a detector. The measurements are made for all directions of the rays. Let $q(x, y)$ be the X-ray absorption coefficient at a point (x, y). Then the relative drop in the radiation intensity over a small interval Δl at (x, y) is equal to $q(x, y)\,\Delta l$. Let J_0 denote the initial radiation intensity, and let J_1 denote the intensity of radiation after it has passed through the body. Then

$$\frac{J_1}{J_0} = \exp\left\{-\int_L q(x, y)\,dl\right\},$$

where L is a straight line in the planar section that coincides with the beam direction. Since J_1/J_0 is measured for different straight lines in the planar section, we arrive at the problem of determining the function $q(x, y)$ from the integrals

$$\int_L q(x, y)\,dl,$$

over all lines L in the planar section.

The problem of reconstructing a function of two variables from the values of its integrals along lines was formulated and solved by J. Radon (see Radon, 1917). Tomography and its applications in medicine started to develop rapidly about half a century after the publication of Radon's paper. We will analyze the Radon problem using the techniques proposed in (Denisov, 1994).

A straight line on a plane can be defined by the equation

$$x\cos\varphi + y\sin\varphi = p, \tag{5.1.1}$$

where $|p|$ is the length of the perpendicular from the origin to the line, and φ is the angle between the x-axis and the perpendicular. A family of lines on the plane is a biparametric family $L(p,\varphi)$, $p \in \mathbb{R}$, $\varphi \in (0, 2\pi)$. From equation (5.1.1) it follows that the pairs (p,φ) and $(-p, \varphi+\pi)$ define the same line.

The *Radon transform* of $q(x,y)$ is the integral $f(p,\varphi)$ of $q(x,y)$ along the line $L(p,\varphi)$:

$$f(p,\varphi) = \int_{L(p,\varphi)} q(x,y)\,dl. \tag{5.1.2}$$

From equation (5.1.1) we pass to the parametric representation of the line

$$x = p\cos\varphi - t\sin\varphi, \quad y = p\sin\varphi + t\cos\varphi, \quad t \in \mathbb{R}.$$

Then the formula (5.1.2) can be written as

$$f(p,\varphi) = \int_{-\infty}^{\infty} q(p\cos\varphi - t\sin\varphi,\, p\sin\varphi + t\cos\varphi)\,dt. \tag{5.1.3}$$

Note that there are functions $q(x,y)$ for which the Radon transform does not exist. For example, as follows from (5.1.3), the Radon transform is not defined for $q(x,y) = \text{const} \neq 0$.

First, consider the problem of inverting the Radon transform for a radially symmetric function $q(x,y)$ vanishing outside the disk of radius R centered at the origin. Let $q(x,y) = q_0(\sqrt{x^2+y^2})$ for $x^2+y^2 \le R^2$, where $q_0(x)$ is continuous on the interval $[0,R]$ and $q(x,y) = 0$ for $x^2+y^2 > R^2$. We will calculate the Radon transform of this function. It is obvious that $f(p,\varphi) = 0$ for $|p| > R$. Let $|p| \le R$. Then

$$\begin{aligned} f(p,\varphi) &= \int_{-\infty}^{\infty} q(p\cos\varphi - t\sin\varphi,\, p\sin\varphi + t\cos\varphi)\,dt \\ &= \int_{-\sqrt{R^2-p^2}}^{\sqrt{R^2-p^2}} q_0(\sqrt{p^2+t^2})\,dt \end{aligned}$$

or

$$f(p,\varphi) = 2\int_0^{\sqrt{R^2-p^2}} q_0(\sqrt{p^2+t^2})\,dt, \quad |p| \le R. \tag{5.1.4}$$

From the representation (5.1.4) for the Radon transform, it follows that in this case it is independent of the variable φ and is even with respect to the variable p. For this reason, in what follows, we will denote the Radon transform by $f(p)$, where $p \in [0,R]$. The change of variables $\theta = \sqrt{p^2+t^2}$ in (5.1.4) yields

$$\int_p^R \frac{2q_0(\theta)\theta\,d\theta}{\sqrt{\theta^2-p^2}} = f(p), \quad 0 \le p \le R. \tag{5.1.5}$$

Thus, the inversion of the Radon transform in the case of a radially symmetric finite function $q(x, y)$ is reduced to the solution of an integral equation of the first kind with a variable limit of integration (5.1.5). The kernel of the equation (5.1.5),

$$G(p,\theta) = \frac{2\theta}{\sqrt{\theta^2 - p^2}} = \frac{2\theta}{\sqrt{\theta - p}\,\sqrt{\theta + p}}, \qquad 0 \le p \le \theta \le R,$$

has an integrable singularity. Such equations were first studied by Niels Abel. Note that the integral in the left-hand side of (5.1.5) is defined for any function $q_0(\theta) \in C[0, R]$.

We now obtain an explicit formula for the solution to equation (5.1.5). Multiply (5.1.5) by $p(p^2 - s^2)^{-1/2}$ and integrate with respect to p from s to R:

$$\int_s^R p(p^2 - s^2)^{-1/2} \int_p^R \frac{2q_0(\theta)\theta\, d\theta}{\sqrt{\theta^2 - p^2}}\, dp = \int_s^R p(p^2 - s^2)^{-1/2} f(p)\, dp. \quad (5.1.6)$$

We denote by $J(q_0)$ the expression in the left-hand side of this equality and transform it by changing the order of integration:

$$J(q_0) = \int_s^R 2q_0(\theta)\theta \Big(\int_s^\theta \frac{p\, dp}{\sqrt{(p^2 - s^2)(\theta^2 - p^2)}} \Big)\, d\theta.$$

Introducing a new variable

$$z = \frac{2p^2 - \theta^2 - s^2}{\theta^2 - s^2}$$

and taking into account that

$$p^2 = \frac{(\theta^2 - s^2)z}{2} + \frac{\theta^2 + s^2}{2},$$

we obtain

$$\begin{aligned}
&\int_s^\theta \frac{p\, dp}{\sqrt{(p^2 - s^2)(\theta^2 - p^2)}} \\
&= \frac{1}{4} \int_{-1}^1 \frac{(\theta^2 - s^2)\, dz}{\sqrt{(\theta^2 - s^2)z/2 + (\theta^2 - s^2)/2}\sqrt{(\theta^2 - s^2)/2 - (\theta^2 - s^2)z/2}} \\
&= \frac{1}{2} \int_{-1}^1 \frac{dz}{\sqrt{1 - z^2}} = \frac{1}{2} \arcsin(z)\Big|_{-1}^1 = \frac{\pi}{2}.
\end{aligned}$$

Consequently,

$$J(q_0) = \pi \int_s^R q_0(\theta)\theta\, d\theta$$

and (5.1.6) can be written in the form

$$\pi \int_s^R q_0(\theta)\theta \, d\theta = \int_s^R p\sqrt{p^2 - s^2} f(p)\, dp.$$

Differentiating this equality with respect to s yields the formula for the solution of equation (5.1.5):

$$q_0(s) = -\frac{1}{\pi s}\frac{d}{ds}\int_s^R p\sqrt{p^2 - s^2} f(p)\, dp. \tag{5.1.7}$$

This representation implies that the solution of equation (5.1.5) is unique in the space of continuous functions, and therefore the radially symmetric function $q(x, y) = q_0(\sqrt{x^2 + y^2})$ is uniquely determined by it Radon transform.

We now turn to the general formulation of the problem of reconstructing the function $q(x, y)$ from it Radon transform $f(p, \varphi)$. Assume that $q(x, y)$ belongs to the set of functions that are infinitely differentiable on the plane and

$$\sup_{(x,y)\in\mathbb{R}^2} \left| x^k y^l \frac{\partial^n q(x, y)}{\partial x^i \, \partial y^j} \right| < \infty \tag{5.1.8}$$

for all nonnegative integers k, l, n, i, and j $(n = i + j)$.

We will show that for any function $q(x, y)$ with the property (5.1.8) there exists a Radon transform $f(p, \varphi)$ that satisfies the estimate

$$|f(p, \varphi)| \le \frac{C}{(1 + p^2)^{3/2}} \tag{5.1.9}$$

for $p \in (-\infty, \infty)$ and $\varphi \in [0, 2\pi]$, where C is a positive constant.

The condition (5.1.8) implies that there exists a positive constant C_1 such that for all $(x, y) \in \mathbb{R}^2$ we have

$$|q(x, y)| \le \frac{C_1}{[(1 + x^2)(1 + y^2)]^2}.$$

From this estimate and the formula (5.1.3) for the Radon transform, it follows that

$$\begin{aligned}|f(p, \varphi)| &\le \int_{-\infty}^{\infty} |q(p\cos\varphi - t\sin\varphi, p\sin\varphi + t\cos\varphi)|\, dt \\ &\le C_1 \int_{-\infty}^{\infty} (1 + (p\cos\varphi - t\sin\varphi)^2)^{-2}(1 + (p\sin\varphi + t\cos\varphi)^2)^{-2}\, dt.\end{aligned}$$

Since

$$\begin{aligned}&(1 + (p\cos\varphi - t\sin\varphi)^2)(1 + (p\sin\varphi + t\cos\varphi)^2) \\ &\quad \ge 1 + (p\cos\varphi - t\sin\varphi)^2 + (p\sin\varphi + t\cos\varphi)^2 = 1 + p^2 + t^2,\end{aligned}$$

we have

$$|f(p,\varphi)| \le C_1 \int_{-\infty}^{\infty} \frac{dt}{(1+p^2+t^2)^2}.$$

Then, using the equality

$$\int \frac{dx}{(x^2+a^2)^2} = \frac{1}{2a^2}\frac{x}{x^2+a^2} + \frac{1}{2a^3}\arctan\left(\frac{x}{a}\right) + \text{const},$$

we obtain

$$|f(p,\varphi)| \le \frac{C}{(1+p^2)^{3/2}},$$

which proves the estimate (5.1.9).

We now establish the relationship between the Fourier transforms of $q(x,y)$ (see Appendix A) and the Radon transform $f(p,\varphi)$ of $q(x,y)$. From the estimate (5.1.8) it follows that there exists a two-dimensional Fourier transform of $q(x,y)$

$$\hat{q}(\omega_1,\omega_2) = \frac{1}{2\pi}\int_{-\infty}^{\infty}\int_{-\infty}^{\infty} \exp(-\mathrm{i}\omega_1 x - \mathrm{i}\omega_2 y) q(x,y)\,dx\,dy.$$

From the estimate (5.1.9) it follows that for any φ there exists a one-dimensional Fourier transform of $f(p,\varphi)$ with respect to the variable p:

$$\hat{f}(\omega,\varphi) = \frac{1}{\sqrt{2\pi}}\int_{-\infty}^{\infty} \exp(-\mathrm{i}\omega p) f(p,\varphi)\,dp.$$

We will show that the Fourier transform $\hat{q}(\omega_1,\omega_2)$ and $\hat{f}(\omega,\varphi)$ are related by the formula

$$\hat{q}(\omega\cos\varphi,\omega\sin\varphi) = \frac{1}{\sqrt{2\pi}}\hat{f}(\omega,\varphi). \tag{5.1.10}$$

Consider

$$\hat{q}(\omega\cos\varphi,\omega\sin\varphi) = \frac{1}{2\pi}\int_{-\infty}^{\infty}\int_{-\infty}^{\infty} \exp(-\mathrm{i}\omega(x\cos\varphi + y\sin\varphi))q(x,y)\,dx\,dy.$$

Using the change of variables $x = p\cos\varphi - t\sin\varphi$ and $y = p\sin\varphi + t\cos\varphi$ in the integral and taking into account the obvious equalities

$$\frac{D(x,y)}{D(p,t)} = \begin{vmatrix} \cos\varphi & -\sin\varphi \\ \sin\varphi & \cos\varphi \end{vmatrix} = 1,$$

$$\begin{aligned} -\mathrm{i}\omega(x\cos\varphi + y\sin\varphi) &= -\mathrm{i}\omega(p\cos^2\varphi - t\sin\varphi\cos\varphi + p\sin^2\varphi + t\sin\varphi\cos\varphi) \\ &= -\mathrm{i}\omega p \end{aligned}$$

we obtain

$$\hat{q}(\omega\cos\varphi, \omega\sin\varphi)$$
$$= \frac{1}{2\pi}\int_{-\infty}^{\infty}\exp(-i\omega p)\int_{-\infty}^{\infty} q(p\cos\varphi - t\sin\varphi, p\sin\varphi + t\cos\varphi)\,dt\,dp.$$

From (5.1.3) it follows that

$$\hat{q}(\omega\cos\varphi, \omega\sin\varphi) = \frac{1}{2\pi}\int_{-\infty}^{\infty}\exp(-i\omega p) f(p,\varphi)\,dp.$$

By the definition of the Fourier transform, we have

$$\hat{q}(\omega\cos\varphi, \omega\sin\varphi) = \frac{1}{\sqrt{2\pi}}\hat{f}(\omega,\varphi),$$

which proves the formula (5.1.10).

From (5.1.10) it follows that $q(x, y)$ is uniquely determined by the Radon transform. Indeed, let $q_1(x, y)$, $q_2(x, y)$ be functions such that $f_1(p,\varphi) = f_2(p,\varphi)$ for $p \in (-\infty,\infty)$ and $\varphi \in [0, 2\pi]$. Then the Radon transform of their difference $q_0(x, y) = q_1(x, y) - q_2(x, y)$ is $f_0(p,\varphi) = 0$ for $p \in (-\infty,\infty)$ and $\varphi \in [0, 2\pi]$. Hence, the Fourier transform of $f_0(p,\varphi)$ with respect to the first argument $\hat{f_0}(\omega,\varphi)$ is equal to zero for all $\omega \in (-\infty,\infty)$ and $\varphi \in [0, 2\pi]$. Then (5.1.10) implies that $\hat{q}_0(\omega\cos\varphi, \omega\sin\varphi) = 0$ for all $\omega \in (-\infty,\infty)$ and $\varphi \in [0, 2\pi]$. Note that for all $(\omega_1, \omega_2) \in \mathbb{R}^2$ the system of equations $\omega\cos\varphi = \omega_1$, $\omega\sin\varphi = \omega_2$ for ω and φ has a solution $\omega \in (-\infty,\infty)$, $\varphi \in [0, 2\pi]$. Consequently, the Fourier transform $\hat{q}_0(\omega_1, \omega_2)$ is identically equal to zero for $(\omega_1, \omega_2) \in \mathbb{R}^2$, and therefore $q_0(x, y) = 0$ for $(x, y) \in \mathbb{R}^2$. Hence $q_1(x, y) = q_2(x, y)$ for $(x, y) \in \mathbb{R}^2$.

Using the formula (5.1.10), given the function $f(p,\varphi)$, one can calculate the two-dimensional Fourier transform $\hat{q}(\omega_1, \omega_2)$ and, subsequently, $q(x, y)$. When applying this algorithm in practice, it is necessary to take into account the problems related to calculating the direct and inverse Fourier transforms of approximate functions. Another characteristic feature of the algorithm is that it is necessary to recalculate the values of the function $\hat{q}(\omega\cos\varphi, \omega\sin\varphi)$ specified in the polar coordinate system to the values of $\hat{q}(\omega_1, \omega_2)$ in the Cartesian coordinate system (ω_1, ω_2).

Consider the formula for the inversion of the Radon transform, i.e., the formula for calculating $q(x, y)$ from $f(p,\varphi)$, which was obtained by Radon. Since for calculating the values of the desired function at a certain point we can assume that this point is the origin, we confine ourselves to determining the value of $q(0, 0)$.

Let D_α denote the exterior of the circle of radius α centered at the origin. Consider the double integral

$$J = \iint_{D_\alpha} \frac{q(x, y)\,dx\,dy}{\sqrt{x^2 + y^2 - \alpha^2}}. \tag{5.1.11}$$

Using the change of variables $x = \alpha\cos\varphi - t\sin\varphi$, $y = \alpha\sin\varphi + t\cos\varphi$, this integral is reduced to the form

$$J = \int_0^{2\pi} d\varphi \int_0^{\infty} q(\alpha\cos\varphi - t\sin\varphi, \alpha\sin\varphi + t\cos\varphi)\, dt$$

or

$$J = \int_0^{2\pi} d\varphi \int_{-\infty}^{0} q(\alpha\cos\varphi - t\sin\varphi, \alpha\sin\varphi + t\cos\varphi)\, dt.$$

Hence,

$$J = \frac{1}{2}\int_0^{2\pi} d\varphi \int_{-\infty}^{\infty} q(\alpha\cos\varphi - t\sin\varphi, \alpha\sin\varphi + t\cos\varphi)\, dt$$

and, in view of (5.1.3), we have

$$J = \frac{1}{2}\int_0^{2\pi} f(\alpha,\varphi)\, d\varphi. \tag{5.1.12}$$

On the other hand, using the change of variables $x = r\cos\varphi$, $y = r\sin\varphi$ in the integral (5.1.11), we obtain

$$J = \int_\alpha^{\infty}\int_0^{2\pi} \frac{q(r\cos\varphi, r\sin\varphi) r\, d\varphi\, dr}{\sqrt{r^2-\alpha^2}}.$$

Setting

$$\bar{q}(r) = \frac{1}{2\pi}\int_0^{2\pi} q(r\cos\varphi, r\sin\varphi)\, d\varphi, \tag{5.1.13}$$

we arrive at the equality

$$J = 2\pi\int_\alpha^{\infty} \frac{\bar{q}(r) r\, dr}{\sqrt{r^2-\alpha^2}}.$$

Comparing the last equality to (5.1.12), we obtain the integral equation for the function $\bar{q}(r)$

$$2\int_\alpha^{\infty} \frac{\bar{q}(r) r\, dr}{\sqrt{r^2-\alpha^2}} = \bar{f}(\alpha), \quad 0 < \alpha < \infty, \tag{5.1.14}$$

where

$$\bar{f}(\alpha) = \frac{1}{2\pi}\int_0^{2\pi} f(\alpha,\varphi)\, d\varphi.$$

Note that equation (5.1.14) coincides with equation (5.1.5) up to the notation, except that the upper limit in (5.1.5) is finite, whereas in (5.1.14) it is infinite. However, this difference occurs only because it was assumed that $q_0(\sqrt{x^2+y^2}) = 0$ for $x^2 + y^2 > R^2$ when deriving the equation (5.1.5).

We now calculate the integral

$$J(\bar{f}) = -\frac{1}{\pi}\int_0^\infty \frac{d\bar{f}(\alpha)}{\alpha} = -\frac{1}{\pi}\lim_{\varepsilon\to 0}\int_\varepsilon^\infty \frac{d\bar{f}(\alpha)}{\alpha}$$
$$= \frac{1}{\pi}\lim_{\varepsilon\to 0}\Big(\frac{\bar{f}(\varepsilon)}{\varepsilon} - \int_\varepsilon^\infty \frac{\bar{f}(\alpha)\,d\alpha}{\alpha^2}\Big).$$

Using (5.1.14), we have

$$J(\bar{f}) = \frac{2}{\pi}\lim_{\varepsilon\to 0}\Big(\frac{1}{\varepsilon}\int_\varepsilon^\infty \frac{\bar{q}(r)r\,dr}{\sqrt{r^2-\varepsilon^2}} - \int_\varepsilon^\infty\int_\alpha^\infty \frac{\bar{q}(r)r\,dr\,d\alpha}{\alpha^2\sqrt{r^2-\varepsilon^2}}\Big).$$

Changing the order of integration in the second integral, we obtain

$$J(\bar{f}) = \frac{2}{\pi}\lim_{\varepsilon\to 0}\Big(\frac{1}{\varepsilon}\int_\varepsilon^\infty \frac{\bar{q}(r)r\,dr}{\sqrt{r^2-\varepsilon^2}} - \int_\varepsilon^\infty \bar{q}(r)\int_\varepsilon^r \frac{r\,d\alpha}{\alpha^2\sqrt{r^2-\varepsilon^2}}\,dr\Big).$$

Since

$$\int \frac{d\alpha}{\alpha^2\sqrt{r^2-\alpha^2}} = -\frac{\sqrt{r^2-\alpha^2}}{r^2\alpha} + \text{const},$$

we have

$$\int_\varepsilon^r \frac{r\,d\alpha}{\alpha^2\sqrt{r^2-\alpha^2}} = -\frac{\sqrt{r^2-\alpha^2}}{r\alpha}\Big|_\varepsilon^r = \frac{\sqrt{r^2-\varepsilon^2}}{r\varepsilon}.$$

Consequently,

$$J(\bar{f}) = \frac{2}{\pi}\lim_{\varepsilon\to 0}\frac{1}{\varepsilon}\int_\varepsilon^\infty \Big(\frac{r}{\sqrt{r^2-\varepsilon^2}} - \frac{\sqrt{r^2-\varepsilon^2}}{r}\Big)\bar{q}(r)\,dr$$
$$= \frac{2}{\pi}\lim_{\varepsilon\to 0}\varepsilon\int_\varepsilon^\infty \frac{\bar{q}(r)\,dr}{r\sqrt{r^2-\varepsilon^2}}$$
$$= \frac{2}{\pi}\lim_{\varepsilon\to 0}\Big(\varepsilon\int_\varepsilon^\infty \frac{\bar{q}(0)\,dr}{r\sqrt{r^2-\varepsilon^2}} + \varepsilon\int_\varepsilon^\infty \frac{\bar{q}(r)-\bar{q}(0)\,dr}{r\sqrt{r^2-\varepsilon^2}}\Big).$$

We calculate the first integral. Since

$$\int \frac{dr}{r\sqrt{r^2-\varepsilon^2}} = \frac{1}{\varepsilon}\arctan\Big(\frac{\sqrt{r^2-\varepsilon^2}}{\varepsilon}\Big) + \text{const}, \tag{5.1.15}$$

we have

$$\varepsilon\int_\varepsilon^\infty \frac{\bar{q}(0)\,dr}{r\sqrt{r^2-\varepsilon^2}} = \frac{\pi}{2}\bar{q}(0).$$

We denote the second term by J_2 and establish that it tends to zero as $\varepsilon \to 0$. Indeed, let $\omega(h)$ be the modulus of continuity of the function $\bar{q}(r)$ on the interval $[0, 1]$. Then for $\varepsilon < 1$ we have

$$|J_2| \le \varepsilon \int_{\varepsilon}^{\sqrt{\varepsilon}} \frac{|\bar{q}(r) - \bar{q}(0)|\, dr}{r\sqrt{r^2 - \varepsilon^2}} + \varepsilon \int_{\sqrt{\varepsilon}}^{\infty} \frac{|\bar{q}(r) - \bar{q}(0)|\, dr}{r\sqrt{r^2 - \varepsilon^2}}$$
$$\le \varepsilon\omega(\sqrt{\varepsilon}) \int_{\varepsilon}^{\sqrt{\varepsilon}} \frac{dr}{r\sqrt{r^2 - \varepsilon^2}} + 2\varepsilon C \int_{\sqrt{\varepsilon}}^{\infty} \frac{dr}{r\sqrt{r^2 - \varepsilon^2}},$$

where the constant C is such that $|\bar{q}(r)| \le C$ for $r \ge 0$. From (5.1.15) it follows that

$$|J_2| \le \omega(\sqrt{\varepsilon}) \arctan\Big(\frac{\sqrt{\varepsilon - \varepsilon^2}}{\varepsilon}\Big) + 2C\Big[\frac{\pi}{2} - \arctan\Big(\frac{\sqrt{\varepsilon - \varepsilon^2}}{\varepsilon}\Big)\Big]$$

and $J_2 \to 0$ for $\varepsilon \to 0$.

As a result,

$$J(\bar{f}) = \bar{q}(0)$$

and the formula obtained by Radon has the form

$$q(0, 0) = \bar{q}(0) = -\frac{1}{\pi} \int_0^{\infty} \frac{d\bar{f}(\alpha)}{\alpha}.$$

Note that in practice it is usually impossible to determine the Radon transform $f(p, \varphi)$ for all values of p and φ. The lack of information can cause non-uniqueness when reconstructing $q(x, y)$. One of the ways to resolve this problem is to take a smaller class of functions $q(x, y)$.

5.2 Reconstructing a function from its spherical means

One of the examples of problems of integral geometry is the problem of reconstructing a function $q(x, y_1, y_2)$ from its mean values on spheres with arbitrary finite radii r and centers on the plane $x = 0$.

We denote a point (x, y_1, y_2) in the three-dimensional space by (x, y), $y = (y_1, y_2)$, and a function $q(x, y_1, y_2)$ by $q(x, y)$. Let (ξ, η) ($\eta = (\eta_1, \eta_2)$) be a variable point on a sphere of radius r centered at the point $(0, y)$. In this notation, the problem of reconstructing the function $q(x, y)$ from its spherical means $f(y, r)$, $r \ge 0$, is reduced to the integral equation

$$\frac{1}{4\pi} \iint_{\xi^2 + |\eta|^2 = r^2} q(\xi, y + \eta)\, d\omega = f(y, r), \tag{5.2.1}$$

where $d\omega$ is the element of a solid angle with vertex $(0, y)$.

This problem will be analyzed in Sections 7.1 and 7.2 (see also (Romanov, 1984)).

5.3 Determining a function of a single variable from the values of its integrals. The problem of moments

The problem of determining a function $q(x)$ from the values of its integrals

$$\langle q, \varphi_n \rangle := \int_a^b q(x)\varphi_n(x)\,dx = f_n, \quad n = 1, 2, \ldots, \tag{5.3.1}$$

where $\{\varphi_n(x)\}$ is a given system of functions, arises in the study of many applied and theoretical problems.

If the system of functions $\{\varphi_n(x)\}$ is complete and orthonormal in $L_2(a,b)$, then the problem of finding $q(x)$ from (5.3.1) consists in reconstructing a function from its Fourier coefficients f_n. From the general theory of the Fourier series in a Hilbert space, it follows that for any sequence $\{f_n\} \in l_2$ there exists a unique function $q(x) \in L_2(a,b)$ such that the equalities (5.3.1) hold for all natural n and Parseval's identity holds:

$$\|q\|^2_{L_2(a,b)} = \sum_{n=1}^{\infty} f_n^2.$$

Consequently, if we reformulate this problem as the problem of solving the operator equation $Aq = f$, $f = \{f_n\}$, where the operator A defined by (5.3.1) acts from the space $L_2(a,b)$ into l_2, then this problem is well-posed because it has a unique solution for any right-hand side $\{f_n\} \in l_2$, and the continuous dependence of $q(x)$ on $\{f_n\}$ follows from Parseval's identity.

We now assume that the operator A acts from the space $C[a,b]$ into l_2, i.e., it is required to find a continuous function $q(x)$ that satisfies the equalities (5.3.1). This problem is ill-posed. Indeed, let $\{\bar{f}_n\} \in l_2$ be the sequence of Fourier coefficients of an arbitrary function $q(x) \in L_2(a,b) \setminus C[a,b]$. Since a function is uniquely determined by its Fourier coefficients, there is no function that is continuous on $[a,b]$ and has the same Fourier coefficients $\{\bar{f}_n\}$. Hence, the well-posedness of the problem of reconstructing $q(x)$ from the values of its integrals (5.3.1) substantially depends on the space that the unknown function $q(x)$ is assumed to belong to. Methods for determining continuous functions from approximations of their Fourier coefficients are described in (Tikhonov and Arsenin, 1986).

Consider the problem of determining $q(x)$ from the values of its integrals in the case of an arbitrary system $\{\varphi_n(x)\}$. The problem of constructing a function $q(x)$ that satisfies the equalities (5.3.1), where $\{\varphi_n(x)\}$ is a given sequence of functions and $\{f_n\}$ is a given sequence of numbers, is called the *problem of moments*. Depending on how many functions $\varphi_n(x)$ and numbers f_n are given, we distinguish finite-dimensional and infinite-dimensional versions of the problem. We now formulate the problem in a more general form. Let $\mathcal{H}$ be a separable Hilbert space. Given a sequence of elements $\{\varphi_n(x)\}$ in $\mathcal{H}$ and and a sequence of numbers $\{f_n\}$, it is required to find an element

$q \in \mathcal{H}$ such that $\langle q, \varphi_n \rangle_{\mathcal{H}} = f_n$, $n = 1, 2, \ldots, N$ (the finite-dimensional problem of moments) or $n \in \mathbb{N}$ (the infinite-dimensional problem of moments). If there is an additional condition $\|q\|_{\mathcal{H}} \leq l$ for the unknown element q, where l is a given number, then the problem is called the *l-problem of moments*. In what follows, we assume that $\langle \cdot, \cdot \rangle := \langle \cdot, \cdot \rangle_{\mathcal{H}}$, $\| \cdot \| := \| \cdot \|_{\mathcal{H}}$.

Consider the infinite-dimensional problem of moments. The uniqueness of the solution of this problem is guaranteed by the completeness of the system of elements $\{\varphi_n\}$. Indeed, if the system $\{\varphi_n\}$ is complete, then the equalities $\langle u, \varphi_n \rangle = 0$, $n = 1, 2, \ldots$, imply that $u = 0$, which means that the solution of the infinite-dimensional problem of moments is unique.

The solution of the finite-dimensional problem of moments is never unique. Indeed, let $\bar{q}$ be a solution to the finite-dimensional problem of moments $\langle \bar{q}, \varphi_n \rangle = f_n$, $n = 1, 2, \ldots, N$. Let $\mathcal{H}_1$ denote the subspace orthogonal to the finite-dimensional subspace generated by the elements $\varphi_1, \ldots, \varphi_N$. Then, in addition to $\bar{q}$, for any element ψ in $\mathcal{H}_1$, the element $\bar{q} + \psi$ is also a solution to this problem.

Note that the solution of the finite-dimensional l-problem of moments may be unique under certain conditions. Indeed, let $\{\varphi_n\}$, $n \in \mathbb{N}$, be an orthonormal basis in $\mathcal{H}$. Consider a finite-dimensional l-problem of moments: $\langle q, \varphi_n \rangle = f_n$, $n = 1, 2, \ldots, N$, $\|q\| \leq l$. Expand the element q in a series with respect to φ_n:

$$q = \sum_{n=1}^{\infty} \langle q, \varphi_n \rangle \varphi_n = \sum_{n=1}^{N} f_n \varphi_n + \sum_{n=N+1}^{\infty} \langle q, \varphi_n \rangle \varphi_n.$$

From Parseval's identity

$$\|q\|^2 = \sum_{n=1}^{N} f_n^2 + \sum_{n=N+1}^{\infty} \langle q, \varphi_n \rangle^2$$

it follows that the finite-dimensional l-problem of moments has a non-unique solution if

$$M = \sum_{n=1}^{N} f_n^2 < l^2,$$

a unique solution if $M = l^2$, and no solutions if $M > l^2$.

We now turn to the solvability of the problem of moments. With the exception of trivial cases, the analysis of the solvability of the infinite-dimensional problem of moments is certainly more difficult that the finite-dimensional one. However, the analysis of the solvability of the infinite-dimensional l-problem of moments can be reduced to the analysis of the solvability of finite-dimensional l-problems of moments. Given a system of elements $\varphi_n \in \mathcal{H}$, $n \in \mathbb{N}$, and a sequence of numbers f_n, it is required to find an element q such that $\langle q, \varphi_n \rangle = f_n$ for $n \in \mathbb{N}$ and $\|q\| \leq l$.

Theorem 5.3.1. *The infinite-dimensional l-problem of moments is solvable if and only if for any natural N there exists a solution to the finite-dimensional l-problem of moments for the same φ_n and f_n.*

Proof. The necessary part of the assertion is obvious, since any element q that is a solution to the infinite-dimensional l-problem of moments is also a solution to the finite-dimensional l-problem of moments for any natural N.

We will prove the sufficient part. Let q_n be a solution to the finite-dimensional l-problem of moments

$$\langle q, \varphi_m \rangle = f_m, \quad m = 1, 2, \ldots, n, \quad \|q\| \leq l.$$

Any element of the sequence $\{q_n\}$, $n = 1, 2, \ldots$, satisfies the condition $\|q_n\| \leq l$. Consequently, the sequence $\{q_n\}$ has a subsequence $\{q_{n_k}\}$ that converges weakly to q. From the weak convergence of q_{n_k} to q and the fact that the norms $\|q_{n_k}\| \leq l$ are bounded, it follows that $\|q\| \leq l$. Since $\langle q_{n_k}, \varphi_m \rangle = f_m$, $m = 1, 2, \ldots, n_k$, passing to the limit as $n_k \to \infty$ in these equalities and using the weak convergence of q_{n_k} to q, we obtain

$$\langle q, \varphi_n \rangle = f_n, \quad n \in \mathbb{N}.$$

Hence, q is a solution to the infinite-dimensional l-problem of moments, which is the desired conclusion. □

We will show that the finite-dimensional problem of moments is always solvable in the case of a linearly independent system of elements $\{\varphi_n\}$. Given a linearly independent system of elements φ_n, $n = 1, 2, \ldots, N$, and numbers f_n, $n = 1, 2, \ldots, N$, consider an orthonormal system of elements ψ_j, $j = 1, 2, \ldots, N$, obtained from φ_n as a result of orthogonalization:

$$\psi_j = \frac{\omega_j}{\|\omega_j\|}, \quad \omega_j = \varphi_j - \sum_{k=1}^{j-1} \langle \varphi_j, \psi_k \rangle \psi_k.$$

Then

$$\varphi_n = \sum_{j=1}^{n} \langle \varphi_n, \psi_j \rangle \psi_j, \quad \langle \varphi_n, \psi_n \rangle \neq 0, \quad n = 1, 2, \ldots, N.$$

Consider the system of linear algebraic equations

$$\sum_{j=1}^{n} \langle \varphi_n, \psi_j \rangle c_j = f_n, \quad n = 1, 2, \ldots, N,$$

for c_j, $j = 1, 2, \dots, N$. This is a system with triangular matrix with nonzero diagonal elements $\langle \varphi_n, \psi_n \rangle$. Hence, it has a unique solution $\bar{c}_j$, $j = 1, 2, \dots, N$. We will show that the element

$$\bar{q}_N = \sum_{j=1}^{N} \bar{c}_j \psi_j$$

is a solution to the finite-dimensional problem of moments. Indeed, since $\langle \varphi_n, \psi_j \rangle = 0$ for $j > n$, it follows that for all $n = 1, 2, \dots, N$

$$\langle \bar{q}_N, \varphi_n \rangle = \Big\langle \sum_{j=1}^{N} \bar{c}_j \psi_j, \varphi_n \Big\rangle = \sum_{j=1}^{N} \bar{c}_j \langle \psi_j, \varphi_n \rangle = f_n.$$

Thus, in the case of a linearly independent system of elements φ_n, the finite-dimensional problem of moments has a solution for any numbers f_n.

We will now find the conditions for the existence of solutions to the finite-dimensional l-problem of moments. Let $\bar{q}_N$ be the solution to the finite-dimensional problem of moments constructed above, and let $B_N : \mathcal{H}^N \to \mathcal{H}^N$ be a linear operator that transforms $\psi = (\psi_1, \dots, \psi_n)$ to $\varphi = (\varphi_1, \dots, \varphi_n)$: $B_N(\psi) = \varphi$. The operator B_N is defined by the matrix with elements $b_{ij} = \langle \varphi_i, \psi_j \rangle$, $i, j = 1, 2, \dots, N$. Since this is a triangular matrix with nonzero diagonal elements, it follows that for all natural N the inverse B_N^{-1} of the operator B_N is bounded and

$$\|\bar{q}_N\|^2 = \sum_{n=1}^{N} \langle \bar{q}_N, \psi_j \rangle^2 \le \|B_N^{-1}\|^2 \sum_{n=1}^{N} f_n^2.$$

Therefore, $\bar{q}_N$ is a solution to the finite-dimensional l-problem of moments if

$$\|B_N^{-1}\|^2 \sum_{n=1}^{N} f_n^2 \le l^2.$$

Let φ_n be a system of linearly independent elements such that $\|B_N^{-1}\| \le B$ for all natural N. Assume that $\{f_n\} \in l_2$ and $\|\{f_n\}\|_{l_2} = \bar{f}$. Then the infinite-dimensional l-problem of moments has a solution if $B\bar{f} \le l$. Indeed, under these conditions, for any natural N there exists a solution $\bar{q}_N$ to the finite-dimensional problem of moments for the set of numbers f_n, $n = 1, 2, \dots, N$. Then

$$\|\bar{q}_N\| \le \|B_N^{-1}\| \Big(\sum_{n=1}^{N} f_n^2 \Big)^{1/2} \le B \|\{f_n\}\|_{l_2} = B\bar{f}.$$

Therefore, if $B\bar{f} \le l$, then the finite-dimensional l-problem of moments has a solution for any natural N, and Theorem 5.3.1 implies that there exists a solution to the infinite-dimensional l-problem of moments.

Remark 5.3.1. The above results are applicable in the study of the original problem (5.3.1) in the case where a solution $q(x)$ is assumed to belong to the space $L_2(a,b)$.

Remark 5.3.2. Other terms used in the scientific literature include the "power problem of moments" ($\varphi_n(x) = x^n$) and the "trigonometric problem of moments". More details on the problem of moments and its applications can be found in (Akhiezer, 1965).

We now return to the problem of reconstructing a function $q(x) \in L_2(a,b)$ from its integrals f_n defined by the equalities (5.3.1), where $\{\varphi_n(x)\}$ is a given system of functions. This problem has a unique solution if the system $\{\varphi_n(x)\}$ is complete in $L_2(a,b)$. However, is usually rather hard to find out whether $\{\varphi_n(x)\}$ is complete. One of the most well-known classes of complete systems in the space $L_2(a,b)$ is represented by the systems of eigenfunctions of the Sturm–Liouville problem.

Another example of such systems is the system of power functions $\{x^n\}$, $n = 0, 1, \ldots$, whose completeness follows from the Weierstrass theorem stating that a continuous function defined on an interval can be uniformly approximated by algebraic polynomials. Consider a more general system of functions $\varphi_n(x) = x^{a_n}$, where a_n is a sequence of positive numbers.

Theorem 5.3.2. *Assume that $q(x) \in L_2(0,1)$, and let a_n, $n \in \mathbb{N}$, be positive numbers such that*

$$\lim_{n\to\infty} a_n = \infty, \qquad \sum_{n=1}^{\infty} \frac{1}{a_n} = \infty, \tag{5.3.2}$$

and

$$\int_0^1 q(x) x^{a_n}\, dx = 0, \qquad n \in \mathbb{N}.$$

Then $q(x) = 0$.

Theorem 5.3.2 implies the uniqueness of solutions to the problem of reconstructing $q(x) \in L_2(0,1)$ from equalities (5.3.1) for $\varphi_n(x) = x^{a_n}$ if the sequence $\{a_n\}$ satisfies the conditions (5.3.2).

Remark 5.3.3. The analysis of the uniqueness of solutions to Fredholm linear integral equations of the first kind can be reduced to establishing the completeness of a system of functions. Indeed, if for the kernel $K(t,x)$ of the equation

$$\int_a^b K(t,x) q(x)\, dx = f(t), \qquad c \le t \le d, \tag{5.3.3}$$

there exists a sequence of points t_n on the interval $[c,d]$ such that the system of functions $\varphi_n(x) = K(t_n, x)$ is complete in the space $L_2(a,b)$, then the solution of equation (5.3.3) is unique in $L_2(a,b)$.

5.4 Inverse kinematic problem of seismology

The inverse kinematic problem of seismology arose in the study of the Earth's interior structure based on surface observations of seismic waves generated by earthquakes.

The physical pattern of seismic wave propagation on the surface of the Earth resembles the pattern of wave propagation on the surface of still water after throwing a stone. The seismic wave front that separates the perturbed and unperturbed regions of the medium moves away from the epicenter. The shape of the wave front and the speed of its points on the surface depend on the speed of the seismic wave in the Earth's interior. For example, for the spherically symmetric model of the Earth, where the speed depends only on depth, the wave fronts have the shape of circles moving away from epicenter. If the speed of the wave depends on all three coordinates, then the shape of its front can be a complex curve. As early as around the beginning of the 20th century, geophysicists faced the question of whether it is possible to determine the speed of seismic waves in the Earth's interior from the pattern of propagation of seismic wave fronts on the surface. Indeed, the speed of a wave depends on the elastic properties of the body in which it propagates. Therefore, information about the speed of propagation of seismic waves in the Earth's interior provides clues about its interior structure.

The mathematical model of the process of seismic wave propagation is formulated as follows (Romanov, 1984). Let D be a domain in the three-dimensional space $\mathbb{R}^3$ with boundary S where signals are transmitted with finite positive speed $c(x)$, $x = (x_1, x_2, x_3)$. Take an arbitrary pair of points $x^0, x \in D$ and consider the functional

$$\tau(L) = \int_{L(x,x^0)} \frac{|dy|}{c(y)}, \tag{5.4.1}$$

where $L(x, x^0)$ is a sufficiently smooth curve that connects the points x^0 and x; $|dy|$ is its Euclidean length element:

$$|dy| = \Big(\sum_{k=1}^{3} dy_k^2\Big)^{1/2}. \tag{5.4.2}$$

In view of formula (5.4.1), the functional $\tau(L)$ represents the time it takes for the signal to travel along the curve $L(x, x^0)$. A *seismic ray* is a curve $L(x, x^0)$ for which the signal travel time is minimal. For media with complex structure, the function $c(x)$ may be substantially different from a constant, which implies that there may be more than one ray (or even infinitely many rays) connecting the points x^0 and x for which the functional (5.4.1) attains its minimum value compared to all other values that correspond to the curves sufficiently close to the given ray. Hence, seismic rays are local extremals of the functional (5.4.1). It is known that the extremals of the functional

(5.4.1) in the n-dimensional space are integral curves of the system of Euler's equations consisting of n ordinary differential equations of the second order. These integral curves satisfy the additional condition that they pass through the points x^0 and x. Let $\Gamma(x, x^0)$ denote the ray that connects x^0 and x, and let $\tau(x, x^0)$ be the signal travel time for this ray. Then

$$\tau(x, x^0) = \int_{\Gamma(x,x^0)} \frac{|dy|}{c(y)}. \tag{5.4.3}$$

The surface defined by the equality $\tau(x, x^0) = \text{const}$ with fixed x^0 is the front of a wave generated by a point source concentrated at x^0. Let $\nu(x, x^0)$ be the unit tangent vector to the ray $\Gamma(x, x^0)$ at the point x in the direction of increasing τ. Then (5.4.3) implies

$$\nabla_x \tau(x, x^0) = \frac{1}{c(x)} \nu(x, x^0), \tag{5.4.4}$$

where $\nabla_x \tau$ is the gradient of the function τ with respect to the variable point x. The equality (5.4.4), which expresses the orthogonality of the fronts and rays, yields the equation

$$|\nabla_x \tau(x, x^0)|^2 = \frac{1}{c^2(x)}. \tag{5.4.5}$$

Note that equation (5.4.4) is closely related to the wave equation

$$u_{tt} = c^2(x)\Delta u, \qquad c(x) > 0, \qquad x \in \mathbb{R}^3, \tag{5.4.6}$$

which describes the propagation of acoustic oscillations in a medium with constant density. Indeed, representing the equation for the characteristic surface of the wave equation (5.4.6) in the form $t = \tau(x)$, we see that the function $\tau(x)$ satisfies the differential equation of the first order

$$|\nabla_x \tau(x)|^2 = c^{-2}(x),$$

which is called the *eikonal equation* (from Greek *eikon*, image). Equation (5.4.5) under the condition

$$\tau(x, x^0) = O(|x - x^0|) \tag{5.4.7}$$

describes the characteristic surfaces $t = \tau(x, x^0)$ that have a conical point at a fixed point x^0. Such surfaces are called *characteristic conoids*. Note that x^0 is a parameter in (5.4.5), and the condition (5.4.7) stands for the Cauchy data in the degenerate case where the surface on which the data is given degenerates to a point x^0.

The algorithm for constructing characteristic conoids $t = \tau(x, x^0)$ from a given function $c(x)$ is based on the construction of their generatrices—lines called *bicharacteristics*. To construct bicharacteristics, we introduce the vector

$$p = (p_1, p_2, p_3) = \nabla_x \tau(x, x^0). \tag{5.4.8}$$

Set $q(x) = 1/c(x)$. Differentiating the eikonal equation $|p|^2 = q^2(x)$ with respect to x_k, we obtain

$$\langle p, p_{x_k} \rangle = q q_{x_k}, \quad k = 1, 2, 3. \tag{5.4.9}$$

From (5.4.8) it follows that $(p_i)_{x_k} = (p_k)_{x_i}$, and therefore (5.4.9) can be written as

$$\langle p, \nabla_x p_k \rangle = q q_{x_k}, \quad k = 1, 2, 3. \tag{5.4.10}$$

Divide (5.4.10) by $q^2(x)$ and rewrite it along the curves $dx/dt = p/q^2(x)$:

$$\frac{dp_k}{dt} = \frac{\partial}{\partial x_k} \ln q(x), \quad k = 1, 2, 3.$$

Along the same curves, $\tau(x, x^0)$ satisfies the equation

$$\frac{d\tau}{dt} = \left\langle \nabla_x \tau, \frac{dx}{dt} \right\rangle = \frac{|p|^2}{q^2(x)} = 1.$$

If the parameter t is chosen so that $\tau = 0$ for $t = 0$, then $t = \tau$ and t is equal to the time it takes for the signal to travel from x^0 to x. We now introduce an arbitrary unit vector ν^0 and solve the Cauchy problem for the system of ordinary differential equations

$$\frac{dx}{dt} = \frac{p}{q^2(x)}, \quad \frac{dp}{dt} = \nabla_x \ln q(x), \tag{5.4.11}$$

$$x\big|_{t=0} = x^0, \quad p\big|_{t=0} = p^0 \equiv q(x^0)\nu^0. \tag{5.4.12}$$

As a result, we find x and p as functions of t and the parameters x^0 and p^0:

$$x = h(t, x^0, p^0), \quad p = \psi(t, x^0, p^0). \tag{5.4.13}$$

For fixed x^0, the first equality in (5.4.13) defines a biparametric family of bicharacteristics that form a characteristic conoid in the space of variables x and t.

A projection of a bicharacteristic onto the space of x is called a *ray*. The equality $x = h(t, x^0, p^0)$ is a parametric definition of a ray, and the first equality in (5.4.11) shows that the tangent to the ray has the same direction as the vector p. Then, as follows from (5.4.8), the rays are orthogonal to the surfaces $\tau(x, x^0) = t$.

As is known, if two points x and x^0 are connected by a smooth curve $L(x, x^0)$ with length element $|dy|$, then the rays are extremals of the functional

$$\tau(L) = \int_{L(x,x^0)} \frac{|dy|}{c(y)},$$

i.e., the rays are curves for which $\tau(L)$ attains its extremum (generally, a local one). Consequently, if a wave is generated by a point source located at x^0 and active since $t = 0$, then the rays we introduced as the projections of bicharacteristics of the eikonal equation onto the space of x are none other than seismic rays that minimize the functional $\tau(L)$. On the other hand, the section of the characteristic conoid by the plane $t = t_0$, $t_0 > 0$, projected onto the space of x (i.e., the solution to the equation $\tau(x, x^0) = t_0$) defines the wave front at $t = t_0$.

Differential properties of solutions to the problem (5.4.11), (5.4.12) for an open domain $D \subset \mathbb{R}^3$ with smooth boundary S are described in detail in (Romanov,1984). We will only present some results that will be necessary in what follows. Put $\overline{D} = D \cup S$.

Lemma 5.4.1. *If $q(x) \in C^{s+1}(\overline{D})$, $s \geq 1$,*

$$q(x) \geq q_0 > 0, \quad x \in \overline{D}, \tag{5.4.14}$$

then for any point $x^0 \in D$ and any unit vector ν^0 there exists a unique solution (5.4.13) *to the problem* (5.4.11), (5.4.12) *for all t such that $h(t, x^0, p^0) \in \overline{D}$; the functions h, h_{tt}, ψ, and ψ_t are continuous in all arguments together with their derivatives up to order s.*

Lemma 5.4.2. *Under the assumptions of Lemma* 5.4.1, *the solution to the problem* (5.4.11), (5.4.12) *can be represented in the form*

$$x = \varphi(\xi, x^0), \quad p = \frac{1}{t}\, q^2(\varphi(\xi, x^0))\, \xi\, \frac{\partial}{\partial \xi}\, \varphi(\xi, x^0), \tag{5.4.15}$$

where $\xi = p^0 t / q^2(x^0)$, the vector function $\varphi(\xi, x^0) = (\varphi_1, \varphi_2, \varphi_3)$ has continuous derivatives up to order s with respect to all its arguments, and $\partial\varphi/\partial\xi$ is the Jacobian matrix

$$\frac{\partial \varphi}{\partial \xi} = \begin{Vmatrix} \frac{\partial\varphi_1}{\partial\xi_1} & \frac{\partial\varphi_2}{\partial\xi_1} & \frac{\partial\varphi_3}{\partial\xi_1} \\ \frac{\partial\varphi_1}{\partial\xi_2} & \frac{\partial\varphi_2}{\partial\xi_2} & \frac{\partial\varphi_3}{\partial\xi_2} \\ \frac{\partial\varphi_1}{\partial\xi_3} & \frac{\partial\varphi_2}{\partial\xi_3} & \frac{\partial\varphi_3}{\partial\xi_3} \end{Vmatrix}, \quad \xi = (\xi_1, \xi_2, \xi_3).$$

The parameter value $t = 0$ corresponds to the point $\xi = \mathbf{0}$. Using the Cauchy data and (5.4.11), the function $\varphi(\xi, x^0)$ can be expanded in a Taylor series in the neighborhood of $\xi = \mathbf{0}$:

$$\varphi(\xi, x^0) = x = x^0 + \left.\frac{dx}{dt}\right|_{t=0} t + O(t^2) = x^0 + \frac{p^0}{q^2(x^0)}\, t + O(t^2).$$

Consequently,

$$\varphi(\xi, x^0) = x^0 + \xi + O(|\xi|^2). \tag{5.4.16}$$

If x^0 is fixed and $|\xi| = t/q(x^0)$, the equality $x = \varphi(\xi, x^0)$ is a parametric definition of the surface $\tau(x, x^0) = t$.

Note that (5.4.16) implies

$$\left.\frac{\partial \varphi}{\partial \xi}\right|_{\xi=\mathbf{0}} = I, \tag{5.4.17}$$

where I is the unit matrix.

Lemma 5.4.3. *Under the assumptions of Lemma* 5.4.1, *in a sufficiently small neighborhood of* $\xi = \mathbf{0}$ *the function* $\varphi(\xi, x^0)$ *has a single-valued inverse* $\xi = g(x, x^0)$, *which is* s *times continuously differentiable with respect to* x *and* x^0. *Moreover,* $g(x^0, x^0) = \mathbf{0}$.

Therefore, if x^0 is fixed, for a set $\Delta(x^0)$ of points $x = \varphi(\xi, x^0)$ such that $|\xi| \le \delta$ and δ is sufficiently small, there exists a one-to-one correspondence between x and ξ. Since $\xi = p^0 t/q^2(x^0)$, each pair x, x^0 corresponds to the unique ray $\Gamma(x, x^0)$ that connects these points and lies in $\Delta(x^0)$. At x^0, the ray $\Gamma(x, x^0)$ has the direction $\nu^0 = p^0/q(x^0)$. The vector ν^0 and the travel time $\tau(x, x^0)$ are determined from the formulas

$$\begin{aligned} t &= \tau(x, x^0) = q(x^0)|g(x, x^0)|, \\ \nu^0 &= \nu^0(x, x^0) = g(x, x^0)/|g(x, x^0)|. \end{aligned} \tag{5.4.18}$$

From (5.4.16) it follows that

$$g(x, x^0) = x - x^0 + O(|x - x^0|^2). \tag{5.4.19}$$

Lemma 5.4.4. *Under the assumptions of Lemma* 5.4.1, *for* $x \ne x_0$ *the functions* $\tau(x, x^0)$ *and* $\nu^0(x, x^0)$ *are* s *times continuously differentiable, and the following estimates hold in the neighborhood of* x^0:

$$|D^\alpha \tau| \le C|x - x^0|^{1-|\alpha|}, \quad |D^\alpha \nu^0| \le C|x - x^0|^{-|\alpha|}, \tag{5.4.20}$$

where $\alpha = (\alpha_1, \alpha_2, \dots, \alpha_6)$ *is a multi-index,* $|\alpha| = \sum_{k=1}^{6} \alpha_k$, $|\alpha| \le s$, $\alpha_k \in \mathbb{N} \cup \{0\}$, *and*

$$D^\alpha = \partial^{|\alpha|}/(\partial x_1)^{\alpha_1} \dots (\partial x_3^0)^{\alpha_6}.$$

We now return to equation (5.4.5). The problem of finding $\tau(x, x^0)$ for a given function $c(x)$ is the direct problem for this equation. It is solved using the Hamilton–Jacobi method presented above, with simultaneous construction of the characteristics of equation (5.4.5) and the function $\tau(x, x^0)$, i.e., the rays and fronts of the wave process. This problem is called the *direct kinematic problem of seismology*. In the *inverse problem*, the function $c(x)$ is unknown, but the law that governs the propagation of wave fronts over the boundary of the domain in which the wave process is taking place is known. We will formulate one of the versions of the inverse kinematic problem. Suppose that a wave process occurs in a domain D of the space $\mathbb{R}^3$ with boundary S, generated by a point source of oscillations at a point x^0. Let $\tau(x, x^0)$ be a given function for all $x^0 \in S$ and $x \in S$. It is required to determine the speed of the signal propagation $c(x)$ for $x \in D$.

The inverse kinematic problem of seismology was first studied by Gustav Herglotz and Emil Wiechert in 1905–1907 under the assumption that the Earth is spherically symmetric ($c = c(r)$, $r = |x|$, the origin of the coordinate system coincides with the Earth's center). This assumption stemmed from experimental data showing that the time of travel of seismic waves from the source to the detector located at a fixed distance from the source is almost independent of the geographic region. It was discovered that for $c = c(r)$ the seismic ray $\Gamma(x, x^0)$ lies in the plane that passes through the points (x, x^0) and the center of the Earth, and therefore it suffices to consider an arbitrary section of the Earth that contains its diameter and select the points x^0 and x on the corresponding great circle. Herglotz and Wiechert showed that the point x^0 can be assumed to be fixed while x is moved along the circle. Consequently, if the function

$$m(r) = \frac{r}{c(r)} \tag{5.4.21}$$

is continuously differentiable and increases monotonically as r increases, then it is uniquely determined by the function $\tau = \tau(\alpha)$, which is called the wave *traveltime curve*. Above, α is the angular distance between the points x^0 and x, and $\tau(\alpha)$ is the time it takes for the signal to travel between these points. It turned out that the function $c(r)$ can be reconstructed locally. Namely, if $\tau(\alpha)$ is known for $\alpha \in [0, \alpha_0]$, $\alpha_0 > 0$, then $c(r)$ is within a layer $r_0 \le r \le R$, where R is the Earth's radius, r_0 is determined from α_0, $r_0(\alpha_0)$ decreases monotonically as α_0 increases, and $r_0(\alpha_0) \to R$ as $\alpha_0 \to 0$.

The solution of the inverse kinematic problem for the spherically symmetric model of the Earth had a considerable impact on the development of the theories about the Earth's interior structure. Calculations based on the measurement data for seismic wave travel times revealed characteristic features of the Earth's structure: its crust, mantle, and core. The crust of the Earth is a relatively thin layer with average thickness of about 30 km. Seismic wave speeds in the Earth's mantle are much higher than in its crust. They increase rather sharply at the boundary between the crust and the mantle. The thickness of the mantle is approximately 3000 km. The physical properties of the core are substantially different from those of the mantle. The core does not transmit shear waves, whereas they do propagate in the mantle. The absence of shear waves is characteristic of fluids. For this reason, the core of the Earth is presumed to be liquid.

Currently there exist several dozen wave speed models of the Earth's interior, but the differences between them are insignificant (10% on average).

Linearized inverse problem of seismology. Consider the inverse problem of reconstructing the speed $c(x)$ from the observations of the seismic waves propagation on the surface of the Earth in the case where $c(x)$ depends on all coordinates.

In other words, assume that the direct problem

$$\begin{aligned} |\nabla_x \tau(x,x^0)|^2 &= c^{-2}(x), \quad x \in D, \\ \tau(x,x^0) &= O(|x-x^0|) \quad \text{for } x \to x^0, \end{aligned} \tag{5.4.22}$$

includes the following additional information about its solution:

$$\tau(x,x^0) = f(x,x^0), \quad x, x^0 \in S. \tag{5.4.23}$$

It is required to find $c(x)$ from (5.4.22) and (5.4.23).

In the linearized formulation of the inverse problem, we assume that the unknown function $q(x) = 1/c(x)$ has the form

$$q(x) = q_0(x) + q_1(x),$$

where $q_0(x)$ is given, $q_0(x) > 0$, and $q_1(x) \in C^2(\overline{D})$ is small compared to $q_0(x)$ in the norm of $C^2(D)$.

First, we seek a solution $\tau_0(x,x^0)$ to the direct problem

$$\begin{aligned} |\nabla_x \tau(x,x^0)|^2 &= q_0^2(x), \quad x \in D, \\ \tau(x,x^0) &= O(|x-x_0|). \end{aligned}$$

Set

$$\begin{aligned} \tau_1(x,x^0) &= \tau(x,x^0) - \tau_0(x,x^0), \quad x^0 \in S,\ x \in D; \\ f_1(x,x^0) &= f(x,x^0) - \tau_0(x,x^0), \quad x, x^0 \in S. \end{aligned}$$

The linearization of the inverse problem (5.4.22), (5.4.23) consists in substituting the right-hand sides of the equalities

$$\tau = \tau_0 + \tau_1, \quad q = q_0 + q_1, \quad f = \tau_0 + f_1$$

for τ, q, and f in (5.4.22) and (5.4.23) with subsequent omission of pairwise products of the functions τ_1, q_1, and f_1 as quantities whose order of smallness is higher than that of q_1. Thus, from (5.4.22) we obtain the equality (linear approximation)

$$\langle \nabla_x \tau_1(x, x^0), \nabla_x \tau_0(x, x^0) \rangle = q_1(x) q_0(x), \quad x^0 \in S, \ x \in D, \tag{5.4.24}$$

and (5.4.23) implies

$$\tau_1(x, x^0) = f_1(x, x^0), \quad x, x^0 \in S. \tag{5.4.25}$$

Divide both sides of (5.4.24) by $q_0(x)$ and integrate along the ray $\Gamma_0(x, x^0)$ constructed from the function $q_0(x)$. Since the unit vector $\nabla_x \tau_0(x, x^0)/q_0(x)$ is tangent to the ray $\Gamma_0(x, x^0)$ at the point x, we have the following integral equation for the unknown function $q_1(x)$:

$$\int_{\Gamma_0(x,x^0)} q_1(y)\,|dy| = \tau_1(x, x^0), \quad x^0 \in S, \quad x \in D, \tag{5.4.26}$$

where $|dy|$ is the Euclidean length element of the curve $\Gamma_0(x, x^0)$:

$$|dy| = \Big(\sum_{i=1}^{3} dy_i^2\Big)^{1/2}.$$

Since the functions $\Gamma_0(x, x^0)$ are known, we obtain a problem of integral geometry: reconstruct $q_1(x)$, $x \in D$, from its integrals (5.4.26) along the given family of curves $\{\Gamma_0(x, x^0)\}$ for all $x, x^0 \in S$.

We first consider the case where D is a ball of radius r_1 centered at zero,

$$D = B(\mathbf{0}, r_1) = \{x \in \mathbb{R}^3 : |x| \le r_1\},$$

and $q_0(x) = q_0(r)$, $r = |x|$. It can be shown (Romanov, 1984) that in this case it suffices to consider the "equatorial" section of the ball produced by the plane $x_3 = 0$, setting $x = (x_1, x_2)$, and specify the function $\tau(x, x^0)$ for the points x and x^0 on the boundary of this section.

Theorem 5.4.1. *Let $q_0(r)$ belong to $C^2[r_0, r_1]$, $r_0 \in (0, r_1)$, and satisfy the conditions*

$$q_0(r) > 0, \quad q_0(r) + r q_0'(r) > 0, \quad r \in [r_0, r_1].$$

Then any continuous function $q_1(x_1, x_2)$ in the domain $r_0 \le \sqrt{x_1^2 + x_2^2} \le r_1$ is uniquely determined from the function $\tau_1(x, x^0)$ specified for the points x and x^0 on the boundary $r = r_1$ such that the rays $\Gamma_0(x, x^0)$ lie within this domain.

We now consider the case where $D \subset \mathbb{R}^3$ is a simply connected bounded domain with boundary S defined by the equation $F(x) = 0$. Let the function F satisfy the conditions

$$F \in C^3(\overline{D}), \quad F(x) < 0 \text{ for } x \in D, \quad |\nabla F|_S = 1, \tag{5.4.27}$$

and let the surface S be convex with respect to the rays $\Gamma_0(x, x^0)$, i.e., any two points $x, x^0 \in S$ can be connected by a ray that lies within D, and none of the rays $\Gamma_0(x, x^0)$ $(x^0 \neq x;\, x^0, x \in S)$ meet the boundary S at any point.

Assume also that the family of rays $\Gamma_0(x, x^0)$ is generated by the function $q_0(x) \geq \lambda_0 > 0$ of class $C^2(\overline{D})$ and is regular in $\overline{D}$ in the following sense. As shown above (see Lemma 5.4.2), the equation of a ray passing through any point $x^0 \in \overline{D}$ in the direction of the unit vector ν^0 can be written as $x = \varphi(\xi, x^0)$, $\xi = \nu^0 t / q_0(x^0)$, where $\varphi \in C^1(\overline{D} \times \overline{D})$, $\varphi(\mathbf{0}, x^0) = x^0$, $(\partial\varphi/\partial\xi)|_{\xi=\mathbf{0}} = I$, and I is the unit matrix. The regularity condition means that there exists a single-valued inverse function $\xi = g(x, x^0)$ in the space $C^1(\overline{D} \times \overline{D})$. In particular, from the regularity condition it follows that the determinant $|\partial\varphi/\partial\xi|$ never vanishes, and is therefore positive.

Note that, given the function $F(x)$ together with $\tau_1(x, x^0)$ and $\tau_0(x, x^0)$ for $x^0, x \in S$, it is possible to determine the function

$$\Phi(\tau_1, \tau_0) = \begin{vmatrix} 0 & 0 & \tau_{1x_1} & \tau_{1x_2} & \tau_{1x_3} \\ 0 & 0 & F_{x_1}(x) & F_{x_2}(x) & F_{x_3}(x) \\ \tau_{1x_1^0} & F_{x_1^0}(x^0) & \tau_{0x_1x_1^0} & \tau_{0x_2x_1^0} & \tau_{0x_3x_1^0} \\ \tau_{1x_2^0} & F_{x_2^0}(x^0) & \tau_{0x_1x_2^0} & \tau_{0x_2x_2^0} & \tau_{0x_3x_2^0} \\ \tau_{1x_3^0} & F_{x_3^0}(x^0) & \tau_{0x_1x_3^0} & \tau_{0x_2x_3^0} & \tau_{0x_3x_3^0} \end{vmatrix}.$$

Theorem 5.4.2. *Assume that the functions $F(x)$, $q_0(x)$, and therefore $\Gamma_0(x, x^0)$ and $\tau_0(x, x^0)$ satisfy the conditions given above. If $q_1(y) \in C^1(\overline{D})$, then the following stability estimate holds for the problem* (5.4.26), (5.4.25)*:*

$$\int_D q_1^2(x)\, dx \leq \frac{1}{8\pi\lambda_0} \int_S \int_S \Phi(\tau_1, \tau_0)\, dS_x\, dS_{x^0}. \tag{5.4.28}$$

Complete formulation of the inverse kinematic problem of seismology. Consider the general case, where $q(x)$ as well as the family of rays $\Gamma(x, x^0)$ generated by $q(x)$ are the unknowns in

$$\int_{\Gamma(x,x^0)} q(y)\, |dy| = \tau(x, x^0), \quad x^0 \in S,\ x \in D, \tag{5.4.29}$$

The function $\tau(x, x^0)$ is specified for all $x, x^0 \in S$.

The function $q(x)$ is said to belong to the class $\Lambda(c_1, c_2)$, $0 < c_1 < c_2 < \infty$, if it satisfies the conditions

1) $c_1 \le q(x)$, $x \in D$;
2) $q \in C^2(D)$, $\|q\|_{C^2(D)} \le c_2$;
3) the family of rays $\Gamma(x, x^0)$ associated with $q(x)$ is regular in D.

Given the function $\tau(x, x^0)$ on $S \times S$, using the eikonal equation, one can calculate the partial derivatives τ_{x_j}, $\tau_{x_j^0}$, $\tau_{x_j x_i^0}$ on $S \times S$ for any i, j. Indeed, the derivatives along the tangent directions can be found directly from $\tau(x, x^0)$, while the derivatives along the normals to $S \times S$ can be constructed using the equations

$$|\nabla_x \tau(x, x^0)| = q(x), \quad |\nabla_x \tau(x^0, x)| = q(x^0).$$

Set

$$\|\tau\|^2 = \int_S \int_S \Big[|\nabla_x \tau|^2 + |\nabla_{x^0} \tau|^2 + |x - x^0|^2 \sum_{i,j=1}^{3} \tau^2_{x_i x_j^0}\Big] \frac{1}{|x - x^0|}\, dS_x\, dS_{x^0}.$$

Theorem 5.4.3. *For any functions $q_1, q_2 \in \Lambda(c_1, c_2)$ and the corresponding functions τ_1, τ_2, we have*

$$\|q_1 - q_2\|^2_{L_2(D)} \le C\,\|\tau_1 - \tau_2\|^2, \tag{5.4.30}$$

where the constant C depends only on the constants c_1 and c_2.

The proofs of Theorems 5.4.1–5.4.3 can be found in (Romanov, 1984).

Chapter 6
Inverse spectral and scattering problems

A little more than 100 years have passed since Max Planck introduced the concept of a quantum. About 75 years ago, Schrödinger formulated his famous equation, which enables one to determine the wave function, and therefore all the properties of the object under study, from the known potential (the direct problem). Equations for solving inverse spectral problems were obtained in the 1950s by I. M. Gelfand, B. M. Levitan, M. G. Krein, and V. A. Marchenko (Gelfand and Levitan, 1955; Krein, 1951; Marchenko, 1952). This glorious achievement of the Russian mathematics gave an all-permeating "mathematical" vision (Zakhariev and Chabanov, 2007). An eye equipped with a microscope is not enough to study atoms, nuclei, and particles. However, powerful computers and methods of the theory of inverse problems made it possible to expand the "quantum" horizon.

Practical importance of inverse spectral and scattering problems cannot be overestimated, since everything we know about the world on a microscopic level (molecules, atoms, microscopic particles) comes from experiments that deal with scattering. The same is true for many problems emerging in medicine, geophysics, and chemistry. On the other hand, our knowledge about macroworld, information about other planets, stars, and galaxies, is also based mostly on the scattering data, namely, on spectral analysis.

This chapter deals with the inverse problems that were the subject of the pioneering studies in this direction. Inverse spectral and scattering problems are so deeply interrelated that any results obtained for the former can usually be extended to the latter, and vice versa. In this respect, the works of Yu. M. Berezanskii are the most significant (Berezanskii, 1953, 1958). In addition to the proof of one of the first uniqueness theorems for a multidimensional inverse spectral problem, these works also present an approach to solving some multidimensional inverse scattering problems based on their relationship with inverse spectral problems.

One chapter is certainly not enough to give an adequate description of the diverse set of ideas, methods, and applications of inverse spectral and scattering problems. We will confine ourselves to one-dimensional spectral problems and a brief description of the Gelfand–Levitan method (Sections 6.1–6.3), a one-dimensional version of the inverse scattering problem (Section 6.4), and inverse scattering problems in the time domain (Section 6.5), focusing on the connections between all the mentioned problems.

Inverse problems in spectral analysis consist in determining operators from certain characteristics of their spectrum (see Section 6.2). The most complete results in the spectral theory of differential operators were obtained for the Sturm–Liouville differential operator. The main results in the field of inverse spectral problems obtained in the second half of the 20th century paved the way for the development of methods for solving some nonlinear evolution equations and for the study of inverse problems of reconstructing an operator from incomplete spectral information and an additional a priori information about the operator. These results are also useful in solving inverse problems with operators of arbitrary order and of other types (for example, see (Yurko, 2007)).

In Sections 6.1–6.3 we consider the spectral theory of Sturm–Liouville operators on a finite interval. The proofs and the main methods for solving inverse one-dimensional spectral problems (the transformation operator method, the method of spectral mappings, the standard models method, and the Borg method) are presented in (Yurko, 2007).

In Section 6.4 we give several formulations of the inverse scattering problem and discuss in detail the one-dimensional case where the potential $q(x)$ depends only on the norm of x: $q(x) = q(|x|)$. In this case the scattering problem is reduced to the Sturm–Liouville problem on a half-axis:

$$-y'' + q(x)y = \lambda y, \quad x \in \mathbb{R}_+, \tag{6.1}$$

$$y(0) = 0. \tag{6.2}$$

Problem (6.1), (6.2) is studied in the case of rapidly decreasing potential $q(x)$, i.e., such that

$$\int_0^\infty x|q(x)|\,dx < \infty. \tag{6.3}$$

In this section we introduce the notions of the Jost solution $e(x, \sqrt{\lambda}\,)$, Jost function $j(\sqrt{\lambda}\,)$, spectral function $\rho(\lambda)$ and define scattering data. We also formulate the main integral equation of the inverse problem of the quantum scattering theory and give a formula for the eigenfunction expansion of the operator of problem (6.1)–(6.3).

In Section 6.5 we consider inverse scattering problems in the time domain. Using Fourier transform, these problems can be reduced to the Sturm–Liouville problem on a half-axis, namely,

$$-y'' + q(x)y = \lambda y, \quad x \in \mathbb{R}_+, \tag{6.4}$$

$$(y' - hy)\big|_{x=0} = 0, \tag{6.5}$$

where $q(x)$ is a continuous finite function. The data of inverse scattering problems are shown to be interrelated by means of the Jost function $j(\sqrt{\lambda}\,)$ of problem (6.4), (6.5). On the other hand, we notice that there exists a one-to-one correspondence between the Jost function ($j(\sqrt{\lambda}\,)$ and the spectral function $\rho(\lambda)$ of problem (6.4), (6.5). Thus the considered inverse problems are shown to be equivalent.

As to the current state of the theory and applications of inverse spectral and scattering problems, it will take a special multi-volume publication to describe it.

6.1 Direct Sturm–Liouville problem on a finite interval

A differential expression defined on a set of functions that satisfy some boundary value conditions generates a differential operator. The differential operator generated by the differential expression

$$l_q y(x) := -y''(x) + q(x)y(x),$$

is called the *Sturm–Liouville operator*. Henceforth the prime denotes the derivative with respect to x. The function $q(x)$ is said to be the *potential* of the operator l_q.

A value of the parameter λ such that the operator equation

$$l_q y(x) = \lambda y(x)$$

has a nontrivial solution is called an *eigenvalue* of the operator l_q, and the corresponding solution is said to be its *eigenfunction*. The set of all eigenvalues of an operator is called its *spectrum*.

The direct Sturm–Liouville problem is the problem of determining the spectrum of the Sturm–Liouville operator. We will consider the Sturm–Liouville problem for operators defined on the set $D(l_q)$ of functions $y(x) \in W_2^2(0,\pi)$ that satisfy the boundary conditions

$$U(y) := y'(0) - hy(0) = 0, \quad V(y) := y'(\pi) + Hy(\pi) = 0,$$

where $h, H \in \mathbb{R}$. Assume that $q(x) \in L_2(0,\pi)$. The definition of the spaces $W_2^2(0,\pi) = H^2(0,\pi)$ and $L_2(0,\pi)$ can be found in Appendix A (Subsection A.1.5).

Remark 6.1.1. The boundary value problem

$$\begin{aligned} &-y'' + q(x)y = \lambda y, \quad x \in (0,\pi), \\ &y'(0) - hy(0) = 0, \quad y'(\pi) + Hy(\pi) = 0, \end{aligned}$$

is sometimes called the classical Sturm–Liouville problem (with split boundary conditions).

In this section we describe the basic properties of eigenvalues and eigenfunctions of the Sturm–Liouville operator on a finite interval.

Spectral properties. To study the properties of the Sturm–Liouville operator, we introduce the notion of a characteristic function. Let $\varphi(x,\lambda)$ and $\psi(x,\lambda)$ be solutions to the equation

$$-y'' + q(x)y = \lambda y, \quad 0 < x < \pi, \tag{6.1.1}$$

that satisfy the boundary conditions

$$\begin{aligned} \varphi(0,\lambda) &= 1, \quad \varphi'(0,\lambda) = h, \\ \psi(\pi,\lambda) &= 1, \quad \psi'(\pi,\lambda) = -H. \end{aligned} \tag{6.1.2}$$

For any fixed x, the functions $\varphi(x,\lambda)$ and $\psi(x,\lambda)$ are entire analytic functions in λ. It is easy to verify that

$$\begin{aligned} U(\varphi) &:= \varphi'(0,\lambda) - h\varphi(0,\lambda) = 0, \\ V(\psi) &:= \psi'(\pi,\lambda) + H\psi(\pi,\lambda) = 0. \end{aligned}$$

Note that any solutions $\psi(x,\lambda)$ and $\varphi(x,\hat{\lambda})$ to equation (6.1.1) satisfy the equality

$$\varphi''(x,\hat{\lambda})\psi(x,\lambda) - \psi''(x,\lambda)\varphi(x,\hat{\lambda}) = (\lambda - \hat{\lambda})\psi(x,\lambda)\varphi(x,\hat{\lambda}),$$

which implies that

$$\frac{d}{dx}\left(\psi(x,\lambda)\varphi'(x,\hat{\lambda}) - \psi'(x,\lambda)\varphi(x,\hat{\lambda})\right) = (\lambda - \hat{\lambda})\psi(x,\lambda)\varphi(x,\hat{\lambda}). \tag{6.1.3}$$

Therefore, for $\lambda = \hat{\lambda}$ the Wronskian $\psi(x,\lambda)\varphi'(x,\lambda) - \psi'(x,\lambda)\varphi(x,\lambda)$ does not depend on x.

Put

$$\Delta(\lambda) = \psi(x,\lambda)\varphi'(x,\lambda) - \psi'(x,\lambda)\varphi(x,\lambda). \tag{6.1.4}$$

The function $\Delta(\lambda)$ is called the *characteristic function* of the operator l_q. Substituting $x = 0$ and $x = \pi$ into (6.1.4), we obtain

$$\Delta(\lambda) = V(\varphi) = -U(\psi). \tag{6.1.5}$$

The function $\Delta(\lambda)$ is an entire function in λ with at most countable set of zeros $\{\lambda_n\}$, $n = 0, 1, \dots$. From (6.1.5) it follows that the functions $\varphi(x,\lambda_n)$ and $\psi(x,\lambda_n)$ are solutions to the equation $l_q y(x) = \lambda_n y(x)$, i.e., eigenfunctions of the operator l_q. Thus, the zeros $\{\lambda_n\}$ of the characteristic function (6.1.4) coincide with the eigenvalues of the operator l_q. The eigenfunctions $\varphi(x,\lambda_n)$ and $\psi(x,\lambda_n)$ are proportional to each other, i.e., there is a sequence of numbers $\{\beta_n\}$ such that

$$\psi(x,\lambda_n) = \beta_n\varphi(x,\lambda_n), \quad \beta_n \neq 0. \tag{6.1.6}$$

Substituting $x = \pi$ into (6.1.6), we have $\beta_n = 1/\varphi(\pi,\lambda_n)$.

Each eigenvalue is associated with a unique eigenfunction (up to a constant factor).

The numbers

$$\alpha_n := \int_0^\pi \varphi^2(x,\lambda_n)\,dx \tag{6.1.7}$$

are called *weights*, and the collection $\{\lambda_n, \alpha_n\}$ is called the *spectral data* of the operator l_q.

The weights α_n and the numbers β_n occurring in the formulas (6.1.6) satisfy the relations

$$\beta_n \alpha_n = -\dot{\Delta}(\lambda_n), \tag{6.1.8}$$

where $\dot{\Delta}(\lambda) = d\Delta(\lambda)/d\lambda$. Indeed, (6.1.3) and (6.1.5) imply

$$\begin{aligned}(\lambda - \lambda_n) \int_0^\pi & \psi(x,\lambda)\varphi(x,\lambda_n)\,dx \\ &= \big(\psi(x,\lambda)\varphi'(x,\lambda_n) - \psi'(x,\lambda)\varphi(x,\lambda_n)\big)\big|_0^\pi = -\Delta(\lambda).\end{aligned}$$

For $\lambda \to \lambda_n$, this yields

$$\int_0^\pi \psi(x,\lambda_n)\varphi(x,\lambda_n)\,dx = -\dot{\Delta}(\lambda_n).$$

Using (6.1.6) and (6.1.7), we obtain (6.1.8). From (6.1.8) it is clear that all zeros of $\Delta(\lambda)$ are simple, i.e., $\dot{\Delta}(\lambda_n) \neq 0$.

The eigenvalues $\{\lambda_n\}_{n\geq 0}$ of the Sturm–Liouville operator l_q are real, and their condensation point is $+\infty$. Note that in the general case the parameter λ in equation (6.1.1), which is called the *spectral parameter*, belongs to the space of complex numbers $\mathbb{C}$. Since the eigenvalues of the Sturm–Liouville operator on a finite interval are real, in Sections 6.1–6.3 we will assume that $\lambda \in \mathbb{R}$.

For large n, the following asymptotic formulas hold for λ_n and the corresponding eigenfunctions $\varphi(x,\lambda_n)$:

$$\sqrt{\lambda_n} = n + \frac{\omega}{\pi n} + \frac{\eta_n}{n}, \quad \{\eta_n\} \in l_2, \tag{6.1.9}$$

$$\varphi(x,\lambda_n) = \cos\sqrt{\lambda_n}\,x + \frac{\xi_n(x)}{n}, \quad |\xi_n(x)| \leq C, \tag{6.1.10}$$

where

$$\omega = h + H + \frac{1}{2}\int_0^\pi q(t)\,dt.$$

Here and in what follows, the symbol $\{\eta_n\}$ stands for different sequences in l_2, while the symbol C stands for different positive constants that do not depend on x, λ, and η.

If the spectrum $\{\lambda_n\}_{n\geq 0}$ is known, the characteristic function $\Delta(\lambda)$ can be uniquely determined by the formula

$$\Delta(\lambda) = \pi(\lambda_0 - \lambda)\prod_{n=1}^{\infty}\frac{\lambda_n - \lambda}{n^2}. \tag{6.1.11}$$

Remark 6.1.2. Similar results can be obtained for Sturm–Liouville operators with other kinds of split boundary conditions. For example, consider the operator $l_q^1 y(x) = -y''(x) + q(x)y(x)$ with the domain of definition $D(l_q^1) = \{y \in W_2^2(0,\pi)\colon U(y) = 0,\ y(\pi) = 0\}$. All eigenvalues $\{\mu_n\}_{n\geq 0}$ of the operator l_q^1 are simple, and they coincide with the zeros of the characteristic function $d(\lambda) := \varphi(\pi,\lambda)$. Furthermore,

$$d(\lambda) = \prod_{n=0}^{\infty} \frac{\mu_n - \lambda}{(n+1/2)^2}. \tag{6.1.12}$$

The following asymptotic formulas hold for the spectral data $\{\mu_n, \alpha_n^1\}_{n\geq 0}$, $\alpha_n^1 := \int_0^\pi \varphi^2(x,\mu_n)\,dx$, of the operator l_q^1:

$$\sqrt{\mu_n} = n + \frac{1}{2} + \frac{\omega^1}{\pi n} + \frac{\eta_n}{n}, \quad \{\eta_n\} \in l_2, \tag{6.1.13}$$

$$\alpha_n^1 = \frac{\pi}{2} + \frac{\eta_n^1}{n}, \quad \{\eta_n^1\} \in l_2, \tag{6.1.14}$$

where

$$\omega^1 = h + \frac{1}{2}\int_0^\pi q(t)\,dt.$$

The spectra of the operators l_q and l_q^1 satisfy the relations

$$\lambda_n < \mu_n < \lambda_{n+1}, \quad n \geq 0, \tag{6.1.15}$$

which means that the eigenvalues of the two operators l_q and l_q^1 alternate.

Properties of the eigenfunctions. Consider the eigenfunctions $\varphi(x,\lambda_n)$ of the Sturm–Liouville operator l_q on the finite interval $(0,\pi)$ associated with the eigenvalues λ_n and normalized by the conditions $\varphi(0,\lambda_n) = 1$, $\varphi'(0,\lambda_n) = h$. The eigenfunctions of the operator l_q are real. Eigenfunctions associated with different eigenvalues are orthogonal in $L_2(0,\pi)$:

$$\int_0^\pi \varphi(x,\lambda_n)\varphi(x,\lambda_m)\,dx = 0, \quad \lambda_m \neq \lambda_n.$$

The set of eigenfunctions $\{\varphi(x,\lambda_n)\}_{n\geq 0}$ of the Sturm–Liouville operator turns out to be not only orthogonal, but also complete, i.e., it constitutes an orthogonal basis in $L_2(0,\pi)$. The completeness theorem for the set of eigenfunctions of the Sturm–Liouville operator on a finite interval was proved by V. A. Steklov at the end of the 19th century.

Theorem 6.1.1. 1. *The set of eigenfunctions* $\{\varphi(x,\lambda_n)\}_{n\geq 0}$ *of the operator* l_q *is complete in* $L_2(0,\pi)$.

2. *Let* $f(x)$, $x \in [0,\pi]$, *be a completely continuous function. Then*

$$f(x) = \sum_{n=0}^{\infty} f_n \varphi(x,\lambda_n), \quad f_n = \frac{1}{\alpha_n} \int_0^{\pi} f(t)\varphi(t,\lambda_n)\,dt, \tag{6.1.16}$$

where the series converges uniformly on $[0,\pi]$.

3. *For* $f(x) \in L_2(0,\pi)$ *the series* (6.1.16) *converges in* $L_2(0,\pi)$, *and Parseval's identity holds:*

$$\int_0^{\pi} |f(x)|^2\,dx = \sum_{n=0}^{\infty} \alpha_n |f_n|^2. \tag{6.1.17}$$

The proof of Theorem 6.1.1 was given in (Yurko, 2007).

It should be noted that completeness and expansion theorems play an important role in solving problems of mathematical physics using the method of separation of variables (for example, see Section 5.3).

Consider the Sturm–Liouville operator l_q with $h = H = 0$ and $q(x) \equiv 0$. In this case $\varphi(x,\lambda_n) = \cos nx$ for $x \in (0,\pi)$, and therefore the nth eigenfunction has exactly n zeros in the interval $(0,\pi)$. It turns out that this property of eigenfunctions also holds in the general case. This is expressed by the following theorem (the Sturm oscillation theorem).

Theorem 6.1.2. *The eigenfunction* $\varphi(x,\lambda_n)$ *of the operator* l_q *has exactly* n *zeros in the interval* $0 < x < \pi$.

Before proving this theorem, we prove a few auxiliary statements.

Lemma 6.1.1. *Let* $u_j(x)$ *be solutions to the equations*

$$u'' + g_j(x)u = 0, \quad j = 1,2,\ x \in [a,b], \tag{6.1.18}$$

where $g_1(x) < g_2(x)$. *Assume that there are* $x_1, x_2 \in [a,b]$ *such that* $u_1(x_1) = u_1(x_2) = 0$ *and* $u_1(x) \neq 0$ *for* $x \in (x_1,x_2)$. *Then there exists an* $x^* \in (x_1,x_2)$ *such that* $u_2(x^*) = 0$. *In other words,* $u_2(x)$ *has at least one zero between any two zeros of* $u_1(x)$.

Proof. Assume the contrary: $u_2(x) \neq 0$ for all $x \in (x_1,x_2)$. Without loss of generality, we can assume that $u_j(x) > 0$ for $x \in (x_1,x_2)$, $j = 1,2$. From (6.1.18) it follows that

$$\frac{d}{dx}(u_1' u_2 - u_2' u_1) = (g_2 - g_1)u_1 u_2.$$

Integrating this equality from x_1 to x_2 yields

$$\int_{x_1}^{x_2} (g_2 - g_1)u_1 u_2 \, dx = (u_1' u_2 - u_2' u_1)\big|_{x_1}^{x_2}$$
$$= u_1'(x_2)u_2(x_2) - u_1'(x_1)u_2(x_1). \tag{6.1.19}$$

The integral in (6.1.19) is strictly positive. On the other hand, since $u_1'(x_1) > 0$, $u_1'(x_2) < 0$, and $u_2(x) \geq 0$ for $x \in [x_1, x_2]$, the right-hand side of (6.1.19) is nonpositive. The obtained contradiction proves the lemma. □

Consider the function $\varphi(x, \lambda)$ for real λ. The zeros of $\varphi(x, \lambda)$ with respect to x are functions of λ. We will show that they are continuous functions of λ.

Lemma 6.1.2. *Let $\varphi(x_0, \hat{\lambda}) = 0$. Then there exists an $\varepsilon_0 > 0$ such that for all $0 < \varepsilon < \varepsilon_0$ there is a $\delta > 0$ such that $\varphi(x, \lambda)$ has a unique zero in the interval $|x - x_0| < \varepsilon$ if $|\lambda - \hat{\lambda}| < \delta$.*

Proof. The zero x_0 of the solution $\varphi(x, \hat{\lambda})$ to the differential equation (6.1.2) is simple. Indeed, if $\varphi'(x_0, \hat{\lambda}) = 0$, then the theorem on the uniqueness of solutions to the Cauchy problem would imply the identity $\varphi(x, \hat{\lambda}) \equiv 0$. Therefore, $\varphi'(x_0, \hat{\lambda}) \neq 0$. Suppose for definiteness that $\varphi'(x_0, \hat{\lambda}) > 0$. Choose $\varepsilon_0 > 0$ so that $\varphi'(x, \hat{\lambda}) > 0$ for $|x - x_0| \leq \varepsilon_0$. Then $\varphi(x, \hat{\lambda}) < 0$ for $x \in [x_0 - \varepsilon_0, x_0)$ and $\varphi(x, \hat{\lambda}) > 0$ for $x \in (x_0, x_0 + \varepsilon_0]$. Take $\varepsilon \leq \varepsilon_0$. Since $\varphi(x, \lambda)$ and $\varphi'(x, \lambda)$ are continuous, there exists a $\delta > 0$ such that for $|\lambda - \hat{\lambda}| < \delta$ and $|x - x_0| < \varepsilon$ we have $\varphi'(x, \lambda) > 0$, $\varphi(x_0 - \varepsilon_0, \lambda) < 0$, $\varphi(x_0 + \varepsilon_0, \lambda) > 0$. Hence, the function $\varphi(x, \lambda)$ has a unique zero in the interval $|x - x_0| < \varepsilon$, which is the desired conclusion. □

Lemma 6.1.3. *Suppose that for some real $\hat{\lambda}$ the function $\varphi(x, \hat{\lambda})$ has m zeros in the interval $0 < x \leq a$. Let $\lambda > \hat{\lambda}$. Then the function $\varphi(x, \lambda)$ has at least m zeros on the same interval, and the kth zero of $\varphi(x, \lambda)$ is smaller than the kth zero of $\varphi(x, \hat{\lambda})$.*

Proof. Let $x_1 > 0$ be the smallest positive zero of $\varphi(x, \hat{\lambda})$. By Lemma 6.1.1, it suffices to show that $\varphi(x, \lambda)$ has at least one zero in the interval $0 < x < x_1$.

Assume the contrary, i.e., $\varphi(x, \lambda) \neq 0$ for all $x \in (0, x_1)$. Since $\varphi(0, \lambda) = 1$, we have $\varphi(x, \lambda) > 0$ for $x \in [0, x_1)$. By the choice of x_1, $\varphi(x, \hat{\lambda}) > 0$ for $x \in [0, x_1)$, $\varphi(x_1, \hat{\lambda}) = 0$, and $\varphi'(x_1, \hat{\lambda}) < 0$. From (6.1.3) it follows that

$$(\lambda - \hat{\lambda}) \int_0^{x_1} \varphi(x, \lambda)\varphi(x, \hat{\lambda}) \, dx = \big(\varphi(x, \lambda)\varphi'(x, \hat{\lambda}) - \varphi'(x, \lambda)\varphi(x, \hat{\lambda})\big)\big|_0^{x_1}$$
$$= \varphi(x_1, \lambda)\varphi'(x_1, \hat{\lambda}) \leq 0.$$

But the integral on the left-hand side is strictly positive. The resulting contradiction proves the lemma. □

Proof of Theorem 6.1.2. Consider the function $\varphi(x,\lambda)$ for real λ. From (6.1.10) it follows that $\varphi(x,\lambda)$ has no zeros for negative λ with sufficiently large absolute values: $\varphi(x,\lambda) > 0$, $\lambda \le -\lambda^* < 0$, $x \in [0,\pi]$. On the other hand, $\varphi(\pi,\mu_n) = 0$, where $\{\mu_n\}_{n\ge 0}$ are eigenvalues of the operator l_q^1.

Applying Lemmas 6.1.2 and 6.1.3, we see that with λ moving from $-\infty$ to ∞, the zeros of $\varphi(x,\lambda)$ on the interval $[0,\pi]$ continuously move to the left. New zeros can only appear in transition through the point $x=\pi$. This means that

1) $\varphi(x,\mu_n)$ has exactly n zeros in the interval $x \in [0,\pi)$;
2) if $\lambda \in (\mu_{n-1},\mu_n)$, $n \ge 0$, $\mu_{-1} := -\infty$, then $\varphi(x,\lambda)$ has exactly n zeros in the interval $x \in [0,\pi]$.

By (6.1.15), we have

$$\lambda_0 < \mu_0 < \lambda_1 < \mu_1 < \lambda_2 < \mu_2 < \cdots$$

and therefore $\varphi(x,\lambda_n)$ has exactly n zeros in the interval $(0,\pi)$. □

Transformation operators. The transformation operators that associate the eigenfunctions of two different Sturm–Liouville operators for all λ were first introduced in the theory of generalized shift operators by Delsarte and Levitan (Levitan, 1973) In the theory of inverse problems, the transformation operators were used by I. M. Gelfand, B. M. Levitan, and V. A. Marchenko (see Section 6.3 and the monographs by (Marchenko, 1986) and (Levitan, 1987)).

Let functions $C(x,\lambda) \in W_2^2(0,\pi)$ and $S(x,\lambda) \in W_2^2(0,\pi)$ be solutions to the differential equation (6.1.1) with initial conditions

$$C(0,\lambda) = 1, \quad C'(0,\lambda) = 0, \quad S(0,\lambda) = 0, \quad S'(0,\lambda) = 1.$$

The function $C(x,\lambda)$ can be represented as

$$C(x,\lambda) = \cos\sqrt{\lambda}x + \int_0^x K(x,t)\cos\sqrt{\lambda}t\,dt, \tag{6.1.20}$$

where $K(x,t)$ is a real-valued continuous function such that

$$K(x,x) = \frac{1}{2}\int_0^x q(t)\,dt. \tag{6.1.21}$$

The proof of the representation (6.1.20) for $C(x,\lambda)$ can be found in (Yurko, 2007).

An operator T defined by the formula

$$Tf(x) = f(x) + \int_0^x K(x,t)f(t)\,dt,$$

maps the function $\cos\sqrt{\lambda}x$ (which is a solution to the equation $-y'' = \lambda y$ with zero potential) to the function $C(x,\lambda)$, which is a solution to equation (6.1.1) with some potential $q(x)$ (i.e., $C(x,\lambda) = T(\cos\sqrt{\lambda}x)$). The operator T is called the *transformation operator* for $C(x,\lambda)$. It is important to note that the kernel $K(x,t)$ does not depend on λ.

The transformation operators for the functions $S(x,\lambda)$ and $\varphi(x,\lambda)$ can be obtained in a similar way:

$$S(x,\lambda) = \frac{\sin\sqrt{\lambda}x}{\sqrt{\lambda}} + \int_0^x P(x,t)\frac{\sin\sqrt{\lambda}t}{\sqrt{\lambda}}\,dt, \tag{6.1.22}$$

$$\varphi(x,\lambda) = \cos\sqrt{\lambda}x + \int_0^x G(x,t)\cos\sqrt{\lambda}t\,dt, \tag{6.1.23}$$

where $P(x,t)$ and $G(x,t)$ are real-valued continuous functions that are as smooth as the function $\int_0^x q(t)\,dt$ and

$$G(x,x) = h + \frac{1}{2}\int_0^x q(t)\,dt, \tag{6.1.24}$$

$$P(x,x) = \frac{1}{2}\int_0^x q(t)\,dt. \tag{6.1.25}$$

A proof of the representations (6.1.22) and (6.1.23) is also given in (Yurko, 2007).

6.2 Inverse Sturm–Liouville problems on a finite interval

Inverse Sturm–Liouville problems consist in reconstructing the operator l_q (i.e., its potential $q(x) \in L_2(0,\pi)$ and the coefficients h and H in the boundary conditions) from its spectral characteristics.

V. A. Ambartsumyan (see Ambartsumyan, 1929) was the first to obtain results in the theory of inverse spectral problems.

Consider the equation $l_q y = \lambda y$ under the condition that $h = H = 0$, i.e.,

$$-y'' + q(x)y = \lambda y, \quad y'(0) = y'(\pi) = 0. \tag{6.2.1}$$

Clearly, if $q(x) = 0$ almost everywhere on $(0,\pi)$, then the eigenvalues of the problem (6.2.1) have the form $\lambda_n = n^2$, $n \geq 0$. The converse is also true.

Theorem 6.2.1 (Ambartsumyan's theorem). *If the eigenvalues of the problem* (6.2.1) *are* $\lambda_n = n^2$, $n \geq 0$, *then* $q(x) = 0$ *almost everywhere on* $(0,\pi)$.

Proof. From (6.1.9) it follows that $\omega = 0$, i.e., $\int_0^\pi q(x)\,dx = 0$. Let $y_0(x)$ be an eigenfunction associated with the smallest eigenvalue $\lambda_0 = 0$. Then

$$y_0''(x) - q(x)y_0(x) = 0, \quad y_0'(0) = y_0'(\pi) = 0.$$

By Theorem 6.1.2, the function $y_0(x)$ has no zeros in the interval $x \in [0, \pi]$. Taking into account the relation

$$q(x) = \frac{y_0''(x)}{y_0(x)} = \Big(\frac{y_0'(x)}{y_0(x)}\Big)^2 + \Big(\frac{y_0'(x)}{y_0(x)}\Big)',$$

we obtain

$$0 = \int_0^\pi q(x)\,dx = \int_0^\pi \frac{y_0''(x)}{y_0(x)}\,dx = \int_0^\pi \Big(\frac{y_0'(x)}{y_0(x)}\Big)^2\,dx.$$

Thus, $y_0'(x) \equiv 0$, i.e., $y_0(x) \equiv \text{const}$, $q(x) = 0$ almost everywhere on $(0, \pi)$, which completes the proof. □

In the general case, the operator l_q is not uniquely determined from its spectrum $\{\lambda_n\}$. Ambartsumyan's result is an exception.

The uniqueness of reconstruction of the differential operator l_q from its spectral data. We now consider the problem of reconstructing the operator l_q from its spectral data $\{\lambda_n, \alpha_n\}_{n\geq 0}$.

In addition to l_q, consider an operator $\hat{l}_q$ of the same form with potential $\hat{q}(x)$ and coefficients $\hat{h}$ and $\hat{H}$. For any symbol, for instance, γ that denotes an object related to the operator l_q, we will denote by $\hat{\gamma}$ the corresponding object related to the operator $\hat{l}_q$.

Theorem 6.2.2. *If $\lambda_n = \hat{\lambda}_n$ and $\alpha_n = \hat{\alpha}_n$ for all $n \geq 0$, then $l_q = \hat{l}_q$, i.e., $q(x) = \hat{q}(x)$ almost everywhere on $(0, \pi)$, $h = \hat{h}$ and $H = \hat{H}$.*

Thus, the spectral data $\{\lambda_n, \alpha_n\}_{n\geq 0}$ uniquely determine the potential and coefficients of the boundary conditions.

There are two ways to prove Theorem 6.2.2. We will present the proof given by Marchenko (1952) that involves the transformation operator and Parseval's identity (6.1.17). This method also works for the Sturm–Liouville operators on a half-axis and can be used to prove the theorem on the uniqueness of reconstruction of an operator from its spectral function. The second proof, which can be found in (Yurko, 2007), was first given by Levinson (1949b). It employs the contour integral method. N. Levinson was the first to apply the ideas of the contour integral method in the study of inverse spectral problems.

Proof. By (6.1.23), the eigenfunctions $\varphi(x, \lambda)$ and $\hat{\varphi}(x, \lambda)$ of the operators l_q and $\hat{l}_q$, respectively, can be represented in the form

$$\varphi(x, \lambda) = \cos\sqrt{\lambda}x + \int_0^x G(x, t)\cos\sqrt{\lambda}t\,dt,$$

$$\hat{\varphi}(x, \lambda) = \cos\sqrt{\lambda}x + \int_0^x \hat{G}(x, t)\cos\sqrt{\lambda}t\,dt.$$

In other words,

$$\varphi(x,\lambda) = (E+G)\cos\sqrt{\lambda}x, \quad \hat{\varphi}(x,\lambda) = (E+\hat{G})\cos\sqrt{\lambda}x,$$

where

$$(E+G)f(x) = f(x) + \int_0^x G(x,t)f(t)\,dt,$$
$$(E+\hat{G})f(x) = f(x) + \int_0^x \hat{G}(x,t)f(t)\,dt.$$

Solving $\hat{\varphi}(x,\lambda) = (E+\hat{G})\cos\sqrt{\lambda}x$ for $\cos\sqrt{\lambda}x$, we have

$$\cos\sqrt{\lambda}x = \hat{\varphi}(x,\lambda) + \int_0^x \hat{H}(x,t)\hat{\varphi}(t,\lambda)\,dt,$$

where $\hat{H}(x,t)$ is a continuous function that is the kernel of the inverse operator

$$(E+\hat{H}) = (E+\hat{G})^{-1}, \quad \hat{H}f(x) = \int_0^x \hat{H}(x,t)f(t)\,dt.$$

Consequently,

$$\varphi(x,\lambda) = \hat{\varphi}(x,\lambda) + \int_0^x Q(x,t)\hat{\varphi}(t,\lambda)\,dt, \tag{6.2.2}$$

where $Q(x,t)$ is a continuous real-valued function.

Take an arbitrary function $f(x) \in L_2(0,\pi)$. From (6.2.2) it follows that

$$\int_0^\pi f(x)\varphi(x,\lambda)\,dx = \int_0^\pi g(x)\hat{\varphi}(x,\lambda)\,dx,$$

where

$$g(x) := f(x) + \int_x^\pi Q(t,x)f(t)\,dt.$$

Consequently, for all $n \geq 0$ we have

$$f_n = \hat{g}_n,$$
$$f_n := \int_0^\pi f(x)\varphi(x,\lambda_n)\,dx, \quad \hat{g}_n := \int_0^\pi g(x)\hat{\varphi}(x,\lambda_n)\,dx.$$

Using Parseval's identity (6.1.17), we obtain

$$\int_0^\pi |f(x)|^2\,dx = \sum_{n=0}^\infty \frac{|f_n|^2}{\alpha_n} = \sum_{n=0}^\infty \frac{|\hat{g}_n|^2}{\alpha_n} = \sum_{n=0}^\infty \frac{|\hat{g}_n|^2}{\hat{\alpha}_n} = \int_0^\pi |g(x)|^2\,dx,$$

i.e.,

$$\|f\|_{L_2} = \|g\|_{L_2}. \tag{6.2.3}$$

Consider the operator

$$Bf(x) = f(x) + \int_x^\pi Q(t,x)f(t)\,dt$$

and note that $Bf = g$. In view of (6.2.3), $\|Bf\|_{L_2} = \|f\|_{L_2}$ for all $f(x) \in L_2(0,\pi)$. Therefore, $B^* = B^{-1}$, which is possible only if $Q(x,t) \equiv 0$. Hence, $\varphi(x,\lambda) \equiv \hat{\varphi}(x,\lambda)$, i.e., $q(x) = \hat{q}(x)$ almost everywhere on $(0,\pi)$, $h = \hat{h}$, and $H = \hat{H}$. □

In the symmetrical case with $q(x) = q(\pi - x)$ and $H = h$, knowing the spectrum $\{\lambda_n\}_{n\geq 0}$ is sufficient to determine the potential $q(x)$ and the coefficient h.

Theorem 6.2.3. *If $q(x) = q(\pi - x)$, $H = h$, $\hat{q}(x) = \hat{q}(\pi - x)$, $\hat{H} = \hat{h}$ and $\lambda_n = \hat{\lambda}_n$ for $n \geq 0$, then $q(x) = \hat{q}(x)$ almost everywhere on $(0,\pi)$ and $h = \hat{h}$.*

Proof. If $q(x) = q(\pi - x)$, $H = h$, and $y(x)$ is a solution to equation (6.1.1), then $y_1(x) := y(\pi - x)$ also satisfies (6.1.1). In particular, $\psi(x,\lambda) = \varphi(\pi - x,\lambda)$. Using (6.1.6), we obtain

$$\psi(x,\lambda_n) = \beta_n\varphi(x,\lambda_n) = \beta_n\psi(\pi - x,\lambda_n) = \beta_n^2\varphi(\pi - x,\lambda_n) = \beta_n^2\psi(x,\lambda_n).$$

Consequently, $\beta_n^2 = 1$.

On the other hand, (6.1.6) implies that $\beta_n\varphi(\pi,\lambda_n) = 1$. Applying Theorem 6.1.2, we conclude that

$$\beta_n = (-1)^n.$$

Then (6.1.8) implies that $\alpha_n = (-1)^{n+1}\dot{\Delta}(\lambda_n)$ and $\hat{\alpha}_n = (-1)^{n+1}\dot{\hat{\Delta}}(\lambda_n)$. By the theorem on the zeros of an entire function and the assumptions of Theorem 6.2.3, $\alpha_n = \hat{\alpha}_n$ for $n \geq 0$. It remains to apply Theorem 6.2.2 to obtain the desired result: $q(x) = \hat{q}(x)$ almost everywhere on $(0,\pi)$, and $h = \hat{h}$. □

6.3 The Gelfand–Levitan method on a finite interval

In the preceding section we discussed the uniqueness of reconstruction of a Sturm–Liouville operator from different sets of its spectral characteristics. An existence theorem formulating the necessary and sufficient conditions for the solvability of the inverse Sturm–Liouville problem and proposing a practical method for constructing the operator was proved by I. M. Gelfand and B. M. Levitan (Gelfand and Levitan, 1955). The proposed method for the reconstruction of a Sturm–Liouville operator from its spectral characteristics was named the *Gelfand–Levitan method.* A detailed explanation of the Gelfand–Levitan method is given in (Levitan, 1987), (Marchenko, 1986), and (Yurko, 2007). Our purpose here is to explain the method to the reader without going into too much technical detail.

Reconstruction of a differential operator from its spectral data. Consider the Sturm–Liouville operator on a finite interval

$$l_q y(x) = -y''(x) + q(x)y(x), \quad 0 < x < \pi, \tag{6.3.1}$$

$$D(l_q) = \{y(x) \in W_2^2(0,\pi) : \ y'(0) - hy(0) = 0, \ y'(\pi) + Hy(\pi) = 0\}, \tag{6.3.2}$$

and assume that the potential $q(x) \in L_2(0,\pi)$ and the coefficients h and H in the boundary conditions are unknown.

Let $\{\lambda_n, \alpha_n\}_{n\geq 0}$ be the spectral data of l_q. We will solve the inverse problem of reconstruction of l_q from the given spectral data $\{\lambda_n, \alpha_n\}_{n\geq 0}$.

The spectral data of the operator l_q have asymptotic properties (Yurko, 2007)

$$\sqrt{\lambda_n} = n + \frac{\omega}{\pi n} + \frac{\eta_n}{n}, \quad \alpha_n = \frac{\pi}{2} + \frac{\eta_n^1}{n}, \quad \{\eta_n\}, \{\eta_n^1\} \in l_2, \tag{6.3.3}$$

$$\alpha_n > 0, \quad \lambda_n \neq \lambda_m \quad (n \neq m). \tag{6.3.4}$$

More precisely,

$$\eta_n = \frac{1}{2\pi} \int_0^\pi q(t) \cos 2nt \, dt + O\Big(\frac{1}{n}\Big),$$

$$\eta_n^1 = -\frac{1}{2} \int_0^\pi (\pi - t) q(t) \sin 2nt \, dt + O\Big(\frac{1}{n}\Big),$$

$$\omega = h + H + \frac{1}{2} \int_0^\pi q(t) \, dt,$$

i.e., the dependence of the principal parts on the potential is linear.

To determine the potential $q(x)$ and the coefficients h and H, we will use the transformation operator for the eigenfunctions of the operator l_q

$$\varphi(x,\lambda) = \cos\sqrt{\lambda}x + \int_0^x G(x,t) \cos\sqrt{\lambda}t \, dt, \tag{6.3.5}$$

where $G(x,t)$ is a continuous real-valued function of the same smoothness as $\int_0^x q(t)dt$ that satisfies the following equality for $t = x$:

$$G(x,x) = h + \frac{1}{2} \int_0^x q(t) \, dt. \tag{6.3.6}$$

This equality means that finding the kernel $G(x,t)$ of the transformation operator will enable us to easily reconstruct the potential $q(x) = 2\frac{d}{dx} G(x,x)$ and the coefficients $h = G(+0,+0)$ and $H = \omega - G(\pi - 0, \pi - 0)$.

To determine the kernel $G(x,t)$, we introduce the function

$$F(x,t) = \sum_{n=0}^\infty \Big(\frac{\cos\sqrt{\lambda_n}\, x \cos\sqrt{\lambda_n}\, t}{\alpha_n} - \frac{\cos nx \cos nt}{\alpha_n^0} \Big), \tag{6.3.7}$$

where

$$\alpha_n^0 = \begin{cases} \pi/2, & n > 0, \\ \pi, & n = 0. \end{cases}$$

The series in the right-hand side of (6.3.7) converges, the function $F(x,t)$ is continuous, and $\frac{d}{dx} F(x,x) \in L_2(0,\pi)$ (Levitan, 1987)

Theorem 6.3.1. *For any fixed $x \in (0,\pi]$, the kernel $G(x,t)$ in the representation* (6.3.5) *satisfies the linear integral equation*

$$G(x,t) + F(x,t) + \int_0^x G(x,s)F(s,t)\,ds = 0, \quad 0 < t < x. \tag{6.3.8}$$

Equation (6.3.8) is called the *Gelfand–Levitan equation*. Thus, Theorem 6.3.1 makes it possible to reduce the inverse problem (6.3.1), (6.3.2) to the problem of solving the Gelfand–Levitan equation.

Proof. Solving the equality (6.3.5) for $\cos\sqrt{\lambda}x$, we obtain

$$\cos\sqrt{\lambda}x = \varphi(x,\lambda) + \int_0^x H(x,t)\varphi(t,\lambda)\,dt, \tag{6.3.9}$$

where $H(x,t)$ is a continuous function. Using the representation (6.3.5), we calculate

$$\sum_{n=0}^{N} \frac{\varphi(x,\lambda_n)\cos\sqrt{\lambda_n}\,t}{\alpha_n} = \sum_{n=0}^{N} \Big(\frac{\cos\sqrt{\lambda_n}\,x \cos\sqrt{\lambda_n}\,t}{\alpha_n} + \frac{\cos\sqrt{\lambda_n}\,t}{\alpha_n} \int_0^x G(x,s)\cos\sqrt{\lambda_n}\,s\,ds \Big).$$

On the other hand, (6.3.9) implies

$$\sum_{n=0}^{N} \frac{\varphi(x,\lambda_n)\cos\sqrt{\lambda_n}\,t}{\alpha_n} = \sum_{n=0}^{N} \Big(\frac{\varphi(x,\lambda_n)\varphi(t,\lambda_n)}{\alpha_n} + \frac{\varphi(x,\lambda_n)}{\alpha_n} \int_0^t H(t,s)\varphi(s,\lambda_n)\,ds \Big).$$

Equating the right-hand sides of the last two equalities and performing simple calculations, we arrive at the equality

$$\Phi_N(x,t) = I_{N_1}(x,t) + I_{N_2}(x,t) + I_{N_3}(x,t) + I_{N_4}(x,t),$$

where

$$\Phi_N(x,t) = \sum_{n=0}^{N} \Big(\frac{\varphi(x,\lambda_n)\varphi(t,\lambda_n)}{\alpha_n} - \frac{\cos nx \cos nt}{\alpha_n^0} \Big),$$

$$I_{N_1}(x,t) = \sum_{n=0}^{N} \Big(\frac{\cos\sqrt{\lambda_n}\, x \cos\sqrt{\lambda_n}\, t}{\alpha_n} - \frac{\cos nx \cos nt}{\alpha_n^0} \Big),$$

$$I_{N_2}(x,t) = \sum_{n=0}^{N} \frac{\cos nt}{\alpha_n^0} \int_0^x G(x,s) \cos ns\, ds,$$

$$I_{N_3}(x,t) = \sum_{n=0}^{N} \int_0^x G(x,s) \Big(\frac{\cos\sqrt{\lambda_n}\, t \cos\sqrt{\lambda_n}\, s}{\alpha_n} - \frac{\cos nt \cos ns}{\alpha_n^0} \Big)\, ds,$$

$$I_{N_4}(x,t) = -\sum_{n=0}^{N} \frac{\varphi(x,\lambda_n)}{\alpha_n} \int_0^t H(t,s)\varphi(s,\lambda_n)\, ds.$$

Let $f(x)$ be an arbitrary function that is absolutely continuous on the interval $[0,\pi]$. By Theorem 6.1.1,

$$\lim_{N\to\infty} \max_{0\le x\le\pi} \int_0^\pi f(t)\Phi_N(x,t)\, dt = 0.$$

Moreover, convergence in each of the following limits is uniform with respect to $x \in [0,\pi]$:

$$\lim_{N\to\infty} \int_0^\pi f(t) I_{N_1}(x,t)\, dt = \int_0^\pi f(t) F(x,t)\, dt,$$

$$\lim_{N\to\infty} \int_0^\pi f(t) I_{N_2}(x,t)\, dt = \int_0^x f(t) G(x,t)\, dt,$$

$$\lim_{N\to\infty} \int_0^\pi f(t) I_{N_3}(x,t)\, dt = \int_0^\pi f(t) \Big(\int_0^x G(x,s) F(s,t)\, ds \Big)\, dt,$$

$$\begin{aligned} \lim_{N\to\infty} &\int_0^\pi f(t) I_{N_4}(x,t)\, dt \\ &= -\lim_{N\to\infty} \sum_{n=0}^{N} \frac{\varphi(x,\lambda_n)}{\alpha_n} \int_0^\pi \varphi(s,\lambda_n) \Big(\int_s^\pi H(t,s) f(t)\, dt \Big)\, ds \\ &= -\int_x^\pi f(t) H(t,x)\, dt. \end{aligned}$$

Put $G(x,t) = H(x,t) = 0$ for $x < t$. Since $f(x)$ is arbitrary, we conclude that

$$G(x,t) + F(x,t) + \int_0^x G(x,s)F(s,t)\, ds - H(t,x) = 0.$$

Since $H(t,x) = 0$ for $t < x$, we arrive at the desired equality (6.3.8). □

Thus, we have shown that the kernel $G(x,t)$ of the transformation operator for the eigenfunctions $\varphi(x,\lambda)$ of the operator l_q is a solution to the Gelfand–Levitan equation (6.3.8). For any fixed $x \in (0,\pi]$, equation (6.3.8) has a unique solution $G(x,t)$ in $L_2(0,x)$. The function $G(x,t)$ is continuous and is as smooth as $F(x,t)$. In particular, $\frac{\partial}{\partial x} G(x,x) \in L_2(0,\pi)$.

We will now establish the uniqueness of solutions to the Gelfand–Levitan equation for fixed $x \in (0,\pi]$.

Note that equation (6.3.8) is a Fredholm equation of the second kind with parameter x. Therefore, to prove the uniqueness of its solutions, it suffices to verify that $g(t) \equiv 0$ is the only solution of the homogeneous equation

$$g(t) + \int_0^x F(s,t)g(s)\,ds = 0. \tag{6.3.10}$$

Let $g(t)$ be a solution to (6.3.10). Then

$$\int_0^x g^2(t)\,dt + \int_0^x \int_0^x F(s,t)g(s)g(t)\,ds\,dt = 0$$

or

$$\int_0^x g^2(t)\,dt + \sum_{n=0}^{\infty} \frac{1}{\alpha_n}\Big(\int_0^x g(t)\cos\sqrt{\lambda_n}\,t\,dt\Big)^2 - \sum_{n=0}^{\infty} \frac{1}{\alpha_n^0}\Big(\int_0^x g(t)\cos nt\,dt\Big)^2 = 0.$$

Applying Parseval's identity

$$\int_0^x g^2(t)\,dt = \sum_{n=0}^{\infty} \frac{1}{\alpha_n^0}\Big(\int_0^x g(t)\cos nt\,dt\Big)^2$$

to the function $g(t)$ extended by zero for $t > x$, we have

$$\sum_{n=0}^{\infty} \frac{1}{\alpha_n}\Big(\int_0^x g(t)\cos\sqrt{\lambda_n}\,t\,dt\Big)^2 = 0.$$

Since $\alpha_n > 0$, we arrive at the equality

$$\int_0^x g(t)\cos\sqrt{\lambda_n}\,t\,dt = 0, \quad n \geq 0.$$

The system of functions $\{\cos\sqrt{\lambda_n}\,t\}_{n\geq 0}$ is complete in $L_2(0,\pi)$; consequently, $g(t) \equiv 0$. Thus we have shown that equation (6.3.8) has a unique solution $G(x,t)$ in $L_2(0,x)$ for any fixed $x \in (0,\pi]$.

The following theorem, which is the focus of Section 6.3, provides an algorithm for solving the inverse problem as well as necessary and sufficient conditions for its solvability.

Theorem 6.3.2. *For real numbers* $\{\lambda_n, \alpha_n\}_{n\geq 0}$ *to be the spectral data of an operator* l_q *of the form* (6.3.1), (6.3.2) *with* $q(x) \in L_2(0,\pi)$, *it is necessary and sufficient that the conditions* (6.3.3), (6.3.4) *hold.*

The operator l_q *is constructed according to the following algorithm:*

1) *construct* $F(x,t)$ *from the given numbers* $\{\lambda_n, \alpha_n\}_{n\geq 0}$ *using formula* (6.3.7);
2) *determine* $G(x,t)$ *from equation* (6.3.8);
3) *calculate* $q(x)$, h, *and* H *using the formulas*

$$q(x) = 2\frac{d}{dx}G(x,x-0), \quad h = G(+0,+0), \tag{6.3.11}$$

$$H = \omega - h - \frac{1}{2}\int_0^\pi q(t)\,dt = \omega - G(\pi-0,\pi-0). \tag{6.3.12}$$

As noted before, a detailed explanation of this theorem, which is the basis for the Gelfand–Levitan method, is given in (Gelfand and Levitan, 1955), (Marchenko, 1986), (Levitan, 1987), and (Yurko, 2007).

Consider the Gelfand–Levitan method in the following simple example.

Example 6.3.1. Take $\lambda_n = n^2$ $(n \geq 0)$, $\alpha_n = \pi/2$ $(n \geq 1)$, and an arbitrary number $\alpha_0 > 0$ Set $a := 1/\alpha_0 - 1/\pi$. Applying the algorithm of Theorem 6.3.2, we have

1) by (6.3.7), $F(x,t) \equiv a$;
2) solving (6.3.8), we find
$$G(x,t) = -\frac{a}{1+ax};$$
3) in view of (6.3.11), (6.3.12),
$$q(x) = \frac{2a^2}{(1+ax)^2}, \quad h = -a, \quad H = \frac{a}{1+a\pi} = \frac{a\alpha_0}{\pi}.$$

By (6.3.5),
$$\varphi(x,\lambda) = \cos\sqrt{\lambda}x - \frac{a}{1+ax}\frac{\sin\sqrt{\lambda}x}{\sqrt{\lambda}}.$$

Reconstruction of the differential operator l_q from two spectra. Together with the operator l_q, we consider the operator

$$l_q^1 y(x) = -y''(x) + q(x)y(x), \quad x \in (0,\pi),$$
$$D(l_q^1) = \{y \in W_2^2(0,\pi) : \; y'(0) - hy(0) = 0, \; y(\pi) = 0\}.$$

The following theorem holds for the operator l_q^1.

Theorem 6.3.3. *For real numbers $\{\mu_n, \alpha_n^1\}_{n\geq 0}$ to be the spectral data of an operator l_q^1 with potential $q(x) \in L_2(0,\pi)$, it is necessary and sufficient that $\mu_n \neq \mu_m$ $(n \neq m)$, $\alpha_n^1 > 0$, and the conditions* (6.1.13), (6.1.14) *hold.*

Let $\{\lambda_n\}_{n\geq 0}$ and $\{\mu_n\}_{n\geq 0}$ be the eigenvalues of l_q and l_q^1, respectively. Then the asymptotic formulas (6.1.9) and (6.1.13) hold, and the representations (6.1.11) and (6.1.12) hold for the characteristic functions $\Delta(\lambda)$ and $d(\lambda)$, respectively. Note that (6.1.8) and (6.1.6) imply

$$\alpha_n = -\dot{\Delta}(\lambda_n)/\beta_n = -\dot{\Delta}(\lambda_n)\varphi(\pi,\lambda_n) = -\dot{\Delta}(\lambda_n)d(\lambda_n).$$

Theorem 6.3.4. *For real numbers $\{\lambda_n, \mu_n\}_{n\geq 0}$ to be the spectra of operators l_q and l_q^1 with $q(x) \in L_2(0,\pi)$, it is necessary and sufficient that* (6.1.9), (6.1.13) *and* (6.1.15) *hold. The function $q(x)$ and the numbers h and H are constructed according to the following algorithm:*

1) *determine the weights α_n of the operator l_q from the given spectra $\{\lambda_n\}_{n\geq 0}$ and $\{\mu_n\}_{n\geq 0}$ using the formula*

$$\alpha_n = -\dot{\Delta}(\lambda_n)d(\lambda_n),$$

where $\Delta(\lambda)$ and $d(\lambda)$ are calculated from (6.1.11) *and* (6.1.12)*;*

2) *use the numbers $\{\lambda_n, \alpha_n\}_{n\geq 0}$ to find $q(x)$, h and H using the algorithm of Theorem* 6.3.2.

More detailed proofs are presented in (Yurko, 2007).

6.4 Inverse scattering problems

Inverse scattering problems represent one of the most important and large areas in the theory and applications of inverse problems. A lot of books on this subject have been written. In this section we will confine ourselves to just a few formulations (mostly one-dimensional) of stationary inverse scattering problems and discuss their relationship with inverse spectral problems. Complete exposition of the relevant results can be found in (Berezanky, 1958), (Faddeev, 1956, 1959), (Marchenko, 1952), (Chadan and Sabatier, 1989), (Levitan, 1987), (Ramm, 1992), and (Colton and Kress, 1983).

Consider the *stationary Schrödinger equation* for two particles in dimensionless variables

$$-\Delta u + q(x)u = k^2 u, \quad x \in \mathbb{R}^3, \tag{6.4.1}$$

where $k > 0$ is a fixed constant representing energy, and $x = (x_1, x_2, x_3)$ is a point in $\mathbb{R}^3$. Assume that $q(x) \in L_2(\overline{B(\mathbf{0}, r_0)})$ is a real-valued measurable function with support in the ball $\overline{B(\mathbf{0}, r_0)} = \{x \in \mathbb{R}^3\colon |x| \leq r_0\}$, where $|x| = \sqrt{x_1^2 + x_2^2 + x_3^2}$

and r_0 is the radius of the smallest ball that contains the support of $q(x)$. We seek a solution to equation (6.4.1) in the form

$$u(x,\theta,k) = \mathrm{e}^{\mathrm{i}k\langle\theta,x\rangle} + u_{\mathrm{sc}}(x,\theta,k), \tag{6.4.2}$$

where θ is a given unit vector that belongs to the unit sphere $S^2 \subset \mathbb{R}^3$, the first term $u_{\mathrm{in}}(x,\theta,k) = \mathrm{e}^{\mathrm{i}k\langle\theta,x\rangle}$ represents the incident field, and the function $u_{\mathrm{sc}}(x,\theta,k)$ represents the scattered field.

The direct scattering problem consists in finding a solution u of (6.4.1) of the form (6.4.2) satisfying the following condition at infinity:

$$u_{\mathrm{sc}}(x,\theta,k) = \mathcal{A}(\theta',\theta,k)\,\frac{\mathrm{e}^{\mathrm{i}kr}}{r} + o\Big(\frac{1}{r}\Big), \quad r = |x| \to \infty; \quad \theta' = \frac{x}{|x|}. \tag{6.4.3}$$

The function $\mathcal{A}(\theta',\theta,k)$ is called the *scattering amplitude*. Note that once we find a solution $u(x,\theta,k)$ to the problem (6.4.1)–(6.4.3), the scattering amplitude can be calculated using the formula

$$\mathcal{A}(\theta',\theta,k) = \lim_{|x|\to\infty,\ x/r=\theta'} r\mathrm{e}^{-\mathrm{i}kr} u_{\mathrm{sc}}(x,\theta,k). \tag{6.4.4}$$

It also worth pointing out that the above conditions imposed on the potential $q(x)$, namely, $q(x) \in L_2(\overline{B(\mathbf{0},r_0)})$, ensure the existence and uniqueness of solution of the direct scattering problem.

For the direct problem (6.4.1)–(6.4.3), we formulate several inverse scattering problems.

Inverse problem 1. Find $q(x)$ assuming that $\mathcal{A}(\theta',\theta,k)$ is given for all $\theta',\theta \in S^2$ and for all $k > 0$.

Inverse problem 2 (for fixed energy). Find $q(x)$ assuming that $\mathcal{A}(\theta',\theta,k)$ is given for all $\theta',\theta \in S^2$ and a fixed $k > 0$.

Inverse problem 3 (for fixed incident wave). Find $q(x)$ assuming that $\mathcal{A}(\theta',\theta,k)$ is given for all $\theta' \in S^2$, all $k > 0$, and a fixed $\theta \in S^2$.

Inverse problem 4. Find $q(x)$ assuming that $\mathcal{A}(-\theta,\theta,k)$ is given for all $\theta \in S^2$ and for all $k > 0$.

Note that $\mathcal{A}(\theta',\theta,k)$ is an analytic function of the variables θ' and θ in $\mathbb{C}^3$, and S^2 is an analytic algebraic manifold in $\mathbb{C}^3$. Therefore, specifying the function $\mathcal{A}(\theta',\theta,k)$ on any open subset of S^2 determines this function uniquely on the entire S^2.

Proofs of the uniqueness theorems for the inverse problems 1–4 can be found, for example, in (Ramm, 1992).

One-dimensional formulation. Consider the case where the potential $q(x)$, $x = (x_1,x_2,x_3) \in \mathbb{R}^3$, depends only on $r = |x| = \sqrt{x_1^2+x_2^2+x_3^2}$. Then we can apply the method of separation of variables can be applied to equation (6.4.1) putting

$u(x_1, x_2, x_3) = \frac{\varphi(r)}{\sqrt{r}} Y_m^l(\beta, \gamma)$, $l = 0, 1, 2, \ldots$, where $x_1 = r \sin\beta \cos\gamma$, $x_2 = r \sin\beta \sin\gamma$, $x_3 = r\cos\beta$, and $Y_m^l(\beta, \gamma)$ are spherical harmonics. Then for every $l = 0, 1, 2, \ldots$ we have a differential equation for $\varphi(r)$:

$$\varphi_{rr} + \frac{1}{r}\varphi_r - \frac{(l + 1/2)^2}{r^2}\varphi - q(r)\varphi + k^2\varphi = 0. \qquad (6.4.5)$$

We will assume that the potential $q(r)$ satisfies the condition

$$\int_0^\infty r|q(r)|\,dr < \infty. \qquad (6.4.6)$$

Then, for a normalized solution to equation (6.4.5) that is regular at zero, the following asymptotic formula holds as $r \to \infty$ with fixed l and k:

$$\sqrt{r}\,\varphi\Big(r, k, l + \frac{1}{2}\Big) = \mathcal{A}\Big(k, l + \frac{1}{2}\Big) \sin\Big[kr - \frac{\pi l}{2} + \delta\Big(k, l + \frac{1}{2}\Big)\Big] + o(1). \qquad (6.4.7)$$

As before, the function $\mathcal{A}(k, l + 1/2)$ represents the scattering amplitude, while $\delta(k, l + 1/2)$ is called the *scattering phase* or the *phase shift*.

In this case, by the *inverse problem of the quantum scattering theory* we mean the problem of determining the potential function $q(r)$ from the given values of the phase shift $\delta(k, l + 1/2)$ for some values of k and l.

The two most thoroughly studied formulations of the inverse quantum scattering problem are: a) the case where the phase shift is specified for a fixed value $l = l_0$ and all $k^2 > 0$; b) the case where the phase shift is given for a fixed value of energy $E = E_0 = k^2$ and all $l = 0, 1, 2, \ldots$. The first problem is called the inverse problem of the quantum scattering theory with fixed angular momentum. The second problem is called the inverse problem of the quantum scattering theory with fixed energy.

We will focus briefly on the first formulation where $l_0 = 0$ and so the equation (6.4.5) has the form

$$\varphi_{rr} + \frac{1}{r}\varphi_r - \frac{1}{4r^2}\varphi - q(r)\varphi + k^2\varphi = 0. \qquad (6.4.8)$$

For the function $\psi = \sqrt{r}\varphi$, equation (6.4.8) assumes the simplified form

$$\psi_{rr} - q(r)\psi + k^2\psi = 0. \qquad (6.4.9)$$

A typical boundary condition that arises in quantum mechanics is

$$\psi(0, k) = 0.$$

Thus, the inverse problem of the quantum scattering theory with the fixed angular momentum $l_0 = 0$ is reduced to the inverse Sturm–Liouville problem on the half-axis $r \in \mathbb{R}_+$.

The Sturm–Liouville problem on a half-axis. We now return to the previous notation and consider the Sturm–Liouville problem

$$-y'' + q(x)y = \lambda y, \quad x \in \mathbb{R}_+, \tag{6.4.10}$$

$$y(0) = 0, \tag{6.4.11}$$

where $q(x)$ is assumed to be a real-valued measurable function satisfying the condition

$$\int_0^\infty x|q(x)|\,dx < \infty. \tag{6.4.12}$$

Let L denote the operator

$$y(x) \to -y''(x) + q(x)y(x),$$

from $L_2(0,\infty)$ into $L_2(0,\infty)$. The domain of definition of this operator is the set of all functions $y(x) \in W_2^2(0,\infty)$ that satisfy the boundary condition (6.4.11), i.e., $y(0) = 0$. Note that the space $W_2^2(0,\infty)$ is everywhere dense in $L_2(0,\infty)$ (see Definition A.1.23).

Since the operator L is self-adjoint, its spectrum lies on the real axis $\operatorname{Im}\lambda = 0$ of the complex plane. Moreover, the condition (6.4.12) guarantees that the entire half-axis $\lambda > 0$ is contained in the continuous spectrum, and the number n of negative eigenvalues is finite (Naimark, 1967):

$$n \le \int_0^\infty x|q(x)|\,dx < \infty.$$

The condition (6.4.12) also implies the existence of solutions $e(x,k)$ to equation (6.4.10) with real $k = \sqrt{\lambda}$ such that

$$e(x,k) = \mathrm{e}^{\mathrm{i}kx} + o(1), \quad x \to +\infty. \tag{6.4.13}$$

In the scattering theory, $e(x,k)$ is called the *Jost solution.* The function $e(x,k)$ admits analytic continuation to the upper half-plane with respect to k so that $e(x,k) \in L_2(0,\infty)$ for $\operatorname{Im} k > 0$.

For real $k \neq 0$, the functions $e(x,k)$ and $e(x,-k)$ represent a pair of linearly independent solutions to equation (6.4.10). These solutions are complex conjugates, i.e., $e(x,-k) = \overline{e(x,k)}$, and their Wronskian $e'(x,k)e(x,-k) - e'(x,-k)e(x,k)$ is independent of x and is equal to $2ik$.

Substituting $e(x,k)$ into the boundary condition (6.4.11), we obtain the function $\hat{j}(k) = e(0,k)$, which is analytic in $\mathbb{C}_+ := \{k\colon \operatorname{Im} k > 0\}$. This function is called the *Jost function* (corresponding to the boundary condition (6.4.11)).

Let $\varphi(x,k)$ ($k = \sqrt{\lambda}$) denote a solution to (6.4.10) that satisfies the boundary condition (6.4.11) and the initial condition

$$\varphi'(0,k) = k. \tag{6.4.14}$$

The function $\varphi(x,k)$ satisfies the following asymptotic formula:

$$\varphi(x,k) = |\hat{j}(k)| \sin(kx + \delta(k)) + o(1), \quad x \to \infty. \tag{6.4.15}$$

The function $\delta(k)$ is called the *scattering phase* or the *phase shift* for the problem (6.4.10)–(6.4.12). The scattering phase $\delta(k)$ and the Jost function $\hat{j}(k)$ are related by the formula

$$\hat{j}(k) = |\hat{j}(k)| e^{-i\delta(k)}. \tag{6.4.16}$$

Since $\overline{\hat{j}(k)} = \hat{j}(-k)$, from (6.4.16) it follows that the scattering phase $\delta(k)$ is odd:

$$\delta(-k) = -\delta(k), \quad k > 0.$$

The Jost function $\hat{j}(k)$ has the following properties (Levitan and Sargasyn, 1970):

1) $\hat{j}(k)$ does not vanish for real $k \neq 0$;
2) $\hat{j}(k)$ and $d\hat{j}(k)/dk$ do not vanish together;
3) the number of zeros of $\hat{j}(k)$ in the upper half-plane is finite, and all of them are simple and purely imaginary.

We introduce an S-function, also called the *scattering function* of the problem (6.4.10)–(6.4.12):

$$S(k) := \frac{\hat{j}(-k)}{\hat{j}(k)} = e^{2i\delta(k)}. \tag{6.4.17}$$

Note that $|S(k)| = 1$ and $S(k)S(-k) = 1$ for $k \in \mathbb{R}$, and

$$S(k) = 1 + O\Big(\frac{1}{k}\Big) \quad \text{for} \quad k \to \infty. \tag{6.4.18}$$

The asymptotic formula (6.4.18) means that a natural normalization for the phase is

$$\delta(\infty) = 0. \tag{6.4.19}$$

The negative eigenvalues of the problem (6.4.10)–(6.4.12) will be denoted by $-\lambda_1, -\lambda_2, \dots, -\lambda_n$, $-\lambda_j = (ik_j)^2$, $k_j > 0$. The *normalization constants* m_j can be defined as the numbers characterizing the asymptotic behavior of eigenfunctions ψ_j associated with the discrete spectrum of the operator L:

$$\begin{aligned} \psi_j(x) &= \mathrm{e}^{-k_j x} + o(1), \quad x \to +\infty, \\ m_j &= \int_0^{\infty} \psi_j^2(x)\,dx. \end{aligned} \tag{6.4.20}$$

We have

$$m_j = -2\mathrm{i}k_j \Big[\frac{\partial e}{\partial x}(0, \mathrm{i}k_j)\,\frac{\partial e}{\partial k}(0, \mathrm{i}k_j)\Big]^{-1}.$$

The *scattering data* of the boundary value problem (6.4.10)–(6.4.12) is the set $s := \{S(k);\ k_1, k_2, \dots, k_n;\ m_1, m_2, \dots, m_n\}$, where $S(k)$ is the scattering function, $-k_j^2$ ($1 \le j \le n$) are the negative eigenvalues of the operator L, and m_j are the normalization constants. It is usually assumed that $k_1 > k_2 > \dots > k_n > 0$. By virtue of (6.4.17), the scattering datas can also be defined by the scattering phase: $s = \{\delta(k); k_1, k_2, \dots, k_n; m_1, m_2, \dots, m_n\}$.

Set

$$F(x) = \sum_{j=1}^{n} \frac{1}{m_j}\,\mathrm{e}^{-k_j x} + \frac{1}{2\pi}\int_{-\infty}^{\infty} [1 - S(k)]\mathrm{e}^{\mathrm{i}kx}\,dk. \tag{6.4.21}$$

The integral equation

$$F(x+y) + K(x,y) + \int_x^{\infty} K(x,t)F(t+y)\,dt = 0, \quad y \ge x, \tag{6.4.22}$$

is called *the main integral equation* of the inverse problem of the quantum scattering theory. The following theorem is valid (Levitan, 1987).

Theorem 6.4.1. *Assume that the function $F(x)$ defined by* (6.4.21) *satisfies the condition*

$$\int_x^{\infty} x|F'(x)|\,dx < \infty, \tag{6.4.23}$$

and $\delta(k)$ satisfies (6.4.19). *Then the main integral equation* (6.4.22) *has a unique solution $K(x,y) \in L(x,\infty)$ for any $x \ge 0$, and the function*

$$f(x,k) = \mathrm{e}^{\mathrm{i}kx} + \int_x^{\infty} K(x,t)\mathrm{e}^{\mathrm{i}kt}\,dt \tag{6.4.24}$$

satisfies the equation

$$-y'' + q(x)y = k^2 y, \quad 0 \le x < \infty,$$

where

$$q(x) = -2\,\frac{d}{dx}\,K(x,x).$$

Moreover,

$$\int_0^\infty x|q(x)|\,dx < \infty.$$

Exercise 6.4.1. Compare (6.4.24) and (6.4.13).

The main integral equation (6.4.22) can be used to derive a formula for the expansion in terms of the eigenfunctions of the Sturm–Liouville problem (6.4.10)–(6.4.12). Under the assumptions of Theorem 6.4.1, let $K(x,y)$ be a solution to the main integral equation (6.4.22). Put

$$\psi_j(x) = \mathrm{e}^{-k_j x} + \int_x^\infty K(x,t)\mathrm{e}^{-k_j t}\,dt, \tag{6.4.25}$$

$$\psi(x,k) = \sin(kx + \delta(k)) + \int_x^\infty K(x,t)\sin(kt + \delta(k))\,dt. \tag{6.4.26}$$

The functions (6.4.25) and (6.4.26) are solutions to the equation

$$-y'' + q(x)y = \lambda y, \quad 0 < x < \infty,$$

where $q(x) = -2\,\frac{d}{dx}\,K(x,x)$, $\lambda = k^2$.

An arbitrary finite smooth function $f(x)$ can be expanded with respect to the eigenfunctions (6.4.25) and (6.4.26) as follows:

$$\begin{aligned} f(x) = {} & \sum_{j=1}^{n} \psi_j(x)\,\frac{1}{m_j}\int_0^\infty f(y)\psi_j(y)\,dy \\ & + \frac{2}{\pi}\int_0^\infty \psi(x,k)\Big[\int_0^\infty f(y)\psi(y,k)\,dy\Big]\,dk. \end{aligned}$$

It should be noted that fundamental results on the inverse scattering problem for the problem (6.4.10), (6.4.11) are presented in (Agranovich and Marchenko, 1963). The problem (6.4.10)–(6.4.12) can be formulated in a more general form if the boundary condition $y(0) = 0$ is replaced by $y(0)\cos\alpha + y'(0)\sin\alpha = 0$, $\alpha \in \mathbb{R}$. Such a formulation of the problem is discussed in (Levitan, 1987). Results analogous to those presented above (for $\sin\alpha = 0$) were also obtained for the general case $\sin\alpha \neq 0$. Several different approaches to the inverse scattering problems as well as their relationship with the inverse Sturm–Liouville problem were discussed in the survey paper by Faddeev (1959), which also contains an extensive bibliography.

The spectral distribution function of the Sturm–Liouville operator. Consider the Sturm–Liouville problem on a half-axis:

$$-y'' + q(x)y = \lambda y, \quad 0 \le x < \infty, \tag{6.4.27}$$

with the boundary condition

$$y(0)\cos\alpha + y'(0)\sin\alpha = 0, \tag{6.4.28}$$

where $q(x)$ is a real-valued continuous function, λ is a spectral parameter, and α is a real number. Clearly, if $\sin\alpha = 0$, we arrive at the problem (6.4.10), (6.4.11). As was mentioned before, the problem (6.4.27), (6.4.28) determines a self-adjoint operator L_α in $L_2(0,\infty)$.

Let $\varphi_\alpha(x,\lambda)$ denote a solution to equation (6.4.27) that satisfies the initial conditions

$$\varphi_\alpha(0,\lambda) = \sin\alpha, \quad \varphi'_\alpha(0,\lambda) = -\cos\alpha.$$

It is obvious that $\varphi_\alpha(0,\lambda)$ satisfies the boundary condition (6.4.28).

One of the main theorems in the spectral analysis of the problem (6.4.27), (6.4.28) is Weyl's theorem formulated below (Levitan and Sargasyn, 1970).

Theorem 6.4.2. *There is at least one nondecreasing function $\rho(\lambda)$, $\lambda \in \mathbb{R}$, such that for any real-valued function $f(x) \in L_2(0,\infty)$ there exists a limit*

$$F_\alpha(\lambda) = \underset{n\to\infty}{\overset{(\rho)}{\lim}} \int_0^n f(x)\varphi_\alpha(x,\lambda)\,dx$$

in the norm of the space $L_{2,\rho}(-\infty,\infty)$, and Parseval's identity holds:

$$\int_0^\infty f^2(x)\,dx = \int_{-\infty}^\infty F_\alpha^2(\lambda)\,d\rho(\lambda). \tag{6.4.29}$$

The function $\rho(\lambda)$ is called the *spectral function* of the problem (6.4.27), (6.4.28). Note that if $g(x) \in L_2(0,\infty)$ is another function such that

$$G_\alpha(\lambda) = \underset{n\to\infty}{\overset{(\rho)}{\lim}} \int_0^n g(x)\varphi_\alpha(x,\lambda)\,dx,$$

then, after applying (6.4.29) to the functions $f(x) + g(x)$ and $f(x) - g(x)$ and subtracting one result from the other, we obtain

$$\int_0^\infty f(x)g(x)\,dx = \int_{-\infty}^\infty F_\alpha(\lambda)G_\alpha(\lambda)\,d\rho(\lambda). \tag{6.4.30}$$

Equality (6.4.30) is called the *generalized Parseval's identity*.

From (6.4.30) it follows that if the integral

$$\int_{-\infty}^{\infty} F_\alpha(\lambda)\varphi_\alpha(x,\lambda)\,d\rho(\lambda)$$

converges uniformly on any finite interval and $f(x)$ is a continuous function, then

$$f(x) = \int_{-\infty}^{\infty} F_\alpha(\lambda)\varphi_\alpha(x,\lambda)\,d\rho(\lambda). \tag{6.4.31}$$

Thus, the integral transformation

$$F_\alpha(\lambda) = \int_0^{\infty} f(x)\varphi_\alpha(x,\lambda)\,dx \tag{6.4.32}$$

generated by $\varphi_\alpha(x,\lambda)$ is an isometric mapping from $L_2(0,\infty)$ onto the weight space $L_{2,\rho}(-\infty,\infty)$ of all ρ-measurable functions $F_\alpha(\lambda)$ such that

$$\int_{-\infty}^{\infty} |F_\alpha(\lambda)|^2\,d\rho(\lambda) < \infty.$$

The image of the operator L_α under the transformation (6.4.32) is multiplication by λ, while the transformation inverse to (6.4.32) is given by (6.4.31). The set of points of increase for the spectral function $\rho(\lambda)$ coincides with the spectrum of the operator L_α. The spectral function $\rho(\lambda)$ for an operator with continuous spectrum is continuous. If the spectrum of the operator L_α is discrete, then the spectral function is piecewise-constant. The behavior of the spectral function in the case of a discrete spectrum is described in Subsection 10.1.4.

The classical inverse spectral Sturm–Liouville problem consists in the reconstruction of an operator L_α (i.e., finding $q(x)$ and α) from a given spectral function $\rho(\lambda)$. Several effective methods for solving this problem were developed by V. A. Marchenko, M. G. Krein, I. M. Gelfand, and B. M. Levitan.

6.5 Inverse scattering problems in the time domain

Unlike the problems for the Schrödinger equation (6.4.1) considered above, inverse scattering problems in the time domain have more intuitive interpretation. Suppose that the segment $(0, x_0)$ of an infinite string whose transverse oscillations are described by the equation

$$u_{tt} = u_{xx} - q(x)u, \quad x \in \mathbb{R},$$

is unknown.

Assume that the string is uniform for $x < 0$ and $x \geq x_0$, i.e., $q(x) = 0$. In inverse scattering problems, it is required to determine the function $q(x)$ by measuring the response of the string to the waves at some of its points. The waves on the string may be induced both outside and within the segment under study. In the latter case, the form of the wave may also be unknown.

A large number of practical problems are reduced to similar problems. For example, consider the inverse problem of seismology following (Alekseev and Belonosov, 1998).

Assume that flat SH-waves are propagating in the elastic half-space $z \geq 0$ whose mechanical properties depend only on depth z, and these waves are polarized along some line parallel to the plane $z = 0$. Suppose that there is no normal stress at the boundary $z = 0$ Under these conditions, the shift w of points of the medium depends only on time t and depth z and satisfies the equation

$$(\mu w_z)_z = \rho w_{tt}, \tag{6.5.1}$$

where $\rho(z)$ is the density of the medium and $\mu(z)$ is the shear modulus at depth z. In what follows, we assume that $\rho(z)$ and $\mu(z)$ are positive and twice continuously differentiable. They are also assumed to be constant for $z \geq z_0 > 0$, being equal to given values $\rho_0 > 0$ and $\mu_0 > 0$, respectively.

The following direct problems consist in finding the medium oscillation $w(z,t)$ induced by a given wave povided the medium properties $g(z)$ and $\mu(z)$ are known.

Direct problem 1 (with internal sources). Assume that there are no external forces on the surface $z = 0$, i.e.,

$$w_z\big|_{z=0} = 0, \tag{6.5.2}$$

and oscillations are induced by a wave $\varphi_1(tc_0+z)$ that moves from the domain $z > z_0$ with the speed $c_0 = \sqrt{\mu_0/\rho_0}$. Let $\varphi_1(z) = 0$ for $z < z_0$ and

$$w\big|_{t\leq 0} = \varphi_1(tc_0 + z). \tag{6.5.3}$$

The equalities (6.5.1)–(6.5.3) will be called the problem with internal sources.

Direct problem 2 (with external sources). If the sources of oscillations are outside the half-space $\mathbb{R}_+$, then the boundary condition (6.5.2) must be replaced with the condition

$$2\mu w_z\big|_{z=0} = \varphi_2(t), \tag{6.5.4}$$

where $\varphi_2(t)$ vanishes for $t \leq 0$. It is assumed that no oscillations occur before the action of (6.5.4):

$$w\big|_{t\leq 0} \equiv 0. \tag{6.5.5}$$

The equations (6.5.1), (6.5.4), (6.5.5) will be called the problem with external sources.

The oscillations described by equation (6.5.1) propagate at depth z with the speed $c(z) = \sqrt{\mu(z)/\rho(z)}$. Equation (6.5.1) and the analysis thereof are substantially simplified if we substitute a new independent variable $x(z)$ for z, where $x(z)$ is equal to the travel time of the wave from depth z to the boundary of the half-space $z = 0$:

$$x(z) := \int_0^z \frac{dy}{c(y)}.$$

The quantity

$$\sigma(x) = \sqrt{\mu(z(x))\rho(z(x))} = \rho(z(x))\, c(z(x)) \tag{6.5.6}$$

is called the *acoustic impedance*.

We now introduce new functions

$$u(x,t) = \sqrt{\sigma(x)}\, w(z(x),t), \quad q(x) = (\ln\sqrt{\sigma(x)}\,)'' + [(\ln\sqrt{\sigma(x)}\,)']^2,$$

$$g_1(x) = \sqrt{\sigma_0}\varphi_1(c_0(x - x_0) + z_0), \quad g_2(t) = \frac{1}{2\sqrt{\sigma(0)}}\varphi_2(t),$$

where $x_0 = x(z_0)$, $\sigma_0 = \sqrt{\mu_0\rho_0}$. Set

$$h = \frac{\sigma'(0)}{2\sigma(0)}.$$

Under the assumptions on the properties of the functions ρ and μ, the coefficient $q(x)$ is continuous on $\overline{\mathbb{R}}_+$ and it vanishes for $x \ge x_0$. The support of $g_1(x)$ is contained in $[x_0, \infty)$.

In the new notation, problem 1 becomes as follows

$$u_{tt} = u_{xx} - q(x)u, \quad x > 0,\ t \in \mathbb{R}, \tag{6.5.7}$$

$$(u_x - hu)\big|_{x=0} = 0, \quad u\big|_{t\le 0} = g_1(t + x), \tag{6.5.8}$$

whereas in problem 2 the conditions (6.5.8) are replaced by the boundary and initial conditions

$$(u_x - hu)\big|_{x=0} = g_2(t), \quad u\big|_{t\le 0} \equiv 0. \tag{6.5.9}$$

In inverse problems of geophysics, it is required to determine the acoustic impedance σ (which becomes the function $q(x)$ after transformation), from the measurements of the wave $u(x,t)$ at a fixed point $\xi \notin (0, x_0)$.

Inverse problem 1.1. Assume that the following additional information about solutions to the direct problem 1 is given:

$$u(0,t) = f_{11}(t), \quad t > 0. \tag{6.5.10}$$

Given the functions $g_1(x)$ and $f_{11}(t)$, it is required to determine the function $q(x)$ and the constant h from (6.5.7), (6.5.8), and (6.5.10).

Inverse problem 1.2. Assume that the following additional information about solutions to the direct problem 1 is given:

$$u(\xi, t) = f_{12}(t), \quad \xi \geq x_0, \ \xi = \text{const}, \ t > 0. \tag{6.5.11}$$

Given the functions $g_1(x)$ and $f_{12}(t)$, it is required to determine $q(x)$ and h from (6.5.7), (6.5.8), and (6.5.11).

Inverse problem 2.1. Assume that the following additional information about solutions to the direct problem 2 is given:

$$u(0, t) = f_{21}(t), \quad t > 0. \tag{6.5.12}$$

Given $g_2(t)$ and $f_{21}(t)$, it is required to determine $q(x)$ and h from (6.5.7), (6.5.9), and (6.5.12).

Inverse problem 2.2. Assume that the following additional information about solutions to the direct problem 2 is given:

$$u(\xi, t) = f_{22}(t), \quad \xi = \text{const}, \ \xi \geq x_0, \ t > 0. \tag{6.5.13}$$

Given $g_2(t)$ and $f_{22}(t)$, it is required to determine $q(x)$ and h from (6.5.7), (6.5.9), and (6.5.13).

Aside from geophysics, the inverse problem formulated above are studied in scattering theory, tomography, nondestructive testing, and many other fields. For example, the positioning problem 1.2 is equivalent to the classical inverse problem of scattering theory. Problems 1.1 and 2.2 can be interpreted as scanning problems, while problems 1.2 and 2.1 can be interpreted as positioning problems. All four inverse problems are equivalent in the sense that they can be reduced to each other by means of explicit formulas and the Fourier transform. Therefore, any result obtained in the study of any one of these problems would apply to the others.

The Sturm–Liouville problem on a half-axis in the case of finite potential. Consider the spectral problem

$$-U'' + q(x)U = k^2 U, \quad x > 0, \tag{6.5.14}$$

$$(U' - hU)\big|_{x=0} = 0 \tag{6.5.15}$$

obtained from equation (6.5.7) and the first condition in (6.5.8) using the formal Fourier transform

$$U(x,k) = \frac{1}{\sqrt{2\pi}} \int_{-\infty}^{\infty} e^{-ikt} u(x,t)\, dt.$$

Note that the boundary condition (6.5.15) coincides with (6.4.28) for $\sin\alpha \neq 0$. Recall that the potential $q(x)$ is continuous on $\overline{\mathbb{R}}_+$ and is finite: $q(x) \equiv 0$ for $x \geq x_0$.

We denote by L_h the operator

$$y(x) \to -y''(x) + q(x)y(x),$$

acting in the space $L_2(0,\infty)$ and defined for all functions $y(x) \in W_2^2(0,\infty)$ such that $y'(0) - hy(0) = 0$.

We now examine the spectrum of L_h. Put $\lambda = k^2$. Since L_h is self-adjoint, its spectrum lies on the real axis $\operatorname{Im}\lambda = 0$ of the complex plane. Since $q(x)$ is finite, the entire half-axis $\lambda > 0$ belongs to the continuous spectrum, while the half-axis $\lambda < 0$ can contain only eigenvalues. In this case, the operator L_h has no eigenvalues $\lambda \leq 0$. Indeed, if we assume the converse and return to the original physical variables—density ρ, shear modulus μ, depth z, and shift w—then we would find eigenvalues $\lambda \leq 0$ of the spectral problem

$$-(\mu w_z)_z = \lambda \rho w, \quad z > 0; \quad w_z\big|_{z=0} = 0,$$

which is a contradiction.

Note that in this case the inversion formula (6.4.31) becomes

$$f(x) = \int_0^\infty F(\lambda)\varphi(x,\lambda)\,d\rho(\lambda),$$

where $\rho(\lambda)$ is the spectral distribution function for the operator L_h, $f(x) \in L_2(0,\infty)$, $F(\lambda) \in L_{2,\rho}(-\infty,\infty)$, and $\varphi(x,\lambda)$ is a solution to the Cauchy problem

$$-\varphi''(x,\lambda) + q(x)\varphi(x,\lambda) = \lambda\varphi(x,\lambda), \quad \varphi(0,\lambda) = 1,\ \varphi'(0,\lambda) = h,\ x \geq 0.$$

We now establish the connection between the spectral function of the operator L_h and the Jost function $j(k)$ of the problem (6.5.14), (6.5.15).

Since $q(x) = 0$ for $x \geq x_0$, equation (6.5.14) has a Jost solution $e(x,k)$ that coincides with e^{ikx} for $x \geq x_0$. This function is holomorphic with respect to the variable k in the entire complex plane and, together with its derivatives $e_x = e^{(1)}$ and $e_{xx} = e^{(2)}$ with respect to x, it can be represented in the form

$$e^{(m)}(x,k) = e^{ikx}[(ik)^m + \psi_m(x,k)], \quad m = 0,1,2. \tag{6.5.16}$$

Here $\psi_m(x,k) = 0$ for $x \geq x_0$ and

$$|\psi_m(x,k)| \leq C_m(|k|+1)^{m-1} e^{\alpha(|\operatorname{Im} k| - \operatorname{Im} k)}$$

for $x \in (0,x_0)$, where C_m and α are positive constants.

Substituting $e(x,k)$ into the boundary condition (6.5.15), we obtain the entire function $j(k) = e'(0,k) - he(0,k)$, which is the Jost function of the problem (6.5.15), (6.5.16).

Lemma 6.5.1. *The Jost function $j(k)$ of the problem* (6.5.15), (6.5.16) *does not vanish for* $\operatorname{Im} k \geq 0$, $k \neq 0$, *but has zero of the first order at* $k = 0$.

The proof is given in (Alekseev and Belonosov, 1998).

The most important role of the Jost function in inverse problems consists in the fact that, on one hand, there is a one-to-one correspondence between it and the spectral distribution function of the operator L_h, while, on the other hand, it is related to the data of the inverse scattering problems 1.1, 1.2, 2.1, 2.2.

The following theorem reveals the relationship between the spectral function $\rho(\lambda)$ and the Jost function $j(\sqrt{\lambda})$.

Theorem 6.5.1. *The spectral function $\rho(\lambda)$ of the operator L_h vanishes for $\lambda \leq 0$ and satisfies the equation*

$$\rho'(\lambda) = \frac{\sqrt{\lambda}}{\pi |j(\sqrt{\lambda})|^2}$$

for $\lambda > 0$.

This formula was first obtained for Sturm–Liouville operators with the Dirichlet boundary condition (Faddeev, 1959; Levitan, 1987; Chadan and Sabatier, 1989). The case of the boundary condition $y'(0,\lambda) - hy(0,\lambda) = 0$ was discussed in (Alekseev, 1962). The complete proof of the theorem is given in (Alekseev and Belonosov, 1998).

As before, let $e(x,k)$ be the Jost solution to equation (6.5.14), and $j(k) = e'(0,k) - he(0,k)$ be the Jost function. If $g_1(x)$ in the direct problem 1 belongs to the space $W_2^2(\mathbb{R})$ and $g_1(x) = 0$ outside of $[x_0, \infty)$, then it can be proved that this problem has a solution $u(x,t)$ such that for any $k \neq 0$, $\operatorname{Im} k \leq 0$, its Fourier image $U(x,k)$ can be represented in the form

$$U(x,k) = G_1(k)\Big[e(x,k) - \frac{j(k)}{j(-k)}\, e(x,-k)\Big], \tag{6.5.17}$$

where $G_1(k)$ is the Fourier image of the function g_1. A similar statement holds for the direct problem 2 in the case where $g_2(t) \in W_2^1(\mathbb{R})$ and $g_2(t) = 0$ for $t < 0$. If we denote by $V(x,k)$ the Fourier image of the derivative u_t of a solution to the problem, then

$$V(x,k) = G_2(k)\Big[\frac{ik}{j(-k)}\Big] e(x,-k), \tag{6.5.18}$$

where $G_2(k)$ is the Fourier image of g_2, $k \neq 0$, and $\operatorname{Im} k \leq 0$.

Formulas (6.5.16) and (6.5.18) make it possible to associate the Fourier images $F_{ij}(k)$ of the data f_{ij}, $i,j = 1,2$, of the inverse problems with the Fourier images $G_i(k)$ of the sources g_i, $i = 1,2$, using the Jost function:

$$U(0,k) = F_{11}(k) = P_{11}(k)G_1(k), \quad U(\xi,k) = F_{12}(k) = P_{12}(\xi,k)G_1(k),$$
$$V(0,k) = F_{21}(k) = P_{21}(k)G_2(k), \quad V(\xi,k) = F_{22}(k) = P_{22}(\xi,k)G_2(k).$$

In the above relations, $\xi > x_0$ is the observation point in the inverse problems 1.2 and 2.2,

$$P_{11}(\xi,k) = -\frac{2\mathrm{i}k}{j(-k)}, \qquad P_{12}(\xi,k) = \mathrm{e}^{\mathrm{i}k\xi} - \frac{j(k)}{j(-k)}\,\mathrm{e}^{-\mathrm{i}k\xi},$$
$$P_{21}(\xi,k) = \frac{\mathrm{i}k}{j(-k)}\,e(0,k), \quad P_{22}(\xi,k) = \frac{\mathrm{i}k}{j(-k)}\,\mathrm{e}^{-\mathrm{i}k\xi}.$$

The functions $P_{ij}(\xi,k)$ are called *transition functions.*

Theorem 6.5.2. *On the real axis, the Jost function can be expressed in terms of any of the transition functions, and vice versa.*

The proof can be found in (Alekseev and Belonosov, 1998).

Note that the transition functions represent none other than the response of the medium to the impulse function $g_j(t) = \delta(t)$, $j = 1, 2$.

Thus, the inverse problems 1.1, 1.2, 2.1, and 2.2 are equivalent in the following sense: any of the four transition functions determines the Jost function and, consequently, the other three transition functions.

In turn, by Theorem 6.5.1, the Jost function determines the spectral function $\rho(\lambda)$. Given $\rho(\lambda)$, one can reconstruct the coefficients $q(x)$ and h of the original equations (6.5.7)–(6.5.9) using well-known effective methods for solving the spectral Sturm–Liouville problem.

Chapter 7
Linear problems for hyperbolic equations

In the beginning of this chapter, we deal with the Cauchy problem on a time-like surface. In Sections 7.1 and 7.2, we analyze the problem of the reconstruction of a function from its spherical means and establish the relationship between this problem and the ill-posed Cauchy problem (Courant and Hilbert, 1951; Romanov, 1973). Section 7.3 provides an example of a numerical method for solving the inverse problem of thermoacoustics. Section 7.4 is devoted to the linearized inverse problem for the wave equation.

7.1 Reconstruction of a function from its spherical means

Consider the problem of reconstruction of a function $q(z, y_1, y_2)$ from the mean values of this function on spheres of arbitrary finite radius r with centers in the plane $z = 0$. A point in the three-dimensional space will be denoted by (z, y), $y = (y_1, y_2)$, and the function $q(z, y_1, y_2)$ by $q(z, y)$. Let (ξ, η) ($\eta = (\eta_1, \eta_2)$) be a variable point on a sphere of radius r centered at $(0, y)$. In this notation, the problem of determining the function $q(z, y)$ from its spherical means can be formulated as a problem of solving the integral equation

$$\frac{1}{4\pi} \iint_{\xi^2+|\eta|^2=r^2} q(\xi, y+\eta)\, d\omega = f(y, r) \qquad (7.1.1)$$

with a given function $f(y, r)$ ($r \geq 0$). In this equation, $d\omega$ denotes an element of the solid angle with vertex at $(0, y)$. Any odd function $q(z, y)$ with respect to z satisfies equation (7.1.1) for $f(y, r) \equiv 0$, and therefore we will assume that $q(z, y)$ is an even function with respect to z and continuously differentiable with respect to y_1 and y_2.

We will use the Courant method (Courant and Hilbert, 1951). Consider the result of applying the operation

$$B_i f := \frac{\partial}{\partial y_i} \int_0^R r^2 f(y, r)\, dr, \quad i = 1, 2,$$

to equation (7.1.1).

Using the Gauss–Ostrogradsky formula, we obtain

$$\begin{aligned} B_i f &= \frac{1}{4\pi}\frac{\partial}{\partial y_i}\int_0^R r^2\Big[\iint_{\xi^2+|\eta|^2=r^2} q(\xi, y+\eta)\,d\omega\Big]\,dr \\ &= \frac{1}{4\pi}\frac{\partial}{\partial y_i}\iiint_{\xi^2+|\eta|^2\le R^2} q(\xi, y+\eta)\,d\xi\,d\eta \\ &= \frac{1}{4\pi}\iiint_{\xi^2+|\eta|^2\le R^2} q_{\eta_i}(\xi, y+\eta)\,d\xi\,d\eta \\ &= \frac{1}{4\pi}\iint_{\xi^2+|\eta|^2=R^2} q(\xi, y+\eta)\cos(n,\eta_i)\,dS \\ &= \frac{1}{4\pi}R\iint_{\xi^2+|\eta|^2=R^2} q(\xi, y+\eta)\eta_i\,d\omega. \end{aligned}$$

In these calculations, dS denotes the surface area element of the sphere $\xi^2+|\eta|^2 = R^2$ equal to $R^2\,d\omega$, while $\cos(n,\eta_i)$ denotes the cosine of the angle between the normal n to the sphere and the axis η_i, and the following equality is used:

$$\cos(n,\eta_i) = \frac{\eta_i}{R}.$$

We now consider the operator

$$L_i f = \Big[\frac{1}{R}B_i f + y_i f(y,R)\Big]_{R=r}.$$

When applied to equation (7.1.1), it yields

$$L_i f = \frac{1}{4\pi}\iint_{\xi^2+|\eta|^2=r^2} q(\xi, y+\eta)(y_i+\eta_i)\,d\omega, \quad i=1,2, \tag{7.1.2}$$

i.e., the action of L_i on f is equivalent to calculating the spherical mean of the function $q(z,y)y_i$, which is the first moment of $q(z,y)$ with respect to the ith coordinate. Therefore, applying the operator L_j to $L_i f$, we have

$$L_j L_i f = \frac{1}{4\pi}\iint_{\xi^2+|\eta|^2=r^2} q(\xi, y+\eta)(y_i+\eta_i)(y_j+\eta_j)\,d\omega, \quad i,j=1,2.$$

We now consider a fixed sphere of radius r centered at $(0,y)$. Having determined the operators L_i $(i=1,2)$ from the given function $f(y,r)$, we find the first moments for the function $q(z,y)$ on this sphere. Then the second application of the operators L_i $(i=1,2)$ will give us all second moments on this sphere with respect to y_1 and y_2, and so on. Thus we can calculate the moments of any order with respect to y_1 and y_2. For example, the spherical mean of the function $q(z,y)y_1^{k_1}y_2^{k_2}$ is given by

$$L_1^{k_1}L_2^{k_2}f = \frac{1}{4\pi}\iint_{\xi^2+|\eta|^2=r^2} q(\xi, y+\eta)(y_1+\eta_1)^{k_1}(y_2+\eta_2)^{k_2}\,d\omega,$$

$$k_1, k_2 = 0, 1, 2, \ldots. \quad (7.1.3)$$

At the same time, the function $q(\xi, y+\eta)$ on our sphere is clearly a function of two variables because $\xi = \pm\sqrt{r^2-|\eta|^2}$ on the sphere. Since $q(\xi, y+\eta)$ is odd with respect to ξ, the relation (7.1.3) can be written in the form

$$\begin{aligned} L_1^{k_1}L_2^{k_2}f &= \frac{1}{2\pi r}\iint_{|\eta|\le r} q(\sqrt{r^2-|\eta|^2}, y+\eta)(y_1+\eta_1)^{k_1}(y_2+\eta_2)^{k_2}\frac{d\eta}{\sqrt{r^2-|\eta|^2}} \\ &= \frac{1}{2\pi r}\iint_{|\zeta-y|\le r} q(\sqrt{r^2-|\zeta-y|^2}, \zeta)\zeta_1^{k_1}\zeta_2^{k_2}\,\frac{d\zeta}{\sqrt{r^2-|\zeta-y|^2}}, \end{aligned}$$

where $\zeta = (\zeta_1, \zeta_2)$ denotes the variable $y+\eta$. The preceding formula shows that for the function of two variables (with fixed (y,r)!)

$$\phi(\zeta_1,\zeta_2) = \frac{q(\sqrt{r^2-|\zeta-y|^2}, \zeta)}{\sqrt{r^2-|\zeta-y|^2}}$$

we know all power moments over the interior of a disk of fixed radius:

$$L_1^{k_1}L_2^{k_2}f = \frac{1}{2\pi r}\iint_{|\zeta-y|\le r} \phi(\zeta)\zeta_1^{k_1}\zeta_2^{k_2}\,d\zeta, \quad k_1, k_2 = 0, 1, 2, \ldots. \quad (7.1.4)$$

It is clear that the function $\phi(\zeta)$ is uniquely determined by these moments. To find its an approximate solution, from the system of functions $\zeta_1^{k_1}\zeta_2^{k_2}$ $(k_1, k_2 = 0, 1, 2, \ldots)$ we construct a complete system of polynomials $P_n(\zeta) = P_n(\zeta_1, \zeta_2)$ orthogonal inside the disk $|\zeta - y| \le r$ using the orthogonalization method. Then the Fourier coefficients s_n for the function $\phi(\zeta)$ are calculated as follows:

$$s_n = \iint_{|\zeta-y|\le r} \phi(\zeta)P_n(\zeta)\,d\zeta = 2\pi r P_n(L_1, L_2)f,$$

which yields the function $\phi(\zeta)$ in the form of a Fourier series. Note that the singularity of $\phi(\zeta)$ on the boundary of the disk is integrable.

We have shown the way to construct the function $q(z,y)$ on any fixed sphere by calculating certain operators for the function $f(y,r)$. It is clear from the construction that in order to uniquely determine the function $q(z,y)$ everywhere (in the class indicated above), it is sufficient that the function $f(y,r)$ be given inside an arbitrarily thin cylinder

$$|y - y_0| \le \varepsilon, \quad 0 \le r < \infty,$$

whose axis passes through a fixed point y_0 (where ε is a positive number that is as small as desired). Assume that this cylinder has finite hight: $0 \le r \le r_0$ ($r_0 > 0$). Then the procedure described above can be used to determine the function $q(z, y)$ on any sphere whose center lies inside the disk $|y - y_0| \le \varepsilon$, $z = 0$, and whose radius r is between 0 and r_0. Consequently, the function $q(z, y)$ is determined inside a certain three-dimensional domain D. Given $q(z, y)$ inside D, we can calculate its spherical means over any sphere whose center lies in the plane $z = 0$ that lies within the domain D. To this end, it is sufficient that the radius r and the center $(0, y)$ of the sphere satisfy the conditions

$$\begin{aligned} r + |y - y_0| &\le r_0 + \varepsilon, \quad \text{if } |y - y_0| \le \varepsilon, \\ r &\le r_0, \qquad\quad\;\; \text{if } |y - y_0| \le \varepsilon. \end{aligned} \tag{7.1.5}$$

Consequently, the function $f(y, r)$ can be determined in the domain defined by (7.1.5).

We have established that the values of the function $f(y, r)$ inside the cylinder $|y - y_0| \le \varepsilon$, $0 \le r \le r_0$ determine $f(y, r)$ inside the circular truncated cone such that its base is the top of the cylinder and the angle between its generatrices and the plane $r = 0$ is 45°. This means that the function $f(y, r)$ cannot be specified in an arbitrary way. Moreover, $f(y, r)$ that can be represented in the form (7.1.1) has a property of analytic type: it is uniquely determined by its values in an arbitrarily small cylinder $|y - y_0| \le \varepsilon$, $0 \le r < \infty$. At the same time, it is obvious that nonanalytic functions $q(z, y)$ correspond to nonanalytic functions $f(y, r)$.

7.2 The Cauchy problem for a hyperbolic equation with data on a time-like surface

We will show that the problem of solving equation (7.1.1) is ill-posed. Consider the function

$$v(z, y, t) = \frac{t}{4\pi} \iint_{\xi^2 + |\eta|^2 = t^2} q(z + \xi, y + \eta)\, dw, \tag{7.2.1}$$

which is a spherical mean over the sphere of radius t centered at (z, y) up to a multiplier for t. The function $q(z, y)$ is assumed to be twice continuously differentiable. It is easy to verify that in the half-space $t \ge 0$ the function $v(z, y, t)$ satisfies the wave equation

$$v_{zz} + v_{y_1 y_1} + v_{y_2 y_2} = v_{tt}. \tag{7.2.2}$$

The function $v(z, y, t)$ vanishes on the boundary of the half-space $t > 0$:

$$v(z, y, 0) = 0. \tag{7.2.3}$$

In addition,

$$v(z,y,t)\big|_{z=0} = tf(y,t). \tag{7.2.4}$$

Since $q(z,y)$ is odd with respect to z, we have

$$\frac{\partial}{\partial z}\, v(z,y,t)\bigg|_{z=0} = 0. \tag{7.2.5}$$

The mixed problem (7.2.2), (7.2.3), (7.2.5) is obviously equivalent to the Cauchy problem with data with respect to the spatial variable z if we take the odd extension of $f(y,t)$ with respect to t and consider the problem in the entire space z, y, t. We will show that the problem (7.2.2)–(7.2.5) is equivalent to equation (7.1.1). To this end, it suffices to show that (7.1.1) follows from the equalities (7.2.2)–(7.2.5) where $v(z,y,t)$ and $q(z,y)$ are related by formula (7.2.1). Note that, given $v(z,y,t)$, the function $q(z,y)$ can be determined from the formula

$$q(z,y) = \lim_{t\to 0} \frac{1}{t}\, v(z,y,t) = \frac{\partial}{\partial t}\, v(z,y,t)\bigg|_{t=0}, \tag{7.2.6}$$

which follows from (7.2.4).

Now let $v(z,y,t)$ be a solution to the problem (7.2.2)–(7.2.5). We denote by $q(z,y)$ the limit of the derivative of $v(z,y,t)$ with respect to t as $t \to 0$. Then $v(z,y,t)$ satisfies equation (7.2.2) and the conditions (7.2.3), (7.2.6), i.e., it is a solution to the Cauchy problem with data with respect to t. As is known, the solution to this problem is given by the Kirchhoff formula, which coincides with formula (7.2.1). Then condition (7.2.5) leads to the equality

$$\iint_{\xi^2+|\eta|^2=t^2} q_\xi(\xi, y+\eta)\, dw = 0,$$

which implies that the even part of $q_z(z,y)$ is equal to zero, i.e., $q(z,y) = q(-z,y)$ because of the uniqueness of solutions to the problem of integral geometry. At the same time, condition (7.2.4) leads to equation (7.1.1), whose solutions should be sought in the class of even functions. We have thus established the equivalence of the problem of integral geometry and the problem (7.2.2)–(7.2.5). In particular, this implies the uniqueness of solutions to the problem (7.2.2)–(7.2.5).

We now show that this problem is unstable. Consider a solution to the problem (7.2.2)–(7.2.5) for

$$f(y,t) = \frac{1}{n^s}\frac{\sin nt}{t} \sin ny_1 \sin ny_2.$$

For sufficiently large n and s, the function $f(y,t)$ is small together with any finite number of its derivatives. In this case, it can be directly verified that a solution to the problem (7.2.2)–(7.2.5) is given by the formula

$$v(z,y,t) = \frac{1}{n^s} \sin nt \sin ny_1 \sin ny_2 \cosh nz. \tag{7.2.7}$$

For any fixed z, the solution tends to infinity as $n \to \infty$. The function $q(z, y)$, which is a solution to the problem (7.2.1), has the same property:

$$q(z, y) = \frac{1}{n^{s-1}} \sin ny_1 \sin ny_2 \cosh nz.$$

Hence, the problem is unstable and therefore classically ill-posed. From the uniqueness of solutions to the problem (7.1.1), it follows that it is conditionally ill-posed. Indeed, consider the set M of even functions $q(z, y)$ with respect to z that is compact in the space C (for example, a bounded set of functions with first derivatives bounded by a given constant). Then, if the problem (7.1.1) has a solution $q(z, y) \in M$, it will be stable by Theorem 2.2.1.

7.3 The inverse thermoacoustic problem

Physical formulation. Consider a domain $\Omega \subset \mathbb{R}^2$ of an elastic medium. Assume that the domain Ω is subjected to electromagnetic radiation starting from the instant $t = 0$ with intensity $I(t)$, and the radiation is partially absorbed by the medium (Kabanikhin et al., 2005). The absorbed energy is transformed into heat, which leads to the increase in the temperature of the medium, then to its expansion, and, ultimately, to acoustic pressure waves. Propagating through the medium, the acoustic pressure waves reach the boundary Γ of the domain, where they can be measured on a part Γ_1 of the boundary. It is required to determine the coefficient of absorption of electromagnetic radiation in the domain Ω from the acoustic pressure measured on the part Γ_1 of the boundary.

Using the model of non-viscous fluid and neglecting diffuse heat flows, the process of propagation of waves of the acoustic pressure u can be described by the equation

$$\frac{1}{c^2}\frac{\partial^2 u}{\partial t^2} - \Delta u = \alpha(z, y)\frac{\beta}{c_p}\frac{\partial I}{\partial t}, \quad (z, y) \in \Omega, \quad t \in \mathbb{R}, \tag{7.3.1}$$

where $c = 1/\sqrt{\rho\varkappa}$ is the speed of propagation of acoustic waves (ρ is density; $\varkappa$ is compressibility); $\alpha(z, y)$ is the coefficient of absorption of the electromagnetic radiation; β is the thermal expansion coefficient; c_p is specific heat capacity at constant pressure.

The condition that there are no acoustic pressure waves before the beginning of irradiation is taken as an initial condition:

$$u\big|_{t<0} = 0. \tag{7.3.2}$$

As a rule, in thermoacoustic problems the duration of the electromagnetic radiation is very short, which makes it possible to (approximately) specify $I(t)$ in the form of Dirac's delta

$$I(t) = I_0\delta(t). \tag{7.3.3}$$

Taking (7.3.3) into account, integrating twice with respect to t from $-\varepsilon$ to $+\varepsilon$ and passing to the limit as $\varepsilon \to +0$, we arrive at the problem

$$\frac{1}{c^2}\frac{\partial^2 u}{\partial t^2} - \Delta u = 0, \tag{7.3.4}$$

$$u\big|_{t=0} = \alpha(z,y)\frac{\beta}{c_p} I_0, \quad u_t\big|_{t=0} = 0. \tag{7.3.5}$$

Formulation of the direct and inverse problems. Consider the case where

$$\Omega = \{(z,y) : z \in (0,a),\ y \in (-b,b)\}, \quad 0 < a < b,$$
$$\Gamma_1 = \{(z,y) : z = 0,\ y \in [-b,b]\}.$$

Assume that $\alpha(z,y)$ is sufficiently smooth and finite:

$$\operatorname{supp}\alpha(z,y) \subset (0,a) \times (-b+a, b-a). \tag{7.3.6}$$

Set $c = 1$, $q(z,y) = \alpha(z,y)\beta I_0/c_p$, $\Pi = \{(z,y) \mid z \in (-a,a),\ y \in (-b,b)\}$ and consider the direct problem

$$u_{tt} = u_{zz} + u_{yy}, \quad (z,y) \in \Pi, \quad t \in (0,a), \tag{7.3.7}$$
$$u(z,y,0) = q(z,y), \quad (z,y) \in \Pi, \tag{7.3.8}$$
$$u_t(z,y,0) = 0, \quad (z,y) \in \Pi, \tag{7.3.9}$$
$$u\big|_{\partial\Pi} = 0, \quad t \in (0,a). \tag{7.3.10}$$

The inverse problem is formulated as follows: find $q(z,y) = u(z,y,0)$ from (7.3.7), (7.3.9), (7.3.10), and the additional information

$$u(0,y,t) = f(y,t), \quad (y,t) \in (-b,b) \times (0,a). \tag{7.3.11}$$

Exercise 7.3.1. Write the objective functional for the problem (7.3.7)–(7.3.11) and determine its gradient.

7.4 Linearized multidimensional inverse problem for the wave equation

In this section we consider the problem of determining the speed of wave propagation in the half-space $(z,y) \in \mathbb{R}_+ \times \mathbb{R}^n$ in the case where the speed can be represented in the form $c^2(z,y) = c_0^2(z) + c_1(z,y)$, $c_1 \ll c_0^2$, and c_1 is nonzero only in a finite domain of the half-space (Kabanikhin and Shishlenin, 2002). We will study the linearized version of the inverse problem (Lavrentiev and Romanov, 1966; Romanov, 1969b). For this version, we will prove the uniqueness theorem, obtain a conditional stability estimate, and construct a regularizing sequence converging to the exact solution of the linearized inverse problem.

Problem formulation. Assume that the speed of wave propagation in the half-space $(z, y) \in \mathbb{R}_+ \times \mathbb{R}^n$, $y = (y_1, \dots, y_n)$, has the structure

$$c^2(z, y) = c_0^2(z) + c_1(z, y). \tag{7.4.1}$$

In addition, assume that the functions c_0 and c_1 satisfy **condition $\mathbf{A_0}$**:

1) $c_0 \in C^2(\overline{\mathbb{R}}_+)$, $c_0'(+0) = 0$;

2) there are constants $M_1, M_2, M_3 \in \mathbb{R}_+$ such that for all $z \in \mathbb{R}_+$

$$0 < M_1 \le c_0(z) \le M_2, \quad \|c_0\|_{C^2(\mathbb{R}_+)} < M_3; \tag{7.4.2}$$

3) the function $c_1(z, y)$ is nonzero only in the domain $(z, y) \in (0, h) \times \mathcal{K}(\mathcal{D}_1)$,

$$\mathcal{K}(\mathcal{D}_1) = \{y \in \mathbb{R}^n : |y_j| < \mathcal{D}_1, \ j = \overline{1, n}\},$$

where $h, \mathcal{D}_1 \in \mathbb{R}_+$ are fixed numbers;

4) $c_1(z, y) \in C^2((0, h) \times \mathcal{K}(\mathcal{D}_1))$,

$$\alpha = \|c_1\|_{C^2((0,h)\times\mathcal{K}(\mathcal{D}_1))} \ll M_1. \tag{7.4.3}$$

The medium is assumed to be at rest before $t = 0$:

$$u\big|_{t<0} \equiv 0, \quad (z, y) \in \mathbb{R}_+ \times \mathbb{R}^n, \tag{7.4.4}$$

and a wave of the form

$$\frac{\partial u}{\partial z}\Big|_{z=0} = r(y)\delta(t), \quad y \in \mathbb{R}^n, \tag{7.4.5}$$

arrives at the boundary $z = 0$ of the half-space $\mathbb{R}_+ \times \mathbb{R}^n$ at the instant $t = 0$ and generates the following wave process in $\mathbb{R}_+ \times \mathbb{R}^n$:

$$\frac{\partial^2 u}{\partial t^2} = c^2(z, y)\Delta_{z,y} u, \quad (z, y) \in \mathbb{R}_+ \times \mathbb{R}^n, \ t \in \mathbb{R}_+. \tag{7.4.6}$$

Assume that the trace of a solution to the direct problem (7.4.4)–(7.4.6) at the boundary $z = 0$ is given:

$$u\big|_{z=0} = f(y, t), \quad y \in \mathbb{R}^n, \quad t \in \mathbb{R}_+. \tag{7.4.7}$$

By the *inverse* problem we will mean the problem of determining $c(z, y)$ from (7.4.4)–(7.4.7) (under the conditions (7.4.1) and A_0).

Linearization. Before linearizing the inverse problem (7.4.4)–(7.4.7), we will show that it can be localized, i.e., it suffices to specify the additional information (7.4.7) only on a bounded subset of the hyperplane $z = 0$ for a finite time interval. Such localization can be done using the finiteness of the domain where a solution to the hyperbolic equation depends on its coefficients and on the initial and boundary conditions.

In view of the assumption (7.4.1) and condition A_0, the minimal time it can take for a perturbation generated by the incident wave (7.4.5) to reach the depth h for all $y \in \mathbb{R}^n$ and return to the surface $z = 0$ is $T_h = 2h/(M_1 - \alpha)$.

Consequently, since the function $c_1(z, y)$ is finite (see condition A_0), the waves reflected from the inhomogeneity represented by $c_1(z, y)$ cannot reach the hyperplanes $|y_j| = \mathcal{D}$, $j = \overline{1, n}$, in time T_h, where $\mathcal{D} = \mathcal{D}_1 + T_h(M_2 + \alpha)$.

Suppose that the incident wave (7.4.5) is flat on an area of the surface $z = 0$ that lies above the domain of an $(n + 1)$-dimensional inhomogeneity, i.e.,

$$r(y)\big|_{y \in \mathcal{K}(\mathcal{D})} \equiv r_0, \quad r_0 = \text{const}, \ r_0 \neq 0. \tag{7.4.8}$$

Then, as follows from the previous arguments, the inverse problem (7.4.4)–(7.4.7) can be replaced with the following one:

$$\frac{\partial^2 u}{\partial t^2} = c^2(z, y)\Delta_{z,y} u, \quad (z, y) \in (0, h) \times \mathcal{K}(\mathcal{D}), \ t \in (0, T_h); \tag{7.4.9}$$

$$u\big|_{t<0} \equiv 0, \quad \frac{\partial u}{\partial z}\Big|_{z=0} = r_0 \delta(t); \tag{7.4.10}$$

$$u\big|_{y_j = \mathcal{D}} = u\big|_{y_j = -\mathcal{D}}, \quad z \in (0, h), \ t \in (0, T_h), \ j = \overline{1, n}; \tag{7.4.11}$$

$$u\big|_{z=0} = f(y, t), \quad y \in \mathcal{K}(\mathcal{D}), \ t \in (0, T_h). \tag{7.4.12}$$

Using the assumption (7.4.3) that c_1 is small, we will linearize the inverse problem (7.4.9)–(7.4.12). To this end, we represent a solution $u(z, y, t)$ to the initial boundary problem (7.4.9)–(7.4.11) in the form

$$u(z, y, t) = u_0(z, t) + u_1(z, y, t),$$

where $u_0(z, t)$ is a solution to the initial boundary value problem

$$\frac{\partial^2 u_0}{\partial t^2} = c_0^2(z)\frac{\partial^2 u_0}{\partial z^2}, \quad z \in (0, h), \ t \in (0, T_h); \tag{7.4.13}$$

$$u_0\big|_{t<0} \equiv 0, \quad \frac{\partial u_0}{\partial z}\Big|_{z=0} = r_0 \delta(t). \tag{7.4.14}$$

Note that for $z = 0$ the solution to the problem (7.4.13), (7.4.14) is identically equal to that of (7.4.4)–(7.4.6) in the case where

$$r(y) = r_0, \quad t \in (0, T_h), \ y \in \mathbb{R}^n \setminus \mathcal{K}(2\mathcal{D}).$$

Neglecting the term $c_1\,\Delta_{z,y}u_1$ of the second infinitesimal order,we arrive at the following problem for $u_1(z,y,t)$:

$$\frac{\partial^2 u_1}{\partial t^2} = c_0^2(z)\Delta_{z,y}u_1 + c_1(z,y),\frac{\partial^2 u_0}{\partial z^2}, \tag{7.4.15}$$

$$(z,y)\in(0,h)\times\mathcal{K}(\mathcal{D}),\ t\in(0,T_h);$$

$$u_1\big|_{t<0}\equiv 0,\quad \frac{\partial u_1}{\partial z}\bigg|_{z=0}=0; \tag{7.4.16}$$

$$u_1\big|_{\partial\mathcal{K}(\mathcal{D})}=0,\quad z\in(0,h),\ t\in(0,T_h), \tag{7.4.17}$$

where $\partial\mathcal{K}(\mathcal{D})$ is the boundary of the domain $\mathcal{K}(\mathcal{D})$.

For the additional information for determining $c_0(z)$ we can take

$$u_0\big|_{z=0} = f(\hat{y},t),\quad t\in(0,T_h), \tag{7.4.18}$$

where $\hat{y}\in\partial\mathcal{K}(\mathcal{D})$ is a fixed point because under our assumptions $f(y,t)$ is independent of y in a neighborhood of the boundary of the domain $\mathcal{K}(\mathcal{D})$ for $t\in(0,T_h)$.

The additional information on the solution to the problem (7.4.15)–(7.4.17) can be written as

$$u_1\big|_{z=0} = g(y,t),\quad y\in\mathcal{K}(\mathcal{D}),\ t\in(0,T_h), \tag{7.4.19}$$

where $g(y,t)=f(y,t)-u_0(0,t)$.

Thus, solution of the inverse problem (7.4.4)–(7.4.7) for $t\in(0,T_h)$ can be subdivided into localized stages:

1) solving the inverse problem (7.4.13), (7.4.14), (7.4.18) up to depth h and determining $c_0(z)$;
2) solving the direct problem (7.4.13), (7.4.14) up to depth h and determining u_{0zz};
3) solving the inverse problem (7.4.15)–(7.4.17), (7.4.19), i.e., the problem of determining $c_1(z,y)$ from the given u_{0zz} and $g(y,t)$.

In what follows, it is assumed that the function $c_0(z)$ is given and satisfies the first two items of condition A_0.

The analysis of the one-dimensional direct problem (7.4.13), (7.4.14). To bring the problem (7.4.13), (7.4.14) to a more convenient form for analysis, we introduce the new variable

$$x=\psi(z),\quad \psi(z)=\int_0^z\frac{d\xi}{c_0(\xi)}, \tag{7.4.20}$$

and the new functions

$$S(x) = \sqrt{c_0(\psi^{-1}(x))}/\sqrt{c_0(+0)}, \quad v(x,t) = u_0(\psi^{-1}(x),t)/S(x),$$
$$a(x) = S''(x)/S(x) - 2[S'(x)]^2/S^2(x). \tag{7.4.21}$$

By condition A_0, the function $\psi(z)$ has the inverse $z = \psi^{-1}(x)$ and therefore S, v, and q are well defined.

Remark 7.4.1. Consider the one-dimensional inverse acoustic problem

$$c^{-2}(z)v_{tt} = v_{zz} - \rho'(z)v_z/\rho(z), \quad z \in \mathbb{R}_+,\ t \in \mathbb{R}_+;$$
$$v\big|_{t<0} = 0, \quad v_z\big|_{z=+0} = \delta(t), \quad v(+0,t) = f(t),$$

where $\rho(z)$ is the density of the medium and $c(z)$ is the speed of wave propagation in the medium. It will be shown in Section 10.2 that it is impossible to find both functions $\rho(z)$ and $c(z)$ at the same time, but it is possible to find $c(z)\rho(z)$. After the change of variables (7.4.20) the problem is reduced to the problem of determining the acoustic impedance of the medium $\sigma(x) = c(\psi^{-1}(x))\rho(\psi^{-1}(x))$:

$$\bar{v}_{tt} = \bar{v}_{xx} - \sigma'(x)\bar{v}_x/\sigma(x), \quad x \in \mathbb{R}_+, \quad t \in \mathbb{R}_+; \tag{7.4.22}$$

$$\bar{v}\big|_{t<0} = 0, \quad \bar{v}_x\big|_{x=+0} = c(+0)\delta(t), \quad \bar{v}(+0,t) = f(t), \tag{7.4.23}$$

where $\bar{v}(x,t) = v(\psi^{-1}(x),t)$ is acoustic pressure. The inverse problem (7.4.22), (7.4.23) consists in finding the functions $\bar{v}(x,t)$ and $\sigma(x)$ from the given function $f(t)$.

We now return to the problem (7.4.13), (7.4.14) and perform the change of variables $x = \psi(z)$. If we use the even extensions of the functions (7.4.21) to $\mathbb{R}_-$ with respect to x, the function $v(x,t)$ becomes a solution to the problem

$$\frac{\partial^2 v}{\partial t^2} = \frac{\partial^2 v}{\partial x^2} + a(x)v, \quad x \in (-h_1, h_1),\ t \in (0, T_{h_1}); \tag{7.4.24}$$

$$v\big|_{t=0} = 0, \quad \frac{\partial v}{\partial t}\bigg|_{t=0} = 2\gamma\delta(x), \tag{7.4.25}$$

where $\gamma = -r_0/c_0(+0)$ and $h_1 = \psi(h)$.

A solution to the problem (7.4.24), (7.4.25) can be represented as

$$v(x,t) = \gamma\theta(t - |x|) + p(x,t).$$

Substituting this representation into (7.4.24), we see that $p(x,t)$ is a solution to the problem

$$\frac{\partial^2 p}{\partial t^2} = \frac{\partial^2 p}{\partial x^2} + a(x)p + \gamma\theta(t - |x|)a(x), \quad x \in (-h_1, h_1),\ t \in (0, T_{h_1}); \tag{7.4.26}$$

$$p\big|_{t=0} = 0, \quad \frac{\partial p}{\partial t}\bigg|_{t=0} = 0. \tag{7.4.27}$$

For $p(x,t)$ we put

$$\mathcal{A}_{x,t}[p(\xi,\tau)] \equiv \mathcal{A}_{x,t}[p] = \frac{1}{2}\int_0^t \int_{x-t+\tau}^{x+t-\tau} p(\xi,\tau)\,d\xi\,d\tau.$$

Then, using D'Alembert's formula for the representation of solutions to the Cauchy problem (7.4.26), (7.4.27), we arrive at the Volterra integral equation of the second kind for $p(x,t)$:

$$p(x,t) = I(x,t) + \mathcal{A}_{x,t}[ap], \quad x \in (-h_1,h_1),\ t \in (0,T_{h_1}), \tag{7.4.28}$$

where

$$I(x,t) = (\gamma/2)\mathcal{A}_{x,t}[a(\xi)\theta(\tau-|\xi|)].$$

Taking into account that $a(x)$ is even and using the properties of the Heaviside function, after obvious transformations we obtain the following useful equalities:

$$\frac{\partial}{\partial x}I(x,t) = \frac{\gamma}{2}\int_0^t a(\xi)\theta(t-|\xi|)\Big|_{\xi=x-t+\tau}^{\xi=x+t-\tau}\,d\tau; \tag{7.4.29}$$

$$\begin{aligned}\frac{\partial^2}{\partial x^2}I(x,t) &= \frac{\gamma}{2}\theta(t)[\delta(t+x)+\delta(t-x)]\int_0^t a(\xi)\,d\xi \\ &\quad + \frac{\gamma}{4}\theta(t-|x|)\Big[a\Big(\frac{t+x}{2}\Big)+a\Big(\frac{t-x}{2}\Big)-4a(x)\Big].\end{aligned} \tag{7.4.30}$$

Lemma 7.4.1. *If $a \in C(\mathbb{R})$, then the integral equation* (7.4.28) *has a unique solution in $C((-h_1,h_1)\times(0,T_{h_1}))$ and for any $x_0 \in (-h_1,h_1)$ and $t_0 \in (0,T_{h_1})$*

$$\|p\|_{C(\Delta(x_0,t_0))} \le (\gamma/2)t_0^2 M_4 e^{t_0\sqrt{M_4}}, \tag{7.4.31}$$

where $\Delta(x_0,t_0) = \{(\xi,\tau)\colon \tau \in (0,t_0),\ |x_0-\xi| < t_0-\tau\}$ and M_4 is a constant that depends only on M_1, M_2, and M_3.

Proof. We will seek a solution to the integral equation (7.4.28) in the form of a series

$$p(x,t) = \sum_{k=0}^{\infty} p_k(x,t), \quad (x,t) \in \Delta(x_0,t_0), \tag{7.4.32}$$

whose terms are defined by induction. For $k=0$ we put $p_0(x,t) = I(x,t)$. Suppose that $p_k(x,t)$ is defined. Then we set

$$p_{k+1}(x,t) = \mathcal{A}_{x,t}[ap_k], \quad (x,t) \in \Delta(x_0,t_0).$$

Let

$$P_k(t) = \sup_{x\in[x_0-t_0+t,\,x_0+t_0-t]} |p_k(x,t)|, \quad t \in (0,t_0).$$

We will use induction to prove the estimate

$$P_k(t) \le (\gamma/2)t_0^2 M_4(\sqrt{M_4}\,t_0)^{2k}/(2k)!, \quad t \in (0, t_0). \tag{7.4.33}$$

Indeed, for $k = 0$ the estimate (7.4.33) clearly holds. Let (7.4.33) hold for some $k > 0$. We will prove that it also holds for $k + 1$.

Estimating $|p_{k+1}(x,t)|$, we have

$$\begin{aligned}
|p_{k+1}(x,t)| \le \mathcal{A}_{x,t}[|ap_k|] &\le \frac{M_4}{2}\int_0^{t_0}\int_0^{t-\tau_0} P_k(\tau)\,d\xi\,d\tau \\
&\le \frac{\gamma}{4}t_0^2 M_4^2 \int_0^{t_0}(t_0-\tau)\frac{(\sqrt{M_4}\,\tau)^{2k}}{(2k)!}\,d\tau \\
&= \frac{\gamma}{2}t_0^2 M_4 \frac{(\sqrt{M_4}\,t_0)^{2(k+1)}}{(2(k+1))!}, \quad t \in (0, t_0).
\end{aligned}$$

Since the right-hand side of this inequality is independent of x, we conclude that the estimate (7.4.33) is true for all k, which implies the uniform convergence of the series (7.4.32) with respect to $(x,t) \in \Delta(x_0, t_0)$ and the inequality (7.4.31).

Setting

$$S_n(x,t) = \sum_{k=0}^{n} p_k(x,t),$$

we arrive at the obvious equality

$$S_{n+1}(x,t) = I(x,t) + \mathcal{A}_{x,t}[aS_n], \quad (x,t) \in \Delta(x_0,t_0).$$

Passing to the limit as $n \to \infty$ in this equality, we conclude that $p(x,t)$ is a solution to the integral equation (7.4.28) that is continuous in $\Delta(x_0, t_0)$. □

Lemma 7.4.2. *Let $a \in C(\mathbb{R})$ and let $\bar{C}((-h_1,h_1)\times(0,T_{h_1}))$ be the set of functions $p(x,t)$ continuous in $(-h_1,h_1)\times(0,T_{h_1})$ everywhere except maybe for the line $t = |x|$. Then the solution to the integral equation* (7.4.28) *has partial derivatives of the first order with respect to t and x in the class $\bar{C}((-h_1,h_1)\times(0,T_{h_1}))$ that satisfy the inequalities*

$$\sup_{\substack{(\xi,\tau)\in\Delta(x,t)\\ |\xi|\ne\tau}} |p'_{(m)}(\xi,\tau)| \le \gamma t M_4[1 + (\gamma/2)t^2 M_4 e^{t\sqrt{M_4}}], \quad t \in (0, T_{h_1}),\ m = 1, 2,$$

$$p'_{(1)}(x,t) = \frac{\partial p}{\partial x}(x,t), \quad p'_{(2)}(x,t) = \frac{\partial p}{\partial t}(x,t).$$

Proof. From Lemma 7.4.1 and the form of the function $I(x,t)$, it follows that the right-hand side of (7.4.28) can be differentiated once with respect to x and t, for example:

$$\frac{\partial p}{\partial x}(x,t) = \frac{\partial}{\partial x} I(x,t) + \frac{1}{2}\int_0^t a(\xi)p(\xi,\tau)\Big|_{\xi=x-t+\tau}^{\xi=x+t-\tau} d\tau, \quad t \in \mathbb{R}_+. \tag{7.4.34}$$

The assertion of the lemma follows from (7.4.29), (7.4.31) and (7.4.34). □

The following lemma follows from Lemmas 7.4.1 and 7.4.2.

Lemma 7.4.3. *If $a \in C(\mathbb{R})$, then the solution to equation* (7.4.28) *has a partial derivative of the second order with respect to x in the class $C((-h_1, h_1) \times (0, T_{h_1}))$ such that*

$$\begin{aligned}\frac{\partial^2 p}{\partial x^2}(x,t) = {} & \frac{\partial^2}{\partial x^2} I(x,t) - a(x)p(x,t) \\ & + \frac{1}{2}\int_0^t [a(x+t-\tau)p'_{(2)}(x+t-\tau,\tau) \\ & \qquad + a(x-t+\tau)p'_{(2)}(x-t+\tau,\tau)]\, d\tau\end{aligned}$$

and

$$\sup_{\substack{(\xi,\tau)\in\Delta(x,t)\\ |\xi|\neq\tau}} \left|\frac{\partial^2 p}{\partial \xi^2}(\xi,\tau)\right| \le \gamma M_5,$$

where the constant M_5 depends on M_1, M_2, M_3, and h.

Theorem 7.4.1. *Let c_0 satisfy condition* A_0. *Then the problem* (7.4.13), (7.4.14) *has a solution in the class C^2 $(t > \psi(z) > 0)$. that has the following structure:*

$$u_0(z,t) = S(\psi(z))(\gamma\theta(t-\psi(z)) + p(\psi(z),t)), \quad t \in (0,T_h),\ z \in (0,h). \tag{7.4.35}$$

The existence of a solution to the direct problem (7.4.15)–(7.4.17). We now turn to the analysis of the direct problem (7.4.15)–(7.4.17). Set

$$\begin{aligned} w(x,y,t) &= u_1(\psi^{-1}(x), y, t)/S(x), \\ q(x,y) &= c_1(\psi^{-1}(x), y), \quad b(x) = c_0(\psi^{-1}(x)) \end{aligned} \tag{7.4.36}$$

(see (7.4.20), (7.4.21)). Using the even extension with respect to x, we arrive at the initial boundary value problem for w

$$Lw = q(x,y)\frac{\partial^2 v}{\partial x^2}, \quad x \in (-h_1,h_1), \quad y \in \mathcal{K}(\mathcal{D}), \quad t \in (0,T_{h_1}); \tag{7.4.37}$$

$$w\big|_{t=0} = 0, \quad \frac{\partial w}{\partial t}\Big|_{t=0} = 0; \tag{7.4.38}$$

$$w\big|_{\partial\mathcal{K}(\mathcal{D})} = 0, \tag{7.4.39}$$

where

$$L = \frac{\partial^2}{\partial t^2} - \frac{\partial^2}{\partial x^2} - b^2(x)\Delta_y - a(x), \quad h_1 = \psi(h), \tag{7.4.40}$$

and the functions $v(x,t)$ and $a(x)$ are defined in (7.4.21). Substituting $v(x,t) = \gamma\theta(t-|x|) + p(x,t)$ into (7.4.37) and applying D'Alembert's formula, it is easy to calculate the limit of $w(x,y,t)$ as $|x| \to t-0$:

$$\lim_{|x|\to t-0} w(x,y,t) = \gamma q(t,y)/4. \tag{7.4.41}$$

Consequently, instead of the problem (7.4.37)–(7.4.39) we can consider the problem

$$Lw = q(x,y)\frac{\partial^2 v}{\partial x^2}, \quad (x,t) \in \Delta(T), \ y \in \mathcal{K}(\mathcal{D}); \tag{7.4.42}$$

$$w(x,y,t)\big|_{t=|x|} = \gamma q(x,y)/4; \tag{7.4.43}$$

$$w\big|_{\partial\mathcal{K}(\mathcal{D})} = 0, \tag{7.4.44}$$

where $T \in (0, T_{h_1}/2)$; $\Delta(T) = \{(x,t) : x \in (-T,T), t \in (|x|, 2T-|x|)\}$.

Let $\Omega(T) = \{(x,y,t)\colon (x,t) \in \Delta(T),\ y \in \mathcal{K}(\mathcal{D})\}$. Assume that there exists a classical solution to the problem (7.4.42)–(7.4.44), i.e., the function

$$w(x,y,t) \in C^2(\Omega(T)) \cap C^1(\overline{\Omega(T)})$$

satisfying the equation (7.4.42) and the boundary conditions (7.4.43) and (7.4.44). We multiply both sides of (7.4.42) by w_t and integrate over the domain

$$\Omega(T,t) = \Omega(T) \cap \{(x,y,t') : t' < t\}.$$

After standard transformations, the application of the Ostrogradsky formula yields the identity

$$\frac{1}{2}\Big[\Big\|\frac{\partial w}{\partial t}\Big\|^2(t) + \Big\|\frac{\partial w}{\partial x}\Big\|^2(t) + \|b|\nabla_y w|\|^2(t)\Big]$$
$$= \int_{S_t} \langle \Phi, \bar{n}\rangle\, ds + \int_{\Omega(T,t)} \frac{\partial w}{\partial \tau}\Big(aw + q\frac{\partial^2 v}{\partial x^2}\Big)\, dx\, dy\, d\tau, \tag{7.4.45}$$

where $|\nabla_y w| = \sqrt{\sum_{i=1}^n (\frac{\partial w}{\partial y_i})^2}$, $S_t = \partial(\Omega(T)) \cap \{(x,y,t) : t' < t\}$; $\bar{n}$ is the external normal vector to S_t;

$$\Phi = \Big(-\frac{\partial w}{\partial t}\frac{\partial w}{\partial x}, -b^2\frac{\partial w}{\partial t}\nabla_y w, \frac{1}{2}\Big[\Big(\frac{\partial w}{\partial t}\Big)^2 + \Big(\frac{\partial w}{\partial x}\Big)^2 + b^2|\nabla_y w|^2\Big]\Big),$$

$$\|w\|^2(t) = \begin{cases} \int_{-t}^{t}\int_{\mathcal{K}(\mathcal{D})} w^2(x,y,t)\, dy\, dx, & t \in [0,T], \\ \int_{t-2T}^{2T-t}\int_{\mathcal{K}(\mathcal{D})} w^2(x,y,t)\, dy\, dx, & t \in [T,2T]. \end{cases}$$

Lemma 7.4.4. *Assume that c_0 and c_1 satisfy condition* A_0 *and the coefficients of the operator L and the initial boundary conditions of the problem* (7.4.42)–(7.4.44) *are constructed from c_0 and c_1 using the formulas* (7.4.21), (7.4.36). *Then the classical solution to the problem* (7.4.42)–(7.4.44) *is unique and*

$$\|w\|_1(t) \leq M_6 \|q\|_{W_2^1([-T,T]\times\mathcal{K}(\mathcal{D}))}, \quad t \in [0, 2T], \tag{7.4.46}$$

where the constant M_6 depends only on M_1, M_2, M_3, and h;

$$\|w\|_1^2(t) := \|w\|^2(t) + \||\nabla_{x,y,t}\, w|\|^2(t), \quad t \in [0, 2T].$$

Proof. The identity (7.4.45) implies the inequality

$$\begin{aligned} &\left\|\frac{\partial w}{\partial t}\right\|^2(t) + \left\|\frac{\partial w}{\partial x}\right\|^2(t) + M_1^2 \||\nabla_y w|\|^2(t) \\ &\quad \leq \tilde{M}_6 \left\{ \left\|\frac{\partial q}{\partial x}\right\|^2(T) + \||\nabla_y q|\|^2(T) \right. \\ &\qquad \left. + \int_0^t \left[\|w\|(\tau) \left\|\frac{\partial w}{\partial \tau}\right\|(\tau) + \|q\|(\tau) \left\|\frac{\partial w}{\partial \tau}\right\|(\tau) \right] d\tau \right\}, \quad t \in [0, 2T], \end{aligned} \tag{7.4.47}$$

where the constant $\tilde{M}_6$ depends on M_1, M_2, M_3, and h. From (7.4.41) it follows that

$$w^2(x, y, t) = \frac{\gamma^2}{8} q^2(x, y) + 2 \int_{|x|}^t w(x, y, \tau) \frac{\partial}{\partial \tau} w(x, y, \tau)\, d\tau.$$

Integrating this equality with respect to x and y within the domain

$$\Omega(T) \cap \{(x, y, t') \mid t' = t\}$$

and applying Hölder's inequality, we obtain

$$\|w\|^2(t) \leq \frac{\gamma^2}{8} \|q\|^2(t) + 2 \int_0^t \|w\|(\tau) \left\|\frac{\partial w}{\partial \tau}\right\|(\tau)\, d\tau. \tag{7.4.48}$$

Combining (7.4.47) with (7.4.48) and applying the inequalities

$$\frac{m_3^2(m_1^2 + m_2^2)}{1 + m_3^2} \leq m_1^2 + m_2^2 m_3^2 \leq (m_1^2 + m_2^2)(1 + m_3^2)$$

and Gronwall's inequality, we arrive at (7.4.46). □

To define a generalized solution to the problem (7.4.42)–(7.4.44) and prove that it exists, we will use a version of the Fourier method: instead of the usual separation of variables into spatial and time variables, we will seek a solution in the form of the sum of functions of the type $X(x, t)Y(y)$.

A function $w(x, y, t)$ is said to belong to the class $\mathcal{P}(T, \mathcal{D})$ if $w(x, y, t)$ is continuous with respect to the variables $(x, t) \in \overline{\Delta(T)}$ in the norm of L_2, i.e., if for any pair $(x, t) \in \overline{\Delta(T)}$ there holds

$$\lim_{(x',t')\to(x,t)} \|w(x', y, t') - w(x, y, t)\|_{L_2(\mathcal{K}(\mathcal{D}))} = 0.$$

In the proof of the theorem on the existence of a generalized solution to the problem (7.4.42)–(7.4.44) we will use modified versions of well-known statements.

Lemma 7.4.5. *Let a sequence* $\{u_m(x, y, t)\}$, $u_m \in \mathcal{P}(T, \mathcal{D})$, *converge to* $u(x, y, t)$ *uniformly with respect to* $(x, t) \in \overline{\Delta(T)}$ *in the norm of* L_2. *Then* $u \in \mathcal{P}(T, \mathcal{D})$.

Lemma 7.4.6. *Let a sequence of functions* $\{u_m(x, y, t)\}$, $u_m \in \mathcal{P}(T, \mathcal{D})$, *converge in itself uniformly with respect to* $(x, t) \in \overline{\Delta(T)}$ *in the norm of* L_2. *Then there exists a function* $u \in \mathcal{P}(T, \mathcal{D})$ *such that the sequence* $\{u_m(x, y, t)\}$ *converges to* $u(x, y, t)$ *uniformly with respect to* $(x, t) \in \overline{\Delta(T)}$ *in the norm of* L_2.

Lemma 7.4.7. *Let* $c_0(\psi^{-1}(x))$ *and* $c_1(\psi^{-1}(x), y)$ *satisfy condition* A_0. *Assume that there exists a sequence of functions* $\{\tilde{q}_m(x, y)\}$, $\tilde{q}_m \in C^1([-T, T] \times \mathcal{K}(\mathcal{D}))$, $m = 1, 2, \ldots$, *such that*

1) $\lim_{m\to\infty} \|q - \tilde{q}_m\|_{W_2^1([-T,T]\times\mathcal{K}(\mathcal{D}))} = 0$;

2) *for any* $m = 1, 2, \ldots$ *there exists a classical solution* $\tilde{w}_m(x, y, t)$ *to the problem* (7.4.42)–(7.4.44) *for* $q = \tilde{q}_m(x, y)$.

Then there exists a function $w \in \mathcal{P}(T, \mathcal{D})$ *such that*

$$\lim_{m\to\infty} \|w - \tilde{w}_m\| = 0. \tag{7.4.49}$$

Proof. Applying the estimate (7.4.46) from Lemma 7.4.4 to the difference $\tilde{w}_m - \tilde{w}_k$ and using the Sobolev embedding theorems, we conclude that the sequence $\{\tilde{w}_m\}$ converges uniformly in itself with respect to $(x, t) \in \overline{\Delta(T)}$ in the norm of L_2.

Hence, by Lemma 7.4.6, there exists a function $w \in \mathcal{P}(T, \mathcal{D})$ such that the condition (7.4.49) holds. □

The function $w(x, y, t)$ from Lemma 7.4.7 will be called a *generalized solution* to the problem (7.4.42)–(7.4.44).

Using (7.4.47) and Lemma 7.4.5, one can show that the generalized solution $w(x, y, t)$ satisfies equation (7.4.42) in the generalized sense.

Theorem 7.4.2. *Let* c_0 *and* c_1 *satisfy condition* A_0. *Then there exists a unique generalized solution to the problem* (7.4.42)–(7.4.44) *for any* $T \in (0, T_{h_1}/2)$.

Proof. We represent $q(x, y)$ in the form

$$q(x, y) = \sum_k q_k(x) Y_k(y), \tag{7.4.50}$$

where $Y_k(y) = \exp\left(\frac{i\pi}{\mathcal{D}} \langle k, y\rangle\right)$, $k = (k_1, \dots, k_n) \in \mathbb{Z}^n$, $\langle k, y\rangle = \sum_{j=1}^n k_j y_j$, and $\sum_k$ means the sum with respect to all $k = (k_1, \dots, k_n)$, $k_j \in \mathbb{Z}$, $j = 1, 2, \dots, n$.

By condition A_0, the series in (7.4.50) converges to $q(x, y)$ uniformly with respect to $x \in [-T, T]$ in the norm of L_2.

We will seek a generalized solution to the problem (7.4.42)–(7.4.44) in the form of a series

$$w(x, y, t) = \sum_k w_k(x, t) Y_k(y), \quad (x, t) \in \Delta(T), \tag{7.4.51}$$

where $w_k(x, t)$ is a classical solution to the problem

$$L_k w_k = q_k(x) \frac{\partial^2 v}{\partial x^2}, \quad (x, t) \in \Delta(T); \tag{7.4.52}$$

$$w_k(x, t)\big|_{t=|x|} = \gamma q_k(x)/4; \tag{7.4.53}$$

$$L_k = \frac{\partial^2}{\partial t^2} - \frac{\partial^2}{\partial x^2} + b^2(x)|k|^2 - a(x), \quad |k|^2 = \sum_{j=1}^n k_j^2. \tag{7.4.54}$$

Lemma 7.4.8. *Under the assumptions of Theorem* 7.4.2, *there exists a unique classical solution to the problem* (7.4.52), (7.4.53) *for any* $k \in \mathbb{Z}^n$, *and the solution satisfies the inequality*

$$\|w_k\|_{C(\overline{\Delta(T)})} \le M_7 \|q_k\|_{C^1[-T,T]} e^{(1+M_2^2|k|^2)T_h},$$

where the constant M_7 *depends only on* M_1, M_2, M_3, *and* h.

The proof of Lemma 7.4.8 is based on the fact that the change of variables $\xi = x + t$, $\eta = x - t$ reduces the problem (7.4.52), (7.4.53) to the Goursat problem.

Hence, the sequence of functions

$$\tilde{q}_m(x, y) = \sum_{|k| \le m} q_k(x) Y_k(y), \quad \tilde{w}_m(x, y, t) = \sum_{|k| \le m} w_k(x, t) Y_k(y)$$

satisfies all the assumptions of Lemma 7.4.7, which completes the proof of Theorem 7.4.2. □

Regularization and the uniqueness of solutions to the inverse problem

Theorem 7.4.3. *Let $c_0(z)$ satisfy items* 1) *and* 2) *of condition* A_0. *Then the solution of the inverse problem* (7.4.15)–(7.4.17), (7.4.19) *is unique in the class of functions $c_1(z, y)$ that satisfy items* 3) *and* 4) *of condition* A_0.

The proof of Theorem 7.4.3 follows from Theorem 7.4.4 as the assumptions 1)–3) of Theorem 7.4.4 follow from the condition A_0 for $c_0(z)$ and $c_1(z, y)$.

Theorem 7.4.4. *Let $c_0(z)$ satisfy condition* A_0. *Assume that functions $g^{(m)}(y,t)$, $m = 1, 2$, satisfy the following conditions:*

1) $g^{(m)}(y,t) = \sum_k g_k^{(m)}(t) Y_k(y)$, $m = 1, 2$;
2) $g^{(m)}(y,t)$ *are continuous in t in the norm of L_2 on the interval* $[0, T_h]$;
3) *for $g^{(m)}(y,t)$, $m = 1, 2$, there exists a solution $c_1^{(m)}(z, y)$, $m = 1, 2$, to the inverse problem* (7.4.15)–(7.4.17), (7.4.19) *that satisfies condition* A_0.

Then for any $k \in \mathbb{Z}^n$

$$\|q_k^{(1)} - q_k^{(2)}\|_{C[-T,T]} \le \omega(k) \|g_k^{(1)} - g_k^{(2)}\|_{C[0,T_h]}, \tag{7.4.55}$$

where $T \in (0, T_{h_1}/2)$; $q^{(m)}(x, y) = c_1^{(m)}(\psi^{-1}(x), y) = \sum_k q_k^{(m)}(x) Y_k(y)$; $\omega(k) = M_8 \exp(M_9 \exp(M_{10}|k|))$, the constants M_8, M_9, and M_{10} depend only on M_1, M_2, M_3, and h.

Proof. By Theorem 7.4.2, for

$$q^{(m)}(x, y) = \sum_k q_k^{(m)}(x) Y_k(y), \quad m = 1, 2,$$

there exists a generalized solution

$$w^{(m)}(x, y, t) = \sum_k w_k^{(m)}(x, t) Y_k(y), \quad m = 1, 2,$$

to the problem (7.4.42)–(7.4.44) whose Fourier coefficients are classical solutions to the problems

$$L_k w_k^{(m)} = q_k^{(m)}(x) \frac{\partial^2 v}{\partial x^2}, \quad (x, t) \in \Delta(T); \tag{7.4.56}$$

$$w_k^{(m)}(x, t)\big|_{t=|x|} = \gamma q_k^{(m)}(x)/4; \tag{7.4.57}$$

$$m = 1, 2, \quad k \in \mathbb{Z}^n, \quad T \in (0, T_{h_1}/2).$$

By the uniqueness of the solution of the direct problem (7.4.42)–(7.4.44),

$$w_k^{(m)}(0,t) = g_k^{(m)}(t), \quad m = 1,2, \quad t \in [0,2T]. \tag{7.4.58}$$

Since $w_k^{(m)}(x,t)$ is even with respect to x, we have the condition

$$\frac{\partial w_k^{(m)}}{\partial x}(0,t) = 0, \quad m = 1,2. \tag{7.4.59}$$

We introduce the operator

$$B_{x,t}[w_k] = \frac{1}{2}\int_0^x \int_{t-x+\xi}^{t+x-\xi} w_k(\xi,\tau)\,d\tau\,d\xi, \quad x \in \mathbb{R}_+, \quad t \in \mathbb{R}_+,$$

and apply D'Alembert's formula for the representation of a solution to the Cauchy problem for equation (7.4.56) with the Cauchy data (7.4.58), (7.4.59). As a result, we have

$$w_k^{(m)}(x,t) = Q_k^{(m)}(x,t) + B_{x,t}[(|k|^2b^2 - a)w_k^{(m)}], \quad (x,t) \in \Delta_1(T), \ m = 1,2, \tag{7.4.60}$$

where

$$\begin{aligned}
Q_k^{(m)}(x,t) &= G_k^{(m)}(x,t) + \Phi_k^{(m)}(x,t), \quad \Phi_k^{(m)}(x,t) = -B_{x,t}[q_k^{(m)} v''_{(1)}],\\
G_k^{(m)}(x,t) &= [g_k^{(m)}(t+x) + g_k^{(m)}(t-x)]/2,\\
v''_{(1)}(x,t) &= \frac{\partial^2 v}{\partial x^2}(x,t),\\
\Delta_1(T) &= \{(x,t):\ x \in (0,T),\ x < t < 2T - x\}.
\end{aligned}$$

For any multi-index k and $m = 1,2$, the Volterra integral equation (7.4.60) has a unique solution $w_k^{(m)} \in C(\Delta_1(T))$, which can be found using the method of successive approximations. Indeed, by the assumptions of Theorem 7.4.4, the functions $Q_k^{(m)}$, $|k|^2b^2 - a$ are continuous. Therefore, a solution $w_k^{(m)}$ can be written as

$$w_k^{(m)}(x,t) = Q_k^{(m)}(x,t) + \frac{1}{2}\int_0^x \int_{t-x+\xi}^{t+x-\xi} R_k(x,t,\xi,\tau) Q_k^{(m)}(\xi,\tau)\,d\tau\,d\xi,$$

$$(x,t) \in \Delta_1(T), \ m = 1,2.$$

In this case, the resolvent $R_k(x,t,\xi,\tau)$ is continuous on $\Delta_1(T)\times\Delta_1(T)$ and

$$\|R_k\|_{C^4(\Delta_1(T)\times\Delta_1(T))} \le e^{T_{h_1}\sqrt{|k|^2 M_2^2+M_4}}. \tag{7.4.61}$$

Let $x = t-0$ and use the condition (7.4.57):

$$\frac{\gamma}{4}q_k^{(m)}(t) = Q_k^{(m)}(t-0,t) + \frac{1}{2}\int_0^t\int_\xi^{2t-\xi} R_k(t-0,t,\xi,\tau)Q_k^{(m)}(\xi,\tau)\,d\tau\,d\xi,$$
$$t\in[0,T],\ m=1,2. \tag{7.4.62}$$

Write (7.4.62) in the form of a linear Volterra integral equation of the second kind for $q_k^{(m)}(t)$:

$$\frac{\gamma}{4}q_k^{(m)}(t) = -B_{t,t}[q_k^{(m)}v''_{(1)}] + \frac{1}{2}\int_0^t\int_\xi^{2t-\xi} R_k(t-0,t,\xi,\tau)\Phi_k^{(m)}(\xi,\tau)\,d\tau\,d\xi$$
$$+ G_k^{(m)}(t-0,t) + \frac{1}{2}\int_0^t\int_\xi^{2t-\xi} R_k(t-0,t,\xi,\tau)G_k^{(m)}(\xi,\tau)\,d\tau\,d\xi,$$
$$t\in[0,T],\ m=1,2. \tag{7.4.63}$$

The absolute value of the kernel of the integral equation (7.4.63) is bounded from above by M_9, while the absolute value of the right-hand side of (7.4.63) is bounded by $M_8\|g_k^{(m)}\|_{C[0,2T]}$. Since (7.4.63) is linear with respect to $q_k^{(m)}$, we conclude that the inequality (7.4.55) holds. □

Theorem 7.4.5. *Let c_0 satisfy condition A_0. Assume that for $g(y,t)$ that is continuous in t in the norm of L_2 on the interval $[0,T_h]$ there exists a solution to the inverse problem* (7.4.15)–(7.4.17), (7.4.19) *that satisfies condition A_0. Let $g_\varepsilon(y,t)$ be continuous in t in the norm of L_2 on $[0,T_h]$, and let*

$$\max_{t\in[0,T_h]}\|g-g_\varepsilon\|_{L_2(\mathcal{K}(\mathcal{D}))} \le \varepsilon.$$

Set

$$\tilde{q}_N^{(\varepsilon)}(x,y) = \sum_{|k|\le N} q_k^{(\varepsilon)}(x)Y_k(y),$$

where $q_k^{(\varepsilon)}(x)$ is a solution to the integral equation (7.4.63) *with the Fourier coefficient $g_k^{(\varepsilon)}(t)$ of the function $g_\varepsilon(y,t)$ substituted for $g_k^{(m)}(t)$:*

$$g_\varepsilon(y,t) = \sum_k g_k^{(\varepsilon)}(t)Y_k(y).$$

Then

$$\|q - \tilde{q}_N^{(\varepsilon)}\|_{L_2(\mathcal{K}(\mathcal{D}))} \leq \omega(N)\varepsilon + M/N, \quad t \in [0, T], \tag{7.4.64}$$

where

$$M = \max_{x \in [0, T_{h_1}]} \|q\|_{W_2^1(\mathcal{K}(\mathcal{D}))}.$$

The proof follows from Theorem 7.4.4.

The estimate (7.4.64) shows that $\tilde{q}_N^{(\varepsilon)}$ is a regularizing family for the inverse problem (7.4.15)–(7.4.17), (7.4.19).

Chapter 8
Linear problems for parabolic equations

Linear problems for parabolic equations are very important for practical applications and are extremely diverse in terms of the degree of complexity. Aside from simple ones, there are problems as complex as the Cauchy problem for the Laplace equation and even more complex ones.

In the beginning of this chapter, we will discuss different formulations of linear inverse problems of heat conduction and compare them with analogous formulations for hyperbolic equations based on the relationship between the solutions of the corresponding direct problems (Section 8.1). In Section 8.2, we will discuss in detail one of the best-studied ill-posed problems—the Cauchy problem with reverse time. We will consider several methods for the approximate solution of this problem (the Lattès–Lions method, the method of separation of variables, and the method of integral equations), obtain a conditional stability estimate, and reduce the problem to an operator equation $Aq = f$. The properties of the operator A and its adjoint operator A^* are analogous to the properties of the corresponding operators for the Laplace equation (see Chapter 9). They will be explained in the next chapter. The same is true for extension problems and inverse boundary-value problems (Section 8.3); nevertheless, we will present several proofs for the reader to be aware of the similarities and differences between them. In conclusion (Section 8.4), we give several examples of linear inverse problems. Note that a more complete analysis of the covered problems can be found in (Alifanov et al., 1988), (Beck et al., 1985) and (Hào, 1995).

8.1 On the formulation of inverse problems for parabolic equations and their relationship with the corresponding inverse problems for hyperbolic equations

Inverse and ill-posed problems for parabolic equations represent one of the most complex and important types of inverse problems for practical applications. The stability of these problems is worse than that of the corresponding inverse problems for hyperbolic equations (see Chapters 7 and 10). Figuratively speaking, it is "worse by one inversion of the Laplace transform". This means that if we want to apply the methods developed for the hyperbolic case to inverse problems for parabolic equations, it can be done by inverting the Laplace transform (see (8.1.8) below), which obviously adds to instability. Therefore, the main methods for the numerical solution of inverse problems for parabolic equations are optimization methods and combinations thereof with methods that use the properties of the operator of the direct problem.

Linear inverse and ill-posed problems for parabolic equations are categorized into the following main groups:

1) problems with reverse time or, equivalently, problems of reconstruction of the initial state from the results of measurements at a certain fixed instant of time (retrospective inverse problems);

2) problems of extension of a solution inside the domain under study from its boundary (or a part thereof);

3) inverse boundary-value problems of reconstruction of a solution on an inaccessible part of the boundary;

4) problems of reconstruction of sources, i.e., determining the right-hand side of a parabolic equation;

5) interior inverse problems, i.e., problems of reconstruction of a solution from its values at interior points of the domain under study.

This classification can certainly be refined further by subdividing each of the groups into classes, but we will confine ourselves to the simplest examples. Further details about the classification and application of the theory and numerical methods for solving inverse problems of heat conduction can be found in (Alifanov et al., 1988), (Lavrentiev et al., 1986) and (Beck et al., 1985).

Note that inverse boundary-value problems, extension problems, and problems of reconstructing sources are interrelated. For example, solving a problem of extension to the inaccessible boundary, we end up solving an inverse boundary-value problem. Extending a solution in the direction of a source, we get more information on its location, shape, and intensity.

Based on the above classification, we can illustrate the differences in the solution of inverse problems for hyperbolic and parabolic equations. In the case of two variables (x, t), the extension problem, the inverse boundary-value problem, and the problem with reverse time for the hyperbolic equation $u_{tt} = u_{xx}$ are well posed, whereas all these problems for parabolic equations are strongly ill-posed. In the multidimensional case, the problem with reverse time for the hyperbolic equation $u_{tt} = Lu$, where the operator L is defined by formula (8.1.4) below, is well posed; moreover, it is reduced to an identical problem by the change of variables $\tau = -t$.

Ill-posedness. Consider a one-dimensional problem of extending a solution of the parabolic equation

$$u_t = u_{xx}, \quad x > 0, \ t \in \mathbb{R}, \tag{8.1.1}$$

to the half-space $x > 0$ based on the measurements on the boundary $x = 0$:

$$u(0,t) = f(t), \quad u_x(0,t) = g(t). \tag{8.1.2}$$

We will show that this problem is ill-posed (Pucci, 1959). Indeed, let $f_n(t) = 2n^{-1}\sin(2n^2t)$, $g(t) \equiv 0$. Then the function

$$u_n(x,t) = n^{-1}\big[e^{nx}\sin(2n^2t + nx) + e^{-nx}\sin(2n^2t - nx)\big]$$

is clearly a solution to equation (8.1.1) satisfying the conditions (8.1.2). But f_n tends to zero as $n \to \infty$, whereas $u_n(x,t)$ increases for any $x > 0$, which means that the problem (8.1.1), (8.1.2) is unstable. (Selection of specific function spaces to complete the analysis of ill-posedness is left to the reader.)

For comparison, consider a similar extension problem for a hyperbolic equation

$$u_{tt} = u_{xx}, \quad x > 0,\ t \in \mathbb{R},$$
$$u(0,t) = f(t), \quad u_x(0,t) = g(t).$$

This problem has an explicit solution

$$u(x,t) = \frac{1}{2}\big[f(t+x) + f(t-x)\big] + \frac{1}{2}\int_{t-x}^{t+x} g(\tau)\,d\tau,$$

and is obviously well posed if considered in any finite domain $\Delta(T) = \{x \in (0,T), t \in (-T+x, T-x)\}$.

Relationship between inverse problems for hyperbolic and parabolic equations. When studying inverse problems for parabolic and elliptic equations, it is possible to consider equivalent inverse problems for hyperbolic equations. In a connected bounded domain $\mathcal{D} \subset \mathbb{R}^n$ with smooth boundary S, consider the mixed problem for a parabolic equation

$$u_t = Lu + \varphi(x,t), \quad u\big|_{t=0} = f(x), \quad \frac{\partial u}{\partial \bar{n}}\Big|_S = \psi(x,t), \tag{8.1.3}$$

where $\bar{n}$ is the conormal to S, and L is a uniformly elliptic operator with continuous coefficients that depend only on the spatial variable $x = (x_1, \dots, x_n)$:

$$Lu = \sum_{i,j=1}^{n} a_{ij}(x)u_{x_i x_j} + \sum_{i=1}^{n} b_i(x)u_{x_i} + c(x)u, \tag{8.1.4}$$

$$\mu\sum_{i=1}^{n}\nu_i^2 \le \sum_{i,j=1}^{n} a_{ij}(x)\nu_i\nu_j \le \frac{1}{\mu}\sum_{i=1}^{n}\nu_i^2 \quad \text{for any } \nu_i \in \mathbb{R},\ 0 < \mu < \infty.$$

We associate the problem (8.1.3) with the following problem for a hyperbolic equation:

$$\begin{gathered}\tilde{u}_{tt} = L\tilde{u} + \tilde{\varphi}(x,t),\\ \tilde{u}\big|_{t=0} = f(x), \quad \tilde{u}_t\big|_{t=0} = 0, \quad \frac{\partial \tilde{u}}{\partial \bar{n}}\Big|_S = \tilde{\psi}(x,t).\end{gathered} \tag{8.1.5}$$

Assume that the functions $\tilde{\varphi}$ and $\tilde{\psi}$ are sufficiently smooth, increase as $t \to \infty$ not faster than $C\,\mathrm{e}^{\alpha t}$, and are related to φ and ψ by the formulas

$$\varphi(x,t) = \int_0^\infty \tilde{\varphi}(x,\tau)G(t,\tau)\,d\tau, \quad \psi(x,t) = \int_0^\infty \tilde{\psi}(x,\tau)G(t,\tau)\,d\tau, \tag{8.1.6}$$

where

$$G(t,\tau) = \frac{1}{\sqrt{\pi t}}\,\mathrm{e}^{-\tau^2/(4t)}.$$

We will show that the function

$$u(x,t) = \int_0^\infty \tilde{u}(x,\tau)G(t,\tau)\,d\tau \tag{8.1.7}$$

is a solution to the problem (8.1.3) if $\tilde{u}$ is a solution to (8.1.5) and (8.1.6) holds.

Indeed, since $G_t = G_{\tau\tau}$ for $t > 0$, we have

$$\begin{aligned}u_t &= \int_0^\infty \tilde{u}(x,\tau)G_t(t,\tau)\,d\tau = \int_0^\infty \tilde{u}(x,\tau)G_{\tau\tau}(t,\tau)\,d\tau\\ &= \tilde{u}G_\tau\big|_0^\infty - \int_0^\infty \tilde{u}_\tau(x,\tau)G_\tau(t,\tau)\,d\tau\\ &= \tilde{u}G_\tau\big|_0^\infty - \tilde{u}_\tau G\big|_0^\infty + \int_0^\infty \tilde{u}_{\tau\tau}(x,\tau)G(t,\tau)\,d\tau\\ &= \int_0^\infty G(t,\tau)\tilde{u}_{\tau\tau}(x,\tau)\,d\tau\\ &= \int_0^\infty G(t,\tau)\big[L\tilde{u}(x,\tau) + \tilde{\varphi}(x,\tau)\big]\,d\tau = \int_0^\infty G(t,\tau)L\tilde{u}(x,\tau)\,d\tau + \varphi(x,t).\end{aligned}$$

However, since

$$Lu = \int_0^\infty G(t,\tau)L\tilde{u}(x,\tau)\,d\tau,$$

for $t > 0$ we have

$$u_t = Lu + \varphi.$$

It is easy to verify that $u(x,t)$ satisfies the initial and boundary conditions:

$$\begin{aligned} u\big|_{t=0} &= \lim_{t\to 0}\int_0^\infty \tilde{u}(x,\tau)G(t,\tau)\,d\tau \\ &= \lim_{t\to 0}\frac{2}{\sqrt{\pi}}\int_0^\infty \tilde{u}(x,2\sqrt{t}\tau)\mathrm{e}^{-\tau^2}\,d\tau = \tilde{u}(x,0) = f(x), \\ \frac{\partial u}{\partial \bar{n}}\Big|_S &= \int_0^\infty \frac{\partial \tilde{u}}{\partial \bar{n}}\Big|_S G(t,\tau)\,d\tau = \int_0^\infty \tilde{\psi}(x,\tau)G(t,\tau)\,d\tau = \psi(x,t). \end{aligned}$$

Thus, formula (8.1.7) enables to find a solution of the problem (8.1.3) through the solution of (8.1.5) if the data of these problems are related by the formulas (8.1.6). It is known that the solution to the problem (8.1.3) is unique under fairly general assumptions on the operator L and the surface S, and it is related to $\tilde{u}$ by formula (8.1.7). Note that (8.1.7) is invertible for any fixed x, since it can be represented in the form of the Laplace transform with parameter $p = 1/(4t)$:

$$u\Big(x,\frac{1}{4p}\Big) = \sqrt{\frac{p}{\pi}}\int_0^\infty \mathrm{e}^{-zp}\,\tilde{u}(x,\sqrt{z}\,)\frac{dz}{\sqrt{z}}. \tag{8.1.8}$$

Therefore, formula (8.1.7) establishes a one-to-one correspondence between the solutions to the problems (8.1.3) and (8.1.5).

We now consider the inverse problem of determining $\varphi(x,t)$ or coefficients of the operator L in the parabolic equation $u_t = Lu + \varphi(x,t)$ from the additional information about a solution to the problem (8.1.3) on the boundary S:

$$u\big|_S = g(x,t), \quad t \geq 0. \tag{8.1.9}$$

To the function $g(x,t)$ we associate a function $\tilde{g}(x,t)$, $x \in S$, that is uniquely determined from the equation

$$g(x,t) = \int_0^\infty \tilde{g}(x,\tau)G(t,\tau)\,d\tau, \quad x \in S, \quad t \geq 0. \tag{8.1.10}$$

Setting

$$\tilde{u}\big|_S = \tilde{g}(x,t), \quad t \geq 0, \tag{8.1.11}$$

we obtain the inverse problem (8.1.5), (8.1.11).

Under rather general assumptions on the coefficients of the operator L, the solution to the problem (8.1.5) satisfies the estimate

$$|\tilde{u}(x,t)| \leq C\,\mathrm{e}^{\alpha t}, \quad x \in S.$$

Remark 8.1.1. Formula (8.1.7) and the equivalent formula (8.1.8) show that $u(x,t)$ is an analytic function of t in the domain $t > 0$. Therefore, it suffices to specify the function $g(x,t)$ from (8.1.9) in an arbitrarily small neighborhood of the point $t = 0$, for example, in an interval $0 \leq t < \delta$, $\delta > 0$.

The established relationship between the problems (8.1.3), (8.1.9) and (8.1.5), (8.1.11) makes it possible to prove the uniqueness of solutions to linear inverse problems for parabolic equations using the results of analysis of the corresponding inverse problems for hyperbolic equations.

8.2 Inverse problem of heat conduction with reverse time (retrospective inverse problem)

In many practical problems it is required to reconstruct the distribution of the temperature of a body at a certain instant of time $t \in (0, T)$ from the temperature measured at $t = T$. Such problems are called retrospective or initial boundary value problems for the heat conduction equation with reverse time (Alifanov et al., 1988). For example, for a bar of length l with fixed temperature at the ends (without loss of generality we can assume that the temperature at both ends is zero) we have the following retrospective problem:

$$\frac{\partial v}{\partial \tau} = \frac{\partial^2 v}{\partial x^2}, \quad x \in (0, l), \quad \tau \in (0, T), \tag{8.2.1}$$

$$v(0, \tau) = v(l, \tau) = 0, \quad \tau \in (0, T), \tag{8.2.2}$$

$$v(x, T) = f(x), \quad x \in (0, l). \tag{8.2.3}$$

It is required to determine $v(x, \tau)$ in the domain $(0, l) \times (0, T)$.

Performing the change of variables $\tau = T - t$ in (8.2.1)–(8.2.3) and setting $u(x, t) = v(x, \tau)$, we arrive at an equivalent (adjoint) problem

$$\frac{\partial u}{\partial t} = -\frac{\partial^2 u}{\partial x^2}, \quad x \in (0, l), \quad t \in (0, T), \tag{8.2.4}$$

$$u(0, t) = u(l, t) = 0, \quad t \in (0, T), \tag{8.2.5}$$

$$u(x, 0) = f(x), \quad x \in (0, l), \tag{8.2.6}$$

in which it is required to determine $u(x, t)$ in the domain $(0, l) \times (0, T)$.

Note that the problem (8.2.4)–(8.2.6) is ill-posed because arbitrarily small variations in the right-hand side f may cause arbitrarily large variations in u. Indeed, for $f(x) = \frac{1}{n} \sin \frac{\pi n x}{l}$ the solution $u(x, t) = \frac{1}{n} \mathrm{e}^{\frac{\pi^2 n^2 t}{l^2}} \sin \frac{\pi n x}{l}$ increases indefinitely as n increases, while the data $f(x)$ of the problem tends to zero.

Consider several examples of regularization of the problem (8.2.4)–(8.2.6). Assume that the functions $u(x, t)$ and $f(x)$ are sufficiently smooth.

The Lattès–Lions quasi-inversion method. We will describe the regularization procedure for the problem (8.2.4)–(8.2.6) proposed by Lattès and Lions (1967).

Let $u_\alpha(x,t)$ be a solution to the problem

$$\frac{\partial u}{\partial t} = -\frac{\partial^2 u}{\partial x^2} - \alpha \frac{\partial^4 u}{\partial x^4},$$
$$u(0,t) = u(l,t) = \frac{\partial^2}{\partial x^2} u(0,t) = \frac{\partial^2}{\partial x^2} u(l,t) = 0, \quad 0 \le t \le T, \tag{8.2.7}$$
$$u(x,0) = f(x).$$

It is easy to verify that the solution $u_\alpha(x,t)$ to the problem (8.2.7) can be constructed using the Fourier method:

$$u_\alpha(x,t) = \sum_{k=1}^{\infty} f_k \exp\Big\{\Big(\frac{\pi k}{l}\Big)^2 \Big(1 - \alpha\Big(\frac{\pi k}{l}\Big)^2\Big)t\Big\} \sin\Big(\frac{\pi k x}{l}\Big), \tag{8.2.8}$$

where f_k are the Fourier coefficients for $f(x)$. The family of operators R_α transforming $f(x)$ into a solution $u_\alpha(x,t)$ of the form (8.2.8), i.e., defined by the formula $R_\alpha f = u_\alpha$, is a regularizing family for the problem (8.2.4)–(8.2.6).

We put

$$\|u(\cdot,t)\| := \|u(\cdot,t)\|_{L_2(0,l)} = \Big(\int_0^l u^2(x,t)\,dx\Big)^{1/2},$$
$$\|f\| := \|f\|_{L_2(0,l)}, \quad \|R_\alpha\|(t) = \sup_{\|f\|=1} \|R_\alpha f(\cdot,t)\|.$$

The norm of the operator R_α satisfies the estimate

$$\|R_\alpha\|(t) \le e^{t/(4\alpha)}. \tag{8.2.9}$$

We can estimate the deviation of u_α from the exact solution u to the problem (8.2.4)–(8.2.6) under the condition $\|u(\cdot,T)\| \le C$, where $C = \text{const}$. Note that

$$\|u(\cdot,T)\|^2 = \sum_{k=1}^{\infty} f_k^2 e^{2(\pi k/l)^2 T}.$$

In this case, we have

$$\|u_\alpha(\cdot,t) - u(\cdot,t)\| = \Big(\int_0^l (u_\alpha(x,t) - u(x,t))^2\,dx\Big)^{1/2} \le \frac{4Ct}{(T-t)^2}\,\alpha e^{-2}.$$

The proof of this estimate together with (8.2.9) is left to the reader as an exercise.

Regularization based on the separation of variables. Consider the problem (8.2.4)–(8.2.6) where it is required to find $u(x,t)$.

We denote by f_k the Fourier coefficients of $f(x)$:

$$f_k = \frac{1}{l}\int_0^l f(x)\sin\frac{\pi k x}{l}\,dx.$$

If $f(x)$ is such that there is a solution $u(x,t)$ of the problem under study, then

$$u(x,t) = \sum_{k=1}^{\infty} f_k \mathrm{e}^{(\pi k/l)^2 t}\sin\frac{\pi k x}{l}. \tag{8.2.10}$$

Consider the sequence of linear operators R_n:

$$(R_n f)(x,t) = \sum_{k=1}^{n} f_k \mathrm{e}^{(\pi k/l)^2 t}\sin\frac{\pi k x}{l}. \tag{8.2.11}$$

The operators of determining the Fourier coefficients f_k and of finite summation are continuous, which implies the continuity of the operators R_n. The fact that the sequence $\{R_n f\}$ converges to the exact solution $u(x,t)$ follows from the representation (8.2.10) and the convergence of the Fourier series in the space L_2.

We now estimate the effectiveness of the application of this regularizing family to approximate data. Assume that the problem under consideration is conditionally well-posed, and the well-posedness set M is defined by the inequality

$$\|u(\cdot,T)\| := \Big(\int_0^l u^2(x,T)\,dx\Big)^{1/2} \le C. \tag{8.2.12}$$

It is obvious that

$$\|R_n\|(t) = \mathrm{e}^{(\pi n/l)^2 t}.$$

From (8.2.10) and (8.2.11) it follows that

$$\|(R_n f)(\cdot,t) - u(\cdot,t)\|^2 = \sum_{k=n+1}^{\infty} f_k^2 \mathrm{e}^{2(\pi k/l)^2 t}. \tag{8.2.13}$$

On the other hand, (8.2.12) implies

$$\sum_{k=1}^{\infty} f_k^2 \mathrm{e}^{2(\pi k/l)^2 T} \le C^2. \tag{8.2.14}$$

It is easy to see that the sum in the right-hand side of (8.2.13) achieves the maximum value under the condition (8.2.14) when

$$f_k = 0, \quad k \ne n+1, \quad f_{n+1} = C\,\mathrm{e}^{-(\pi(n+1)/l)^2 T},$$

and therefore

$$\|(R_n f)(\cdot,t) - u(\cdot,t)\| \leq C\,\mathrm{e}^{-(\pi(n+1)/l)^2(T-t)}.$$

Consequently, if there is an error δ in the given data, i.e., $\| f_\delta - f \| \leq \delta$, then

$$\begin{aligned}\|(R_n f_\delta)(\cdot,t) - u(\cdot,t)\| &\leq \|R_n(f_\delta - f)(\cdot,t)\| + \|(R_n f)(\cdot,t) - u(\cdot,t)\| \\ &\leq \mathrm{e}^{(\pi n/l)^2 t}\delta + C\,\mathrm{e}^{-(\pi(n+1)/l)^2(T-t)}. \end{aligned} \tag{8.2.15}$$

Exercise 8.2.1. Assume that the error in the given data f is fixed. Find the index $n = n(\delta)$ such that the right-hand side of the estimate (8.2.15) is minimal.

A detailed discussion of similar matters within the framework of the theory of operator differential equations can be found in Krein (1957).

Reduction to an integral equation. The method of integral equations is one of the main methods in the study of problems of mathematical physics. Problems of mathematical physics that are classically well-posed can usually be reduced to Fredholm or Volterra integral equations of the second kind. Ill-posed problems can usually be reduced to integral equations of the first kind. It is easy to verify that solution of the problem (8.2.4)–(8.2.6) is equivalent to solution of an integral equation of the form

$$\int_0^l K(x,\xi)q(\xi)\,d\xi = f(x), \tag{8.2.16}$$

where

$$K(x,\xi) = \frac{1}{l}\sum_{k=1}^{\infty} \mathrm{e}^{-(\pi k/l)^2 T} \sin\frac{\pi k x}{l} \sin\frac{\pi k \xi}{l}.$$

Conditional stability estimate. In the initial boundary value problem for the heat conduction equation with reverse time, it is natural to assume that the solution $u(x,t)$ belongs to the set

$$M = \{u(x,t) : C_1 \leq u(x,t) \leq C_2\},$$

where C_1 and C_2 are constants. Indeed, if $u(x,t)$ describes the concentration of a substance in the process of diffusion, i.e., $u(x,t)$ is the ratio of the mass of the substance in a unit layer to the total mass of the layer, then, obviously,

$$0 \leq u(x,t) \leq 1.$$

If $u(x,t)$ is the temperature distribution in a bar, then, for example,

$$C_1 \leq u(x,t) \leq C_2,$$

where C_1 is the absolute zero temperature and C_2 is the melting point of the bar.

Theorem 2.2.1 implies that if a well-posedness set M is compact, then a conditional stability estimate $\beta(\delta)$ (see Definition 2.8.1) can be found on the set M. The estimate $\beta(\delta)$ is of high importance in the study of ill-posed problems. As noted above, in the initial boundary value problem for the heat conduction equation, it is natural to take the set of bounded functions for the well-posedness set. We will obtain the conditional stability estimate $\beta(\delta)$ on the set of functions $u(x,t)$ that are bounded in $L_2(0,l)$ for any $t \in (t,0)$.

Let $u(x,t)$ be a solution to the equation

$$\frac{\partial u}{\partial t} = -\frac{\partial^2 u}{\partial x^2}, \quad x \in (0,l),\ t \in (0,T), \tag{8.2.17}$$

that satisfies the boundary conditions

$$u(0,t) = u(l,t) = 0, \quad 0 \le t \le T. \tag{8.2.18}$$

We introduce the function

$$\rho(t) = \int_0^l u^2(x,t)\,dx.$$

Differentiating $\rho(t)$ twice, we obtain

$$\rho'(t) = 2\int_0^l u\,u_t\,dx,$$

$$\rho''(t) = 2\int_0^l u_t^2\,dx + 2\int_0^l u\,u_{tt}\,dx.$$

Transforming the second term in the expression for $\rho''(t)$, we have

$$\int_0^l u\,u_{tt}\,dx = \int_0^l u\,u_{xxxx}\,dx = \int_0^l u_{xx}^2\,dx = \int_0^l u_t^2\,dx.$$

Consequently,

$$\rho''(t) = 4\int_0^l u_t^2\,dx.$$

Consider the function

$$\varphi(t) = \ln \rho(t).$$

Differentiating $\varphi(t)$ twice, we have

$$\begin{aligned}\varphi''(t) &= \frac{1}{\rho^2(t)}\left[\rho''(t)\,\rho(t) - (\rho'(t))^2\right] \\ &= \frac{1}{\rho^2(t)}\Big[4\int_0^l u^2\,dx \int_0^l u_t^2\,dx - 4\Big(\int_0^l u u_t\,dx\Big)^2\Big].\end{aligned}$$

Hence, by the Cauchy–Bunyakovsky inequality,

$$\varphi''(t) \geq 0. \tag{8.2.19}$$

The inequality (8.2.19) means that $\varphi(t)$ is convex, which implies that on the interval $[0, T]$ the function $\varphi(t)$ does not exceed the linear function that has the same values at the ends of the interval as $\varphi(t)$:

$$\varphi(t) \leq \frac{T-t}{T}\,\varphi(0) + \frac{t}{T}\,\varphi(T). \tag{8.2.20}$$

Exponentiating the inequality (8.2.20), we have

$$\rho(t) \leq [\rho(0)]^{(T-t)/T}[\rho(T)]^{t/T}. \tag{8.2.21}$$

Let $u(x,t)$ be a solution of the problem (8.2.4)–(8.2.6) and let $u_\delta(x,t)$ solve the same problem but with $f_\delta(x)$ instead of $f(x)$. Setting $\tilde{u}(x,t) = u(x,t) - u_\delta(x,t)$, we note that $\tilde{u}(x,t)$ satisfies (8.2.17), (8.2.18) and $\tilde{u}(x,0) = f(x) - f_\delta(x)$. Assume that

$$\Big(\int_0^l \tilde{u}^2(x,T)\,dx\Big)^{1/2} \leq C, \tag{8.2.22}$$

$$\Big(\int_0^l \tilde{u}^2(x,0)\,dx\Big)^{1/2} \leq \delta, \tag{8.2.23}$$

where C and δ are constants. Clearly, from (8.2.21)–(8.2.23) it follows that

$$\Big(\int_0^l \tilde{u}^2(x,t)\,dx\Big)^{1/2} \leq (C)^{t/T}(\delta)^{(T-t)/T} = \beta(\delta). \tag{8.2.24}$$

Thus, in the initial boundary value problem for the heat conduction equation with reverse time, the conditional stability estimate (8.2.24) holds in the case where the set of well-posedness consists of functions $u(x,t)$ bounded in $L_2(0,l)$ for $t = T$.

Formulation of the initial boundary value problem for the heat conduction equation in the form of an inverse problem. Reduction to an operator equation. The ill-posed initial boundary value problem for the heat conduction equation

$$\frac{\partial u}{\partial t} = -\frac{\partial^2 u}{\partial x^2}, \quad x \in (0,l), \quad t \in (0,T), \tag{8.2.25}$$

$$u(x,0) = f(x), \quad x \in (0,l), \tag{8.2.26}$$

$$u(0,t) = u(l,t) = 0, \quad t \in (0,T), \tag{8.2.27}$$

can be formulated in the form of the inverse to a certain direct (well-posed) problem. Indeed, let us consider the problem

$$\frac{\partial u}{\partial t} = -\frac{\partial^2 u}{\partial x^2}, \quad x \in (0,l), \quad t \in (0,T), \tag{8.2.28}$$

$$u(x,T) = q(x), \quad x \in (0,l), \tag{8.2.29}$$

$$u(0,t) = u(l,t) = 0, \quad t \in (0,T), \tag{8.2.30}$$

where it is required to determine $u(x,t)$ from a given function $q(x)$.

The problem (8.2.28)–(8.2.30) is well posed. Therefore, the problem (8.2.25)–(8.2.27) can be reduced to the following inverse problem: determine $q(x)$ from the additional information about the solution to the direct problem (8.2.28)–(8.2.30)

$$u(x,0) = f(x), \quad x \in (0,l). \tag{8.2.31}$$

Consider the operator

$$A: \quad q(x) := u(x,T) \to f(x) := u(x,0),$$

where $u(x,t)$ is a solution to the problem (8.2.28)–(8.2.30).

Then the inverse problem can be written in the operator form

$$Aq = f,$$

where $f(x)$ is given and $q(x)$ is the unknown.

The steepest descent method in the initial boundary value problem for the heat conduction equation. Consider the inverse problem (8.2.28)–(8.2.31) in the operator form

$$Aq = f. \tag{8.2.32}$$

To solve this problem, we will minimize the objective functional $J(q) = \|Aq - f\|^2_{L_2(0,l)}$ using the steepest descent method. This method consists in the successive calculation of approximations q_n according to the procedure

$$q_{n+1} = q_n - \alpha_n J'(q_n), \quad n = 0, 1, \dots,$$

where q_0 is an initial approximation, $J'(q)$ is the gradient of the objective functional, and the descent parameter α_n is determined from the condition

$$\alpha_n = \arg\min_{\alpha>0} (J(q_n - \alpha J'(q_n))). \tag{8.2.33}$$

The gradient $J'(q)$ in this inverse problem is calculated as follows. First, calculate the increment of the functional $J(q)$:

$$\begin{aligned}J(q+\delta q)-J(q) &= \|A(q+\delta q)-f\|^2_{L_2(0,l)} - \|Aq-f\|^2_{L_2(0,l)}\\ &= 2\langle Aq-f, A\,\delta q\rangle + \|A\,\delta q\|^2_{L_2(0,l)}\\ &= 2\int_0^l [u(x,0)-f(x)]\delta u(x,0)\,dx + \int_0^l (\delta u(x,0))^2\,dx,\end{aligned}$$

where $u(x,t)$ is a solution to the direct problem (8.2.28)–(8.2.30) and $\delta u(x,t)$ is a solution to the problem

$$\frac{\partial u}{\partial t} = -\frac{\partial^2 u}{\partial x^2}, \quad x\in(0,l), \quad t\in(0,T), \tag{8.2.34}$$

$$u(x,T) = \delta q(x), \quad x\in(0,l), \tag{8.2.35}$$

$$u(0,t) = u(l,t) = 0, \quad t\in(0,T). \tag{8.2.36}$$

Let $\psi(x,t)$ be a solution to the adjoint problem

$$\frac{\partial \psi}{\partial t} = \frac{\partial^2 \psi}{\partial x^2}, \tag{8.2.37}$$

$$\psi(x,0) = \mu(x), \tag{8.2.38}$$

$$\psi(0,t) = \psi(l,t) = 0. \tag{8.2.39}$$

Using (8.2.34)–(8.2.39), we obtain the formulas relating the solutions of the problems (8.2.34)–(8.2.36) and (8.2.37)–(8.2.39):

$$\begin{aligned}0 &= \int_0^l\int_0^T \psi(x,t)\big(\delta u_t(x,t)+\delta u_{xx}(x,t)\big)\,dx\,dt\\ &= \int_0^l\int_0^T \psi(x,t)\delta u_t(x,t)\,dx\,dt + \int_0^l\int_0^T \psi(x,t)\delta u_{xx}(x,t)\,dx\,dt\\ &= \int_0^l \psi(x,t)\delta u(x,t)\,dx\Big|_0^T - \int_0^l\int_0^T \psi_t(x,t)\delta u(x,t)\,dx\,dt\\ &\quad + \int_0^T \psi(x,t)\delta u_x(x,t)\,dt\Big|_0^l - \int_0^l\int_0^T \psi_x(x,t)\delta u_x(x,t)\,dx\,dt\\ &= \int_0^l \psi(x,t)\delta u(x,t)\,dx\Big|_0^T - \int_0^l\int_0^T \psi_t(x,t)\delta u(x,t)\,dx\,dt\\ &\quad - \int_0^T \psi_x(x,t)\delta u(x,t)\,dt\Big|_0^l + \int_0^l\int_0^T \psi_{xx}(x,t)\delta u(x,t)\,dx\,dt\end{aligned}$$

$$= \int_0^l (\psi(x,T)\delta u(x,T) - \psi(x,0)\delta u(x,0))\,dx$$
$$- \int_0^l \int_0^T \big(\psi_t(x,t) - \psi_{xx}(x,t)\big)\delta u(x,t)\,dx\,dt$$
$$= \int_0^l \psi(x,T)\delta u(x,T)\,dx - \int_0^l \psi(x,0)\delta u(x,0)\,dx.$$

Hence

$$\int_0^l \psi(x,T)\delta q(x)\,dx - \int_0^l \psi(x,0)\delta u(x,0)\,dx = 0.$$

Put $\mu(x) = 2(u(x,0) - f(x))$ in the adjoint problem (8.2.37)–(8.2.39). Then

$$\begin{aligned} J(q+\delta q) - J(q) &= \int_0^l \psi(x,0)\delta u(x,0)\,dx + \int_0^l (\delta u(x,0))^2\,dx \\ &= \int_0^l \psi(x,T)\delta q(x)\,dx + \int_0^l (\delta u(x,0))^2\,dx \\ &= \langle \psi(\cdot,T), \delta q(\cdot)\rangle + \int_0^l (\delta u(x,0))^2\,dx. \end{aligned}$$

Comparing the obtained expression for the increment of the functional to the definition of its gradient

$$J(q+\delta q) - J(q) = \langle J'(q), \delta q\rangle + o(\|\delta q\|)$$

and applying Theorem 8.2.2 (see below), we conclude that $J'(q) = 2A^*(Aq - f) = \psi(x,T)$, where ψ is a solution to the adjoint problem (8.2.37)–(8.2.39) with the function $2(Aq - f) = 2(u(x,0) - f(x))$ standing for $\mu(x)$.

Multidimensional formulation. In a connected bounded domain $D \subset \mathbb{R}^n$ with smooth boundary ∂D, consider the initial boundary value problem for the heat conduction equation with reverse time

$$u_t = -Lu, \quad (x,t) \in D \times (0,T), \tag{8.2.40}$$
$$u(x,0) = f(x), \quad x \in D, \tag{8.2.41}$$
$$u\big|_{\partial D} = 0, \quad t \in [0,T], \tag{8.2.42}$$

where it is required to determine $u(x,t) \in L_2(D \times (0,T))$ from a given function $f(x) \in L_2(D)$. In the above formulation, L is an elliptic operator such that

$$Lu = \sum_{i,j=1}^{n} D_i(a_{ij}(x)D_j u) - c(x)u, \quad D_i u = \frac{\partial u}{\partial x_i}, \quad i = \overline{1,n}, \tag{8.2.43}$$

$$\sum_{i,j=1}^{n} a_{ij}(x)\nu_i \nu_j \geq \sum_{1}^{n} \nu_i^2 \quad \text{for any } \nu_i \in \mathbb{R}, \quad a_{ij}(x) = a_{ji}(x), \ x \in D; \tag{8.2.44}$$

$$0 \leq c(x) \leq C_1. \tag{8.2.45}$$

Note that the coefficients of the operator L depend only on the spatial variable $x \in D \subset \mathbb{R}^n$.

The problem (8.2.40)–(8.2.42) is ill-posed because arbitrarily small variations in the right-hand side f may cause arbitrarily large variations in the solution u. The problem can be formulated as the inverse to the direct (well-posed) problem

$$u_t = -Lu, \quad (x,t) \in D \times (0,T), \tag{8.2.46}$$

$$u(x,T) = q(x), \quad x \in D, \tag{8.2.47}$$

$$u\big|_{\partial D} = 0, \quad t \in [0,T]. \tag{8.2.48}$$

The function $q(x)$ is assumed to satisfy the concordance condition

$$q\big|_{\partial D} = 0. \tag{8.2.49}$$

In the direct problem (8.2.46)–(8.2.49) it is required to determine $u(x,t)$ in $D \times (0,T)$ from the given function $q(x)$.

The inverse problem consists in finding $q(x)$ from (8.2.46)–(8.2.48) and the additional information

$$u(x,0) = f(x), \quad x \in D. \tag{8.2.50}$$

The function $u \in L_2(D \times (0,T))$ will be called a *generalized solution to the direct problem* (8.2.46)–(8.2.48) if any $w \in H^{2,1}(D \times (0,T))$ such that

$$w(x,0) = 0, \quad x \in D, \tag{8.2.51}$$

$$w\big|_{\partial D} = 0, \quad t \in [0,T], \tag{8.2.52}$$

satisfies the equality

$$\int_0^T \int_D u(w_t - Lw)\,dx\,dt - \int_D q(x)w(x,T)\,dx = 0. \tag{8.2.53}$$

The direct problem (8.2.46)–(8.2.48) is well-posed in the generalized sense, which follows from the next theorem.

Theorem 8.2.1. *If $q \in L_2(D)$, then the problem* (8.2.46)–(8.2.48) *has a unique generalized solution $u \in L_2(D \times (0,T))$ that depends continuously on the initial data:*

$$\|u\|_{L_2(D\times(0,T))} \le \sqrt{T}\,\|q\|_{L_2(D)}. \tag{8.2.54}$$

Proof. Consider a function $\tilde{q} \in C_0^\infty(D)$ (see Definition A.5.2). For any such function, the problem (8.2.46)–(8.2.48) has a unique infinitely smooth solution $\tilde{u}(x,t) \in C_0^\infty(D \times (0,T))$. The solution to the auxiliary problem

$$\tilde{w}_t - L\tilde{w} = \tilde{u}, \quad (x,t) \in D \times (0,T), \tag{8.2.55}$$

$$\tilde{w}(x,0) = 0, \tag{8.2.56}$$

$$\tilde{w}\big|_{\partial D} = 0 \tag{8.2.57}$$

has the same property.

Using (8.2.56), from the equality

$$\tilde{w}(x,T) = \tilde{w}(x,0) + \int_0^T \tilde{w}_t(x,t)\,dt$$

we obtain

$$\begin{gathered} |\tilde{w}(x,T)|^2 \le T \int_0^T \tilde{w}_t^2(x,t)\,dt, \\ \|\tilde{w}(\,\cdot\,,T)\|^2_{L_2(D)} \le T\,\|\tilde{w}_t\|^2_{L_2(D\times(0,T))}. \end{gathered} \tag{8.2.58}$$

Consider the identity

$$\begin{aligned} \tilde{w}_t\tilde{u} = \tilde{w}_t(\tilde{w}_t - L\tilde{w}) = \tilde{w}_t^2 &- \sum_{i,j=1}^n D_i(\tilde{w}_t a_{ij}(x) D_j\tilde{w}) \\ &+ \sum_{i,j=1}^n a_{ij}(x) D_i\tilde{w}_t D_j\tilde{w} + \frac{1}{2} c(x)(\tilde{w}^2)_t. \end{aligned} \tag{8.2.59}$$

Since $a_{ij}(x) = a_{ji}(x)$ for $i,j = \overline{1,n}$, we have

$$\sum_{i,j=1}^n a_{ij}(x) D_i\tilde{w}_t D_j\tilde{w} = \frac{1}{2}\sum_{i,j=1}^n a_{ij}(x)(D_i\tilde{w} D_j\tilde{w})_t. \tag{8.2.60}$$

Using (8.2.60), integrate (8.2.59) over the domain $D \times (0, T)$:

$$\int_0^T \int_D \tilde{w}_t \tilde{u} \, dx \, dt = \int_0^T \int_D \tilde{w}_t^2 \, dx \, dt - \int_0^T \sum_{i,j=1}^n \tilde{w}_t a_{ij}(x) D_j \tilde{w} \Big|_{\partial D} dt$$
$$+ \frac{1}{2} \int_D \sum_{i,j=1}^n a_{ij}(x) D_i \tilde{w} D_j \tilde{w}(x, T) \, dx$$
$$- \frac{1}{2} \int_D \sum_{i,j=1}^n a_{ij}(x) D_i \tilde{w} D_j \tilde{w}(x, 0) \, dx$$
$$+ \frac{1}{2} \int_D c(x)[\tilde{w}^2(x, T) - \tilde{w}^2(x, 0)] \, dx. \quad (8.2.61)$$

By condition (8.2.45),

$$\int_D c(x) \tilde{w}^2(x, T) \geq 0. \quad (8.2.62)$$

By virtue of (8.2.44), (8.2.56), (8.2.57), and (8.2.62), from (8.2.61) we obtain

$$\begin{aligned} \|\tilde{w}_t\|^2_{L_2(D\times(0,T))} &\leq \|\tilde{w}_t\|^2_{L_2(D\times(0,T))} \|\tilde{u}\|_{L_2(D\times(0,T))}, \\ \|\tilde{w}_t\|_{L_2(D\times(0,T))} &\leq \|\tilde{u}\|_{L_2(D\times(0,T))}. \end{aligned} \quad (8.2.63)$$

From (8.2.63) and (8.2.58) it follows that

$$\|\tilde{w}(\cdot, t)\|^2_{L_2(D)} \leq T \|\tilde{u}\|^2_{L_2(D\times(0,T))}. \quad (8.2.64)$$

Take a solution $\tilde{w}(x, t)$ to the auxiliary problem (8.2.55)–(8.2.57) as a test function and substitute it into (8.2.53). As a result, using (8.2.64) we obtain

$$\|\tilde{u}\|^2_{L_2(D\times(0,T))} \leq \|\tilde{q}\|^2_{L_2(D)} \|\tilde{w}(\cdot, T)\|_{L_2(D)} \leq \sqrt{T} \, \|\tilde{q}\|_{L_2(D)} \|\tilde{u}\|_{L_2(D\times(0,T))},$$
$$\|\tilde{u}\|_{L_2(D\times(0,T))} \leq \sqrt{T} \, \|\tilde{q}\|_{L_2(D)}.$$

Thus, approximating $q \in L_2(D)$ with a sequence of functions $\tilde{q}_j \in C_0^\infty(D)$, we obtain the sequence of solutions $\tilde{u}_j \in C_0^\infty(D \times (0, T))$ converging to the solution $u \in L_2(D \times (0, T))$ because of the continuity of the scalar product, and the estimate (8.2.54) holds. □

Theorem 8.2.2. *If $q \in L_2(D)$, then the generalized solution to the direct problem* (8.2.46)–(8.2.48) *has the trace $u(x, 0) \in L_2(D)$ and*

$$\|u(\cdot, 0)\|_{L_2(D)} < \|q\|_{L_2(D)}. \quad (8.2.65)$$

The proof of this theorem is similar to that of Theorem 8.2.1.

Using arguments similar to those in the proof of the estimate (8.2.24), we can obtain an analogous estimate for the solution $u(x,t)$ to the problem (8.2.40)–(8.2.42):

$$\int_D u^2(x,t)\,dx \le \|u(\cdot,T)\|_{L_2(D)}^{2t/T}\|f\|_{L_2(D)}^{2(T-t)/T}. \tag{8.2.66}$$

This estimate implies that the problem (8.2.40)–(8.2.42) is conditionally well posed on the set M of functions $u(x,t)$ such that $\|u(\cdot,T)\|_{L_2(D)} \le C$.

Similar assertions and estimates hold for the adjoint problem. A detailed theoretical analysis of the direct and the adjoint problem is provided in Kabanikhin et al. (2006).

We now introduce an operator

$$A : q(x) \mapsto u(x,0),$$

where $u(x,t)$ is a solution to the direct problem (8.2.46)–(8.2.48).

The inverse problem (8.2.46)–(8.2.50) in the operator form is written as

$$Aq = f.$$

Note that from Theorem 8.2.2 it follows that A: $L_2(D) \to L_2(D)$ and $\|A\| \le 1$. We will solve the problem by minimizing the functional $J(q) = \|Aq - f\|_{L_2(D)}^2$ using the steepest descent method. The following theorem provides the rate of convergence for this method.

Theorem 8.2.3. *Assume that there exists a sufficiently smooth solution q_e to the problem $Aq = f$ with $f \in L_2(D)$, and the initial approximation satisfies the condition $\|q_0 - q_e\| \le C$. Then the sequence of solutions $\{u_n\}$ to the direct problems for the corresponding iterations q_n converges to the exact solution u to the prboblem* (8.2.40)–(8.2.42) *and the following estimate holds:*

$$\|u_n - u\|_{L_2(D\times(0,t))}^2 \le C^2T(n^{t/T-1}/\ln n), \quad t \in (0,T). \tag{8.2.67}$$

Proof. Using the estimate (8.2.66), for $\tau \in (0,T)$ we obtain

$$\begin{aligned}\|u_n(\cdot,\tau) - u(\cdot,\tau)\|_{L_2(D)}^2 &= \int_D [u_n(x,\tau) - u(x,\tau)]^2\,dx\\ &\le \|q_n - q_e\|_{L_2(D)}^{2\tau/T}\|Aq_n - f\|_{L_2(D)}^{2(T-\tau)/T}.\end{aligned}$$

From Theorem 2.7.1 it follows that

$$\|q_n - q_e\| \le \|q_0 - q_e\| \le C,$$

$$J(q_n) = \|Aq_n - f\|^2 \le \frac{\|A\|^2\|q_0 - q_e\|^2}{n} \le \frac{C^2}{n}.$$

Hence

$$\|u_n(\cdot,\tau) - u(\cdot,\tau)\|^2_{L_2(D)} \le C^{2\tau/T}\Big(\frac{C^2}{n}\Big)^{(T-\tau)/T} = C^2 n^{\tau/T-1}. \qquad (8.2.68)$$

Integrating (8.2.68) with respect to τ from 0 to t, we arrive at the inequality

$$\|u_n - u\|^2_{L_2(D\times(0,t))} \le \frac{C^2 T}{n}\frac{n^{t/T}-1}{\ln n} \le C^2 T\,\frac{n^{t/T-1}}{\ln n},$$

which is the desired conclusion. □

8.3 Inverse boundary-value problems and extension problems

Consider the process of heat propagation in a medium (diffusion process). Heat flow (substance flow) and temperature (concentration) are measured on one part of the boundary, while the other part of the boundary is not available or too difficult to access for measurements. It is required to determine the temperature (or the substance concentration) inside the domain up to the inaccessible part of the boundary. Problems of this type arise in geophysics (Ballani et al., 2002), the theory of nuclear reactors, aeronautics, etc. (Becket al., 1985; Griedt, 1995).

Consider a mathematical model of this physical process in the two-dimensional case.

Let $D := \{(x,y) \in \mathbb{R}^2\colon x \in (0,1),\ y \in (0,1)\}$ be the domain under study. The process of heat propagation in the domain D during the time $t \in [0,T]$, $T \in \mathbb{R}^+$, is described by the initial boundary value problem for a parabolic equation

$$u_t = u_{xx} + u_{yy}, \quad (x,y) \in D,\ t \in (0,T), \qquad (8.3.1)$$

$$u(0,y,t) = f(y,t), \quad y \in (0,1),\ t \in (0,T), \qquad (8.3.2)$$

$$u_x(0,y,t) = 0, \quad y \in (0,1),\ t \in (0,T), \qquad (8.3.3)$$

$$u(x,0,t) = u(x,1,t) = 0, \quad x \in (0,1),\ t \in (0,T), \qquad (8.3.4)$$

$$u(x,y,0) = 0, \quad (x,y) \in D. \qquad (8.3.5)$$

In the above formulation, $u(x,y,t)$ represents the temperature of the medium at the point $(x,y) \in D$ at the instant of time $t \in [0,T]$. It is required to find $u(x,y,t)$ in $D \times (0,T)$ from the temperature $f(y,t)$ measured on a part of the boundary of D.

Let us replace the condition (8.3.2) in the problem (8.3.1)–(8.3.5) by the condition

$$u(1,y,t) = q(y,t), \quad y \in (0,1),\ t \in (0,T), \qquad (8.3.6)$$

where $q(y,t)$ is a given function. The problem of finding $u(x,y,t)$ from (8.3.1), (8.3.3)–(8.3.6) will be called the direct problem. Then the problem (8.3.1)–(8.3.5) is inverse to the problem (8.3.1), (8.3.3)–(8.3.6), since it is reduced to determining $q(y,t)$ from (8.3.1), (8.3.3)–(8.3.6) using the additional information about the solution to the direct problem

$$u(0,y,t) = f(y,t), \quad y \in (0,1),\ t \in (0,T).$$

As before, we introduce the operator

$$A : q(y,t) = u(1,y,t) \mapsto f(y,t) = u(0,y,t),$$

where $u(x,y,t)$ is a solution to the direct problem (8.3.1), (8.3.3)–(8.3.6), and write the problem (8.3.1)–(8.3.6) in the operator form $Aq = f$. It can be proved (Kabanikhin et al., 2006) that the operator A acts from the space $L_2((0,1)\times(0,T))$ into $L_2((0,1)\times(0,T))$ and is bounded: $\|A\| \le 1$. Consequently, the problem (8.3.1)–(8.3.6) can be solved using the gradient method (see Chapter 2) and, in particular, using the steepest descent method.

8.4 Interior problems and problems of determining sources

In this section, we will present several formulations of linear ill-posed inverse problems for parabolic equations such that gradient methods are applicable in the numerical solution of these problems.

Finding the initial condition from an additional measurement at a point. Consider the boundary-value problem

$$u_t = u_{xx}, \quad x \in (0,1), \quad t \in (0,T), \tag{8.4.1}$$

$$u(x,0) = q(x), \quad x \in (0,1), \tag{8.4.2}$$

$$u_x(0,t) = u_x(1,t) = 0, \quad t \in (0,T), \tag{8.4.3}$$

where $q(x)$ is a given function and $u(x,t)$ is the unknown. This problem is well posed.

We now suppose that $q(x)$ is unknown, and the following additional information about the solution to the problem (8.4.1)–(8.4.3) is known at a fixed point $x_0 \in [0,1]$:

$$u(x_0,t) = f(t), \quad t \in (t_0,t_1), \quad 0 < t_0 < t_1 < T. \tag{8.4.4}$$

The problem of reconstruction of the function $q(x)$ from the function $f(t)$ and (8.4.1)–(8.4.4) is inverse to the problem (8.4.1)–(8.4.3).

The solution to the direct problem can be represented in the form of a Fourier series

$$u(x,t) = \frac{1}{\sqrt{2}}q_0 + \sum_{k=1}^{\infty} q_k\, e^{-\pi^2k^2t}\sqrt{2}\cos(\pi kx),$$

$$q_k = \sqrt{2}\int_0^1 q(\xi)\cos(\pi k\xi)\,d\xi, \quad k \in \mathbb{N}\cup\{0\}.$$

Substituting this representation of $u(x,t)$ into (8.4.4), we obtain a Fredholm integral equation of the first kind for $q(x)$:

$$f(t) = \int_0^1 K(t, x_0, \xi) q(\xi)\, d\xi, \quad t \in (t_0, t_1), \tag{8.4.5}$$

where

$$K(t, x_0, \xi) = 1 + 2 \sum_{k=1}^{\infty} e^{-\pi^2 k^2 t} \cos(\pi k x_0) \cos(\pi k \xi). \tag{8.4.6}$$

Theorem 8.4.1. *If* $x_0 = 0$, *then the solution of equation* (8.4.5) *is unique in the space* $L_2(0,1)$.

Proof. Since equation (8.4.5) is linear, it suffices to prove that the homogeneous equation $\int_0^1 K(t,0,\xi) q(\xi)\, d\xi = 0$ has only zero solution (Denisov, 1994).

Substituting (8.4.6) into (8.4.5), we have

$$\int_0^1 q(\xi)\, d\xi + 2 \sum_{k=1}^{\infty} e^{-\pi^2 k^2 t} \int_0^1 q(\xi) \cos(\pi k \xi)\, d\xi = 0, \quad t \in (t_0, t_1). \tag{8.4.7}$$

Let $\alpha \in (0, t_0)$. Then the function

$$\Phi(z) = \int_0^1 q(\xi)\, d\xi + 2 \sum_{k=1}^{\infty} e^{-\pi^2 k^2 z} \int_0^1 q(\xi) \cos(\pi k \xi)\, d\xi \tag{8.4.8}$$

is analytic in the complex half-plane $\operatorname{Re} z \geq \alpha$, which follows from the Weierstrass theorem and the estimate

$$|e^{-\pi^2 k^2 z}| \leq e^{-\pi^2 k^2 \alpha}, \quad \operatorname{Re} z \geq \alpha.$$

Equality (8.4.7) means that the analytic function $\Phi(z)$ vanishes on the interval (t_0, t_1) of the real axis that lies in the domain of analyticity of $\Phi(z)$. Consequently, by the uniqueness theorem for analytic functions, $\Phi(z) = 0$ for all z such that $\operatorname{Re} z \geq \alpha$. Passing to the limit as $t \to +\infty$ in (8.4.7), we have $\int_0^1 q(\xi)\, d\xi = 0$.

Hence,

$$e^{-\pi^2 t} \int_0^1 q(\xi) \cos(\pi \xi)\, d\xi + \sum_{k=2}^{\infty} e^{-\pi^2 k^2 t} \int_0^1 q(\xi) \cos(\pi k \xi)\, d\xi = 0.$$

Multiplying the last equality by $e^{\pi^2 t}$ and again passing to the limit as $t \to \infty$, we obtain

$$\int_0^1 q(\xi) \cos(\pi \xi)\, d\xi = 0.$$

Using similar arguments, we arrive at the system of equalities

$$\int_0^1 q(\xi)\cos(\pi k\xi)\,d\xi = 0, \quad k = 0, 1, 2, \dots. \tag{8.4.9}$$

Since the system of functions $\{\cos(\pi k x)\}_{k\in\mathbb{N}\cup\{0\}}$ is complete in the space $L_2(0,1)$, from (8.4.9) it follows that $q(x) \equiv 0$. □

Exercise 8.4.1. Analyze the case of $x_0 \neq 0$.

Determining the boundary condition. For the direct problem we take the problem of finding the function $u(x,t)$ for $x \in (0,1)$ and $t \in [0,T]$ such that

$$u_t = u_{xx}, \quad x \in (0,1),\ t \in (0,T], \tag{8.4.10}$$

$$u(x,0) = 0, \quad x \in (0,1), \tag{8.4.11}$$

$$u(0,t) = q(t), \quad t \in (0,T], \tag{8.4.12}$$

$$u_x(1,t) = 0, \quad t \in (0,T], \tag{8.4.13}$$

where $q(t)$ is a given function. In the inverse problem, $q(t)$ is the unknown, and the following additional information is given for the solution to the direct problem (8.4.10)–(8.4.13):

$$u(1,t) = f(t), \quad t \in [0,T]. \tag{8.4.14}$$

The uniqueness theorem for the solution of this inverse problem was proved in (Landis, 1959).

Theorem 8.4.2. *Let a function $u(x,t) \in C^{2,1}([0,1]\times[0,T])$ satisfy the equation* (8.4.10) *in* $[0,1]\times[0,T]$. *Let $u(1,t) = 0$ and $u_x(1,t) = 0$ for $t \in [0,T]$. Then $u(x,t) \equiv 0$ in* $[0,1]\times[0,T]$.

Note that the initial condition is not required in 8.4.2. The uniqueness theorem still holds if the additional information (8.4.14) results from the measurements at an interior point of the interval: $u(x_0,t) = f(t)$, $x_0 \in (0,1)$ (Denisov, 1994)

Reconstruction of sources. The problem of determining the function $u(x,t)$, $x \in (0,1)$, $t \in (0,T]$, from given functions $q_1(x)$, $q_2(t)$ and the relations

$$u_t = u_{xx} + q_1(x)q_2(t), \quad x \in (0,1),\ t \in (0,T], \tag{8.4.15}$$

$$u(x,0) = 0, \quad x \in [0,1], \tag{8.4.16}$$

$$u_x(0,t) = u_x(1,t) = 0, \quad t \in (0,T], \tag{8.4.17}$$

will be called the direct problem.

Let $q_1(x)$ be given and $u(x,t)$ be measured at a fixed point $x_0 \in [0,1]$:

$$u(x_0,t) = f(t), \quad t \in [0,T]. \tag{8.4.18}$$

The problem of finding $q_2(t)$ from the given x_0, $q_1(x)$, and $f(t)$ and (8.4.15)–(8.4.18) is inverse to the problem (8.4.15)–(8.4.17).

The direct problem can be solved using the method of separation of variables:

$$u(x,t) = q_{1,0}\int_0^t q_2(\tau)\,d\tau + \sum_{k=1}^{\infty} q_{1,k}\sqrt{2}\cos(\pi k x)\int_0^t q_2(\tau)\mathrm{e}^{-\pi^2k^2(t-\tau)}\,d\tau, \tag{8.4.19}$$

$$\begin{aligned} q_{1,k} &= \sqrt{2}\int_0^l q_1(\xi)\cos(\pi k\xi)\,d\xi, \quad k = 1,2,\dots, \\ q_{1,0} &= \int_0^l q_1(\xi)\,d\xi. \end{aligned} \tag{8.4.20}$$

For $x = x_0$, the equality (8.4.18) becomes

$$q_{1,0}\int_0^t q_2(\tau)\,d\tau + \sum_{k=1}^{\infty} q_{1,k}\sqrt{2}\cos(\pi k x_0)\int_0^t q_2(\tau)\mathrm{e}^{-\pi^2k^2(t-\tau)}\,d\tau = f(t),$$
$$t \in [0,T].$$

Assume that $q_1 \in C^2[0,1]$ and $q_2 \in C[0,1]$. Changing the order of integration and summation, we obtain the Volterra integral equation of the first kind for q_2

$$\int_0^t K(t,\tau)q_2(\tau)\,d\tau = f(t), \quad t \in [0,T], \tag{8.4.21}$$

with the kernel

$$K(t,\tau) = q_{1,0} + \sum_{k=1}^{\infty} q_{1,k}e^{-\pi^2k^2(t-\tau)}\sqrt{2}\cos(\pi k x_0).$$

The following theorem appears in Denisov (1994).

Theorem 8.4.3. *Let the data of the inverse problem* (8.4.15)–(8.4.18) *satisfy the conditions*

1) $q_1(x) \in C^4[0,1]$, $q_1'(+0) = q_1'(1-0) = 0$, $q_1(x_0) \neq 0$;
2) $f(t) \in C^4[0,T]$, $f(+0) = 0$.

Then there exists a unique solution $q_2(t)$ *to equation* (8.4.21) *in* $C[0,T]$.

Proof. Since $q_1 \in C^4[0,T]$, there exists a constant $c > 0$ such that $|q_{1,k}| < Ck^{-4}$ for all $k \in \mathbb{N}$. Consequently, the kernel $K(t,\tau)$ of equation (8.4.21) is continuous and has a continuous partial derivative $K_t(t,\tau)$ for $0 \le \tau \le t \le T$. Differentiating (8.4.21) with respect to t, we have

$$K(t,t)q_2(t) + \int_0^t K_t(t,\tau)q_2(\tau)\,d\tau = f'(t), \quad t \in [0,T]. \tag{8.4.22}$$

Note that $K(t,t) = q_1(x_0) \neq 0$. Therefore, the Volterra equation of the second kind (8.4.22) with continuous kernel $K_t(t,\tau)$ and continuous right-hand side $f'(t)$ has a unique solution $q_2(t)$ in $C[0,T]$, which is also a solution to equation (8.4.21). □

If $q_1(x_0) = 0$, then the solution of the inverse problem (8.4.15)–(8.4.18) is not necessarily unique. This can be proved by choosing $x_0 = 1/2$ and $q_1(x)$ so that $q_1(x) = -q_1(1/2 - x)$ for all $x \in [0,1]$.

Exercise 8.4.2. Analyze the direct problem (8.4.15)–(8.4.17) in the case $q_2(t) \equiv 1$.

Chapter 9
Linear problems for elliptic equations

The Cauchy problem for the Laplace equation was the first one whose ill-posedness was noticed by mathematicians (Hadamard, 1902). This problem, however, turned out to be so important for practical applications that it had been constantly studied throughout the 20th century. The uniqueness and conditional stability estimates and the first algorithms for constructing approximate solutions (Carleman, 1926; Lavrentiev, 1955, 1956, 1962; Pucci, 1955; Mergelyan, 1956; Landis, 1959; Aronszain, 1957; Calderon, 1958) were followed by a great number of publications devoted to the numerical solution and different generalizations and modifications of this problem. Such were the early publications by Douglas (1960) and by Chudov (1962), as well as more recent works (Kozlov et al., 1991; Kabanikhin and Karchevsky, 1995; Hào and Lesnic, 2000; Elden, 2002; Johansson, 2003; Kabanikhin et al., 2003) that developed iterative methods for minimizing the corresponding residual functional.

In this chapter, we will use a simple example to demonstrate the methods for proving the uniqueness and conditional stability theorems (Section 9.1). In the next section, we will reduce the initial boundary value problem for the Laplace equation to an inverse problem and then to an operator equation $Aq = f$ (Section 9.2). Studying the properties of the linear operator A and its adjoint A^*, we will show, in particular, that they are bounded (Section 9.3). As follows from the results of Chapter 2, if A is bounded, the conditional stability estimate enables not only to prove the convergence but also to estimate the rate of convergence of gradient methods for minimizing the objective functional $J(q) = \|Aq - f\|^2$. We will extend all the obtained results to the multidimensional case of an elliptic equation with divergent principal part (Section 9.4) and conclude the chapter by giving estimates for the rate of convergence with respect to the objective functional and the rate of strong convergence of the steepest descent method.

There are certainly many more formulations of linear inverse and ill-posed problems for elliptic equations, for example, the problem of determining the right-hand side of the Poisson equation $\Delta u = q$. But the study of the uniqueness and conditional stability of solutions in the cases that are of importance for practical applications is considerably more difficult.

9.1 The uniqueness theorem and a conditional stability estimate on a plane

Consider the following initial boundary value problem in the domain $\Omega := \{(x, y) \in \mathbb{R}^2 : x \in (0, 1),\ y \in (0, 1)\}$:

$$\Delta u = 0, \quad (x, y) \in \Omega, \tag{9.1.1}$$

$$u(0, y) = f(y), \quad y \in (0, 1), \tag{9.1.2}$$

$$u_x(0, y) = 0, \quad y \in (0, 1), \tag{9.1.3}$$

$$u(x, 0) = u(x, 1) = 0, \quad x \in (0, 1). \tag{9.1.4}$$

Two-dimensional Cauchy problems for the Laplace equation that have many applications, for example, in finding the potential of electrostatic field in the Earth's interior, are reduced to the above problem.

This problem is ill-posed in the sense of Hadamard. The solution is unique (see Theorem 9.1.1), but unstable. Indeed, for

$$f(y) = \frac{1}{n}\sin(ny)$$

the solution to the problem (9.1.1)–(9.1.4) is given by the formula

$$u(x, y) = \frac{1}{n}\sin(ny)\,\frac{e^{nx} + e^{-nx}}{2}.$$

Consequently, the function $f(y)$ can become as small as desired as n increases, whereas the solution $u(x, y)$ increases indefinitely as $n \to \infty$.

Exercise 9.1.1. Using the obtained particular solutions, determine the functional spaces where the problem (9.1.1)–(9.1.4) is ill-posed.

Theorem 9.1.1 (on the uniqueness of solution). *The solution of the problem* (9.1.1)–(9.1.4) *is unique in the class of functions $u \in C^4(\overline{\Omega})$ that can be represented in the form of a Fourier series*

$$u(x, y) = \sum_{k=1}^{\infty} u_k(x)\sin\left(\frac{ky}{\pi}\right) \tag{9.1.5}$$

with Fourier coefficients $u_k \in C^4[0, 1]$.

Proof. If $u \in C^4(\overline{\Omega})$, then $f(y) \in C^4[0, 1]$. Substituting (9.1.5) into (9.1.1)–(9.1.4), we obtain

$$u_k'' - k^2 u_k = 0, \tag{9.1.6}$$

$$u_k(0) = f_k, \tag{9.1.7}$$

$$u_k'(0) = 0, \tag{9.1.8}$$

where f_k are Fourier coefficients of the function $f(y)$.

Integrating (9.1.6) twice from 0 to x, we have

$$u_k(x) = f_k + k^2 \int_0^x \int_0^\xi u_k(\zeta) d\zeta d\xi. \tag{9.1.9}$$

As in Chapter 4, we introduce an auxiliary norm

$$\|u_k\|_{k,m} := \sup_{x\in[0,1]} \{|u_k(x)| e^{-kmx}\}, \quad m \in \mathbb{R}^+,$$

and note that for any $m > 0$

$$\sup_{x\in[0,1]} |u_k(x)| = \|u_k\|_{C[0,1]} \le \|u_k\|_{k,m} e^{km}. \tag{9.1.10}$$

Using (9.1.9), estimate $|u_k(x)|$ for any $x \in (0, 1)$:

$$\begin{aligned}
|u_k(x)| &\le |f_k| + k^2 \int_0^x \int_0^\xi |u_k(\zeta)| e^{-\zeta km} e^{\zeta km} \, d\zeta \, d\xi \\
&\le |f_k| + k^2 \|u_k\|_{k,m} \int_0^x \int_0^\xi e^{\zeta km} \, d\zeta \, d\xi \\
&\le |f_k| + k^2 \|u_k\|_{k,m} \int_0^x \frac{e^{\xi km} - 1}{km} \, d\xi \\
&\le |f_k| + k^2 \|u_k\|_{k,m} \frac{e^{xkm} - 1}{k^2 m^2}.
\end{aligned}$$

Consequently, for any $x \in (0, 1)$

$$|u_k(x)| \le |f_k| + \|u_k\|_{k,m} \frac{e^{xkm}}{m^2}. \tag{9.1.11}$$

Multiplying (9.1.11) by e^{-xkm}, we obtain

$$|u_k| e^{-xkm} \le |f_k| e^{-xkm} + \frac{\|u_k\|_{k,m}}{m^2},$$

$$\|u_k\|_{k,m} \le |f_k| + \frac{\|u_k\|_{k,m}}{m^2}.$$

Then for any $m > 1$ we have

$$\|u_k\|_{k,m} \le \frac{m^2}{m^2 - 1} |f_k|. \tag{9.1.12}$$

Combining (9.1.10) and (9.1.12), we arrive at

$$\|u_k\|_{C[0,1]} \le \frac{m^2}{m^2 - 1} |f_k| e^{km}, \tag{9.1.13}$$

which implies the uniqueness of the solution of the problem (9.1.1)–(9.1.4). □

Theorem 9.1.2 (on the conditional stability). *Assume that for $f \in C^2(0,1)$ there exists a solution $u \in C^2(\overline{\Omega})$ to the problem* (9.1.1)–(9.1.4). *Then the following conditional stability estimate holds:*

$$\int_0^1 u^2(x,y)\,dy \le \Big(\int_0^1 f^2(y)\,dy\Big)^{1-x}\Big(\int_0^1 u^2(1,y)\,dy\Big)^{x}. \tag{9.1.14}$$

Proof. Consider the auxiliary function

$$g(x) = \int_0^1 u^2(x,y)\,dy. \tag{9.1.15}$$

Differentiating (9.1.15) twice with respect to x, we have

$$g'(x) = 2\int_0^1 u(x,y)u_x(x,y)\,dy, \tag{9.1.16}$$

$$g''(x) = 2\int_0^1 u_x^2(x,y)\,dy + 2\int_0^1 u(x,y)u_{xx}(x,y)\,dy. \tag{9.1.17}$$

We now transform the second term in (9.1.17) using (9.1.1):

$$\begin{aligned}\int_0^1 u(x,y)u_{xx}(x,y)\,dy &= -\int_0^1 u(x,y)u_{yy}(x,y)\,dy\\ &= -(uu_y)(x,1) + (uu_y)(x,0) + \int_0^1 u_y^2(x,y)\,dy\\ &= \int_0^1 u_y^2(x,y)\,dy.\end{aligned}$$

Hence,

$$g''(x) = 2\int_0^1 u_x^2(x,y)\,dy + 2\int_0^1 u_y^2(x,y)\,dy. \tag{9.1.18}$$

Note that

$$\begin{aligned}\frac{d}{dx}\int_0^1 u_y^2(x,y)\,dy &= 2\int_0^1 u_y(x,y)u_{yx}(x,y)\,dy\\ &= 2(u_yu_x)(1,y) - 2(u_yu_x)(0,y) - 2\int_0^1 u_x(x,y)u_{yy}^2(x,y)\,dy\\ &= \frac{d}{dx}\int_0^1 u_x^2(x,y)\,dy.\end{aligned} \tag{9.1.19}$$

Integrating (9.1.19) with respect to x from 0 to an $x^0 \in (0, 1)$, we obtain

$$\int_0^1 u_y^2(x^0, y)\,dy - \int_0^1 u_y^2(0, y)\,dy = \int_0^1 u_x^2(x^0, y)\,dy - \int_0^1 u_x^2(0, y)\,dy$$
$$= \int_0^1 u_x^2(x^0, y)\,dy. \tag{9.1.20}$$

From the boundary conditions (9.1.2), it follows that

$$\int_0^1 u_y^2(0, y)\,dy = \int_0^1 [f'(y)]^2\,dy.$$

By (9.1.20) and the arbitrary choice of $x^0 \in (0, 1)$, from (9.1.18) we obtain

$$g''(x) = 4\int_0^1 u_x^2(x, y)\,dy + 2\int_0^1 [f'(y)]^2\,dy.$$

We introduce the function

$$G(x) = \ln g(x).$$

Differentiating $G(x)$ twice, we have

$$G''(x) = \frac{1}{g^2(x)}\,[g''(x)g(x) - [g'(x)]^2].$$

Consequently,

$$G''(x) = \frac{1}{g^2(x)}\Big\{\Big[4\int_0^1 u_x^2(x, y)\,dy + 2\int_0^1 [f'(y)]^2\,dy\Big]\int_0^1 u^2(x, y)\,dy$$
$$- \Big(2\int_0^1 u(x, y)u_x(x, y)\,dy\Big)^2\Big\}.$$

We will prove that $G''(x) > 0$, $x \in [0, 1]$. Indeed, by the Cauchy–Bunyakovsky inequality,

$$\Big(\int_0^1 u(x, y)u_x(x, y)\,dy\Big)^2 \le \int_0^1 u^2(x, y)\,dy \int_0^1 u_x^2(x, y)\,dy.$$

Therefore,

$$G''(x) \ge \frac{2}{g^2(x)}\int_0^1 u^2(x, y)\,dy \int_0^1 |f'(y)|^2\,dy > 0.$$

This means that $G(x)$ is convex on $[0, 1]$. Then for $x \in (0, 1)$ we have

$$G(x) \le (1 - x)G(0) + xG(1),$$

and exponentiation yields the inequality

$$g(x) \le [g(0)]^{1-x}[g(1)]^x, \quad x \in (0, 1),$$

i.e.,

$$\int_0^1 u^2(x,y)\,dy \le \Big(\int_0^1 f^2(y)\,dy\Big)^{1-x}\Big(\int_0^1 u^2(1,y)\,dy\Big)^x, \quad x \in (0,1),$$

which is the desired conclusion. □

Exercise 9.1.2. Analyze the proof of Theorem 9.1.2 and show that the conditions on the smoothness of $u(x,y)$ and $f(y)$ can be relaxed.

9.2 Formulation of the initial boundary value problem for the Laplace equation in the form of an inverse problem. Reduction to an operator equation

Consider the initial boundary value problem (9.1.1)–(9.1.4). As will be shown in Section 9.3, if the condition (9.1.3) is replaced by the condition

$$u(1,y) = q(y), \quad y \in (0,1),$$

where $q(y) \in L_2(0,1)$ is a given function, then the problem becomes well-posed. This means that the problem (9.1.1)–(9.1.4) of finding $u(x,y)$, $(x,y) \in \Omega$, can be reduced to the problem of finding $q(y) = u(1,y)$, $y \in (0,1)$.

Therefore, the initial boundary value problem (9.1.1)–(9.1.4) can be viewed as inverse to the problem

$$\Delta u = 0, \quad (x,y) \in \Omega, \tag{9.2.1}$$

$$u_x(0,y) = 0, \quad y \in (0,1), \tag{9.2.2}$$

$$u(1,y) = q(y), \quad y \in (0,1), \tag{9.2.3}$$

$$u(x,0) = u(x,1) = 0, \quad x \in (0,1), \tag{9.2.4}$$

where $q(x)$ is a given function (Kabanikhin and Karchevsky, 1995).

Indeed, it is required to reconstruct the unknown function $q(y)$, $y \in (0,1)$, from the additional information

$$u(0,y) = f(y), \quad y \in (0,1), \tag{9.2.5}$$

about the solution to the direct problem (9.2.1)–(9.2.4).

From Theorem 9.1.1 it follows that the inverse problem (9.2.1)–(9.2.5) has at most one solution.

Consider the operator

$$A: \quad q(y) := u(1,y) \mapsto f(y) := u(0,y),$$

where $u(x,y)$ is a solution to the problem (9.2.1)–(9.2.4).

Then the inverse problem (9.2.1)–(9.2.5) can be written as an operator equation

$$Aq = f. \tag{9.2.6}$$

We will show below that the operator A acts from $L_2(0,1)$ into $L_2(0,1)$ and is bounded in norm.

9.3 Analysis of the direct initial boundary value problem for the Laplace equation

This section employs a somewhat unusual definition of a generalized solution to the direct problem (9.2.1)–(9.2.4) in the space L_2 (usually a generalized solution to the Laplace equation is defined in the Sobolev space H^1). However, the definition given below is more convenient for the analysis of gradient methods for solving the corresponding inverse problem.

Definition 9.3.1. A function $u \in L_2(\Omega)$ will be called a *generalized* (or *weak*) *solution to the direct problem* (9.2.1)–(9.2.4) if any $w \in H^2(\Omega)$ such that

$$w_x(0,y) = 0, \quad y \in (0,1), \tag{9.3.1}$$

$$w(1,y) = 0, \quad y \in (0,1), \tag{9.3.2}$$

$$w(x,0) = w(x,1) = 0, \quad x \in (0,1), \tag{9.3.3}$$

satisfies the equality

$$\int_\Omega u \Delta w \, dx\, dy - \int_0^1 q(y) w_x(1,y)\, dy = 0. \tag{9.3.4}$$

A function $w \in H^2(\Omega)$ satisfying (9.3.1)–(9.3.3) will be called a *test function*.

It is easy to verify that if $u \in C^2(\Omega) \cap C^1(\overline{\Omega})$, then the generalized solution to the problem (9.2.1)–(9.2.4) is also its classical solution.

Theorem 9.3.1 (on the well-posedness of the direct problem). *If* $q \in L_2(0,1)$, *then the problem* (9.2.1)–(9.2.4) *has a unique generalized solution* $u \in L_2(\Omega)$ *satisfying the estimate*

$$\|u\|^2_{L_2(\Omega)} \le \sqrt{2}\, \|q\|^2_{L_2(0,1)}. \tag{9.3.5}$$

Proof. Let $\bar q \in C_0^\infty(0,1)$, where $C_0^\infty(0,1)$ is the space of finite infinitely differentiable functions whose support lies within the interval $(0,1)$. For any such function, the problem (9.2.1)–(9.2.4) has an infinitely differentiable solution, which will be denoted by $\bar u$ (Ladyzhenskaya and Uraltseva, 1973)

Consider the auxiliary problem

$$\Delta w = \bar{u}, \quad (x, y) \in \Omega, \tag{9.3.6}$$

$$w_x(0, y) = 0, \quad y \in (0, 1), \tag{9.3.7}$$

$$w(1, y) = 0, \quad y \in (0, 1), \tag{9.3.8}$$

$$w(x, 0) = w(x, 1) = 0, \quad x \in (0, 1). \tag{9.3.9}$$

Integrating the identity

$$w\bar{u} = w\Delta w = (ww_x)_x - w_x^2 + (ww_y)_y - w_y^2$$

over the domain Ω, we have

$$\int_\Omega w\bar{u}\, dx\, dy = -\int_\Omega [w_x^2 + w_y^2]\, dx\, dy.$$

Hence,

$$\int_\Omega [w_x^2 + w_y^2]\, dx\, dy \le \|w\|_{L_2(\Omega)} \|\bar{u}\|_{L_2(\Omega)}. \tag{9.3.10}$$

The equalities

$$0 = w(1, y) = w(x, y) + \int_x^1 w_\xi(\xi, y)\, d\xi,$$

$$w(x, y) = w(x, 0) + \int_0^y w_\eta(x, \eta)\, d\eta = \int_0^y w_\eta(x, \eta)\, d\eta$$

imply that

$$2w(x, y) = \int_0^y w_\eta(x, \eta)\, d\eta - \int_x^1 w_\xi(\xi, y)\, dy.$$

Consequently,

$$|w(x, y)| \le \frac{1}{2}\Big(\int_0^1 |w_\eta(x, \eta)|\, d\eta + \int_0^1 |w_\xi(\xi, y)|\, d\xi\Big).$$

By the Cauchy–Bunyakovsky inequality,

$$\int_0^1 |w_\eta(x, \eta)|\, d\eta \le \Big(\int_0^1 w_\eta^2(x, \eta)\, d\eta\Big)^{1/2}.$$

Since $(a + b)^2 \le 2a^2 + 2b^2$, we arrive at the inequality

$$|w(x, y)|^2 \le \frac{1}{2}\Big(\int_0^1 w_\eta^2(x, \eta)\, d\eta + \int_0^1 w_\xi^2(\xi, y)\, d\xi\Big).$$

Integrating the preceding inequality over Ω, we obtain

$$\|w\|^2_{L_2(\Omega)} \le \frac{1}{2}\int_\Omega (w_x^2 + w_y^2)\,dx\,dy. \tag{9.3.11}$$

Comparing (9.3.11) and (9.3.10), we have

$$\|w\|_{L_2(\Omega)} \le \frac{1}{2}\,\|\bar{u}\|_{L_2(\Omega)},$$
$$\int_\Omega (w_x^2 + w_y^2)\,dx\,dy \le \frac{1}{2}\,\|\bar{u}\|^2_{L_2(\Omega)}. \tag{9.3.12}$$

We now integrate the expression $w_x\bar{u} = w_x(w_{xx} + w_{yy})$ over Ω using the boundary conditions (9.3.7)–(9.3.9):

$$\begin{aligned}
&\int_\Omega w_x(w_{xx} + w_{yy})\,dx\,dy \\
&= \int_\Omega \left[\frac{1}{2}\,(w_x^2)_x + (w_x w_y)_y - w_{xy} w_y\right] dx\,dy \\
&= \frac{1}{2}\int_0^1 w_x^2(1,y)\,dy - \frac{1}{2}\int_0^1 w_x^2(0,y)\,dy + \int_0^1 (w_x w_y)(x,1)\,dx \\
&\quad - \int_0^1 (w_x w_y)(x,0)\,dx - \frac{1}{2}\int_\Omega (w_y^2)_x\,dx\,dy \\
&= \frac{1}{2}\int_0^1 w_x^2(1,y)\,dy - \frac{1}{2}\int_0^1 w_y^2(1,y)\,dy + \frac{1}{2}\int_0^1 w_y^2(0,y)\,dy \\
&= \frac{1}{2}\int_0^1 w_x^2(1,y)\,dy + \frac{1}{2}\int_0^1 w_y^2(0,y)\,dy.
\end{aligned}$$

Consequently,

$$\int_0^1 w_x^2(1,y)\,dy \le 2\int_\Omega |w_x\bar{u}|\,dx\,dy \le 2\|w_x\|_{L_2(\Omega)}\|\bar{u}\|_{L_2(\Omega)}.$$

From (9.3.12) it follows that

$$\|w_x\|^2_{L_2(\Omega)} \le \frac{1}{2}\,\|\bar{u}\|^2_{L_2(\Omega)}.$$

Hence,

$$\int_0^1 w_x^2(1,y)\,dy \le \sqrt{2}\,\|\bar{u}\|^2_{L_2(\Omega)}. \tag{9.3.13}$$

Take a solution $w(x, y)$ to the auxiliary problem (9.3.6)–(9.3.9) as a test function (see Definition 9.3.1) and substitute it into (9.3.4). Changing q and u for $\bar{q}$ and $\bar{u}$ in (9.3.4) and using (9.3.13), we obtain

$$\|\bar{u}\|^2_{L_2(\Omega)} \leq \Big(\int_0^1 w_x^2(1,y)\,dy\Big)^{1/2} \|\bar{q}\|_{L_2(0,1)} \leq \sqrt[4]{2}\,\|\bar{u}\|_{L_2(\Omega)}\|\bar{q}\|_{L_2(0,1)}.$$

Thus, by approximating the function $q \in L_2(0,1)$ by the sequence of functions $\bar{q}_j \in C_0^\infty(0,1)$, we get the sequence of solutions $\bar{u}_j \in C_0^\infty(\Omega)$ that converges to the solution $u \in L_2(\Omega)$ because of the continuity of the scalar product, which yields the estimate (9.3.5). This completes the proof. □

Theorem 9.3.2. *If* $q \in L_2(0,1)$, *then the solution to the direct problem* (9.2.1)–(9.2.4) *has the trace* $u(0, y) \in L_2(0,1)$ *and the following estimate holds:*

$$\|u(0,y)\|_{L_2(0,1)} < \|q\|_{L_2(0,1)}. \tag{9.3.14}$$

Proof. As in Theorem 9.3.1, we take an arbitrary $\bar{q} \in C_0^\infty(0,1)$ and note that for $q = \bar{q}$ there exists a unique infinitely differentiable solution to the problem (9.2.1)–(9.2.4), which will be denoted by $\bar{u}$. Integrate the identity

$$0 = \bar{u}\Delta\bar{u} = (\bar{u}_x\bar{u})_x - \bar{u}_x^2 + (\bar{u}_y\bar{u})_y - \bar{u}_y^2$$

with respect to ξ from 0 to x and with respect to y from 0 to 1. As a result, we have

$$0 = \frac{1}{2}\int_0^1 (\bar{u}^2)_x(x,y)\,dy - \int_0^1\int_0^x (\bar{u}_x^2 + \bar{u}_y^2)\,d\xi\,dy,$$

whence

$$\frac{1}{2}\int_0^1 (\bar{u}^2)_x(x,y)\,dy > 0.$$

Integrating the preceding inequality with respect to x from 0 to 1, we have

$$\int_0^1 \bar{u}^2(1,y)\,dy > \int_0^1 \bar{u}^2(0,y)\,dy,$$

i.e.,

$$\|\bar{u}(0,y)\|^2_{L_2(0,1)} < \|\bar{q}\|^2_{L_2(0,1)}.$$

Approximating $q(x) \in L_2(0,1)$ by a sequence of functions $\tilde{q}_j \in C_0^\infty(0,1)$, we can show that the corresponding sequence $\tilde{u}_n(0,y)$ has a limit that satisfies the estimate (9.3.14). The proof is complete. □

From Theorem 9.3.2 it follows that the operator A (see Section 9.2) acts from $L_2(0,1)$ into $L_2(0,1)$ and $\|A\| \leq 1$. Note that these properties of A make it possible to estimate the rate of convergence of the method of simple iterations (see Section 2.7).

Consider the problem adjoint to the direct problem (9.2.1)–(9.2.4):

$$\Delta\psi = 0, \quad (x,y) \in \Omega, \tag{9.3.15}$$

$$\psi_x(0,y) = \mu(y), \quad y \in (0,1), \tag{9.3.16}$$

$$\psi(1,y) = 0, \quad y \in (0,1), \tag{9.3.17}$$

$$\psi(x,0) = \psi(x,1) = 0, \quad x \in (0,1). \tag{9.3.18}$$

Definition 9.3.2. A function $\psi \in L_2(\Omega)$ is called a *generalized solution to the adjoint problem* (9.3.15)–(9.3.18) if any $v \in H^2(\Omega)$ such that

$$v_x(0,y) = 0, \quad y \in (0,1), \tag{9.3.19}$$

$$v(1,y) = 0, \quad y \in (0,1), \tag{9.3.20}$$

$$v(x,0) = v(x,1) = 0, \quad x \in (0,1), \tag{9.3.21}$$

satisfies the equality

$$\int_\Omega \psi \Delta v \, dx\, dy - \int_0^1 \mu(y) v(0,y)\, dy = 0. \tag{9.3.22}$$

If $\psi \in C^2(\Omega)$, then a generalized solution to (9.3.15)–(9.3.18) is also its classical solution.

Theorem 9.3.3 (on the well-posedness of the adjoint problem). *If $\mu \in L_2(0,1)$, then the problem* (9.3.15)–(9.3.18) *has a unique generalized solution $\psi \in L_2(\Omega)$ and the following estimate holds:*

$$\|\psi\|_{L_2(\Omega)} \leq \|\mu\|_{L_2(0,1)}. \tag{9.3.23}$$

Theorem 9.3.4. *If $\mu \in L_2(0,1)$, then the derivative of a solution to the adjoint problem has the trace $\psi_x(1,y) \in L_2(0,1)$ and the following estimate holds:*

$$\|\psi_x(1,y)\|_{L_2(0,1)} < \|\mu\|_{L_2(0,1)}. \tag{9.3.24}$$

Theorems 9.3.3 and 9.3.4 are similar to Theorems 9.3.1 and 9.3.2, and we leave it to the reader to prove them.

9.4 The extension problem for an equation with self-adjoint elliptic operator

In the domain $\Omega := \{(x,y) \in \mathbb{R}^{n+1} \mid x \in (0,l),\ y \in \mathcal{D} \subset \mathbb{R}^n\}$, consider the extension problem

$$u_{xx} + L(y)u = 0, \quad (x,y) \in \Omega, \tag{9.4.1}$$

$$u(0,y) = f(y), \quad y \in \mathcal{D}, \tag{9.4.2}$$

$$u_x(0,y) = 0, \quad y \in \mathcal{D}, \tag{9.4.3}$$

$$u\big|_{\partial\mathcal{D}} = 0, \quad x \in (0,l), \tag{9.4.4}$$

with the concordance condition

$$f\big|_{\partial\mathcal{D}} = 0, \quad x \in (0,l), \tag{9.4.5}$$

where

$$L(y)u = \sum_{i,j=1}^{n} D_i(a_{ij}(y)D_j u) - c(y)u, \quad D_i u = \frac{\partial u}{\partial y_i}, \quad i = \overline{1,n}. \tag{9.4.6}$$

Assume that $\mathcal{D}$ is a connected bounded domain with Lipschitz boundary, and the operator $L(y)$ has the following properties:

$$C_1 \sum_{j=1}^{n} \nu_j^2 \le \sum_{i,j=1}^{n} a_{ij}(y)\nu_i\nu_j \quad \text{for any } \nu_i \in \mathbb{R}, \quad a_{ij} = a_{ji},\ i,j = \overline{1,n}, \tag{9.4.7}$$

$$0 \le c(y) \le C_2, \tag{9.4.8}$$

$$a_{ij} \in C^1(\bar{\mathcal{D}}), \quad c \in C(\bar{\mathcal{D}}), \tag{9.4.9}$$

where $C_1 > 0$ and $C_2 > 0$ are constants.

The problem (9.4.1)–(9.4.4) is ill-posed. Hadamard's example in the case where $\mathcal{D} = (0,1)$, $c(y) \equiv 0$, and $\{a_{ij}\}$ is a unit matrix was already considered in Section 9.1

Consider the ill-posed problem (9.4.1)–(9.4.4) as an inverse problem with respect to the direct problem

$$u_{xx} + L(y)u = 0, \quad (x,y) \in \Omega, \tag{9.4.10}$$

$$u_x(0,y) = 0, \quad y \in \mathcal{D}, \tag{9.4.11}$$

$$u(l,y) = q(y), \quad y \in \mathcal{D}, \tag{9.4.12}$$

$$u\big|_{\partial\mathcal{D}} = 0, \quad x \in (0,l), \tag{9.4.13}$$

with the concordance condition

$$q\big|_{\partial\mathcal{D}} = 0, \quad x \in (0,l). \tag{9.4.14}$$

In the direct problem (9.4.10)–(9.4.13), it is required to determine $u(x, y)$ in Ω from the function $q(y)$ given on a part of the boundary $x = l$ of the domain Ω.

The inverse problem consists in finding $q(y)$ from (9.4.10)–(9.4.13) and the additional information

$$u(0, y) = f(y), \quad y \in \mathcal{D}. \tag{9.4.15}$$

Definition 9.4.1. A function $u \in L_2(\Omega)$ is called a *generalized solution to the direct problem* (9.4.10)–(9.4.13) if any $w \in H^2(\Omega)$ such that

$$w_x(0, y) = 0, \quad y \in \mathcal{D}, \tag{9.4.16}$$

$$w(l, y) = 0, \quad y \in \mathcal{D}, \tag{9.4.17}$$

$$w|_{\partial\mathcal{D}} = 0, \quad x \in (0, l), \tag{9.4.18}$$

satisfies the equality

$$\int_\Omega u(w_{xx} + L(y)w)\,dx\,dy - \int_\mathcal{D} q(y)w_x(l, y)\,dy = 0. \tag{9.4.19}$$

Theorems 9.4.1–9.4.4 are generalizations of Theorems 9.3.1–9.3.4. Their proofs can be found in Kabanikhin et al. (2006).

Theorem 9.4.1 (on the well-posedness of the direct problem). *Assume that the conditions* (9.4.6)–(9.4.9) *hold, and $\mathcal{D}$ is a connected bounded domain with Lipschitz boundary. If $q \in L_2(\mathcal{D})$, then the problem* (9.4.10)–(9.4.14) *has a unique generalized solution $u \in L_2(\Omega)$ such that*

$$\|u\|_{L_2(\Omega)} \le \sqrt{2l + 2C_3 l^3}\,\|q\|_{L_2(\mathcal{D})}. \tag{9.4.20}$$

Theorem 9.4.2. *If $q \in L_2(\mathcal{D})$, then the solution to the direct problem* (9.4.10)–(9.4.14) *has the trace $u(0, y) \in L_2(\mathcal{D})$ and the following estimate holds:*

$$\|u(0, y)\|_{L_2(\mathcal{D})} < \|q\|_{L_2(\mathcal{D})}. \tag{9.4.21}$$

Together with the direct problem (9.4.10)–(9.4.14), we consider the adjoint problem

$$\psi_{xx} + L(y)\psi = 0, \quad (x, y) \in \Omega, \tag{9.4.22}$$

$$\psi_x(0, y) = \mu(y), \quad y \in \mathcal{D}, \tag{9.4.23}$$

$$\psi(l, y) = 0, \quad y \in \mathcal{D}, \tag{9.4.24}$$

$$\psi|_{\partial\mathcal{D}} = 0, \quad x \in (0, l), \tag{9.4.25}$$

where it is required to determine $\psi(x, y)$ from the given function $\mu(y)$.

Definition 9.4.2. A function $\psi \in L_2(\Omega)$ is called a *generalized solution to the adjoint problem* (9.4.22)–(9.4.25) if any $v \in H^2(\Omega)$ such that

$$v_x(0, y) = 0, \quad y \in \mathcal{D}, \tag{9.4.26}$$

$$v(l, y) = 0, \quad y \in \mathcal{D}, \tag{9.4.27}$$

$$v|_{\partial\mathcal{D}} = 0, \quad x \in (0, l), \tag{9.4.28}$$

satisfies the equality

$$\int_\Omega \psi(v_{xx} + L(y)v)\,dx\,dy - \int_{\mathcal{D}} \mu(y)v(0, y)\,dy = 0. \tag{9.4.29}$$

Theorem 9.4.3 (on the well-posedness of the adjoint problem). *If $\mu \in L_2(\mathcal{D})$, then the problem* (9.4.22)–(9.4.25) *has a unique generalized solution $\psi \in L_2(\Omega)$ such that*

$$\|\psi\|_{L_2(\Omega)} \le l\sqrt{2l}\,\|\mu\|_{L_2(\mathcal{D})}. \tag{9.4.30}$$

Theorem 9.4.4. *Let $\psi \in L_2(\Omega)$ be a generalized solution to the adjoint problem* (9.4.22)–(9.4.25) *with $\mu \in L_2(\mathcal{D})$. Then there exists a trace $\psi_x(l, y) \in L_2(\mathcal{D})$ and the following estimate holds:*

$$\|\psi_x(l, y)\|_{L_2(\mathcal{D})} < \|\mu\|_{L_2(\mathcal{D})}. \tag{9.4.31}$$

Theorem 9.4.5 (conditional stability estimate for the extension problem). *Let $q, f \in L_2(\mathcal{D})$. If the extension problem* (9.4.1)–(9.4.5) *has a solution $u \in C^2(\overline{\Omega})$, then it satisfies the inequality*

$$\int_{\mathcal{D}} u^2(x, y)\,dy \le \|q\|^{2x/l}_{L_2(\mathcal{D})}\|f\|^{2(l-x)/l}_{L_2(\mathcal{D})}. \tag{9.4.32}$$

The proof of this theorem is similar to that of Theorem 9.1.2.

The steepest descent method. We introduce the operator

$$A : q(y) \mapsto u(0, y),$$

where $u(x, y)$ is a solution to the direct problem (9.4.10)–(9.4.13).

Then the adjoint of A has the form

$$A^* : \mu(y) \mapsto \psi_x(l, y), \tag{9.4.33}$$

where $\psi(x, y)$ is a solution to the adjoint problem (9.4.22)–(9.4.25).

From Theorems 9.4.2 and 9.4.4 it follows that the operators A and A^* act from $L_2(\mathcal{D})$ into $L_2(\mathcal{D})$ and $\|A\| \le 1$, $\|A^*\| \le 1$.

Therefore, the inverse problem (9.4.1)–(9.4.4) can be written in the operator form

$$Aq = f. \tag{9.4.34}$$

We will solve the problem (9.4.34) by minimizing the functional

$$J(q) = \|Aq - f\|^2_{L_2(\mathcal{D})}. \tag{9.4.35}$$

To this end, we will use the steepest descent method

$$q_{n+1} = q_n - \alpha_n J'q_n, \quad n = 0, 1, 2, \dots, \tag{9.4.36}$$

where $J'q$ is the gradient of the functional $J(q)$ and α_n is the descent parameter:

$$J'q = 2A^*(Aq - f), \tag{9.4.37}$$

$$\alpha_n = \arg\min_{\alpha \geq 0} J(q_n - \alpha J'q_n). \tag{9.4.38}$$

Note that (9.4.37) implies

$$J'q = \psi_x(l, y), \tag{9.4.39}$$

where $\psi(x, y)$ is a solution to the adjoint problem (9.4.22)–(9.4.25) with $\mu(y) = 2[Aq - f]$.

For $q, \delta q \in L_2(\mathcal{D})$ we have

$$|J(q + \delta q) - J(q) - \langle \delta q, J'(q)\rangle| < \|\delta q\|^2. \tag{9.4.40}$$

To determine the descent parameter α_n, we note that (9.4.37) implies that $J(q_n - \alpha J'q_n)$ is a quadratic polynomial in α. Consequently, it achieves its minimum at the point

$$\alpha_n = \frac{\|J'q_n\|^2}{2\|A(J'q_n)\|^2} = \frac{\|A^*(Aq_n - f)\|^2}{2\|AA^*(Aq_n - f)\|^2}. \tag{9.4.41}$$

If there exists a solution q_e to the inverse problem $Aq = f$ with $f \in L_2(\mathcal{D})$, then the descent parameter α_n and iteration q_{n+1} of the steepest descent method defined by formula (9.4.36) satisfy the inequalities

$$\frac{1}{2\|A\|^2} \leq \alpha_n \leq \frac{\|q_n - q_e\|^2}{2\|Aq_n - f\|^2}, \tag{9.4.42}$$

$$\|q_{n+1} - q_e\| < \|q_n - q_e\|.$$

Theorem 9.4.6 (on the rate of convergence with respect to the functional). *Assume that the problem $Aq = f$ with $f \in L_2(\mathcal{D})$ has a solution $q_e \in L_2(\mathcal{D})$. If the initial approximation satisfies the condition $\|q_0 - q_e\| \le C$, then the iterations q_n of the steepest descent method converge with respect to the objective functional, and the following estimate holds:*

$$J(q_n) \le \frac{C^2}{n}, \qquad n = 1, 2, \ldots.$$

The proof of a more general formulation of this theorem is given in Chapter 2. Using the estimate of the rate of convergence of iterations q_n with respect to the objective functional, one can estimate the rate of convergence of the sequence $\{u_n\}$ of solutions to the direct problems (9.4.10)–(9.4.14) for the corresponding q_n.

Set $\Omega_0 := \{(x, y) \in \mathbb{R}^{n+1} : x \in (0, l_0),\ y \in \mathcal{D}\}$, $l_0 \in (0, l)$.

Theorem 9.4.7. *Assume that the problem $Aq = f$ with $f \in L_2(\mathcal{D})$ has a solution $q_e \in L_2(\mathcal{D})$, and the initial approximation q_0 satisfies the condition $\|q_0 - q_e\| \le C$. Then the sequence $\{u_n\}$ of solutions to the direct problems* (9.4.10)–(9.4.14) *for the corresponding iterations q_n converges to the exact solution $u_e \in L_2(\Omega)$ to the problem* (9.4.1)–(9.4.5)*, and the following estimate holds:*

$$\|u_n - u_e\|^2_{L_2(\Omega_0)} \le l C^2 n^{-(l-l_0)/l}.$$

Proof. Theorems 9.4.5 and 9.4.6 imply the inequality

$$\int_{\mathcal{D}} (u_n - u_e)^2(x, y)\, dy \le C^{2x/l} \Big(\frac{C^2}{n}\Big)^{(l-x)/l} = C^2 n^{x/l-1}. \tag{9.4.43}$$

Integrating (9.4.43) with respect to x from 0 to l_0 ($l_0 \in (0, l)$), we obtain

$$\|u_n - u_e\|^2_{L_2(\Omega_0)} \le \frac{l C^2}{n} \frac{n^{l_0/l} - 1}{\ln n} \le l C^2 n^{-\frac{l-l_0}{l}},$$

which is the desired conclusion. □

Chapter 10

Inverse coefficient problems for hyperbolic equations

The reader may notice that this chapter is more voluminous than the others. A natural question arises of whether this is caused by a particular interest of the author to this subject or some external reasons exist. The answer is "both". In fact, the hyperbolic equations describing wave processes are of great concern for a large number of researchers in abstract and applied mathematics. Physically, it is explained by the capacity of a wave to come through the object under investigation and to deliver information about its internal structure to the surface. For example, a primitive knocking might help to locate a cavity inside a wall. How to achieve the matching results by heating (parabolic equations) or applying pressure (elliptic ones) is difficult to imagine. On the other side, theoretical results for the inverse problems for hyperbolic equations are more obtainable as compared to those for parabolic equations and, a fortiori, elliptic ones. Indeed, the more stable is the operator, the more difficult is to invert it. The direct problems for parabolic and elliptic equations have, as a rule, rather smooth solutions. The same cannot be said of hyperbolic equations. Solutions of the corresponding direct problems may have discontinuities, singular components of complicated structure, etc. That is why the inverse coefficient problems for hyperbolic equations are more amenable to study. Vast amount of literature have been published concerning these problems and different methods for their solution. The enclosed bibliography contains only a small part of the scientific publications.

10.1 Inverse problems for the equation $u_{tt} = u_{xx} - q(x)u + F(x,t)$

The problems we are consider in this chapter can be roughly separated into two types. Some of them are aimed at illustrating the investigative technique; the others are of practical significance. We start with a problem of the first type, namely, an inverse problem for the equation $u_{tt} = u_{xx} - q(x)u + F(x,t)$, $x \in \mathbb{R}$, $t > 0$, with smooth initial data.

Many results obtained for this equation can be extended to the case of several space variables. However, the theory of inverse problems for the equation

$$u_{tt} = \Delta u - q(x)u + F, \quad x \in \mathbb{R}^n,$$

is not so developed as in the one-dimensional case. In fact, the ill-posedness of one-dimensional inverse problems in the large is caused by their non-linearity, which may be remedied by using the fundamental Gelfand–Levitan–Krein approach. The main difficulties experienced in solving multidimensional inverse problems are created by the fact that additional information is often given on a timelike surface. As was shown in Chapter 9, this circumstance leads to a serious problem even in the linear case.

In this section we use results and technique presented in (Romanov, 1984).

10.1.1 An inverse problem for the equation $u_{tt} = u_{xx} - q(x)u + F(x,t)$ with smooth initial data

Denote $L_q u = u_{tt} - u_{xx} + q(x)u$ and consider the Cauchy problem

$$L_q u = F(x,t), \quad x \in \mathbb{R},\ t > 0; \tag{10.1.1}$$

$$u(x,0) = \varphi(x), \quad u_t(x,0) = \psi(x), \quad x \in \mathbb{R}. \tag{10.1.2}$$

The problem of finding the function $u(x,t)$ from the given functions $q(x)$, $F(x,t)$, $\varphi(x)$, $\psi(x)$ will be referred to as a *direct problem*.

The direct problem (10.1.1), (10.1.2) is well-posed if functional spaces for the data and the solution are suitably chosen, for example, if $q, \varphi, \psi \in C(\mathbb{R})$ and $F, u \in C(\mathbb{R} \times \mathbb{R}_+)$. In this case the function $u(x,t)$ is a generalized solution of problem (10.1.1), (10.1.2). It will be shown later that putting more restrictions on the data, namely, $q \in C(\mathbb{R})$, $\varphi \in C^2(\mathbb{R})$, $\psi \in C^1(\mathbb{R})$, F, $F_t \in C(\mathbb{R} \times (0,T])$, provides the existence of a classical solution $u \in C^2(\mathbb{R} \times (0,T])$ to problem (10.1.1), (10.1.2).

Consider now the *inverse problem* of recovering the coefficient $q(x) \in C(\mathbb{R})$ of equation (10.1.1) if an additional information on the solution $u(x,t)$ of the direct problem (10.1.1), (10.1.2) is given in the form

$$u(x_0,t) = f_1(t), \quad u_x(x_0,t) = f_2(t), \quad t \geq 0, \tag{10.1.3}$$

where $x_0 \in \mathbb{R}$ is a fixed point. In other words, it is required to find the functions $q(x)$ and $u(x,t)$ satisfying (10.1.1)–(10.1.3) provided the functions $F(x,t)$, $\varphi(x)$, $\psi(x)$, $f_1(t)$ and $f_2(t)$ are known.

Since the functions f_1 and f_2 are traces of functions u and u_x on the line $x = x_0$, they can not be arbitrary. In particular, the smoothness of these functions depends on that of q, φ, ψ, F. In other words, the necessary conditions on f_1, f_2 follow from the properties of the solution $u(x,t)$ to problem (10.1.1), (10.1.2).

Denote by $\Delta(x_0,t_0)$ a triangle on the half-plane $\mathbb{R} \times \mathbb{R}_+$ which is enclosed by the x-axis and the characteristics of equation (10.1.1) passing through the point (x_0,t_0) (see Figure 1).

Lemma 10.1.1. *If, for some $t_0 > 0$ and $x_0 \in \mathbb{R}$, there hold $q(x) \in C[x_0-t_0, x_0+t_0]$, $\varphi(x) \in C^2[x_0-t_0, x_0+t_0]$, $\psi(x) \in C^1[x_0-t_0, x_0+t_0]$, F, $F_t \in C(\Delta(x_0,t_0))$,*

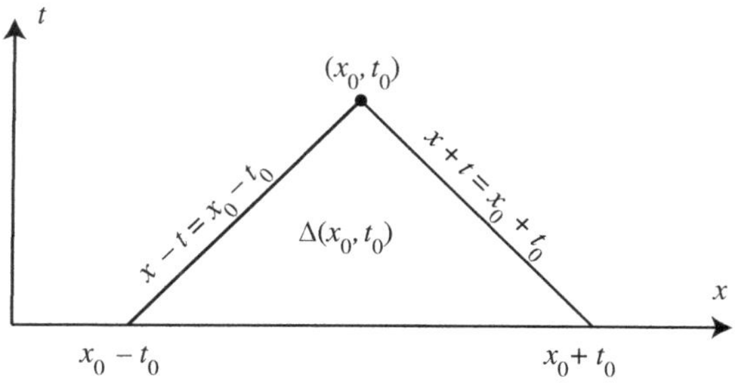

Figure 1 The domain $\Delta(x_0, t_0)$.

then in the domain $\Delta(x_0, t_0)$ *there exists a unique classical solution of the direct problem* (10.1.1), (10.1.2).

Proof. Problem (10.1.1), (10.1.2) can be reduced to a linear integral equation for the function $u(x,t)$. Indeed, rewrite equation (10.1.1) in the form

$$\Big(\frac{\partial^2}{\partial t^2} - \frac{\partial^2}{\partial x^2}\Big)u = F(x,t) - q(x)u. \tag{10.1.4}$$

By d'Alembert's formula, the solution of problem (10.1.4), (10.1.2) is representable as

$$u(x,t) = \frac{1}{2}[\varphi(x-t) + \varphi(x+t)] + \frac{1}{2}\int_{x-t}^{x+t} \psi(\xi)\,d\xi + \frac{1}{2}\iint_{\Delta(x,t)} [F(\xi,\tau) - q(\xi)u(\xi,\tau)]\,d\xi\,d\tau.$$

Therefore, the solution $u(x,t)$ of problem (10.1.1), (10.1.2) satisfies the integral equation

$$u(x,t) = u_0(x,t) - \frac{1}{2}\iint_{\Delta(x,t)} q(\xi)u(\xi,\tau)\,d\xi\,d\tau, \quad (x,t) \in \Delta(x_0,t_0), \tag{10.1.5}$$

where

$$u_0(x,t) = \frac{1}{2}[\varphi(x-t) + \varphi(x+t)] + \frac{1}{2}\int_{x-t}^{x+t} \psi(\xi)\,d\xi + \frac{1}{2}\iint_{\Delta(x,t)} F(\xi,\tau)\,d\xi\,d\tau.$$

Equation (10.1.5) admits a unique continuous solution in $\Delta(x_0, t_0)$. To prove this, we shall use the method of successive approximations assuming

$$u(x,t) = \sum_{n=0}^{\infty} u_n(x,t), \tag{10.1.6}$$

where $u_n(x,t)$, $n \geq 1$, are determined by the formulae

$$u_n(x,t) = -\frac{1}{2}\iint_{\Delta(x,t)} q(\xi)u_{n-1}(\xi,\tau)\,d\xi\,d\tau, \quad (x,t) \in \Delta(x_0,t_0), \quad n \geq 1. \tag{10.1.7}$$

Let us show that the series defined by (10.1.6) converges in $\Delta(x_0,t_0)$.

Note that $u_0 \in C^2(\Delta(x_0,t_0))$ under the lemma assumptions. From (10.1.7) it follows that all the functions u_n are continuous in $\Delta(x_0,t_0)$. Denote

$$\begin{aligned} U_n(t) &= \max_{x_0+(t-t_0)\leq x\leq x_0-(t-t_0)} |u_n(x,t)|, \quad 0 \leq t \leq t_0, \\ \|u\|_k &= \sum_{|\alpha|\leq k} \max_{(x,t)\in\Delta(x_0,t_0)} |D^\alpha u(x,t)|, \quad k = 0,1,2, \end{aligned} \tag{10.1.8}$$

where

$$D^\alpha = \frac{\partial^{|\alpha|}}{\partial^{\alpha_1}\partial^{\alpha_2}}, \quad \alpha = (\alpha_1,\alpha_2), \quad |\alpha| = \alpha_1 + \alpha_2, \quad \alpha_j = 0,1,2, \quad j = 1,2.$$

Then (10.1.7) implies the inequality

$$\begin{aligned} U_n(t) &\leq \frac{1}{2}\int_0^t \max_{x_0+(t-t_0)\leq x\leq x_0-(t-t_0)} \left| \int_{\tau-t+x}^{-\tau+t+x} q(\xi)u_{n-1}(\xi,\tau)\,d\xi \right| d\tau \\ &\leq \|q\|_0 \int_0^t (t-\tau)U_{n-1}(\tau)\,d\tau, \quad n \geq 1,\ 0 \leq t \leq t_0. \end{aligned}$$

Using this inequality successively for $n = 1,2,\ldots$, we come to the estimate

$$U_n(t) \leq (\|q\|_0)^n \frac{t^{2n}}{(2n)!}\|u_0\|_0, \quad n \geq 1. \tag{10.1.9}$$

As seen from (10.1.9), there holds $|u_n(x,t)| \leq U_n(t) \leq \|u_0\|_0 \frac{(\|q\|_0 t_0^2)^n}{(2n)!}$. Since the numerical series $\sum_{n=0}^\infty \|u_0\|_0 \frac{(\|q\|_0 t_0^2)^n}{(2n)!}$ converges, we apply the Weierstrass criterion to conclude that the series in (10.1.6) converges uniformly in the domain $\Delta(x_0,t_0)$. Consequently, there exists a continuous function $u(x,t) = \sum_{n=0}^\infty u_n(x,t)$ in $\Delta(x_0,t_0)$ which solves equation (10.1.5). The solution is unique because the corresponding homogeneous equation

$$u(x,t) = -\frac{1}{2}\iint_{\Delta(x,t)} q(\xi)u(\xi,\tau)\,d\xi\,d\tau \tag{10.1.10}$$

admits only the trivial solution in the class of continuous functions in $\Delta(x_0,t_0)$. Indeed, putting

$$U(t) = \max_{x_0-(t_0-t)\leq x\leq x_0+(t_0-t)} |u(x,t)|,$$

we note that, in view of (10.1.10), the function $U(t)$ satisfies the inequality

$$U(t) \le \|q\|_0 \int_0^t (t-\tau)U(\tau)\,d\tau, \quad t \in [0, t_0],$$

but this is possible only for $U(t) \equiv 0$. Consequently, $u(x,t) \equiv 0$ for $(x,t) \in \Delta(x_0, t_0)$.

Thus, we have shown the existence of a unique continuous solution to equation (10.1.5). Let us prove that the solution has continuous partial derivatives of order ≤ 2 in $\Delta(x_0, t_0)$ and hence is a classical solution to the direct problem (10.1.1), (10.1.2). Rewrite (10.1.5) in the form

$$u(x,t) = u_0(x,t) - \frac{1}{2}\int_{x-t}^{x+t} q(\xi)\Big[\int_0^{t-|x-\xi|} u(\xi,\tau)\,d\tau\Big]\,d\xi, \quad (x,t) \in \Delta(x_0,t_0). \tag{10.1.11}$$

Since $u_0 \in C^2(\Delta(x_0,t_0))$, the expression in the right-hand side of (10.1.11) is differentiable with respect to x and t. Therefore, so is the function $u(x,t)$:

$$u_t(x,t) = \frac{\partial}{\partial t}u_0(x,t) - \frac{1}{2}\int_{x-t}^{x+t} q(\xi)u(\xi, t-|x-\xi|)\,d\xi, \tag{10.1.12}$$

$$u_x(x,t) = \frac{\partial}{\partial x}u_0(x,t) - \frac{1}{2}\int_{x-t}^{x+t} q(\xi)u(\xi, t-|x-\xi|)\,\mathrm{sign}(\xi - x)\,d\xi. \tag{10.1.13}$$

As evident from equalities (10.1.12), (10.1.13), the functions u_t and u_x are continuous and have first-order derivatives with respect to x and t in $\Delta(x_0,t_0)$. Consequently, the function $u(x,t)$ has partial derivatives of the second order in the domain $\Delta(x_0,t_0)$:

$$\begin{aligned} u_{tt}(x,t) = {} & \frac{\partial^2}{\partial t^2}u_0(x,t) - \frac{1}{2}[q(x+t)u(x+t,0) + q(x-t)u(x-t,0)] \\ & - \frac{1}{2}\int_{x-t}^{x+t} q(\xi)u_t(\xi, t-|x-\xi|)\,d\xi, \end{aligned} \tag{10.1.14}$$

$$\begin{aligned} u_{xt}(x,t) = {} & \frac{\partial^2}{\partial x \partial t}u_0(x,t) - \frac{1}{2}[q(x+t)u(x+t,0) - q(x-t)u(x-t,0)] \\ & - \frac{1}{2}\int_{x-t}^{x+t} q(\xi)u_t(\xi, t-|x-\xi|)\,\mathrm{sign}(\xi - x)\,d\xi, \end{aligned} \tag{10.1.15}$$

$$\begin{aligned} u_{xx}(x,t) = {} & \frac{\partial^2}{\partial x^2}u_0(x,t) - \frac{1}{2}[q(x+t)u(x+t,0) + q(x-t)u(x-t,0)] \\ & + q(x)u(x,t) - \frac{1}{2}\int_{x-t}^{x+t} q(\xi)u_t(\xi, t-|x-\xi|)\,d\xi. \end{aligned} \tag{10.1.16}$$

From formulae (10.1.14)–(10.1.16) it follows that $u_{tt}, u_{xx}, u_{xt} \in C(\Delta(x_0,t_0))$, that is, $u \in C^2(\Delta(x_0,t_0))$. Thus, the lemma is proved. □

Turning back to the inverse problem (10.1.1)–(10.1.3), we see that under the assumptions of Lemma 10.1.1 the data $f_1(t)$, $f_2(t)$ are bound to be smooth enough, namely,

$$f_1(t) \in C^2[0, t_0], \quad f_2(t) \in C^1[0, t_0]. \tag{10.1.17}$$

Moreover, the functions $f_1(t)$, $f_2(t)$ have to meet the following conditions of consistency with the direct problem data:

$$\varphi(x_0) = f_1(0), \quad \psi(x_0) = f_1'(0), \quad \varphi'(x_0) = f_2(0), \quad \psi'(x_0) = f_2'(0). \tag{10.1.18}$$

The necessary conditions (10.1.17), (10.1.18) are also sufficient for solvability of the inverse problem in the small.

Theorem 10.1.1. *Let $t_0 > 0$ and $x_0 \in \mathbb{R}$ be fixed. Suppose that the assumptions of Lemma* 10.1.1, *conditions* (10.1.17), (10.1.18) *and the inequality*

$$|\varphi(x)| \geq \alpha > 0, \quad x \in [x_0 - t_0, x_0 + t_0], \tag{10.1.19}$$

hold. Then, for a small enough $h > 0$, a solution $q(x)$ of the inverse problem (10.1.1)–(10.1.3) *exists on the interval $[x_0-h, x_0+h]$ and is unique in the class $C[x_0-h, x_0+h]$.*

Proof. Setting $x = x_0$ in formulae (10.1.14), (10.1.15) and using (10.1.3), we obtain

$$\begin{aligned} f_1''(t) = {} & \frac{\partial^2}{\partial t^2} u_0(x_0, t) - \frac{1}{2} [q(x_0 + t)u(x_0 + t, 0) + q(x_0 - t)u(x_0 - t, 0)] \\ & - \frac{1}{2} \int_{x_0-t}^{x_0+t} q(\xi) u_t(\xi, t - |x_0 - \xi|) \, d\xi, \end{aligned} \tag{10.1.20}$$

$$\begin{aligned} f_2'(t) = {} & \frac{\partial^2}{\partial x \, \partial t} u_0(x, t)|_{x=x_0} - \frac{1}{2} [q(x_0 + t)u(x_0 + t, 0) - q(x_0 - t)u(x_0 - t, 0)] \\ & - \frac{1}{2} \int_{x_0-t}^{x_0+t} q(\xi) u_t(\xi, t - |x_0 - \xi|) \operatorname{sign}(\xi - x_0) \, d\xi, \quad t \in [0, t_0], \end{aligned} \tag{10.1.21}$$

whence, in view of (10.1.2),

$$\begin{aligned} q(x) = {} & \frac{1}{\varphi(x)} \Big\{ -f_1''(x - x_0) - f_2'(x - x_0) + \Big[\Big(\frac{\partial^2}{\partial t^2} + \frac{\partial^2}{\partial x \partial t} \Big) u_0(x, t) \Big]_{x=x_0, t=x-x_0} \\ & - \int_{x_0}^{x} q(\xi) u_t(\xi, x - \xi) \, d\xi \Big\}, \quad x_0 \leq x \leq x_0 + t_0, \end{aligned}$$

$$\begin{aligned} q(x) = {} & \frac{1}{\varphi(x)} \Big\{ -f_1''(x_0 - x) + f_2'(x_0 - x) + \Big[\Big(\frac{\partial^2}{\partial t^2} - \frac{\partial^2}{\partial x \partial t} \Big) u_0(x, t) \Big]_{x=x_0, t=x_0-x} \\ & - \int_{x}^{x_0} q(\xi) u_t(\xi, \xi - x) \, d\xi \Big\}, \quad x_0 - t_0 \leq x \leq x_0. \end{aligned}$$

Combining these two equalities yields

$$q(x) = q_0(x) + \frac{1}{\varphi(x)} \int_{x_0}^{x} q(\xi)u_t(\xi, |x-\xi|)\, d\xi \operatorname{sign}(x_0 - x),$$
$$x \in [x_0 - t_0, x_0 + t_0], \quad (10.1.22)$$

where

$$q_0(x) = \frac{1}{\varphi(x)} \Big\{ -f_1''(|x - x_0|) - f_2'(|x-x_0|) \operatorname{sign}(x - x_0) \qquad (10.1.23)$$
$$+ \Big[\Big(\frac{\partial^2}{\partial t^2} + \operatorname{sign}(x - x_0) \frac{\partial^2}{\partial x\, \partial t} \Big) u_0(x,t) \Big]_{x = x_0, t = |x - x_0|} \Big\}.$$

Though the discontinuous function $\operatorname{sign}(x - x_0)$ is involved in (10.1.23), the function $q_0(x)$ is continuous on $[x_0 - t_0, x_0 + t_0)$ due to the consistency conditions (10.1.18). Indeed, since

$$-f_2'(0) + \frac{\partial^2}{\partial x\, \partial t} u_0(x,t)\big|_{x = x_0, t=0} = -f_2'(0) + \psi'(x_0) = 0,$$

the multiplier at $\operatorname{sign}(x - x_0)$ vanishes for $x = x_0$.

We consider now a system of nonlinear second-kind integral equations (10.1.5), (10.1.12), (10.1.22) for the functions u, u_t, q in the domain $\Delta(x_0, t_0)$. Regarding the measure of the integration domain as a small parameter, we can apply the principle of contracting mappings to the system. Indeed, system (10.1.5), (10.1.12), (10.1.22) can be rewritten in the operator form

$$\boldsymbol{q} = A\boldsymbol{q}, \qquad (10.1.24)$$

where $\boldsymbol{q} = (q_1, q_2, q_3)^\top$ is a vector-function of two variables x and t whose components are as follows:

$$q_1(x,t) = u(x,t), \quad q_2(x,t) = u_t(x,t), \quad q_3(x,t) \equiv q_3(x) = q(x).$$

The operator A in (10.1.24) is defined over the set of vector-functions $\boldsymbol{q} \in C(\Delta(x_0, t_0))$ and acts so that $A\boldsymbol{q} = (A_1\boldsymbol{q}, A_2\boldsymbol{q}, A_3\boldsymbol{q})^\top$ where

$$A_1\boldsymbol{q} = u_0(x,t) - \frac{1}{2} \iint_{\Delta(x,t)} q_3(\xi) q_1(\xi, \tau)\, d\xi\, d\tau,$$
$$A_2\boldsymbol{q} = \frac{\partial}{\partial t} u_0(x,t) - \frac{1}{2} \int_{x-t}^{x+t} q_3(\xi) q_1(\xi, t - |x - \xi|)\, d\xi, \qquad (10.1.25)$$
$$A_3\boldsymbol{q} = q_0(x) + \frac{\operatorname{sign}(x_0 - x)}{\varphi(x)} \int_{x_0}^{x} q_3(\xi) q_2(\xi, |x-\xi|)\, d\xi.$$

Denote

$$\|\boldsymbol{q}\|(t_0) = \max_{1\le k\le 3}\ \max_{(x,t)\in\Delta(x_0,t_0)} |q_k(x,t)|,$$

$$\boldsymbol{q}_0(x,t) = \Big(u_0(x,t), \frac{\partial}{\partial t}u_0(x,t), q_0(x)\Big).$$

Let $0 < h \le t_0$ and $M(h)$ be a set of functions $\boldsymbol{q}(x,t) \in C(\Delta(x_0,h))$ satisfying the inequality

$$\|\boldsymbol{q} - \boldsymbol{q}_0\|(h) \le \|\boldsymbol{q}_0\|(t_0). \tag{10.1.26}$$

Let us show that if h is small enough, then A is a contracting mapping of the set $M(h)$ into itself. Note that for $\boldsymbol{q} \in M(h)$ there holds

$$\|\boldsymbol{q}\|(h) \le 2\|\boldsymbol{q}_0\|(h).$$

Consequently, from formulae (10.1.25) we obtain

$$\|A\boldsymbol{q} - \boldsymbol{q}_0\|(h) \le 4\|\boldsymbol{q}_0\|^2(t_0)\max\Big(\frac{h^2}{2}, h, \frac{h}{\alpha}\Big).$$

Thus, if

$$h \le h^* = \min\Big(\frac{1}{\sqrt{2\|\boldsymbol{q}_0\|(t_0)}}, \frac{\min(\alpha, 1)}{4\|\boldsymbol{q}_0\|(t_0)}, t_0\Big), \tag{10.1.27}$$

then the operator A maps the set $M(h)$ into itself.

Let $\boldsymbol{q}^{(1)}$ and $\boldsymbol{q}^{(2)}$ be arbitrary elements of the set $M(h)$, $h \le h^*$. Then using the evident inequalities

$$\begin{aligned} |q_k^{(1)}q_s^{(1)} - q_k^{(2)}q_s^{(2)}| &\le |q_k^{(1)} - q_k^{(2)}|\,|q_s^{(1)}| + |q_k^{(2)}|\,|q_s^{(1)} - q_s^{(2)}| \\ &\le 4\|\boldsymbol{q}_0\|(t_0)\|\boldsymbol{q}^{(1)} - \boldsymbol{q}^{(2)}\|(h), \end{aligned}$$

we get

$$\begin{aligned} \|A\boldsymbol{q}^{(1)} - A\boldsymbol{q}^{(2)}\|(h) &\le 4\|\boldsymbol{q}_0\|(t_0)\|\boldsymbol{q}^{(1)} - \boldsymbol{q}^{(2)}\|(h)\max\Big(\frac{h^2}{2}, h, \frac{h}{\alpha}\Big) \\ &\le \frac{h}{h^*}\|\boldsymbol{q}^{(1)} - \boldsymbol{q}^{(2)}\|(h). \end{aligned}$$

As evident from the last inequality, for any $h < h^*$ the operator A is a contracting mapping of the set $M(h)$ into itself. Consequently, by the Banach fixed point theorem, the operator A admits one and only one fixed point in $M(h)$, that is, there exists a unique solution of equation (10.1.24). Therefore, solving the system of equations (10.1.5), (10.1.12), (10.1.22) by, for example, the method of successive approximations, we thereby uniquely determine the functions u, u_t, q in the domain $\Delta(x_0,h)$, $h \in (0,h^*)$. The function $q(x)$ thus constructed is a solution of the inverse problem on the interval $[x_0 - h, x_0 + h]$. □

10.1.2 An inverse point-source problem for the equation $u_{tt} = u_{xx} - q(x)u$

The one-dimensional homogeneous equation $u_{tt} = u_{xx} - q(x)u$ seems to be the most demonstrative among hyperbolic equations. First, this equation is closely connected with the classical Sturm–Lioville problem (see Subsection 10.1.4). Secondly, inverse problems for this equation are reducible to a system of nonlinear integral Volterra equations of the second kind, which makes it possible to prove their local well-posedness and conditional well-posedness in the large (Subsection 10.1.2 and Chapter 4). Thirdly, the technique developed for the equation $u_{tt} = u_{xx} - q(x)u$ may be exploited in solving the acoustic equation $c^{-2}(x)v_{tt} = v_{xx} - (\ln \rho(x))'v_x$ (Section 10.2) and others, for example, $c^{-2}(x)v_{tt} = v_{xx}$, $v_{tt} = v_{xx} - (\ln \sigma(x))'v_x$ (see also Subsection A.6.3). Finally, the considered equation is associated with the parabolic equation $v_t = v_{xx} - q(x)v$ (Section 11.2).

The direct and inverse point-source problems for the equation $u_{tt} = u_{xx} - q(x)u$ are formulated as follows.

In *the direct problem* it is required to find a generalized solution $u(x,t)$ to the Cauchy problem

$$u_{tt} = u_{xx} - q(x)u, \quad (x,t) \in \mathbb{R} \times \mathbb{R}_+, \tag{10.1.28}$$

$$u\big|_{t=0} = 0, \quad u_t\big|_{t=0} = \delta(x). \tag{10.1.29}$$

The inverse point-source problem consists in recovering the function $q(x)$ from an additional information about the solution of the direct problem (10.1.28), (10.1.29) given in the form

$$u(0,t) = f_1(t), \quad u_x(0,t) = f_2(t), \quad t \geq 0. \tag{10.1.30}$$

Properties of the direct problem solution. Taking into account that

$$\int_{-\infty}^{\infty} \psi(x)\delta(x)\,dx = \psi(0)$$

for any continuous finite function $\psi(x)$ and applying d'Alembert's formula, we conclude that the generalized solution of the direct problem (10.1.28), (10.1.29) is a piecewise continuous solution of the integral equation

$$u(x,t) = \frac{1}{2}\theta(t - |x|) - \frac{1}{2}\iint_{\Delta(x,t)} q(\xi)u(\xi,\tau)\,d\xi\,d\tau, \quad (x,t) \in \mathbb{R} \times \mathbb{R}_+. \tag{10.1.31}$$

Here $\theta(\cdot)$ is the Heaviside step function:

$$\theta(t) = \begin{cases} 1, & t \geq 0, \\ 0, & t < 0. \end{cases}$$

From (10.1.31) it follows that

$$u(x,t) \equiv 0, \quad t < |x|, \quad (x,t) \in \mathbb{R} \times \mathbb{R}_+. \tag{10.1.32}$$

To prove this, we consider the domain $\Delta(x_1, t_1)$ where $(x_1, t_1) \in \mathbb{R} \times \mathbb{R}_+$ is an arbitrary point such that $t_1 < |x_1|$. In the domain $\Delta(x_1, t_1)$, owing to (10.1.31), there holds

$$u(x,t) = -\frac{1}{2} \iint_{\Delta(x,t)} q(\xi) u(\xi,\tau)\, d\xi\, d\tau, \quad (x,t) \in \Delta(x_1, t_1). \tag{10.1.33}$$

Denote

$$Q_1 = \max_{x_1 - t_1 \le x \le x_1 + t_1} |q(x)|, \quad U(t) = \max_{x_1 - (t_1 - t) \le x \le x_1 + (t_1 - t)} |u(x,t)|.$$

In view of (10.1.33), we obtain

$$U(t) \le Q_1 t_1 \int_0^t U(\tau)\, d\tau, \quad 0 \le t \le t_1.$$

By Gronwall's lemma, $U(t) \equiv 0$ for $t \in [0, t_1]$. Therefore, $u(x,t) \equiv 0$ in the domain $\Delta(x_1, t_1)$. Taking into account that the point (x_1, t_1) was chosen arbitrarily, we obtain (10.1.32).

From (10.1.32) it follows that the support of the function $u(x,t)$ is wholly contained in the domain $\{(x,t) : t \ge |x|\}$.

Let $D = \{(x,t) : t > |x|\}$. In what follows we shall consider the function $u(x,t)$ in the domain D only. Formulae (10.1.31), (10.1.32) imply

$$u(x,t) = \frac{1}{2} - \frac{1}{2} \iint_{\square(x,t)} q(\xi) u(\xi,\tau)\, d\xi\, d\tau, \quad (x,t) \in D. \tag{10.1.34}$$

Here $\square(x,t) = \{(\xi,\tau) : |\xi| \le \tau \le t - |x - \xi|\}$ is a rectangle confined between the characteristics of equation (10.1.28) passing through the points $(0,0)$ and (x,t) (see Figure 2).

Lemma 10.1.2. *If for a point $(x_1, t_1) \in D$ the function $q(x)$ is continuous on the interval $\big[(x_1 - t_1)/2, (x_1 + t_1)/2\big]$, then a solution to equation* (10.1.34) *exists and belongs to the class $C^2\big(\square(x_1, t_1)\big)$.*

We shall give a sketchy proof of the lemma.

Let

$$u_0 = \frac{1}{2}, \quad u_n(x,t) = -\frac{1}{2} \iint_{\square(x,t)} q(\xi) u_{n-1}(\xi,\tau)\, d\xi\, d\tau, \quad n \ge 1,\ (x,t) \in \square(x_1, t_1).$$

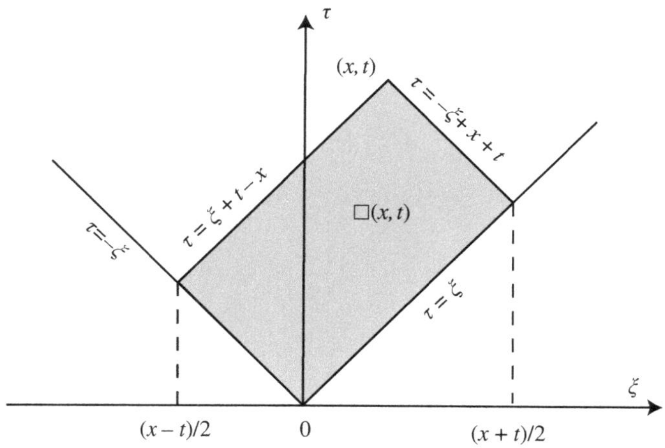

Figure 2 The domain $\square(x,t)$.

It can be shown that the series

$$u(x,t) = \sum_{n=0}^{\infty} u_n(x,t)$$

converges and thus defined function $u(x,t)$ solves equation (10.1.34). This solution is continuous in the domain $\square(x_1,t_1)$ because the functions $u_n(x,t)$ are obviously continuous and the series $u(x,t) = \sum_{n=0}^{\infty} u_n(x,t)$ uniformly converges in $\square(x_1,t_1)$, which follows from the inequalities

$$|u_n(x,t)| \le \frac{1}{2^{n+1}}[\|q\|_0(t_1 - |x_1|)]^n \frac{t^n}{n!}, \quad (x,t) \in \square(x_1,t_1), \quad n = 0,1,2,\ldots.$$

These inequalities can be easily verified by induction. The required smoothness of the solution can be checked by differentiating (10.1.34). To show its uniqueness, it suffices to observe that the corresponding homogeneous equation has no nontrivial solutions. The proof of this fact is similar to that of (10.1.32).

Note that, due to (10.1.34), the function $u(x,t)$ is constant on the boundary of the domain D and equals to $1/2$:

$$u(x,|x|) = 1/2. \tag{10.1.35}$$

Reducing the inverse point-source problem to a system of integral Volterra equations. Let us write down partial derivatives of the function $u(x,t)$ in the domain D. For this purpose we replace the double integral in (10.1.34) by a repeated one:

$$u(x,t) = \frac{1}{2} - \frac{1}{2}\int_{(x-t)/2}^{(x+t)/2} \int_{|\xi|}^{t-|x-\xi|} q(\xi)u(\xi,\tau)\,d\xi\,d\tau, \quad (x,t) \in D.$$

Differentiating this equality, we obtain

$$u_t = \frac{1}{2}\int_{(x-t)/2}^{(x+t)/2} q(\xi)u(\xi, t - |x - \xi|)\, d\xi, \tag{10.1.36}$$

$$u_x = -\frac{1}{2}\int_{(x-t)/2}^{(x+t)/2} q(\xi)u(\xi, t - |x - \xi|)\,\mathrm{sign}\,(\xi - x)\, d\xi, \tag{10.1.37}$$

$$\begin{aligned} u_{xt} = &-\frac{1}{4}\Big[q\Big(\frac{x+t}{2}\Big) - q\Big(\frac{x-t}{2}\Big)\Big] \\ &- \frac{1}{2}\int_{(x-t)/2}^{(x+t)/2} q(\xi)u_t(\xi, t - |x - \xi|)\,\mathrm{sign}\,(\xi - x)\, d\xi, \end{aligned} \tag{10.1.38}$$

$$\begin{aligned} u_{tt} = &-\frac{1}{4}\Big[q\Big(\frac{x+t}{2}\Big) + q\Big(\frac{x-t}{2}\Big)\Big] \\ &- \frac{1}{2}\int_{(x-t)/2}^{(x+t)/2} q(\xi)u_t(\xi, t - |x - \xi|)\, d\xi. \end{aligned} \tag{10.1.39}$$

Lemma 10.1.2 and equalities (10.1.35)–(10.1.38) imply the conditions (see Lemma 10.1.3) which the inverse problem data $f_1(t)$, $f_2(t)$ are to satisfy.

Lemma 10.1.3. *If* $q(x) \in C[-T/2, T/2]$, $T > 0$, *then* $f_1 \in C^2[0,T]$, $f_2 \in C^1[0,T]$ *and*

$$f_1(+0) = 1/2, \quad f_1'(+0) = f_2(+0) = f_2'(+0) = 0. \tag{10.1.40}$$

Summing formulae (10.1.38), (10.1.39) for $x = 0$ and setting then $t = 2|x|$, we come to

$$q(x) = q_0(x) - 4\,\mathrm{sign}(x)\int_0^x q(\xi)u_t(\xi, 2|x| - \xi)\, d\xi. \tag{10.1.41}$$

Here

$$q_0(x) = -4\big[f_1''(2|x|) + f_2'(2|x|)\,\mathrm{sign}(x)\big].$$

Equations (10.1.41), (10.1.34), (10.1.36) form a closed system of integral equations for the functions u, u_t, q in the domain D. Using this system, it is not difficult to prove the following theorems.

Theorem 10.1.2. *Assume that, for some* $t_0 > 0$, *the functions* $f_1(t) \in C^2[0, t_0]$ *and* $f_2(t) \in C^1[0, t_0]$ *meet conditions* (10.1.40). *Then there exists such* $h \in (0, t_0/2)$ *that the inverse problem* (10.1.28)–(10.1.30) *is uniquely solvable in the class of functions* $q(x) \in C[-h, h]$.

Theorem 10.1.3. *On the assumptions of Theorem* 10.1.2 *the function* $q(x) \in C[-t_0/2, t_0/2]$ *is uniquely determined from the functions* f_1, f_2 *given for* $t \in (0, t_0]$.

Theorem 10.1.4. *Let* $q(x)$, $\bar{q}(x)$ *be solutions of the inverse problem* (10.1.28)–(10.1.30) *with the data* f_1, f_2 *and* $\bar{f}_1$, $\bar{f}_2$, *respectively. Then*

$$|q(x) - \bar{q}(x)| \leq c(\|f_1'' - \bar{f}_1''\|_{C[0,t_0]} + \|f_2' - \bar{f}_2'\|_{C[0,t_0]}), \quad x \in [-t_0/2, t_0/2],$$

with the constant c *depending on* $\|q\|_{C[-t_0/2,t_0/2]}$, $\|\bar{q}\|_{C[-t_0/2,t_0/2]}$ *and* t_0 *only.*

These theorems can be proven in a similar way as Theorem 10.1.1.

Let us look now at the case when information about the solution of the Cauchy problem (10.1.28), (10.1.29) is given at a point $x_1 \neq 0$:

$$u(x_1, t) = f_1(t), \quad u_x(x_1, t) = f_2(t), \quad t > 0. \tag{10.1.42}$$

Considering the uniqueness of a solution to the inverse problem (10.1.28), (10.1.29), (10.1.42), we face a different situation here. Indeed, let $x_1 > 0$. Then the following theorem is valid (see (Romanov, 1984)).

Theorem 10.1.5. *The information* (10.1.42) *on the solution of the Cauchy problem* (10.1.28), (10.1.29) *uniquely determines the function* $q(x)$ *in the domain* $x \geq x_1 > 0$.

10.1.3 Solvability in the large of the inverse problem for the equation $u_{tt} = u_{xx} - q(x)u$

In this subsection our main concern is to formulate necessary and sufficient conditions for the inverse problem (10.1.28)–(10.1.30) to be solvable in the large.

Consider the inverse problem in the case when $f_2(t) \equiv 0$:

$$u_{tt} = u_{xx} - q(x)u, \quad (x,t) \in \mathbb{R} \times \mathbb{R}_+; \tag{10.1.43}$$

$$u|_{t=0} = 0, \quad u_t|_{t=0} = \delta(x); \tag{10.1.44}$$

$$u(0,t) = f(t), \quad u_x(0,t) = 0, \quad t \geq 0. \tag{10.1.45}$$

We extend the functions $u(x,t)$ and $f(t)$ over $t < 0$ assuming

$$u(x,t) = -u(x,-t), \quad t < 0;$$
$$f(t) = -f(-t), \quad t < 0.$$

In view of (10.1.40), $f(+0) = 1/2$. Therefore, the extended function $f(t)$ has a jump at $t = 0$:

$$f(+0) = 1/2, \quad f(-0) = -1/2.$$

The function $u(x,t)$ thus continued satisfies the equation

$$u_{tt} = u_{xx} - q(x)u, \quad (x,t) \in \mathbb{R}^2, \tag{10.1.46}$$

and the condition

$$u(x,t) \equiv 0, \quad t < |x|. \tag{10.1.47}$$

Suppose a solution of the inverse problem (10.1.43)–(10.1.45) to exist. Then the function $u(x,t)$ meets additional conditions

$$u(0,t) = f(t), \quad u_x(0,t) = 0, \tag{10.1.48}$$

with $f(t)$ being twice continuously differentiable at $t \neq 0$.

Consider an auxiliary Cauchy problem:

$$w_{tt} = w_{xx} - q(x)w, \quad x > 0, \quad t \in \mathbb{R}; \tag{10.1.49}$$

$$w\big|_{x=0} = \delta(t), \quad w_x\big|_{x=0} = 0. \tag{10.1.50}$$

A solution of this problem must obey the integral equation

$$w(x,t) = \frac{1}{2}[\delta(t-x) + \delta(t+x)] + \frac{1}{2}\iint_{\Delta_1(x,t)} q(\xi)\,w(\xi,\tau)\,d\xi\,d\tau, \tag{10.1.51}$$

where $\Delta_1(x,t) = \{(\xi,\tau) : 0 < \xi \le x, t - x + \xi < \tau < t + x - \xi\}$ is a triangle formed by the t-axis and the characteristics passing through the point (x,t).

It is not difficult to show that

$$w(x,t) \equiv 0, \quad 0 < x < |t|. \tag{10.1.52}$$

From (10.1.52) it follows that the domain of integration in equation (10.1.51) for $(x,t) \in D_1 = \{(x,t) : x \ge |t|\}$ is, in fact, a rectangle $\square_1(x,t) = \{(\xi,\tau) : |\tau| \le \xi \le x - |t-\tau|\}$ formed by the characteristics passing through the points $(0,0)$ and (x,t).

Denote

$$\tilde{w}(x,t) = w(x,t) - \frac{1}{2}[\delta(t-x) + \delta(t+x)]. \tag{10.1.53}$$

In view of (10.1.51), the piecewise continuous function $\tilde{w}(x,t)$ solves the equation

$$\begin{aligned} \tilde{w}(x,t) = {} & \frac{1}{4}\theta(x - |t|)\left[\int_0^{(x+t)/2} q(\xi)\,d\xi + \int_0^{(x-t)/2} q(\xi)\,d\xi\right] \\ & + \frac{1}{2}\iint_{\square_1(x,t)} q(\xi)\,\tilde{w}(\xi,\tau)\,d\xi\,d\tau, \quad x > 0. \end{aligned} \tag{10.1.54}$$

From (10.1.54) it follows that $\tilde{w}(x,t) \equiv 0$ for $x < |t|$, $\tilde{w}$ is even in t and has continuous first derivatives inside D_1. Moreover,

$$\tilde{w}(x, x-0) = \frac{1}{4}\int_0^x q(\xi)\, d\xi, \quad x > 0. \tag{10.1.55}$$

The solution of problem (10.1.46), (10.1.48) is representable in the form

$$u(x,t) = \int_{-\infty}^{\infty} f(\tau)\, w(x, t-\tau)\, d\tau = \int_{-\infty}^{\infty} f(t-\tau)\, w(x,\tau)\, d\tau. \tag{10.1.56}$$

Let us prove this.

It follows from (10.1.49), (10.1.50) that

$$L_q u = \int_{-\infty}^{\infty} f(\tau)\, L_q w(x, t-\tau)\, d\tau = 0,$$

$$u\big|_{x=0} = \int_{-\infty}^{\infty} f(\tau)\, \delta(t-\tau)\, d\tau = f(t), \quad u_x\big|_{x=0} = 0,$$

where $L_q \equiv \frac{\partial^2}{\partial t^2} - \frac{\partial^2}{\partial x^2} + q(x)$. We use (10.1.52) and (10.1.53) to bring equality (10.1.56) to the form:

$$u(x,t) = \frac{1}{2}[f(t-x) + f(t+x)] + \int_{-x}^{x} f(t-\tau)\, \tilde{w}(x,\tau)\, d\tau. \tag{10.1.57}$$

In view of (10.1.47),

$$\frac{1}{2}[f(t-x) + f(t+x)] + \int_{-x}^{x} f(t-\tau)\, \tilde{w}(x,\tau)\, d\tau = 0, \quad x > |t|. \tag{10.1.58}$$

For any fixed $x > 0$, (10.1.58) is a first-kind integral equation for the function $\tilde{w}(x,t)$, $t \in (-x, x)$. Let us look more closely at this equation. Note that its kernel $f(t-\tau)$ has a finite discontinuity at $t = \tau$: $f(+0) = 1/2$, $f(-0) = -1/2$. Therefore, differentiating equation (10.1.58) with respect to t, we come to the following second-kind integral equation of Fredholm type:

$$\tilde{w}(x,t) + \int_{-x}^{x} f'(t-\tau)\, \tilde{w}(x,\tau)\, d\tau = -\frac{1}{2}[f'(t-x) + f'(t+x)],$$
$$x > 0, \quad t \in (-x, x). \tag{10.1.59}$$

The kernel $f'(t-\tau)$ of equation (10.1.59) is continuous and, since $f'(t)$ is even, symmetric.

Taking into account that $\tilde{w}(x,t) = \tilde{w}(x,-t)$, we can put (10.1.59) in a more convenient form, namely,

$$\tilde{w}(x,t) + \int_0^x [f'(t-\tau) + f'(t+\tau)]\,\tilde{w}(x,\tau)\,d\tau = -\frac{1}{2}\,[f'(t-x) + f'(t+x)],$$
$$x > 0, \quad t \in [0,x). \tag{10.1.60}$$

Equation (10.1.60) is equivalent to (10.1.58) under the assumption that its solution is even in t. Indeed, any even in t solution of equation (10.1.58) solves (10.1.60). On the other hand, any solution $\tilde{w}(x,t)$ of (10.1.60) extended so that $\tilde{w}(x,t) = \tilde{w}(x,-t)$ for $t \in (-x,0)$ satisfies equation (10.1.59) (which is derived from (10.1.58) by differentiation with respect to t) and meets the condition

$$\int_{-x}^{x} f(-\tau)\tilde{w}(x,\tau)\,d\tau = 0$$

coinciding with (10.1.58) for $t = 0$. Consequently, any solution of equation (10.1.60) evenly extended in t solves (10.1.58). Thus, we have shown the equivalence of equations (10.1.58) and (10.1.60).

Suppose that, for every $x > 0$, equation (10.1.60) is uniquely solvable in the class of continuous functions. Let $\tilde{w}(x,t)$ satisfy (10.1.60). Setting $\tilde{w}(x,t) = \tilde{w}(x,-t)$ for $t \in (-x,0)$, we obtain a continuous function $\tilde{w}(x,t)$ in the domain D_1. Since the term on the right of (10.1.59) is continuously differentiable in D_1 and the kernel $f'(t-s)$ is piecewise continuously differentiable, the solution of equation (10.1.59) is continuously differentiable in D_1. Note that the solution $\tilde{w}(x,t)$ of (10.1.60) enjoys the property $\tilde{w}_t(x,+0) = 0$. Hence, upon the even continuation of $\tilde{w}(x,t)$ for $t < 0$, its partial derivatives remain continuous.

From (10.1.55) we derive

$$q(x) = 4\,\frac{d}{dx}\,\tilde{w}(x,x-0), \quad x > 0. \tag{10.1.61}$$

For $x < 0$ we can obtain an equation identical to (10.1.60) except for the upper limit of integration which becomes $|x|$ instead of x. This implies $\tilde{w}(-x,t) = \tilde{w}(x,t)$ and, hence, $q(-x) = q(x)$.

Thus, to find a solution of the inverse problem at a point $x > 0$, it suffices to solve equation (10.1.60) and calculate $q(x)$ by formula (10.1.61).

As is known from the theory of Fredholm equations, equation (10.1.60) is uniquely solvable if x is small enough. In the case in question the following assertion holds.

Lemma 10.1.4. *If there exists a solution of the inverse problem* (10.1.43)–(10.1.45), *then equation* (10.1.60) *is uniquely solvable for all* $x > 0$.

Proof. Equation (10.1.60) is uniquely solvable for a fixed $x > 0$ if and only if the homogeneous equation

$$\tilde{w}(x,t) + \int_0^x [f'(t-\tau) + f'(t+\tau)]\,\tilde{w}(x,\tau)\,d\tau = 0, \quad t \in [0,x), \qquad (10.1.62)$$

has only trivial solution $\tilde{w} \equiv 0$. Suppose the contrary, namely, let $x_0 > 0$ exist such that for $x = x_0$ equation (10.1.62) has a solution $\tilde{w}(x_0,t) \not\equiv 0$, $t \in [0,x_0)$. Put $\tilde{w}(x_0,t) = \tilde{w}(x_0,-t)$ for $t \in (-x_0,0)$ and $\tilde{w}(x_0,t) = 0$ for $t \notin (-x_0,x_0)$. Then from (10.1.62) it follows

$$\int_{-\infty}^{\infty} f(t-\tau)\tilde{w}(x_0,\tau)\,d\tau = 0, \quad t \in (-x_0,x_0).$$

Introduce a function

$$\Phi(x,t) = \int_{-\infty}^{\infty} u(x,t-\tau)\tilde{w}(x_0,\tau)\,d\tau. \qquad (10.1.63)$$

Evidently,

$$L_q\Phi = 0, \quad 0 < x < x_0, \qquad (10.1.64)$$

$$\Phi\big|_{x=0} = 0, \quad t \in (-x_0,x_0), \quad \Phi_x\big|_{x=0} = 0. \qquad (10.1.65)$$

Therefore, $\Phi(x,t) \equiv 0$ for $(x,t) \in \Delta_1(x_0,0)$, that is,

$$\int_{-\infty}^{\infty} u(x,t-\tau)\tilde{w}(x_0,\tau)\,d\tau = 0, \quad (x,t) \in \Delta_1(x_0,0). \qquad (10.1.66)$$

Differentiating equality (10.1.66) with respect to t and setting $t = 0$, we obtain

$$\int_{-\infty}^{\infty} u_t(x,-\tau)\tilde{w}(x_0,\tau)\,d\tau = 0, \quad x \in [0,x_0]. \qquad (10.1.67)$$

Taking into account the evenness of the functions $u_t(x,t)$ and $\tilde{w}(x_0,t)$, rewrite (10.1.66) as

$$\int_{-\infty}^{\infty} u_t(x,\tau)\tilde{w}(x_0,\tau)\,d\tau = 0, \quad x \in [0,x_0]. \qquad (10.1.68)$$

The solution of the Cauchy problem (10.1.28), (10.1.29) in the domain $x > 0$, $t > 0$ has the form

$$u(x,t) = \theta(t-x)\Big[\frac{1}{2} + \tilde{u}(x,t)\Big],$$

where $\tilde{u}(x,t)$ is a twice continuously differentiable function in the domain $t \ge x \ge 0$ vanishing at $x = t$: $\tilde{u}(x,x) = 0$. Consequently,

$$u_t(x,t) = \frac{1}{2}\,\delta(t-x) + \theta(t-x)\tilde{u}_t(x,t), \quad x > 0,\ t > 0.$$

Combining this equality with (10.1.68) yields

$$\frac{1}{2}\,\tilde{w}(x_0,x) + \int_x^{x_0} \tilde{u}_t(x,t)\tilde{w}(x_0,t)\,dt = 0, \quad x \in [0,x_0]. \tag{10.1.69}$$

Note that (10.1.69) is a homogeneous integral Volterra equation of second kind for the function $\tilde{w}(x_0,t)$ with a continuous kernel. Therefore, $\tilde{w}(x_0,t) \equiv 0$, $t \in [0,x_0]$, which contradicts the assumption that a nontrivial solution of (10.1.62) exists. Thereby the lemma is proven. □

From Lemma 10.1.4 follows uniqueness of the solution to the inverse problem in question.

The following lemma may be regarded as converse to Lemma 10.1.4.

Lemma 10.1.5. *Let $T > 0$ be fixed, $f(t) \in C^2(0,T]$, $f(+0) = 1/2$, $f'(+0) = 0$. Suppose equation* (10.1.60) *where $f(t)$ is oddly extended for $t \in [-T,0)$ to be uniquely solvable for all $x \in (0,T/2]$. Then there exists a solution of the inverse problem* (10.1.43)–(10.1.45) *on the interval* $[-T/2,T/2]$.

Proof. Let $f(t) \in C^3(0,T]$. Denote by $D_1(T)$ the part of the domain D_1 where $x \le T/2$. Extend a solution $\tilde{w}(x,t)$ of (10.1.60) evenly for $t < 0$ and note that the extended solution is twice continuously differentiable in $D_1(T)$.

Put

$$q(x) = 4\,\frac{d}{dx}\,\tilde{w}(x,x-0), \quad 0 < x \le \frac{T}{2}, \quad q(-x) = q(x),$$

$$w(x,t) = \frac{1}{2}\,[\delta(t-x) + \delta(t+x)] + \theta(x-|t|)\tilde{w}(x,t), \quad 0 < x \le \frac{T}{2}, \tag{10.1.70}$$

$$\hat{w}(x,t) = L_q w(x,t), \quad 0 \le x \le \frac{T}{2}, \quad -\infty < t < \infty. \tag{10.1.71}$$

Note that though the function $w(x,t)$ is defined for $x \in (0,T/2]$, the function $\hat{w}(x,t)$ is piecewise continuous on the closed interval $[0,T/2]$. Indeed,

$$\begin{aligned}
\hat{w}(x,t) &= \frac{1}{2}\,L_q[\delta(t-x) + \delta(t+x)] + L_q[\theta(x-|t|)\tilde{w}(x,t)] \\
&= \frac{1}{2}\,q(x)[\delta(t-x) + \delta(t+x)] \\
&\quad - [\delta(t-x) + \delta(t+x)]\,2\,\frac{d}{dx}\,\tilde{w}(x,x-0) + \theta(x-|t|)L_q\tilde{w}(x,t) \\
&= \theta(x-|t|)L_q\tilde{w}(x,t).
\end{aligned}$$

The function $w(x,t)$ obeys the equation

$$\int_{-\infty}^{\infty} f(t-\tau)w(x,\tau)\,d\tau = 0, \quad (x,t) \in D_1(T). \tag{10.1.72}$$

Rewriting (10.1.72) in the form

$$\int_{-\infty}^{\infty} f(\tau)w(x,t-\tau)\,d\tau = 0, \quad (x,t) \in D_1(T), \tag{10.1.73}$$

and applying L_q to equation (10.1.73), we observe that $\hat{w}(x,t)$ solves the equation

$$\int_{-\infty}^{\infty} f(\tau)\hat{w}(x,t-\tau)\,d\tau = 0, \quad (x,t) \in D_1(T). \tag{10.1.74}$$

Since $\hat{w}(x,t) \equiv 0$ for $x < |t|$, equation (10.1.74) is equivalent to

$$\int_{x}^{-x} f(t-\tau)\hat{w}(x,\tau)\,d\tau = 0, \quad (x,t) \in D_1(T).$$

We differentiate this equation with respect to t and use the conditions

$$\hat{w}(x,-t) = \hat{w}(x,t), \quad f(+0) = \frac{1}{2}, \quad f(-0) = -\frac{1}{2}$$

to get

$$\hat{w}(x,t) + \int_{0}^{x} [f'(t-\tau) + f'(t+\tau)]\hat{w}(x,\tau)\,d\tau = 0, \quad (x,t) \in D_1(T). \tag{10.1.75}$$

Owing to the unique solvability of (10.1.60), equality (10.1.75) implies

$$\hat{w}(x,t) \equiv 0, \quad (x,t) \in D_1(T).$$

Consequently, $\hat{w} \equiv 0$ for $x \in [0, T/2]$. Thus, the function $w(x,t)$ satisfies the equation

$$L_q w = 0.$$

Since $f'(0) = 0$, from (10.1.60) it follows that $\tilde{w}(0,0) = 0$. Assuming $x \to +0$ in (10.1.70), we find

$$w(0,t) = \delta(t), \quad w_x\big|_{x=0} = 0, \quad t \in \mathbb{R}.$$

Therefore, $w(x,t)$ is a solution of the Cauchy problem (10.1.49), (10.1.50) for $x \in (0, T/2]$.

Then the function $u(x,t)$ defined by formula (10.1.56) solves equation (10.1.46) for $x \in [0, T/2]$, meets conditions (10.1.48) for $t \in (-T, T)$ and, in view of (10.1.72), satisfies (10.1.47). Assuming $u(x,t) = u(-x,t)$ for $x \in [-T/2, 0)$, we note that the extended function $u(x,t)$ still solves (10.1.46) and fulfills (10.1.47), (10.1.48). Moreover, it is apparent that

$$u(x,0) = 0, \quad u_t\big|_{t=0} = \delta(x). \tag{10.1.76}$$

So, the inverse problem has a solution $q(x) \in C^1[-T/2, T/2]$.

Suppose now that $f \in C^2[0,T]$. Consider such a sequence of functions $f_n(t) \in C^3[0,T]$ that $f_n(t)$ converge to $f(t)$ in the norm of $C^2[0,T]$ as $n \to \infty$. To the sequence $\{f_n\}$ correspond sequences $\{\tilde{w}_n(x,t)\}$, $\{q_n(x)\}$, $\{u_n(x,t)\}$. Evidently, the functions $\tilde{w}_n \in C^2(D_1(T))$ converge in the norm of $C^1(D_1(T))$ to a function $\tilde{w}(x,t) \in C^1(D_1(T))$ while the functions $q_n(x)$ and $u_n(x,t)$ converge uniformly to $q(x) \in C[-T/2, T/2]$ and to a piecewise continuous function $u(x,t)$, respectively. The limiting functions $q(x)$ and $u(x,t)$ regarded as distributions satisfy (10.1.46), (10.1.48), (10.1.76). So, the lemma is proven. □

From Lemmata 10.1.4 and 10.1.5 follows the next theorem where necessary and sufficient conditions for the unique solvability of the inverse problem on the interval $[-T/2, T/2]$ are formulated.

Theorem 10.1.6. *The inverse problem* (10.1.43), (10.1.44), (10.1.48) *is uniquely solvable in the class* $C[-T/2, T/2]$ *if and only if*

a) *the function* $f(t)$ *fulfills the following conditions:* $f(t) \in C^2[0,T]$, $f(+0) = 1/2$, $f'(0) = 0$;

b) *the integral equation* (10.1.60) (*with* $f(t)$ *extended over* $t \in [-T, 0)$ *by the formula* $f(-t) = -f(t)$, $t \in [0,T]$) *is uniquely solvable for any* $x \in (0, T/2]$.

The second condition of the theorem is equivalent to the condition of positive definiteness of an operator A_x acting by the formula

$$A_x\varphi = \varphi(t) + \int_0^x [f'(t-\tau) + f'(t+\tau)]\,\varphi(\tau)\,d\tau, \quad x \in [0, T/2],$$

that is, instead of condition b), we can verify that $\langle A_x\varphi, \varphi\rangle_{L_2[0,x]} > 0$ for any $x \in [0, T/2]$ and any $\varphi \in L_2[0,x]$.

10.1.4 A connection between the inverse Sturm–Liouville problem and the inverse point-source problem

In this subsection we derive a formula which relates spectral data and the inverse point-source problem data (Romanov, 1984). As noted above, such formulae are of great interest because they permit the results obtained for one problem to be extended to the other.

We remind that the expression

$$l_q y = -y'' + q(x)y, \quad x \in (a,b), \tag{10.1.77}$$

together with the boundary conditions

$$(y' - hy)\big|_{x=a} = 0, \quad (y' + Hy)\big|_{x=b} = 0, \tag{10.1.78}$$

where $q \in C[a,b]$ and h, H, a, b are finite real numbers, defines an operator l_q in $L_2[a,b]$ which is referred to as a *Sturm–Liouville operator* on the interval $[a,b]$. We say a number λ to be an *eigenvalue* of the operator l_q if there exists a nontrivial solution $y(x,\lambda)$ of the equation

$$y'' - q(x)y = -\lambda y, \quad x \in (a,b), \tag{10.1.79}$$

satisfying the boundary condition (10.1.78). The function $y(x,\lambda)$ is called the *eigenfunction* of the operator l_q corresponding to the eigenvalue λ.

The *direct Sturm–Liouville problem* is to find all the eigenvalues and eigenfunctions of the Sturm–Liouville operator.

Note that all the eigenfunctions corresponding to a given eigenvalue λ are proportional. Hence we can assume

$$y(a,\lambda) = 1, \quad y'(a,\lambda) = h.$$

Eigenvalues of the Sturm–Liouville operator form a countable set $\{\lambda_n\} \subset \mathbb{R}$. If we arrange the set so that $\lambda_1 < \lambda_2 < \lambda_3 < \cdots < \lambda_n < \cdots$, then $\lim_{n\to\infty} \lambda_n = \infty$. A sequence of the corresponding eigenfunctions $\{y_n(x)\}$, where $y_n(x) = y(x,\lambda_n)$, contains mutually orthogonal functions and is complete in $L_2[a,b]$. This means that for any $f \in L_2[a,b]$ the Fourier series

$$f(x) = \sum_{n=1}^{\infty} f_n y_n(x), \quad f_n = \frac{1}{\|y_n\|^2} \int_a^b f(x) y_n(x)\, dx, \tag{10.1.80}$$

converges in the norm of $L_2[a,b]$ to $f(x)$. Consequently, for any $f \in L_2[a,b]$ the Parseval equality

$$\int_a^b f^2(x)\, dx = \sum_{n=1}^{\infty} f_n^2 \|y_n\|^2 = \sum_{n=1}^{\infty} \frac{\langle f, y_n\rangle^2}{\|y_n\|^2}$$

holds. The sequences $\{\lambda_n\}$ and $\{\|y_n\|\}$ are called the *spectral data.*

The *inverse Sturm–Liouville problem* consists in recovering the function $q(x)$ from the spectral data.

This problem can be reformulated in terms of the *spectral function* $\rho(\lambda)$ of the Sturm–Liouville operator, which has the form

$$\rho(\lambda) = \sum_{n:\{\lambda_n<\lambda\}} \frac{1}{\|y_n\|^2}\,\theta(\lambda-\lambda_n). \tag{10.1.81}$$

Here we sum over all n such that $\lambda_n < \lambda$.

To illustrate the spectral function, consider the Cauchy problem

$$\begin{aligned} y''(x) - q(x)y(x) &= -\lambda y(x), \quad x \in [a,b], \\ y(a,\lambda) = 1, &\quad y'(a,\lambda) = h. \end{aligned} \tag{10.1.82}$$

Denote by $y(x,\lambda)$ a solution of problem (10.1.82). For an arbitrary $g(x) \in L_2[a,b]$, we introduce a function $\tilde{g}(\lambda)$ putting

$$\tilde{g}(\lambda) = \int_a^b g(x)y(x,\lambda)\,dx, \quad \lambda \in \mathbb{R}. \tag{10.1.83}$$

Since $y(x,\lambda_n) = y_n(x)$, the function $g(x)$ can be represented as the Fourier series

$$g(x) = \sum_{n=1}^{\infty} \frac{\tilde{g}(\lambda_n)}{\|y_n\|^2}\,y_n(x). \tag{10.1.84}$$

Therefore, $g(x)$ is determined by the sequence $\{\tilde{g}(\lambda_n)\}$. Using the spectral function (10.1.81), we rewrite (10.1.84) in the form

$$g(x) = \int_{-\infty}^{\infty} \tilde{g}(\lambda)y(x,\lambda)\,d\rho(\lambda). \tag{10.1.85}$$

The spectral function $\rho(\lambda)$ is displayed in Figure 3. This is a piecewise constant function having finite jumps at the points $\lambda = \lambda_n$. The jump value at $\lambda = \lambda_n$ is equal to $\|y_n\|^{-2}$. Hence, from the spectral function $\rho(\lambda)$ we can recover the spectral data

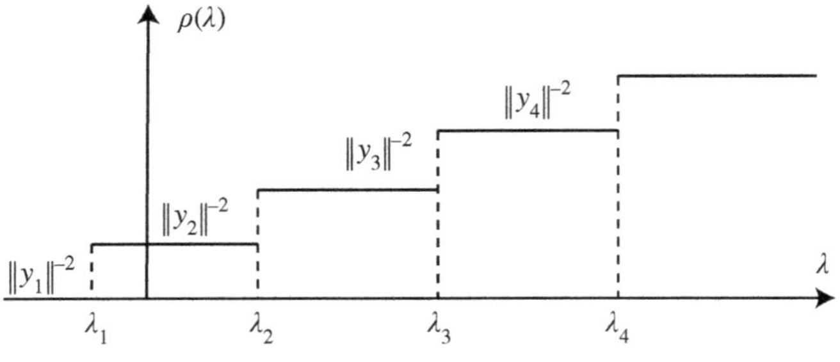

Figure 3 Spectral function $\rho(\lambda)$.

$\{\lambda_n\}$, $\{\|y_n\|\}$. So, the inverse Sturm–Liouville problem can be rephrased as follows: it is required to restore the operator l_q from its spectral function. This formulation is more convenient for considering the problem in the case when $b = \infty$.

The inverse Sturm–Liouville problem is closely connected with the inverse point-source problem. To demonstrate this, we derive a formula relating the spectral function $\rho(\lambda)$ and the data $f(t)$ of the following inverse problem: find $q(x) \in C(a,b)$ from the relations

$$u_{tt} = u_{xx} - q(x)u, \quad (x,t) \in (a,b) \times \mathbb{R}_+, \tag{10.1.86}$$

$$u\big|_{t=0} = 0, \quad u_t\big|_{t=0} = \delta(x-a), \tag{10.1.87}$$

$$(u_x - hu)\big|_{x=a} = 0, \quad (u_x + Hu)\big|_{x=b} = 0, \quad t > 0, \tag{10.1.88}$$

$$u\big|_{x=a} = f(t), \quad t \in (0, 2(b-a)). \tag{10.1.89}$$

One of the ways to study this problem is by reducing it to an integral equation. We shall apply another approach based on the Fourier method of separation of variables.

Let $u(x,t)$ be a solution of the direct problem (10.1.86)–(10.1.88). We present $u(x,t)$ in the form

$$u(x,t) = \sum_{n=1}^{\infty} T_n(t) y_n(x),$$

where $y_n(x)$ are eigenfunctions of the operator l_q such that $y_n(a) = 1$, $y_n'(a) = h$. From (10.1.86), (10.1.87) it follows that the functions $T_n(t)$ satisfy the equation

$$T_n'' + \lambda_n T_n = 0 \tag{10.1.90}$$

and the conditions

$$T_n(0) = 0, \quad T_n'(0) = \|y_n\|^{-2}. \tag{10.1.91}$$

The Cauchy problem (10.1.90), (10.1.91) admits a solution in the form

$$T_n(t) = \frac{1}{\sqrt{\lambda_n}\,\|y_n\|^2} \sin\sqrt{\lambda_n}\, t. \tag{10.1.92}$$

Note that if $\lambda_n < 0$, then in (10.1.92) the equality

$$\frac{\sin\sqrt{\lambda_n}\, t}{\sqrt{\lambda_n}} = \frac{\sinh\sqrt{-\lambda_n}\, t}{\sqrt{-\lambda_n}}$$

should be used.

In view of (10.1.92), the solution of the direct problem (10.1.86)–(10.1.88) takes the form

$$u(x,t) = \sum_{n=1}^{\infty} \frac{\sin\sqrt{\lambda_n}\, t}{\sqrt{\lambda_n}\|y_n\|^2}\, y_n(x). \tag{10.1.93}$$

Rewrite (10.1.93) using the spectral function $\rho(\lambda)$:

$$u(x,t) = \int_{-\infty}^{\infty} \frac{\sin\sqrt{\lambda}\,t}{\sqrt{\lambda}}\, y(x,\lambda)\, d\rho(\lambda). \tag{10.1.94}$$

Setting $x = a$ in (10.1.93) and (10.1.94), we obtain the desired formulae expressing the data $f(t)$ through the spectral data $\{\lambda_n\}$, $\{\|y_n\|\}$ or the spectral function $\rho(\lambda)$, respectively:

$$\begin{aligned} f(t) &= \sum_{n=1}^{\infty} \frac{1}{\|y_n\|^2} \frac{\sin\sqrt{\lambda_n}\,t}{\sqrt{\lambda_n}}, \\ f(t) &= \int_{-\infty}^{\infty} \frac{\sin\sqrt{\lambda}\,t}{\sqrt{\lambda}}\, d\rho(\lambda). \end{aligned} \tag{10.1.95}$$

Given $\rho(\lambda)$, we can calculate $f(t)$ for $t \in (0, 2(b-a))$ using (10.1.95). Conversely, if the data $f(t)$ are known for $t \in (0, 2(b-a))$, we first restore $q(x)$ by solving the inverse problem (10.1.86)–(10.1.89) and then solve the direct Sturm–Liouville problem to find $\rho(\lambda)$.

So, we have shown that there exists a one-to-one correspondence between the inverse problem data $f(t)$, $t \in (0, 2(b-a))$, and the spectral function $\rho(\lambda)$. Solving either the inverse Sturm–Liouville problem or inverse point-source problem (10.1.86)–(10.1.89) with appropriate data, we obtain a function $q(x)$ desired for both of the problems.

10.2 Inverse problems of acoustics

Consider the following one-dimensional inverse problem of acoustics:

$$\frac{1}{c^2(z)} v_{tt} = v_{zz} - \frac{\rho'(z)}{\rho(z)} v_z, \quad z > 0, \quad t > 0, \tag{10.2.1}$$

$$v\big|_{t<0} \equiv 0, \quad z > 0, \tag{10.2.2}$$

$$v_z\big|_{z=0} = \delta(t), \quad t > 0, \tag{10.2.3}$$

$$v(+0,t) = g(t), \quad t > 0. \tag{10.2.4}$$

Here $\rho(z) > 0$ stands for the density of a medium, $c(z) > 0$ is the wave velocity in the medium and $v(z,t)$ denotes the acoustic pressure of the medium at the moment $t > 0$. The direct (generalized initial boundary value) problem (10.2.1)–(10.2.3) consists in determining the function $v(z,t)$ from the given functions $c(z)$ and $\rho(z)$. In the inverse problem (10.2.1)–(10.2.4) it is required to find either $c(z)$ or $\rho(z)$ (or some their combination) from the known function $g(t)$. We will show below that it is impossible to recover both the functions $c(z)$ and $\rho(z)$ from the relations (10.2.1)–(10.2.4) though we can find their product $c(\psi(x))\rho(\psi(x))$, where $x = \varphi(z) := \int_0^z \frac{dz}{c(z)}$ is a new variable, $\psi(\varphi(z)) = z$. Note that the condition $c(z) > 0$ provides the existence of the function $\psi(x)$ inverse to $\varphi(z)$.

Denote

$$u(x,t) = v(\psi(x),t), \quad a(x) = c(\psi(x)), \quad b(x) = \rho(\psi(x)).$$

Exercise 10.2.1. Verify that the inverse problem (10.2.1)–(10.2.4) in the variables (x,t) takes the form

$$\begin{aligned} u_{tt} &= u_{xx} - (a'/a + b'/b)u_x, \quad x > 0, \quad t > 0; \\ u\big|_{t<0} &\equiv 0, \quad x > 0, \\ u_x(+0,t) &= c(+0)\delta(t), \quad t > 0, \\ u(+0,t) &= g(t), \quad t > 0. \end{aligned}$$

Let $\sigma(x) = a(x)b(x)$, $c_0 = c(+0)$. Since

$$a'/a + b'/b = (\ln a)' + (\ln b)' = (\ln(ab))' = (\ln \sigma)' = \sigma'/\sigma,$$

we come to the problem

$$u_{tt} = u_{xx} - \frac{\sigma'(x)}{\sigma(x)} u_x, \quad x > 0, \quad t > 0, \tag{10.2.5}$$

$$u\big|_{t<0} \equiv 0, \quad x > 0, \tag{10.2.6}$$

$$u_x(+0,t) = c_0\delta(t), \quad t > 0, \tag{10.2.7}$$

$$u(+0,t) = g(t), \quad t > 0, \tag{10.2.8}$$

where $u(x,t)$ and $\sigma(x) > 0$ are the desired functions.

Exercise 10.2.2. Show that if $\sigma \in C^1(\mathbb{R}_+)$, then the solution $u(x,t)$ of the direct problem (10.2.5)–(10.2.7) has the form

$$u(x,t) = s(x)\theta(t-x) + \tilde{u}(x,t), \tag{10.2.9}$$

where $\tilde{u}(x,t)$ is a function continuous for $x \geq 0$ and smooth for $t > x > 0$, $s(x) = -c_0\sqrt{\sigma(x)/\sigma(+0)}$, θ is the Heaviside function.

Exercise 10.2.3. Prove that, in view of (10.2.9), the inverse problem (10.2.5)–(10.2.8) is equivalent to the problem of finding the functions $u(x,t)$ and $s(x)$ from the relations

$$u_{tt} = u_{xx} - 2\frac{s'(x)}{s(x)} u_x, \quad t > x > 0, \tag{10.2.10}$$

$$u_x\big|_{x=0} = 0, \quad t > 0, \tag{10.2.11}$$

$$u(x, x+0) = s(x), \quad x > 0, \tag{10.2.12}$$

$$u\big|_{x=+0} = g(t), \quad t > 0. \tag{10.2.13}$$

We prefer to consider the inverse problem (10.2.10)–(10.2.13) instead of (10.2.1)–(10.2.4) for the following reasons. First, unlike problem (10.2.1)–(10.2.3), the direct problem (10.2.10)–(10.2.12) contains no singular functions. Second, we need to restore only one coefficient $s(x)$ while in (10.2.1)–(10.2.4) two coefficients $c(z)$ and $\rho(z)$ are unknown. Third, the inverse problem (10.2.10)–(10.2.13) can be reduced to a system of nonlinear integral Volterra equations for which a local theorem of unique solvability, among other results, will be proven. It is remarkable that from this theorem it follows that the original problem (10.2.1)–(10.2.4) admits a lot of solutions. Indeed, if $\sigma(x) = c(\psi(x))\rho(\psi(x))$ is a solution to problem (10.2.5)–(10.2.8), then any pair $\{\tilde{c}((\psi(x)), \tilde{\rho}(\psi(x))\}$ such that $\tilde{c}((\psi(x)) = Cc(\psi(x))$, $\tilde{\rho}(\psi(x)) = C^{-1}\rho(\psi(x))$, $C = \text{const}$, solves problem (10.2.1)–(10.2.4)

Reduction of problem (10.2.10)–(10.2.13) to a system of integral equations. Let

$$q_1(x,t) = u_x(x,t), \quad q_2(x) = \frac{1}{s(x)}, \quad q_3(x) = 2\,\frac{s'(x)}{s(x)} = \frac{\sigma'(x)}{\sigma(x)}.$$

Note that

$$q_2'(x) = -\frac{s'(x)}{s^2(x)} = -\frac{1}{2}\,q_3(x)q_2(x), \quad s(+0) = -c_0,$$

whence

$$q_2(x) = -\frac{1}{c_0} - \frac{1}{2}\int_0^x q_3(\xi)q_2(\xi)\,d\xi. \tag{10.2.14}$$

Applying the d'Alembert formula to the Cauchy problem (10.2.10)–(10.2.12), we obtain

$$u(x,t) = \frac{1}{2}\,[g(t-x) + g(t+x)] + \frac{1}{2}\int_0^x \int_{t-x+\xi}^{t+x-\xi} q_3(\xi)q_1(\xi,\tau)\,d\tau\,d\xi. \tag{10.2.15}$$

Differentiating (10.2.15) with respect to x yields

$$\begin{aligned} u_x(x,t) = q_1(x,t) = {}& \frac{1}{2}\,[-g'(t-x) + g'(t+x)] \\ & + \frac{1}{2}\int_0^x q_3(\xi)[q_1(\xi,t+x-\xi) + q_1(\xi,t-x+\xi)]\,d\xi. \end{aligned} \tag{10.2.16}$$

Set $t = x + 0$ in (10.2.15) and use (10.2.12) to come to

$$s(x) = \frac{1}{2}\,[g(+0) + g(2x)] + \frac{1}{2}\int_0^x \int_{\xi}^{2x-\xi} q_3(\xi)q_1(\xi,\tau)\,d\tau\,d\xi,$$

$$s'(x) = g'(2x) + \int_0^x q_3(\xi)q_1(\xi,2x-\xi)\,d\xi.$$

The last equality multiplied by (10.2.14) takes the form

$$q_3(x) = \Big[-\frac{1}{c_0} - \frac{1}{2}\int_0^x q_3(\xi)q_2(\xi)\,d\xi\Big]\Big[2g'(2x) + 2\int_0^x q_3(\xi)q_1(\xi, 2x-\xi)\,d\xi\Big]. \tag{10.2.17}$$

Equations (10.2.16), (10.2.14), (10.2.17) form a system of nonlinear integral Volterra equations of the second kind. Note that if system (10.2.16), (10.2.14), (10.2.17) is solvable and $u \in L_2(\Delta(l)) \cap C^2(\Delta(l))$, $s \in C^1(0,l)$, $g \in C^2(0,2l)$, where $\Delta(l) = \{(x,t) \in \mathbb{R}^2\colon 0 < x < t < 2l - x\}$, $l > 0$, then we can find a solution of problem (10.2.5)–(10.2.8) by the formula $\sigma(x) = (\sigma(+0)/c_0^2)s^2(x)$ providing $\sigma(+0)$ is known.

Formulation of problem (10.2.10)–(10.2.13) in an operator form. The system of integral equations (10.2.16), (10.2.14), (10.2.17) can be written in the operator form

$$Aq = f, \tag{10.2.18}$$

where

$$Aq := q + Bq, \tag{10.2.19}$$

$$q(x,t) = (q_1, q_2, q_3)^\top, \tag{10.2.20}$$

$$q_1(x,t) = u_x(x,t), \quad q_2(x) = 1/s(x), \quad q_3(x) = 2s'(x)/s(x),$$

$$\begin{aligned} f(x,t) &= (f_1, f_2, f_3)^\top, \\ f_1(x,t) &= [g'(t+x) - g'(t-x)]/2, \\ f_2(x) &= -1/c_0, \quad f_3(x) = -2g'(2x)/c_0, \end{aligned} \tag{10.2.21}$$

$$\begin{aligned} Bq &= (B_1q, B_2q, B_3q)^\top, \\ B_1q &= -\frac{1}{2}\int_0^x q_3(\xi)[q_1(\xi, t+x-\xi) + q_1(\xi, t-x+\xi)]\,d\xi, \\ B_2q &= \frac{1}{2}\int_0^x q_3(\xi)q_2(\xi)\,d\xi, \\ B_3q &= 2B_2q[g'(2x) + B_4q] + (2/c_0)B_4q, \\ B_4q &= \int_0^x q_3(\xi)q_1(\xi, 2x-\xi)\,d\xi. \end{aligned} \tag{10.2.22}$$

Note that if $\{u(x,t), s(x)\}$ is a solution of problem (10.2.10)–(10.2.13), then the vector-function $q(x,t)$ defined by (10.2.20) solves problem (10.2.18)–(10.2.22).

Definition 10.2.1. A vector-function $q(x,t) = (q_1, q_2, q_3)^\top$ will be said to belong to the space $\vec{L}_2(l)$ if

$$q_1(x,t) \in L_2(\Delta(l)), \quad q_k(x) \in L_2(0,l), \quad k = 2,3,$$

where $\Delta(l) = \{(x,t) \in \mathbb{R}^2 : 0 < x < t < 2l - x\}, l > 0$.

Define an inner product in the space $\vec{L}_2(l)$ by the formula

$$\langle q^{(1)}, q^{(2)} \rangle_{\vec{L}_2(l)} = \iint_{\Delta(l)} q_1^{(1)}(x,t) q_1^{(2)}(x,t)\,dx\,dt + \sum_{k=2}^{3} \int_0^l q_k^{(1)}(x) q_k^{(2)}(x)\,dx. \tag{10.2.23}$$

Then the associated norm in $\vec{L}_2(l)$ has the form

$$\|q\|^2_{\vec{L}_2(l)} := \langle q, q \rangle_{\vec{L}_2(l)} = \|q_1\|^2_{L_2(\Delta(l))} + \sum_2^3 \|q_k\|^2_{L_2(0,l)}.$$

Exercise 10.2.4. Verify that the operation defined in the space $\vec{L}_2(l)$ by formula (10.2.23) is really an inner product and that $\vec{L}_2(l)$ is a Hilbert space.

Exercise 10.2.5. Prove that if $q(x,t) \in \vec{L}_2(l)$ is a solution of problem (10.2.18)–(10.2.22), then the pair of functions $\{u(x,t), s(x)\}$, where $u(x,t)$ is found by formula (10.2.15) and $s(x) = 1/q_2(x)$, solves problem (10.2.10)–(10.2.13).

Features of the operator A. We begin a study of problem (10.2.18)–(10.2.22) with discussing some properties of the operator A.

Lemma 10.2.1 (on boundedness of A). *Let A be an operator defined by* (10.2.19), (10.2.22) *and $g' \in L_2(0, 2l)$. Then the following estimates hold:*

$$\|B_1 q\|^2_{L_2(\Delta(l))} \le \|q\|^4_{\vec{L}_2(l)} l, \quad \|B_2 q\|^2_{L_2(0,l)} \le (1/4)\|q\|^4_{\vec{L}_2(l)} l,$$
$$\|B_3 q\|^2_{L_2(0,l)} \le (3/2)(4/c_0^2 + \beta)\|q\|^4_{\vec{L}_2(l)} + (3/2)\|q\|^8_{\vec{L}_2(l)},$$

where $\beta := \|g'\|^2_{L_2(0,2l)}$. Consequently, $A : \vec{L}_2(l) \to \vec{L}_2(l)$.

Proof. Let $q = (q_1, q_2, q_3)^\top \in \vec{L}_2(l)$, $x \in (0, l)$. Consider the operator $Bq = (B_1 q, B_2 q, B_3 q)^\top$ and estimate its components. Note that

$$|B_1 q| \le \frac{1}{2}\Big(\int_0^x q_3^2(\xi)\,d\xi\Big)^{1/2}\Big(\Big(\int_0^x q_1^2(\xi, t + x - \xi)\,d\xi\Big)^{1/2} + \Big(\int_0^x q_1^2(\xi, t - x + \xi)\,d\xi\Big)^{1/2}\Big),$$

whence

$$|B_1 q|^2 \le \frac{1}{2}\|q\|^2_{\vec{L}_2(l)}\Big[\int_0^x q_1^2(\xi, t + x - \xi)\,d\xi + \int_0^x q_1^2(\xi, t - x + \xi)\,d\xi\Big].$$

Consequently,

$$\begin{aligned}
&\|B_1 q\|^2_{L_2(\Delta(l))} \\
&\quad \le \frac{1}{2}\,\|q\|^2_{\vec{L}_2(l)} \int_0^l \int_x^{2l-x} \int_0^x (q_1^2(\xi, t+x-\xi) + q_1^2(\xi, t-x+\xi)\,d\xi\,dt\,dx \\
&\quad \le \|q\|^2_{\vec{L}_2(l)} \int_0^l \int_0^x \int_\xi^{2l-\xi} q_1^2(\xi, t)\,dt\,d\xi\,dx \le \|q\|^4_{\vec{L}_2(l)} l.
\end{aligned}$$

Similarly, we get

$$|B_2 q|^2 \le \frac{1}{4} \int_0^x q_3^2(\xi)\,d\xi \int_0^x q_2^2(\xi)\,d\xi \le \frac{1}{4}\,\|q\|^4_{\vec{L}_2(l)}$$

and therefore

$$\|B_2 q\|^2_{L_2(0,l)} \le \frac{1}{4}\|q\|^4_{\vec{L}_2(l)} l.$$

Finally,

$$\begin{aligned}
|B_3 q| &\le 2|B_2 q|\,|g'(2x)| + 2|B_4 q|\,(|B_2 q| + 1/c_0), \\
|B_3 q|^2 &\le 12|B_2 q|^2\,|g'(2x)|^2 + 12|B_4 q|^2(|B_2 q|^2 + 1/c_0^2).
\end{aligned}$$

To derive the desired estimate for $B_3 q$, we first evaluate $\|B_4 q\|^2_{L_2(0,l)}$. Since

$$|B_4 q|^2 \le \int_0^x q_3^2(\xi)\,d\xi \int_0^x q_1^2(\xi, 2x-\xi)\,d\xi \le \|q\|^2_{\vec{L}_2(l)} \int_0^x q_1^2(\xi, 2x-\xi)\,d\xi,$$

we obtain

$$\begin{aligned}
\|B_4 q\|^2_{L_2(0,l)} &\le \|q\|^2_{\vec{L}_2(l)} \int_0^l \int_0^x q_1^2(\xi, 2x-\xi)\,d\xi\,dx \\
&= \|q\|^2_{\vec{L}_2(l)} \int_0^l \int_\xi^l q_1^2(\xi, 2x-\xi)\,dx\,d\xi \\
&= \frac{1}{2}\,\|q\|^2_{\vec{L}_2(l)} \int_0^l \int_\xi^{2l-\xi} q_1^2(\xi, t)\,dt\,d\xi \le \frac{1}{2}\,\|q\|^4_{\vec{L}_2(l)}.
\end{aligned}$$

It is therefore concluded that

$$\begin{aligned}
\|B_3 q\|^2_{L_2(0,l)} &\le 3\|q\|^4_{\vec{L}_2(l)} \int_0^l [g'(2x)]^2\,dx + 12\Big(\frac{1}{4}\,\|q\|^4_{\vec{L}_2(l)} + \frac{1}{c_0^2}\Big) \int_0^l |B_4 q|^2 \\
&\le (3/2)\|q\|^4_{\vec{L}_2(l)}\beta + (3/2)(\|q\|^4_{\vec{L}_2(l)} + 4/c_0^2)\|q\|^4_{\vec{L}_2(l)} \\
&= (3/2)(4/c_0^2 + \beta)\|q\|^4_{\vec{L}_2(l)} + (3/2)\|q\|^8_{\vec{L}_2(l)}.
\end{aligned}$$

So, the lemma is proven. □

Lemma 10.2.2 (on existence of $A'(q)$). *If $q \in \vec{L}_2(l)$, $g' \in L_2(0,2l)$, then the operator A defined by* (10.2.19), (10.2.22) *is Frechet differentiable and its Frechet derivative $A'(q) : \vec{L}_2(l) \to \vec{L}_2(l)$ is determined by the formula $A'(q)p = p + B'(q)p$, where $p \in \vec{L}_2(l)$, $B'(q)p = (B_1'(q)p, B_2'(q)p, B_3'(q)p)^\top$,*

$$\begin{aligned}
B_1'(q)p &= -\frac{1}{2}\int_0^x q_3(\xi)[p_1(\xi, t+x-\xi) + p_1(\xi, t-x+\xi)]\,d\xi \\
&\quad - \frac{1}{2}\int_0^x p_3(\xi)[q_1(\xi, t+x-\xi) + q_1(\xi, t-x+\xi)]\,d\xi, \\
B_2'(q)p &= \frac{1}{2}\int_0^x [q_3(\xi)p_2(\xi) + q_2(\xi)p_3(\xi)]\,d\xi, \\
B_3'(q)p &= 2B_2'(q)p[g'(2x) + B_4 q] + 2B_4'(q)p[B_2 q + 1/c_0], \\
B_4'(q)p &= \int_0^x [q_3(\xi)p_1(\xi, 2x-\xi) + p_3(\xi)q_1(\xi, 2x-\xi)]\,d\xi.
\end{aligned}$$

Moreover, the following estimates hold:

$$\begin{aligned}
\|B_1'(q)p\|^2_{L_2(\Delta(l))} &\le 4\|q\|^2_{\vec{L}_2(l)}\|p\|^2_{\vec{L}_2(l)}\, l, \\
\|B_2'(q)p\|^2_{L_2(0,l)} &\le \|p\|^2_{\vec{L}_2(l)}\|q\|^2_{\vec{L}_2(l)}\, l, \\
\|B_3'(q)p\|^2_{L_2(0,l)} &\le 8(\beta + 4/c_0^2)\|q\|^2_{\vec{L}_2(l)}\|p\|^2_{\vec{L}_2(l)} + 16\|q\|^6_{\vec{L}_2(l)}\|p\|^2_{\vec{L}_2(l)}.
\end{aligned}$$

Lemma 10.2.3 (on properties of $[A'(q)]^*$). *Let A be an operator defined by formulae* (10.2.19), (10.2.22) *and let $g' \in L_2(0,2l)$. Then, for any $q \in \vec{L}_2(l)$, the operator $A'(q) : \vec{L}_2(l) \to \vec{L}_2(l)$ has an adjoint $[A'(q)]^* : \vec{L}_2(l) \to \vec{L}_2(l)$ such that $[A'(q)]^* r = r + [B'(q)]^* r$, $r \in \vec{L}_2(l)$, and the components of $[B'(q)]^* r$ have the form*

$$\begin{aligned}
[B_1'(q)]^* r = -\frac{1}{2} q_3(x)\Big\{ &\int_x^{(x+t)/2} r_1(\xi, t+x-\xi)\,d\xi \\
&+ \int_x^{l-(t-x)/2} r_1(\xi, t-x+\xi)\,d\xi \\
&- 2\Big[B_2 q\Big(\frac{t+x}{2}\Big) + \frac{1}{c_0}\Big] r_3\Big(\frac{t+x}{2}\Big)\Big\}, \\
[B_2'(q)]^* r = \frac{1}{2} q_3(x) &\int_x^l \{r_2(\xi + 2r_3(\xi)[g'(2\xi) + B_4 q(\xi)]\}\,d\xi,
\end{aligned}$$

$$[B_3'(q)]^* r = -\frac{1}{2}\int_x^l \Big\{ \int_\xi^{2l-\xi} [q_1(x, t+\xi-x) + q_1(x, t-\xi+x)] r_1(\xi, t)\, dt \\ - q_2(x) r_2(\xi) - 2q_2(x) r_3(\xi)[g'(2\xi) + B_4 q(\xi)] \\ - 4q_1(x, 2\xi - x)\Big[B_2 q(\xi) + \frac{1}{c_0}\Big] r_3(\xi)\Big\}\, d\xi.$$

Furthermore, the estimates

$$\|[B_1'(q)]^* r\|^2_{L_2(\Delta(l))} \le 2l \|r\|^2_{\vec{L}_2(l)} \|q\|^2_{\vec{L}_2(l)} + 2(\|q\|^4_{\vec{L}_2(l)} + 4/c_0^2)\|r\|^2_{\vec{L}_2(l)} \|q\|^2_{\vec{L}_2(l)},$$
$$\|[B_2'(q)]^* r\|^2_{L_2(0,l)} \le (3/4)\|q\|^2_{\vec{L}_2(l)} \|r\|^2_{\vec{L}_2(l)} (l + 2\beta + 2\|q\|^4_{\vec{L}_2(l)}),$$
$$\|[B_3'(q)]^* r\|^2_{L_2(0,l)} \le 3\|r\|^2_{\vec{L}_2(l)} \|q\|^2_{\vec{L}_2(l)} (8/c_0^2 + \beta + 3l/2 + 3\|q\|^4_{\vec{L}_2(l)})$$

hold true.

Lemma 10.2.4. *Let an operator A be defined by* (10.2.19), (10.2.22) *and let $g' \in L_2(0, 2l)$. Then, for any $q \in \vec{L}_2(l)$, the operator $A'(q)$ is invertible and for its inverse $[A'(q)]^{-1} : \vec{L}_2(l) \to \vec{L}_2(l)$ there holds*

$$\|[A'(q)]^{-1} r\|_{\vec{L}_2(l)} \le M_1 \|r\|_{\vec{L}_2(l)}, \tag{10.2.24}$$

where $M_1 = M_1(l, c_0, \beta, \|q\|) =$ const.

Proof. Let $r \in \vec{L}_2(l)$. If

$$[A'(q)]^{-1} r = p, \tag{10.2.25}$$

then

$$A'(q)p = p + B'(q)p = r. \tag{10.2.26}$$

So, to find the image of the element $r \in \vec{L}_2(l)$ under the operator $[A'(q)]^{-1}$, it suffices to solve the linear operator equation (10.2.26), which is a system of linear integral Volterra equations of the second kind. We leave it as an exercise for the reader to show that there exists a unique solution of (10.2.26) in $\vec{L}_2(l)$.

Let $\langle \cdot, \cdot \rangle$ denote the inner product (10.2.23) in $\vec{L}_2(l)$ and $\|\cdot\|$ the associated norm. By Lemma 10.2.2, we get

$$\langle A'(q)p, p\rangle = \langle p, p\rangle + \langle B'(q)p, p\rangle,$$

whence

$$\|p\|^2 = \langle r, p\rangle - \langle B'(q)p, p\rangle \le \|r\|\,\|p\| + \|p\|\,\|[B'(q)]^* p\|.$$

Consequently,

$$\begin{aligned}\|p\| &\le \|r\| + \|[B'(q)]^* p\|,\\ \|p\|^2 &\le 2\|r\|^2 + 2\|[B'(q)]^* p\|^2.\end{aligned} \tag{10.2.27}$$

For $p = (p_1, p_2, p_3)^\top \in \vec{L}_2(l)$ we introduce

$$\|p\|^2(l,x) = \|p_1\|^2(l,x) + \|p_2\|^2(x) + \|p_3\|^2(x),$$

where

$$\begin{aligned}\|p_1\|^2(l,x) &= \int_x^l \int_\xi^{2l-\xi} p_1^2(\xi,\tau)\, d\xi\, d\tau,\\ \|p_k\|^2(x) &= \int_x^l p_k^2(\xi)\, d\xi, \quad k = 2,3.\end{aligned}$$

Using Lemma 10.2.3, one can show that

$$\|[B'(q)]^* p\|^2(l,x) \le \int_x^l \mu_0(\xi)\|p\|^2(l,\xi)\, d\xi, \tag{10.2.28}$$

where

$$\begin{aligned}\mu_0(x) = {} & q_3^2(x)\Big(2l + \|q\|^4 + \frac{4}{c_0^2}\Big) + \frac{3}{2} q_3^2(x)\Big(\frac{l}{2} + \beta + \|q\|^4\Big)\\ & + \frac{7}{2}\|q\|^2 + \frac{7}{4}(l + 2\beta + 2\|q\|^4 q_2^2(x)\\ & + 7\Big(\|q\|^4 + \frac{4}{c_0^2}\Big)\int_x^l q_1^2(x, 2\zeta - x)\, d\zeta.\end{aligned} \tag{10.2.29}$$

From (10.2.27), (10.2.28) it follows that

$$\|p\|^2(l,x) \le 2\|r\|^2 + 2\int_x^l \mu_0(\xi)\|p\|^2(l,\xi)\, d\xi.$$

Applying the Bellman–Gronwall lemma to the last inequality, we obtain the desired estimate (10.2.24) with

$$M_1 = \sqrt{2}\exp\Big(\int_0^l \mu_0(\xi)\, d\xi\Big). \tag{10.2.30}$$

Thus, the lemma is proven. □

10.2.1 Properties of a one-dimensional inverse problem of acoustics

In this subsection we will describe some features of the inverse problem (10.2.5)–(10.2.8).

Conditional stability in H^1. The inverse problem (10.2.5)–(10.2.8) is conditionally stable if its solution $\sigma(x)$ and data $g(t)$ belong to special classes of functions.

Definition 10.2.2. The function $\sigma(x)$ will be said to belong to the class $\Sigma(l, M_0, c_0, \rho_0, \sigma_*)$ if the following conditions are fulfilled:

1) $\sigma(x) \in H^1(0, l)$;
2) $\|\sigma\|_{H^1(0,l)} \leq M_0$, $M_0 = \text{const}$;
3) $0 < \sigma_* \leq \sigma(x)$, $x \in (0, l)$, $\sigma_* = \text{const}$;
4) $\sigma(+0) = c_0 \rho_0$, $c_0 = c(+0)$, $\rho_0 = \rho(+0)$.

Definition 10.2.3. The data $g(t)$ will be said to belong to $\mathcal{G}(l, \beta, c_0)$ if g satisfies the following conditions:

1) $g(t) \in H^1(0, 2l)$;
2) $\|g'\|^2_{L_2(0,2l)} \leq \beta$, $\beta = \text{const}$;
3) $g(+0) = -c_0$.

Theorem 10.2.1 (Kabanikhin et al., 2006). *Suppose that, for some $g^{(1)}$, $g^{(2)}$ from the class $\mathcal{G}(l, \beta, c_0)$, there exist $\sigma^{(1)}, \sigma^{(2)} \in \Sigma(l, M_0, c_0, \rho_0, \sigma_*)$ solving the inverse problems* (10.2.5)–(10.2.8) *with the data $g^{(1)}$, $g^{(2)}$, respectively. Then*

$$\|\sigma^{(1)} - \sigma^{(2)}\|^2_{H^1(0,l)} \leq C \, \|g^{(1)} - g^{(2)}\|^2_{H^1(0,2l)},$$
$$C = C(l, M_0, \beta, c_0, \rho_0, \sigma_*) = \text{const}.$$

Well-posedness for sufficiently small data. Consider the system of nonlinear integral Volterra equations (10.2.16), (10.2.14), (10.2.17) in the operator form:

$$q(x,t) + Bq(x,t) = f(x,t), \quad (x,t) \in \Delta(l). \tag{10.2.31}$$

We remind that the vector-functions $q(x,t)$, $f(x,t)$ and the operator B are specified in (10.2.20)–(10.2.22). Introduce an operator V as

$$Vq = f - Bq, \quad (x,t) \in \Delta(l). \tag{10.2.32}$$

Lemma 10.2.5. *If $f \in \vec{L}_2(l)$, then*

$$V : \vec{L}_2(l) \to \vec{L}_2(l). \tag{10.2.33}$$

Define k-norms ($k \geq 0$) for $q \in \vec{L}_2(l)$ as follows:

$$\|q_1\|_k^2(x) = \int_0^x \|q_1\|_{(1)}^2(\xi)\exp\{-2k\xi\}d\xi, \quad x \in (0,l),$$

$$\|q_1\|_{(1)}^2(x) = \int_x^{2l-x} q_1^2(x,t)\,dt, \quad x \in (0,l),$$

$$\|q_j\|_k^2(x) = \int_0^x q_j^2(\xi)\exp\{-2k\xi\}d\xi, \quad x \in (0,l), \quad j = 2,3,$$

$$\|q\|_k^2(x) = \sum_{j=1}^{3} \|q_j\|_k^2(x), \quad \|q\|_k := \|q\|_k(l).$$

For $f \in \vec{L}_2(l)$, $\varepsilon > 0$, $k \geq 0$, we denote by $B(f,\varepsilon,k)$ the ball of radius ε centered at f:

$$B(f,\varepsilon,k) = \{q \in \vec{L}_2(l) : \|f - q\|_k < \varepsilon\}.$$

Lemma 10.2.6. *Let V be an operator defined by formula* (10.2.32). *Suppose that f, k and ε are such that*

$$16l^2 + 2l + 3/2 < k, \tag{10.2.34}$$

$$\|f\|_{\vec{L}_2(l)}^2 < \frac{1}{2}\exp\{-2kl\}, \tag{10.2.35}$$

$$\varepsilon^2 < \frac{1}{2}\exp\{-2kl\}. \tag{10.2.36}$$

Then

$$V(B(f,\varepsilon,k)) \subset B(f,\varepsilon,k). \tag{10.2.37}$$

Lemma 10.2.7. *Let $f \in \vec{L}_2(l)$. Suppose that $c_0 \neq 0$, $l > 0$, $k \in (M(l,c_0),\infty)$ are fixed, where*

$$M(l,c_0) = \max\{24/c_0^2 + 5l + 4c_0^2 + 16, 16l^2 + 2l + 3/2\}.$$

If f and $\varepsilon > 0$ satisfy (10.2.35), (10.2.36), *then V is a contracting operator in $B(f,\varepsilon,k)$ with the contraction constant $\nu^2 = M(l,c_0)/k$.*

Theorem 10.2.2 (on well-posedness for sufficiently small data). *Let $f \in \vec{L}_2(l)$, $l > 0$, $c_0 \neq 0$,*

$$k \in (M(l,c_0),\infty), \tag{10.2.38}$$

$$\varepsilon^2 \in \Big(0, \frac{1}{2}\mathrm{e}^{-2kl}\Big). \tag{10.2.39}$$

If

$$\|f\|^2_{\vec{L}_2(l)} < \varepsilon^2, \tag{10.2.40}$$

then system (10.2.31) *has a unique solution* $q \in \vec{L}_2(l)$ *continuously dependent on the data* f.

Sketch of proof. Following Lemma 10.2.6 and the Banach fixed point theorem, the operator V has a unique fixed point $\bar{q} \in B(f, \varepsilon, k)$, that is, $\bar{q}$ solves the system

$$q(x,t) = Vq(x,t), \quad (x,t) \in \Delta(l), \tag{10.2.41}$$

which is the same as (10.2.31).

Let us prove that the solution is stable. Suppose $q^{(1)}$ and $q^{(2)}$ to be solutions of the corresponding systems

$$q^{(j)} = f^{(j)} + Bq^{(j)}, \quad j = 1, 2. \tag{10.2.42}$$

Put

$$\tilde{q} = q^{(1)} - q^{(2)}, \quad \tilde{f} = f^{(1)} - f^{(2)}.$$

Then

$$\tilde{q} = \tilde{f} + Bq^{(1)} - Bq^{(2)}. \tag{10.2.43}$$

Using the same reasoning as in Section 4.4, we can deduce from Lemma 10.2.7 that

$$\|\tilde{q}\|_k \le \|\tilde{f}\|_k + \nu\|\tilde{q}\|_k$$

and, since $\nu \in (0, 1)$,

$$\|\tilde{q}\|_k \le \frac{1}{1-\nu}\|\tilde{f}\|_k \le \frac{1}{1-\nu}\|\tilde{f}\|. \tag{10.2.44}$$

Combining (10.2.44) with the inequality

$$\|\tilde{q}\|_{\vec{L}_2(l)} \le e^{2kl}\|\tilde{q}\|_k,$$

which follows from the definition of k-norm, we obtain

$$\|\tilde{q}\|_{\vec{L}_2(l)} \le \frac{1}{1-\nu} e^{2kl} \|\tilde{f}\|_{\vec{L}_2(l)}. \tag{10.2.45}$$

Thus, we have shown that the solution $q(x,t)$ of system (10.2.31) continuously depends on the data $f(x,t)$. □

Theorem 10.2.3 (on well-posedness in the small). *Let* $f \in \vec{L}_2(l_0)$, $l_0 > 0$, $c_0 \neq 0$. *Then there exists such* $l_* \in (0, l_0)$ *that for any* $l \in (0, l_*)$ *system* (10.2.31) *is uniquely solvable in* $\vec{L}_2(l)$ *and its solution* $q(x,t)$ *continuously depends on the data* $f \in \vec{L}_2(l)$.

The proof of this theorem is similar to that of Theorem 10.1.1, with the only difference that now we should use the L_2-norm instead of the C-norm.

Theorem 10.2.4 (on well-posedness in a neighbourhood of exact solution). *Assume that, for some* $l > 0$ *and* $f \in \vec{L}_2(l)$, *there exists an exact solution* $q_e \in \vec{L}_2(l)$ *of problem* (10.2.18)–(10.2.22). *Then such* $\delta > 0$ *can be found that, for any* $f^\delta \in B(f, \delta)$, *the problem* $Aq = f^\delta$ *has a solution* $q^\delta \in \vec{L}_2(l)$ *continuously depending on the data* f^δ. *Here*

$$B(f, \delta) := B(f, \delta, 0) = \{q \in \vec{L}_2(l) : \|q - f\| < \delta\}.$$

This theorem is similar to Theorem 4.5.1.

10.2.2 Methods for solving the inverse problem of acoustics

In this subsection we consider the inverse problem of acoustics in the operator form (10.2.18)–(10.2.22) and describe different iterative methods as applied to this problem.

Note that Theorem 10.2.4 implies the uniqueness of the solution to problem (10.2.18)–(10.2.22). Therefore, we suppose only the existence of the solution q_e in the following theorems.

Method of successive approximations. Let q_0 be an initial approximation. Then we calculate successively the next approximations using the recurrence formula

$$q_{n+1} = f - Aq_n, \quad n = 0, 1, \dots. \tag{10.2.46}$$

Theorem 10.2.5 (on convergence of the method of successive approximations). *Suppose that* $l > 0$ *and, for* $f \in \vec{L}_2(l)$, *there exists a solution* $q_e \in \vec{L}_2(l)$ *of the problem* $Aq = f$. *Then such* $\delta_* > 0$ *and* $\nu_* \in (0, 1)$ *can be found that, if* $q_0 \in B(q_e, \delta_*)$, *then the iterative procedure defined by* (10.2.46) *converges to the solution* q_e *and the following estimate of convergence rate is valid:*

$$\|q_e - q_n\|_{\vec{L}_2(l)} \leq \nu_*^n \delta_*. \tag{10.2.47}$$

Landweber iterations. Given an initial approximation q_0, we compute the next approximations using the formula

$$q_{n+1} = q_n - \alpha[A'(q_n)]^*(Aq_n - f), \quad \alpha > 0,\ n = 0, 1, \dots. \tag{10.2.48}$$

Theorem 10.2.6 (on convergence of Landweber's iterations). *Let $l > 0$ and $f \in \vec{L}_2(l)$. Suppose that there exists a solution $q_e \in \vec{L}_2(l)$ of the problem $Aq = f$. Then one can find such $\nu_* \in (0,1)$, $\delta_* > 0$, $\alpha_* > 0$ that, if $q_0 \in B(q_e, \delta_*)$ and $\alpha \in (0, \alpha_*)$, then the approximations q_n determined by formula* (10.2.48) *converge to the solution q_e as $n \to \infty$ at the rate*

$$\|q_e - q_n\|^2_{\vec{L}_2(l)} \leq \nu_*^n \delta_*^2. \tag{10.2.49}$$

Gradient descent. Consider the functional

$$J(q) = \langle Aq - f, Aq - f \rangle_{\vec{L}_2(l)} = \|Aq - f\|^2_{\vec{L}_2(l)}. \tag{10.2.50}$$

Iterations of the gradient descent method are defined by the rule

$$q_{n+1} = q_n - \alpha_n J' q_n, \quad n = 0, 1, \ldots. \tag{10.2.51}$$

Here q_0 is a fixed initial approximation, $J'q$ is the gradient of the functional $J(q)$, the parameter α_n is chosen so as to satisfy

$$J(q_n - \alpha_n J' q_n) = \inf_{\alpha > 0} (J(q_n - \alpha J' q_n)), \quad \alpha_n > 0. \tag{10.2.52}$$

We remind that

$$J'q = 2[A'(q)]^*(Aq - f). \tag{10.2.53}$$

Theorem 10.2.7 (on uniqueness of the critical point). *Let the problem $Aq = f$ have a unique solution q_e in $\vec{L}_2(l)$. If the equation $J'q = 0$ is also uniquely solvable in $\vec{L}_2(l)$, then its solution is precisely q_e.*

Theorem 10.2.8 (on convergence of gradient descent). *Assume that $l > 0$ and, for $f \in \vec{L}_2(l)$, there exists a solution $q_e \in \vec{L}_2(l)$ of the problem $Aq = f$. Then one can specify such $\delta_* > 0$, $\nu_* \in (0,1)$, $M_* > 0$ that, if $q_0 \in B(q_e, \delta_*)$, then the iterations q_n of the gradient descent method converge to q_e as $n \to \infty$ and there holds*

$$\|q_n - q_e\|^2_{\vec{L}_2(l)} \leq M_* \nu_*^n \delta_*^2. \tag{10.2.54}$$

Newton–Kantorovich method. The Newton–Kantorovich method is based on the following recurrence formula for the iteration sequence:

$$q_{n+1} = q_n + [A'(q_n)]^{-1}(f - Aq_n). \tag{10.2.55}$$

As before, q_0 is supposed to be given.

Theorem 10.2.9 (on convergence of the Newton–Kantorovich method). *Let $l > 0$ and, for $f \in \vec{L}_2(l)$, the problem $Aq = f$ have a solution $q_e \in \vec{L}_2(l)$. Then there exist such $\delta_* > 0$ and $\nu_* \in (0,1)$ that, if $q_0 \in B(q_e, \delta_*)$, then the iterations q_n computed by* (10.2.55) *converge to q_e as $n \to \infty$ and the following estimate holds:*

$$\|q_n - q_e\|_{\vec{L}_2(l)} \le \nu_*^n \|q_0 - q_e\|_{\vec{L}_2(l)}. \tag{10.2.56}$$

10.3 Inverse problems of electrodynamics

In this section we introduce an inverse problem related to Maxwell's equations (Romanov and Kabanikhin, 1994)

$$\begin{cases} \varepsilon \dfrac{\partial}{\partial t} \boldsymbol{E} - \operatorname{rot} \boldsymbol{H} + \sigma \boldsymbol{E} + \boldsymbol{j}^* = \boldsymbol{0}, \\ \mu \dfrac{\partial}{\partial t} \boldsymbol{H} + \operatorname{rot} \boldsymbol{E} = \boldsymbol{0}, \end{cases} \quad x \in \mathbb{R}^3, \quad t \in \mathbb{R}. \tag{10.3.1}$$

Here $\boldsymbol{E} = (E_1, E_2, E_3)^\top$ and $\boldsymbol{H} = (H_1, H_2, H_3)^\top$ are vector-functions depending on the space variable $x \in \mathbb{R}^3$, $x = (x_1, x_2, x_3)$, and the time variable t. The coefficients ε, μ, σ are functions of $x \in \mathbb{R}^3$, $\boldsymbol{j}^* = \boldsymbol{j}^*(x,t)$. The space $\mathbb{R}^3$ is divided by the plane $x_3 = 0$ into the two half-spaces

$$\mathbb{R}^3_- = \{x \in \mathbb{R}^3 : x_3 < 0\}, \quad \mathbb{R}^3_+ = \{x \in \mathbb{R}^3 : x_3 > 0\}.$$

The coefficients ε, μ, σ are supposed to be constants in $\mathbb{R}^3_-$ and smooth functions in $\overline{\mathbb{R}}^3_+$. On the common boundary $x_3 = 0$ of $\mathbb{R}^3_+$ and $\mathbb{R}^3_-$ the functions ε, μ, σ may have finite jumps. Therefore, equations (10.3.1) are to be understood in the following sense: they hold in each of the half-spaces $\mathbb{R}^3_-$, $\mathbb{R}^3_+$ while on the plane $x_3 = 0$ the tangential components of $\boldsymbol{E}$ and $\boldsymbol{H}$ should meet the continuity conditions

$$E_j\big|_{x_3=-0} = E_j\big|_{x_3=+0}, \quad H_j\big|_{x_3=-0} = H_j\big|_{x_3=+0}, \quad j = 1, 2. \tag{10.3.2}$$

The vectors $\boldsymbol{E}$, $\boldsymbol{H}$ are also assumed to vanish identically up to the moment $t = 0$:

$$\boldsymbol{E}\big|_{t<0} \equiv \boldsymbol{0}, \quad \boldsymbol{H}\big|_{t<0} \equiv \boldsymbol{0}, \quad \boldsymbol{j}^*\big|_{t<0} \equiv \boldsymbol{0}. \tag{10.3.3}$$

The problem of finding $\boldsymbol{E}$, $\boldsymbol{H}$ from (10.3.1)–(10.3.3), when the coefficients ε, μ, σ and the function $\boldsymbol{j}^*$ are given, is said to be the *direct problem of electrodynamics*.

Physically, equations (10.3.1)–(10.3.3) describe the electromagnetic field in a medium with parameters ε, μ, σ characterizing permittivity, permeability, and conductivity, respectively. The medium consists of two parts: the Earth $\mathbb{R}^3_+$ with the flat surface $x_3 = 0$ and the air $\mathbb{R}^3_-$. The electromagnetic field is generated by an exterior current $\boldsymbol{j}^*$ at the moment $t = 0$. The direct problem of electrodynamics consists in determining the electric and magnetic strength vectors $\boldsymbol{E}$, $\boldsymbol{H}$ in the medium of a given structure, i.e., with known parameters ε, μ, σ. However the opposite situation is more frequent in practice, namely: it is required to explore a medium by observing the electromagnetic field on its surface. In other words, the coefficients ε, μ, σ are desired to be found in the Earth $\mathbb{R}^3_+$. For this purpose the tangential components of $\boldsymbol{E}$, $\boldsymbol{H}$ are measured on the surface $x_3 = 0$:

$$E_j\big|_{x_3=0} = \varphi_j(x_1, x_2, t), \quad H_j\big|_{x_3=0} = \psi_j(x_1, x_2, t), \quad j = 1, 2. \tag{10.3.4}$$

Thereby we come to the *inverse problem of electrodynamics* which consists in recovering the coefficients ε, μ, σ from information (10.3.4) on the solution $(\boldsymbol{E}, \boldsymbol{H})$ to the direct problem (10.3.1)–(10.3.3), provided the function $\boldsymbol{j}^*$ is given.

Throughout this section the support of $\boldsymbol{j}^*$ is assumed to belong to the domain $D^- = \{(x,t) \in \mathbb{R}^4 : x \in \overline{\mathbb{R}}^3_-, t \ge 0\}$.

Notice that if the coefficients ε, μ, σ are known in $\mathbb{R}^3_-$, it suffices to measure only two components of $\boldsymbol{E}$, $\boldsymbol{H}$ in (10.3.4). Indeed, if, for example, φ_j, $j = 1, 2$, are given, we can solve the initial boundary value problem (10.3.1)–(10.3.3) in the domain D^- with the boundary condition $E_j|_{x_3=0} = \varphi_j$, $j = 1, 2$, and thereby uniquely determine all the components of $\boldsymbol{E}$, $\boldsymbol{H}$ in D^-. Hence, we can calculate ψ_j from the functions φ_j, $j = 1, 2$.

It should be noted that system (10.3.1) is often supplemented by the equations

$$\operatorname{div}(\mu \boldsymbol{H}) = 0, \tag{10.3.5}$$

$$\operatorname{div}(\varepsilon \boldsymbol{E}) = 4\pi\rho. \tag{10.3.6}$$

Strictly speaking, it is system (10.3.1), (10.3.5), (10.3.6) which is usually called a system of Maxwell's equations. However, we shall not use equations (10.3.5), (10.3.6) anywhere because (10.3.5) is a direct consequence of (10.3.1)–(10.3.3) and equation (10.3.6) can be regarded as an independent equation for determining the charge density ρ which is beyond our interest here. At the same time, the vector $\boldsymbol{E}$ can be found from (10.3.1)–(10.3.3). Thus, we can consider system (10.3.1) as an independent object.

Consider now the inverse problem (10.3.1)–(10.3.4) in the half-space $\mathbb{R}^3_+$ and assume that the vector-function has the particular form

$$\boldsymbol{j}^* = (0, 1, 0)^\top g(x_1)\delta(x_3)\theta(t). \tag{10.3.7}$$

This means that the current is switched on instantaneously, directed along the x_2-axis, concentrated on the plane $x_3 = 0$ and distributed over the x_1-axis with the density $g(x_1)$ (for example, an infinite cable may be used). Assume also that the coefficients of Maxwell's equations do not depend on x_2 and satisfy conditions

$$\begin{aligned} \varepsilon = \varepsilon(x_1, x_3) \ge \varepsilon_0 > 0, \quad & \mu = \mu(x_1, x_3) \ge \mu_0 > 0, \\ \sigma = \sigma(x_1, x_3) \ge 0, \quad & \forall (x_1, x_3) \in \mathbb{R} \times \mathbb{R}_+. \end{aligned} \tag{10.3.8}$$

It is not difficult to show that under conditions (10.3.3), (10.3.7), (10.3.8) the components E_1, E_3, H_2 vanish and Maxwell's equations take the form

$$\begin{cases} \varepsilon \dfrac{\partial}{\partial t} E_2 + \dfrac{\partial}{\partial x_1} H_3 - \dfrac{\partial}{\partial x_3} H_1 + \sigma E_2 + g(x_1)\delta(x_3)\theta(t) = 0, \\ \mu \dfrac{\partial}{\partial t} H_1 - \dfrac{\partial}{\partial x_3} E_2 = 0, \quad \mu \dfrac{\partial}{\partial t} H_3 + \dfrac{\partial}{\partial x_1} E_2 = 0. \end{cases} \tag{10.3.9}$$

Let an addition information be given as

$$E_2\big|_{x_3=0} = \varphi_2(x_1, t), \quad x_1 \in \mathbb{R},\ t > 0. \tag{10.3.10}$$

The following condition is necessary for $\varphi_2(x_1, t)$ to be the trace of a smooth solution of problem (10.3.9) with the initial conditions $(E_2, H_1, H_3)|_{t<0} \equiv 0$:

$$\varphi_2(x_1, +0) = g(x_1)/2, \quad \forall x_1 \in \mathbb{R}.$$

If $\varphi_2(x_1, t)$ is given and the coefficients of system (10.3.9) are known for $x_3 \le 0$, we can solve the direct problem for $x_3 \le 0$ using (10.3.10) as a boundary condition. So we can find

$$H_1\big|_{x_3=+0} = \psi_1(x_1, t), \quad \forall (x_1, t) \in \mathbb{R} \times \mathbb{R}_+. \tag{10.3.11}$$

Setting the function $\psi_1(x_1, t)$ allows us to avoid solving the direct problem for $x_3 < 0$ (i.e. in the air).

Assume that the coefficients of (10.3.9) are sufficiently smooth for $x_3 > 0$. Eliminating the partial derivatives of H_1 and H_3 from system (10.3.9), we obtain the second-order equation for E_2:

$$\varepsilon \frac{\partial^2}{\partial t^2} E_2 + \sigma \frac{\partial}{\partial t} E_2 = \frac{\partial}{\partial x_1}\Big(\frac{1}{\mu}\frac{\partial}{\partial x_1} E_2\Big) + \frac{\partial}{\partial x_3}\Big(\frac{1}{\mu}\frac{\partial}{\partial x_3} E_2\Big) - g(x_1)\delta(x_3)\theta'(t),$$
$$x_1 \in \mathbb{R},\ x_3 \in \mathbb{R},\ t \ge 0, \tag{10.3.12}$$

with the initial condition

$$E_2\big|_{t<0} \equiv 0 \tag{10.3.13}$$

and the boundary conditions

$$E_2\big|_{x_3=+0} = \varphi_2(x_1, t), \quad \Big(\frac{1}{\mu}\frac{\partial}{\partial x_3} E_2\Big)\Big|_{x_3=+0} = \frac{\partial}{\partial t}\psi_1(x_1, t). \tag{10.3.14}$$

If the coefficients of (10.3.12) depend on x_3 only, then we can apply the Fourier transform $F_{x_1}[\cdot]$. Denote by $\tilde{E}_2$ the Fourier image of E_2 with respect to x_1, i.e.,

$$\tilde{E}_2(\lambda, x_3, t) = \int_{-\infty}^{\infty} E_2(x_1, x_3, t) e^{i\lambda x_1}\, dx_1,$$

where λ is a fixed parameter. Then problem (10.3.12)–(10.3.14) can be reduced to the following one-dimensional problem for $\tilde{E}_2$:

$$\varepsilon \frac{\partial^2 \tilde{E}_2}{\partial t^2} + \sigma \frac{\partial \tilde{E}_2}{\partial t} = \frac{\partial}{\partial x_3}\Big(\frac{1}{\mu}\frac{\partial \tilde{E}_2}{\partial x_3}\Big) - \frac{\lambda^2}{\mu}\tilde{E}_2 - \tilde{g}\delta(x_3)\delta(t), \tag{10.3.15}$$

$$\tilde{E}_2\big|_{t<0} \equiv 0, \quad \tilde{E}_2\big|_{x_3=0} = \tilde{\varphi}_2, \quad \Big(\frac{1}{\mu}\frac{\partial \tilde{E}_2}{\partial x_3}\Big)\Big|_{x_3=+0} = \frac{\partial \tilde{\psi}_1}{\partial t}. \tag{10.3.16}$$

Here $\tilde{g}$, $\tilde{\varphi}_2$, $\tilde{\psi}_1$ are the Fourier images with respect to x_1 of the functions g, φ_2, ψ_1 with the same parameter λ.

Consider now the problem in the case when μ is constant. Set

$$v(x_3, t) = v(\lambda, x_3, t) = \tilde{E}_2(\lambda, x_3, t), \quad g^\lambda = \tilde{g}(\lambda),$$
$$f_0^\lambda(t) = \tilde{\varphi}_2(\lambda, t), \quad f_1^\lambda(t) = \frac{\partial}{\partial t}\tilde{\psi}_1(\lambda, t).$$

Then (10.3.15), (10.3.16) imply

$$v_{tt} + \frac{\sigma}{\varepsilon} v_t = \frac{1}{\mu\varepsilon}(v_{x_3x_3} - \lambda^2 v) - \frac{1}{\varepsilon} g^\lambda \delta(x_3)\delta(t), \quad x_3 \in \mathbb{R},\ t \geq 0, \tag{10.3.17}$$

$$v\big|_{t<0} \equiv 0, \tag{10.3.18}$$

$$v\big|_{x_3=0} = f_0^\lambda(t), \quad t \geq 0, \tag{10.3.19}$$

$$v_{x_3}\big|_{x_3=+0} = \mu f_1^\lambda(t), \quad t \geq 0. \tag{10.3.20}$$

The *direct problem* is to find the function $v(x_3, t)$ from (10.3.17), (10.3.18), provided $\varepsilon(x_3)$, μ, $\sigma(x_3)$ are given.

The *inverse problem* consists in recovering $\varepsilon(x_3)$ or $\sigma(x_3)$ (or both $\varepsilon(x_3)$ and $\sigma(x_3)$) from (10.3.17)–(10.3.20), provided the functions $f_0^\lambda(t)$ and $f_1^\lambda(t)$ are known for a fixed $\lambda = \lambda_0$ (or for two values of λ).

To study the direct and inverse problems for equation (10.3.17), it is convenient to change from x_3 to a new variable x:

$$x = x(x_3) = \int_0^{x_3} \sqrt{\mu\varepsilon(\xi)}\, d\xi, \quad x_3 = \omega(x).$$

Introduce

$$a(x) = \frac{\sigma(\omega(x))}{\varepsilon(\omega(x))}, \quad b(x) = \frac{1}{\sqrt{\mu\varepsilon(\omega(x))}},$$

$$u(x,t) = v(\omega(x),t), \quad \gamma = -g^{\lambda}\sqrt{\frac{\mu}{\varepsilon(+0)}}.$$

Then (10.3.17), (10.3.18) takes the form

$$u_{tt}(x,t) = u_{xx}(x,t) - Pu(x,t) + \gamma\delta(x)\delta(t), \tag{10.3.21}$$

$$u\big|_{t<0} \equiv 0. \tag{10.3.22}$$

Here

$$Pu(x,t) = a(x)u_t(x,t) + \frac{b'(x)}{b(x)}u_x(x,t) + (\lambda b(x))^2 u(x,t).$$

Lemma 10.3.1 (Romanov and Kabanikhin, 1994). *If $a \in C^1(\mathbb{R})$, then the direct problem* (10.3.21), (10.3.22) *is uniquely solvable in $C^2(t \geq |x|)$ and its solution has the form*

$$u(x,t) = s(x)\theta(t-|x|) + \tilde{u}(x,t),$$

where $\tilde{u}(x,t)$ is a continuous function and $s(x)$ solves the following problem:

$$\begin{cases} s'(x) = \dfrac{1}{2}\Big(\dfrac{b'(x)}{b(x)} - a(x)\operatorname{sign} x\Big)s(x), & x \in \mathbb{R}, \\ s(0) = \gamma/2. \end{cases} \tag{10.3.23}$$

From Lemma 10.3.1 it follows that we can replace the initial condition (10.3.22) by the condition

$$u\big|_{t=|x|} = s(x), \quad x \in \mathbb{R}, \tag{10.3.24}$$

and equation (10.3.21) by the following one:

$$u_{tt}(x,t) = u_{xx}(x,t) - Pu(x,t), \quad 0 < |x| < t. \tag{10.3.25}$$

Since all the coefficients are to be recovered for $x_3 > 0$ and hence for $x > 0$, we will consider the problem in the domain

$$\Delta(l) = \big\{(x,t) \in \mathbb{R}^2 : 0 < x < l,\ 0 < x < t < 2l - x\big\}.$$

Thus, the direct problem (10.3.21), (10.3.22) can be reformulated as a problem of finding the function $u(x,t)$ from the relations

$$u_{tt}(x,t) = u_{xx}(x,t) - Pu(x,t), \quad (x,t) \in \Delta(l), \tag{10.3.26}$$

$$u\big|_{t=x} = s(x), \quad x \in (0,l), \tag{10.3.27}$$

$$u_x(0,t) = \sqrt{\frac{\mu}{\varepsilon(+0)}}\, f_1^\lambda(t), \quad t \in (0,2l), \tag{10.3.28}$$

providing the functions $a(x)$, $b(x)$, $f_1^\lambda(t)$ are known. Notice that the additional condition (10.3.20) of the inverse problem (10.3.17)–(10.3.20) is used to solve the direct problem (10.3.26)–(10.3.28) for $x > 0$.

In the inverse problem it is required to restore the function $a(x)$ from the additional information

$$u(0,t) = f_0^\lambda(t), \quad t \in (0,2l), \tag{10.3.29}$$

on the solution to the direct problem (10.3.26)–(10.3.28). The function $b(x)$, as well as $f_0^\lambda(t)$ and $f_1^\lambda(t)$, is supposed to be known.

Reduction of the inverse problem (10.3.26)–(10.3.29) to a system of integral equations. Consider the inverse problem (10.3.26)–(10.3.29) in the domain $\Delta(l)$.

Note that the solution $s(x)$ of system (10.3.23) for $x \in (0,l)$ satisfies the following second-kind integral Volterra equation:

$$s(x) = \frac{\gamma}{2} + \frac{1}{2}\int_0^x \Big[\frac{b'(\xi)}{b(\xi)} - a(\xi)\Big] s(\xi)\, d\xi, \quad x \in (0,l). \tag{10.3.30}$$

Using the d'Alembert formula and taking conditions (10.3.28), (10.3.29) into account, we obtain the integral equation

$$\begin{aligned} u(x,t) = {} & \frac{1}{2}\big[f_0^\lambda(t+x) + f_0^\lambda(t-x)\big] + \frac{1}{2}\int_{t-x}^{t+x} \sqrt{\frac{\mu}{\varepsilon(+0)}}\, f_1^\lambda(\xi)\, d\xi \\ & + \frac{1}{2}\int_0^x \int_{t-x+\xi}^{t+x-\xi} Pu(\xi,\tau)\, d\tau\, d\xi, \quad (x,t) \in \Delta(l). \end{aligned} \tag{10.3.31}$$

We differentiate (10.3.31) with respect to x and t to get

$$\begin{aligned} u_x(x,t) = {} & \frac{1}{2}[(f_0^\lambda)'(t+x) - (f_0^\lambda)'(t-x)] + \frac{1}{2}\sqrt{\frac{\mu}{\varepsilon(+0)}}\big(f_1^\lambda(t+x) + f_1^\lambda(t-x)\big) \\ & + \frac{1}{2}\int_0^x \Big(Pu(\xi, t+x-\xi) + Pu(\xi, t-x+\xi)\Big)\, d\xi, \end{aligned} \tag{10.3.32}$$

$$\begin{aligned} u_t(x,t) = {} & \frac{1}{2}[(f_0^\lambda)'(t+x) + (f_0^\lambda)'(t-x)]) + \frac{1}{2}\sqrt{\frac{\mu}{\varepsilon(+0)}}(f_1^\lambda(t+x) - f_1^\lambda(t-x)) \\ & + \frac{1}{2}\int_0^x [Pu(\xi, t+x-\xi) - Pu(\xi, t-x+\xi)]\, d\xi. \end{aligned} \tag{10.3.33}$$

Taking the limit in (10.3.31) as $t \to x + 0$ and using (10.3.27), we obtain

$$s(x) = \frac{1}{2}[f_0^\lambda(2x) + f_0^\lambda(0)] + \frac{1}{2}\sqrt{\frac{\mu}{\varepsilon(+0)}} \int_0^{2x} f_1^\lambda(\xi)\, d\xi + \frac{1}{2}\int_0^x \int_\xi^{2x-\xi} Pu(\xi,\tau)\, d\tau\, d\xi.$$

Differentiating this equation with respect to x, in view of (10.3.34), yields

$$s'(x) = \frac{1}{2}\Big[\frac{b'(x)}{b(x)} - a(x)\Big]s(x) = [f_0^\lambda]'(2x) + \sqrt{\frac{\mu}{\varepsilon(+0)}}\, f_1^\lambda(2x) + \int_0^x Pu(\xi, 2x-\xi)\, d\xi.$$

It is not difficult to check that the function $[s(x)]^{-1}$ solves the second-kind integral Volterra equation

$$[s(x)]^{-1} = \frac{2}{\gamma} - \frac{1}{2}\int_0^x \Big[\frac{b'(\xi)}{b(\xi)} - a(\xi)\Big][s(\xi)]^{-1}\, d\xi. \qquad (10.3.34)$$

Combining (10.3.34) and the preceding equality results in

$$a(x) = \frac{b'(x)}{b(x)} - \frac{4}{\gamma}[[f_0^\lambda]'(2x) + \sqrt{\frac{\mu}{\varepsilon(+0)}}\, f_1^\lambda(2x)] - \frac{4}{\gamma}\int_0^x Pu(\xi, 2x-\xi)\, d\xi + \Big[[f_0^\lambda]'(2x) + \sqrt{\frac{\mu}{\varepsilon(+0)}}\, f_1^\lambda(2x) + \int_0^x Pu(\xi, 2x-\xi)\, d\xi\Big] \times \int_0^x \Big[\frac{b'(\xi)}{b(\xi)} - a(\xi)\Big][s(\xi)]^{-1}\, d\xi. \qquad (10.3.35)$$

Thus, relations (10.3.31)–(10.3.35) form a system of integral equations for the functions $u(x,t)$, $u_x(x,t)$, $u_t(x,t)$, $[s(x)]^{-1}$, $a(x)$ in the domain $\Delta(l)$. This system can be written in the operator form

$$Aq := q + Bq = f, \qquad (10.3.36)$$

where

$$q(x,t) = \big(q_1, q_2, q_3, q_4, q_5\big)^\top, \quad f(x,t) = \big(f_1, f_2, f_3, f_4, f_5\big)^\top, \tag{10.3.37}$$

$$Bq = \big(B_1 q,\ B_2 q,\ B_3 q,\ B_4 q,\ B_5 q\big)^\top, \tag{10.3.38}$$

$$q_1(x,t) = u(x,t), \quad q_2(x,t) = u_x(x,t), \quad q_3(x,t) = u_t(x,t),$$

$$q_4(x) = [s(x)]^{-1}, \quad q_5(x) = a(x).$$

$$f_1(x,t) = \frac{1}{2}\,[f_0^\lambda(t+x) + f_0^\lambda(t-x)] + \frac{1}{2}\int_{t-x}^{t+x} \sqrt{\frac{\mu}{\varepsilon(+0)}}\, f_1^\lambda(\xi)\,d\xi,$$

$$\begin{aligned} f_2(x,t) &= \frac{1}{2}[(f_0^\lambda)'(t+x) - (f_0^\lambda)'(t-x)] \\ &\quad + \frac{1}{2}\sqrt{\frac{\mu}{\varepsilon(+0)}}[f_1^\lambda(t+x) + f_1^\lambda(t-x)], \end{aligned}$$

$$\begin{aligned} f_3(x,t) &= \frac{1}{2}[(f_0^\lambda)'(t+x) + (f_0^\lambda)'(t-x)] \\ &\quad + \frac{1}{2}\sqrt{\frac{\mu}{\varepsilon(+0)}}\,[f_1^\lambda(t+x) - f_1^\lambda(t-x)], \end{aligned}$$

$$f_4 = \frac{2}{\gamma},$$

$$f_5(x) = \frac{b'(x)}{b(x)} - \frac{4}{\gamma}\Big[(f_0^\lambda)'(2x) + \sqrt{\frac{\mu}{\varepsilon(+0)}}\, f_1^\lambda(2x)\Big] = b_1(x) + \frac{4}{\gamma}\, g_0(x),$$

$$B_1 q = -\frac{1}{2}\int_0^x \int_{t-x+\xi}^{t+x-\xi} Lq(\xi,\tau)\,d\tau\,d\xi, \tag{10.3.39}$$

$$B_2 q = -\frac{1}{2}\int_0^x [Lq(\xi, t+x-\xi) + Lq(\xi, t-x+\xi)]\,d\xi, \tag{10.3.40}$$

$$B_3 q = -\frac{1}{2}\int_0^x [Lq(\xi, t+x-\xi) - Lq(\xi, t-x+\xi)]\,d\xi, \tag{10.3.41}$$

$$B_4 q = \frac{1}{2}\int_0^x [b_1(\xi) - q_5(\xi)]q_4(\xi)\,d\xi, \tag{10.3.42}$$

$$B_5 q = 2g_0(x)B_4 q + \Big(\frac{4}{\gamma} - 2B_4 q\Big)\int_0^x Lq(\xi, 2x-\xi)\,d\xi. \tag{10.3.43}$$

$$Lq(x,t) = b_0(x)q_1(x,t) + b_1(x)q_2(x,t) + q_5(x)q_3(x,t),$$

$$b_0(x) = (\lambda b(x))^2, \quad b_1(x) = \frac{b'(x)}{b(x)},$$

$$g_0(x) = -(f_0^\lambda)'(2x) - \sqrt{\frac{\mu}{\varepsilon(+0)}}\, f_1^\lambda(2x).$$

We define a space $\vec{L}_2(l)$ as containing all vector-functions

$$q = (q_1(x,t), q_2(x,t), q_3(x,t), q_4(x), q_5(x))^\top$$

such that

$$q_j(x,t) \in L_2(\Delta(l)), \quad j = 1,2,3, \quad q_j(x) \in L_2(0,l), \quad j = 4,5.$$

The space $\vec{L}_2(l)$ when equipped with the inner product

$$\begin{aligned}\langle q^{(1)}, q^{(2)} \rangle_{\vec{L}_2(l)} &= \sum_{j=1}^{3} \int_0^l \int_x^{2l-x} q_j^{(1)}(x,t) q_j^{(2)}(x,t)\, dt\, dx \\ &\quad + \sum_{j=4}^{5} \int_0^l q_j^{(1)}(x) q_j^{(2)}(x)\, dx\end{aligned}$$

and with the associated norm

$$\|q\|^2_{\vec{L}_2(l)} = \langle q, q \rangle_{\vec{L}_2(l)}$$

becomes a Hilbert space.

In the inverse problem it is required to find a vector $q \in \vec{L}_2(l)$ satisfying (10.3.36) provided $f \in \vec{L}_2(l)$ is known. This problem may be studied using the same reasoning as we have applied to the inverse problem of acoustics (see Section 10.2).

10.4 Local solvability of multidimensional inverse problems

Little has been reported regarding the solvability of multidimensional inverse problems. This makes a first step toward studying the solvability of the problems in the class of functions analytic in horizontal variables all the more remarkable. The presented approach invoking the concept of a scale of Banach spaces of analytic functions (Ovsyannikov, 1965) was developed by Romanov (Romanov, 1989a,b).

In this section we deal with the inverse problem of recovering the coefficient $q(x,y)$ of the hyperbolic equation (10.4.3) in the space $(x,y) \in \mathbb{R}^{n+1}_+ := \mathbb{R}_+ \times \mathbb{R}^n$, where $\mathbb{R}_+ := \{x \in \mathbb{R} : x > 0\}$. The initial data are thought of as zero, while the Neumann boundary condition is set on $\partial\mathbb{R}^{n+1}_+ \times [0,T]$, $\partial\mathbb{R}^{n+1}_+ = \{(x,y) \in \mathbb{R}^{n+1}_+ : x = 0\}$, $T > 0$, and has the form

$$u_x\big|_{x=0} = -g(y)\delta'(t),$$

where g is an analytic function. We assume also that the solution to the corresponding initial boundary value problem is fixed on the surface $\partial\mathbb{R}^{n+1}_+ \times [0,T]$ and forms a function analytic in $y \in \mathbb{R}^n$.

Denote by $\mathbb{A}_s(r)$, $r > 0$, $s > 0$, a space of real-valued functions $\varphi(y)$, $y \in \mathbb{R}^n$, such that

$$\|\varphi\|_s(r) := \sup_{|y|\le r} \sum_{|\alpha|=0}^{\infty} \frac{s^{|\alpha|}}{\alpha!} |D^\alpha \varphi(y)| < \infty, \tag{10.4.1}$$

where α is a multiindex,

$$D^\alpha := \frac{\partial^{|\alpha|}}{\partial y_1^{\alpha_1} + \cdots + \partial y_n^{\alpha_n}}, \quad \alpha = (\alpha_1, \ldots, \alpha_n), \quad \alpha_k \in \mathbb{N} \cup \{0\},$$

$$|\alpha| := \alpha_1 + \cdots + \alpha_n, \quad \alpha! := (\alpha_1)! \cdot \ldots \cdot (\alpha_n)!.$$

The space $\mathbb{A}_s(r)$ equipped with the norm (10.4.1) is a Banach space of analytic functions. Assuming r as a fixed parameter and s as a variable one, we obtain the *scale of Banach spaces of analytic functions* $\{\mathbb{A}_s(r)\}$, $s > 0$. Evidently, if $\varphi(y) \in \mathbb{A}_s(r)$, then $\varphi(y) \in \mathbb{A}_{s'}(r)$ for all $s' \in (0, s)$. Hence, $\mathbb{A}_s(r) \subset \mathbb{A}_{s'}(r)$ provided $s' < s$. Moreover, if $\varphi(y) \in \mathbb{A}_s(r)$, then $D^\alpha \varphi(y)$, for any α, belongs to the space $\mathbb{A}_{s'}(r)$ for all $s' \in (0, s)$ and the following inequality holds:

$$\|D^\alpha \varphi\|_{s'}(r) \le C_\alpha \frac{\|\varphi\|_s(r)}{(s - s')^{|\alpha|}}.$$

Here C_α is a constant depending on α only.

Indeed,

$$\begin{aligned}
\|D^\alpha \varphi\|_{s'}(r) &= \sup_{|y|\le r} \sum_{|\beta|=0}^{\infty} \frac{(s')^{|\beta|}}{\beta!} |D^{\alpha+\beta} \varphi(y)| \\
&\le \|\varphi\|_s(r) (s')^{-|\alpha|} \sup_\beta \left[\frac{(\alpha+\beta)!}{\beta!} \left(\frac{s'}{s}\right)^{|\alpha+\beta|} \right] \\
&\le \|\varphi\|_s(r) (s')^{-|\alpha|} \sup_{k \ge |\alpha|} \left[k^{|\alpha|} \left(\frac{s'}{s}\right)^k \right] \le C_\alpha \frac{\|\varphi\|_s(r)}{(s - s')^{|\alpha|}},
\end{aligned}$$

where $C_\alpha = |\alpha|^{|\alpha|}$. In particular,

$$\|\Delta \varphi\|_{s'}(r) \le \frac{4n}{(s - s')^2} \|\varphi\|_s(r), \quad s > s' > 0. \tag{10.4.2}$$

In what follows we will omit the fixed parameter r in the corresponding notations writing $\|\varphi\|_s$, $\mathbb{A}_s$ instead of $\|\varphi\|_s(r)$, $\mathbb{A}_s(r)$.

Introduce now a space of functions analytic in $y \in \mathbb{R}^n$ and dependent on the variables $(x, t) \in G$ where G is a compact domain with piecewise-smooth boundary.

Definition 10.4.1. A function $w = w(x, y, t)$ will be said to belong to $\mathbb{C}(\mathbb{A}_s, G)$ if $w \in \mathbb{A}_s$ for all $(x, t) \in G$, w is continuous in G as an element of the space $\mathbb{A}_s$ and, moreover,

$$\|w\|_{\mathbb{C}(\mathbb{A}_s,G)} := \sup_{(x,t)\in G} \|w\|_s(x,t) < \infty.$$

It is clear that $\mathbb{C}(\mathbb{A}_s, G)$ is a Banach space.

Consider the initial boundary value problem of determining the function $u(x, y, t)$ from the relations

$$u_{tt} - u_{xx} - \Delta u - q(x, y)u = 0, \quad x \in \mathbb{R}_+,\ y \in \mathbb{R}^n,\ t \in \mathbb{R}, \tag{10.4.3}$$

$$u|_{t<0} = 0, \quad u_x|_{x=0} = -g(y)\delta'(t), \tag{10.4.4}$$

provided $q(x, y)$ and $g(y)$ are known functions. This problem is well-posed and will be referred to as the direct problem.

In the inverse problem the functions $q(x, y)$ and $u(x, y, t)$ are to be found from the given function $g(y)$ and from the trace of the solution to problem (10.4.3)–(10.4.4) on the hyperplane $x = 0$:

$$u\big|_{x=0} = F(y, t), \quad y \in \mathbb{R}^n,\ t > 0. \tag{10.4.5}$$

It can be easily shown that the function $F(y, t)$ is representable in the form

$$F(y, t) = g(y)\delta(t) + f(y, t)\theta(t), \tag{10.4.6}$$

where $\theta(t)$ is the Heaviside function.

Indeed, the solution to problem (10.4.3)–(10.4.4) can be written in the form

$$u(x, y, t) = g(y)\delta(t - x) + \alpha(x, y)\theta(t - x) + \sum_{k=1}^{\infty} \alpha_k(x, y)\theta_k(t - x), \tag{10.4.7}$$

where

$$\theta_k(t) := \frac{t^k}{k!}\,\theta(t), \quad k = 1, 2, \ldots.$$

A similar asymptotic expansion of the solution to a more general problem was derived in (Romanov, 2005, Section 2.2).

Therefore, the function $u(x, y, t)$ can be represented as

$$u(x, y, t) = g(y)\delta(t - x) + \bar{u}(x, y, t)\theta(t - x). \tag{10.4.8}$$

It is not difficult to verify that

$$\alpha(x, y) := \bar{u}(x, y, x + 0) = \frac{1}{2}\Big[x\,\Delta g(y) + g(y)\int_0^x q(\xi, y)\,d\xi\Big]. \tag{10.4.9}$$

From formula (10.4.8) follows the validity of (10.4.6). At the same time, (10.4.9) implies

$$f(y, +0) = 0. \tag{10.4.10}$$

Equality (10.4.10) is a necessary condition for the inverse problem to have a solution. Note that $u(x, y, t) \equiv \bar{u}(x, y, t)$ when $x > t$. Therefore, we will write $u(x, y, t)$ instead of $\bar{u}(x, y, t)$ when considering the function in this domain.

Set

$$\begin{aligned} u_0(x, y, t) &= \frac{1}{2}[f(x + t, y) + f(x - t, y)], \\ q_0(x, y) &= \frac{1}{g(y)}\big[-\Delta g(y) + 2 f_t(y, t)|_{t=2x}\big]. \end{aligned} \tag{10.4.11}$$

Theorem 10.4.1. *Assume that condition* (10.4.10) *is fulfilled, the functions* $g(y)$, $1/g(y)$ *belong to* $\mathbb{A}_{s_0}$ *for some* $s_0 > 0$, *and* $f(y, t)$, $f_t(y, t)$ *belong to* $\mathbb{C}(\mathbb{A}_{s_0}, [0, T])$ *for some* $T > 0$. *Denote*

$$R_0 := \max\Big[\|(g(y))^{-1}\|_{s_0}, \max_{0 \le x \le T/2} \|q_0(x, y)\|_{s_0}(x), \max_{0 \le t \le T} \|f(y, t)\|_{s_0}(t)\Big]. \tag{10.4.12}$$

Then such number $a = a(s_0, T, R_0) \in (0, T/(2s_0))$ *can be found that, for any* $s \in (0, s_0)$, *there exists a unique solution* $q_e(x, y)$, $u_e(x, y, t)$ *of the inverse problem* (10.4.3)–(10.4.5) *in* $\mathbb{C}(\mathbb{A}_{s_0}, G_s)$, *where* $G_s = \{(x, t) \in \mathbb{R}^2 : 0 \le x \le a(s_0 - s),$ $x \le t \le T - x\}$. *Moreover, the following estimates hold:*

$$\|u_e - u_0\|_s(x, t) \le R_0, \quad \|q_e - q_0\|_s(x) \le \frac{R_0}{s_0 - s}, \quad (x, t) \in G_s. \tag{10.4.13}$$

Proof. In view of (10.4.9), the inverse problem (10.4.3)–(10.4.5) is equivalent to the problem

$$\begin{aligned} u_{tt} - u_{xx} - \Delta u - qu &= 0, \quad y \in \mathbb{R}^n, \quad t > x > 0, \\ u_x\big|_{x=0} &= 0, \quad u\big|_{t=x} = \alpha(x, y), \\ u\big|_{x=0} &= f(y, t), \end{aligned} \tag{10.4.14}$$

where $f(y, t)$ is related to the data of the original problem by equality (10.4.6).

Since the Cauchy data u, u_x are specified for $x = 0$, we can use the d'Alembert formula to get the integro-differential equation for $u(x, y, t)$:

$$u(x, y, t) = u_0(x, y, t) - \frac{1}{2}\int_{\Delta(x,t)} (\Delta u + qu)(\zeta, y, \tau)\, d\zeta\, d\tau. \tag{10.4.15}$$

Here

$$\Delta(x,t) = \{(\zeta,\tau) : 0 \le \zeta \le x, \zeta + t - x \le \tau \le t + x - \zeta\}$$

is a characteristic triangle in the plane (ζ,τ) whose base lies on the τ-axis and other sides intersect at the point (x,t). Differentiating (10.4.15) with respect to t and with respect to x yields two equalities. We sum these equalities assuming $t = x$ and taking (10.4.9) into account. As a result, we have

$$\begin{aligned}\Big(\frac{\partial u}{\partial t} + \frac{\partial u}{\partial x}\Big)_{t=x} &= \frac{1}{2}(g(y)q(x,y) + \Delta g(y))\\ &= \Big(\frac{\partial u_0}{\partial t} + \frac{\partial u_0}{\partial x}\Big)_{t=x} - \int_0^x (\Delta u + qu)(\zeta, y, 2x-\zeta)\, d\zeta.\end{aligned}$$

From this we infer that, in view of (10.4.11), the function $q(x,y)$ obeys the integro-differential equation

$$q(x,y) = q_0(x,y) - \frac{2}{g(y)} \int_0^x (\Delta u + qu)(\zeta, y, 2x-\zeta)\, d\zeta. \tag{10.4.16}$$

Thus, we have derived the system of integro-differential equations (10.4.15), (10.4.16) for the functions $q(x,y)$, $u(x,y,t)$. This system can be solved using a modified Nirenberg method (Romanov, 2005). Let a sequence $a_0, a_1, \dots, a_m, \dots$ be defined as

$$a_{m+1} = a_m\Big(1 + \frac{1}{(m+1)^2}\Big)^{-1}, \quad m = 0,1,2,\dots, \tag{10.4.17}$$

and let

$$a = \lim_{m\to\infty} a_m = a_0 \prod_{n=0}^{\infty} \Big(1 + \frac{1}{(m+1)^2}\Big)^{-1}. \tag{10.4.18}$$

A way of choosing $a_0 \in (0, T/(2s_0))$ will be specified later. We calculate successively approximations to the functions $u(x,y,t)$, $q(x,y)$ using the recurrent formulae

$$\begin{aligned} u_{m+1}(x,y,t) &= u_0(x,y,t) - \frac{1}{2}\int_{\Delta(x,t)} (\Delta u_m + q_m u_m)(\zeta, y, \tau)\, d\zeta\, d\tau,\\ q_{m+1}(x,y) &= q_0(x,y) - \frac{2}{g(y)}\int_0^x (\Delta u_m + q_m u_m)(\zeta, y, 2x-\zeta)\, d\zeta \end{aligned} \tag{10.4.19}$$

for $m = 0,1,2,\dots$. Then the functions

$$v_m = u_{m+1} - u_m, \quad p_m = q_{m+1} - q_m, \quad m = 0,1,2,\dots \tag{10.4.20}$$

satisfy the relations

$$v_0(x,y,t) = -\frac{1}{2}\int_{\Delta(x,t)} (\Delta u_0 + q_0 u_0)(\zeta, y, \tau)\, d\zeta\, d\tau, \tag{10.4.21}$$

$$p_0(x,y) = -\frac{2}{g(y)}\int_0^x (\Delta u_0 + q_0 u_0)(\zeta, y, 2x-\zeta)\, d\zeta,$$

$$v_{m+1}(x,y,t) = -\frac{1}{2}\int_{\Delta(x,t)} (\Delta v_m + p_m u_{m+1} + q_m v_m)(\zeta, y, \tau)\, d\zeta\, d\tau, \tag{10.4.22}$$

$$p_{m+1}(x,y) = -\frac{2}{g(y)}\int_0^x (\Delta v_m + p_m u_{m+1} + q_m v_m)(\zeta, y, 2x-\zeta)\, d\zeta$$

for $m = 0, 1, 2, \ldots$. Let us prove that, for a suitable a_0, the inequalities

$$\begin{aligned}\lambda_m := \max\Big\{ &\sup_{(x,t,s)\in F_m} \Big[\|v_m\|_s(x,t)\frac{a_m(s_0-s)-x}{x}\Big],\\ &\sup_{(x,t,s)\in F_m} \Big[\|p_m\|_s(x,t)\,\frac{(a_m(s_0-s)-x)^2}{a_m x}\Big]\Big\} < \infty,\end{aligned} \tag{10.4.23}$$

$$\|u_{m+1} - u_0\|_{s_0}(x,t) \le R_0, \quad \|q_{m+1} - q_0\|_{s_0}(x) \le \frac{R_0}{s_0 - s} \tag{10.4.24}$$

are valid for $(x,t,s) \in F_{m+1}$, where

$$\begin{aligned} F_m &:= \{(x,t,s) : (x,t) \in G,\ 0 < x < a_m(s_0 - s),\ 0 < s < s_0\},\\ G &:= \{(x,t) : 0 \le x \le t \le T - x\}.\end{aligned} \tag{10.4.25}$$

Since $u_0(x,y,t) \in \mathbb{C}(\mathbb{A}_{s_0}, G_s)$, from (10.4.21) it follows that

$$\begin{aligned}\|v_0\|_s(x,t) &\le \frac{1}{2}\int_0^x \int_{t-x+\zeta}^{t+x-\zeta} (\|\Delta u_0\|_s(\zeta,\tau) + \|q_0\|_s(\zeta)\|u_0\|_s(\zeta,\tau))\, d\tau\, d\zeta\\ &\le \frac{1}{2}\int_0^x \int_{t-x+\zeta}^{t+x-\zeta} \Big(\frac{4nR_0}{(s(\zeta)-s)^2} + R_0^2\Big)\, d\tau\, d\zeta,\end{aligned} \tag{10.4.26}$$

provided $(x,t,s) \in F_0$ and $s(\zeta) = (s_0 + s - \zeta/a_0)/2$. Here we have used inequality (10.4.2) for the function Δu_0 regarded as an element of $\mathbb{A}_s$ and the estimates $\|u_0\|_s(x,t) < R_0$, $\|q_0\|_s(x) < R_0$, $s \in (0, s_0)$, $(x,t) \in G$.

Note that $(x,t,s) \in F_0$ implies $(\zeta, \tau, s) \in F_0$, whence $\zeta < a_0(s_0 - s)$ or, what is the same, $s < s_0 - \zeta/a_0$. Consequently, $s(\zeta) \in (s, s_0)$ and $s(\zeta) - s = (s_0 - s - \zeta/a_0)/2$. Therefore,

$$\begin{aligned}\|v_0\|_s(x,t) &\le \frac{1}{2}\int_0^x \int_{t-x+\zeta}^{t+x-\zeta} \Big(\frac{16na_0^2R_0}{((s_0-s)a_0-\zeta)^2} + R_0^2\Big)\, d\tau\, d\zeta\\ &\le a_0^2R_0(16n + R_0s_0^2)\int_0^x \frac{(x-\zeta)\, d\zeta}{((s_0-s)a_0-\zeta)^2} \le \frac{xa_0^2R_0(16n+R_0s_0^2)}{(s_0-s)a_0-x}.\end{aligned}$$

In a similar manner we derive

$$\|p_0\|_s(x) \le \Big\|\frac{2}{g(y)}\Big\|_{s_0}\int_0^x \Big(\frac{16na_0^2R_0}{[(s_0-s)a_0-\zeta]^2}+R_0^2\Big)\,d\zeta$$
$$\le 2R_0x\Big(\frac{16na_0^2R_0}{[(s_0-s)a_0-x]^2}+R_0^2\Big) \le \frac{2xa_0^2R_0^2(16n+R_0s_0^2)}{[(s_0-s)a_0-x]^2}.$$

Hence, upon setting $\mu_0 = a_0(16n+R_0s_0^2)\max(a_0, 2R_0)$, we get

$$\lambda_0 \le R_0\mu_0 < \infty.$$

Furthermore, for $(x,t,s)\in F_1$ the following inequalities hold true:

$$\|u_1-u_0\|_s(x,t) = \|v_0\|_s(x,t) \le \frac{\lambda_0 x}{a_0(s_0-s)-x} \le \frac{\lambda_0 a_1}{a_0-a_1}$$
$$= \frac{\lambda_0}{a_0/a_1-1} = \lambda_0 \le \mu_0 R_0,$$
$$\|q_1-q_0\|_s(x) = \|p_0\|_s(x) \le \frac{\lambda_0 a_0 x}{(a_0(s_0-s)-x)^2} \le \frac{\lambda_0 a_0 a_1}{(a_0-a_1)^2(s_0-s)}$$
$$= \frac{\lambda_0 a_0/a_1}{(a_0/a_1-1)^2(s_0-s)} \le \frac{2\lambda_0}{s_0-s} \le \frac{2\mu_0R_0}{s_0-s}.$$

We choose a_0 so that the inequality $2\mu_0 \le 1$ holds. Then relations (10.4.24) are valid for $m=0$. To demonstrate the validity of (10.4.23), (10.4.24) for $m = 1,2,\dots,$ we apply the method of mathematical induction. Suppose that a_0 is so chosen that (10.4.23), (10.4.24) are fulfilled for $m \le k$. Then for $(x,t,s)\in F_{k+1}$ there holds

$$\|v_{k+1}\|_s(x,t) \le \frac{1}{2}\int_0^x\int_{t-x+\zeta}^{t+x-\zeta}\Big(\frac{4n\|v_k\|_{s(\zeta)}(\zeta,\tau)}{(s(\zeta)-s)^2}$$
$$+\|p_k\|_s(\zeta)\|u_{k+1}\|_s(\zeta,\tau)+\|q_k\|_s(\zeta)\|v_k\|_s(\zeta,\tau\Big)d\tau\,d\zeta$$
$$\le \int_0^x (x-\zeta)\Big(\frac{4n\lambda_k\zeta}{(s(\zeta)-s)^2[a_{k+1}(s_0-s(\zeta))-\zeta]}$$
$$+\frac{2R_0^2\lambda_k a_k\zeta}{[a_{k+1}(s_0-s(\zeta))-\zeta]^2}+\frac{\lambda_k\zeta}{a_{k+1}(s_0-s(\zeta))-\zeta}\,\frac{R_0(1+s_0)}{s_0-s}\Big)d\zeta,$$

where $s(\zeta) = (s_0+s-\zeta/a_{k+1})/2 > s$. Evaluating the last integral, we have

$$\|v_{k+1}\|_s(x,t)$$
$$\le 2\lambda_k\int_0^x \frac{16na_{k+1}^2+4R_0a_ka_{k+1}s_0+R_0(1+s_0)s_0a_{k+1}^2}{(a_{k+1}(s_0-s)-\zeta)^3}(x-\zeta)\zeta\,d\zeta$$
$$\le 2\lambda_k a_0^2(16n+R_0s_0(5+s_0))\,\frac{x}{a_{k+1}(s_0-s)-x}$$

for $(x,t,s) \in F_{k+1}$. We derive similarly

$$\|p_{k+1}\|_s(x) \le \lambda_k a_0^2 R_0(16n + R_0 s_0(5+s_0)) \frac{x a_{k+1}}{(a_{k+1}(s_0-s)-x)^2}$$

at the same values of (x,t,s). Thus,

$$\lambda_{k+1} \le \lambda_k \rho < \infty, \tag{10.4.27}$$

where

$$\rho := a_0(16n + R_0 s_0(5+s_0)) \max(2a_0, R_0).$$

On the other hand, if $(x,t,s) \in F_{k+2}$, then

$$\begin{aligned}
\|u_{k+2} - u_0\|_s(x,t) &= \Big\| \sum_{m=0}^{k+1} v_m \Big\|_s (x,t) \le \sum_{m=0}^{k+1} \|v_m\|_s(x,t) \\
&\le \sum_{m=0}^{k+1} \frac{\lambda_m x}{a_m(s_0-s)-x} \le \sum_{m=0}^{k+1} \frac{\lambda_m a_{k+2}}{a_m - a_{k+2}} < \sum_{m=0}^{k+1} \frac{\lambda_m a_{m+1}}{a_m - a_{m+1}} \\
&= \sum_{m=0}^{k+1} \lambda_m (m+1)^2 \le \lambda_0 \sum_{m=0}^{k+1} \rho^m (m+1)^2,
\end{aligned}$$

$$\begin{aligned}
\|q_{k+2} - q_0\|_s(x) &\le \sum_{m=0}^{k+1} \|p_m\|_s(x,t) \le \sum_{m=0}^{k+1} \frac{\lambda_m a_m x}{(a_m(s_0-s)-x)^2} \\
&\le \frac{1}{s_0-s} \sum_{m=0}^{k+1} \frac{\lambda_m a_m a_{k+2}}{(a_m - a_{k+2})^2} \le \frac{2\lambda_0}{s_0-s} \sum_{m=0}^{k+1} \rho^m (m+1)^4.
\end{aligned}$$

We choose $a_0 \in (0, T/(2s_0))$ so as to provide

$$\rho < 1, \quad 2\mu_0 \sum_{m=0}^{\infty} \rho^m (m+1)^4 \le 1. \tag{10.4.28}$$

Then

$$\|u_{k+2} - u_0\|_s(x,t) \le R_0, \quad \|q_{k+2} - q_0\|_s(x) \le \frac{R_0}{s_0-s}, \quad (x,t,s) \in F_{k+2}.$$

Consequently, (10.4.23), (10.4.24) hold for all $m = 0, 1, 2, \dots$.

Since a_0 is independent of the number of iteration m, approximations u_m and q_m belong to the space $\mathbb{A}_s$ as functions of y for all fixed $(x,t,s) \in F = \bigcap_{m=0}^{\infty} F_m$. Therefore, $u_m \in \mathbb{C}(\mathbb{A}_s, G_s)$, $q_m \in \mathbb{C}(\mathbb{A}_s, G_s)$ and u_m, q_m satisfy the relations

$$\|u_m - u_0\|_s(x,t) \le R_0, \quad \|q_m - q_0\|_s(x) \le \frac{R_0}{s_0-s}, \quad (x,t,s) \in F.$$

The series

$$\sum_{m=0}^{\infty} v_m(x,y,t), \quad \sum_{m=0}^{\infty} p_m(x,y)$$

uniformly converge in the variable y in norm of the space $\mathbb{A}_s$, $s \in (0, s_0)$, and in the variables $(x,t) \in G_s$ in norm of $C(G_s)$. Consequently $u_m \to u$, $q_m \to q$ and the limiting functions u, q belong to $\mathbb{C}(\mathbb{A}_s, G_s)$ for all $s \in (0, s_0)$.

The proof of uniqueness of the solution is similar to that offered earlier for one-dimensional inverse problems and we leave to the reader to make appropriate modifications. □

Let F consist of all pairs $(g(y), f(y,t))$ such that $g(y) \in \mathbb{A}_{s_0}$, $f(y,t) \in \mathbb{C}(\mathbb{A}_{s_0}, [0,T])$ and conditions (10.4.10), (10.4.12) are fulfilled for a fixed R_0. Consider the system of equations (10.4.15), (10.4.16) with the functions $u_0(x,y,t), q_0(x,y)$ defined by (10.4.11). The following theorem establishes the stability of the problem in question (Romanov, 2005).

Theorem 10.4.2. *If, for $(g, f) \in F$ and $(\hat{g}, \hat{f}) \in F$, there exist solutions (u, q) and $(\hat{u}, \hat{q})$ of the corresponding system of equations* (10.4.15), (10.4.16), *then they obey the following inequalities*

$$\|u - \hat{u}\|_s(x,t) \le C\varepsilon, \quad \|q - \hat{q}\|_s(x,t) \le \frac{C\varepsilon}{s_0 - s}, \quad (x,t) \in G_s, \quad s \in (0, s_0), \tag{10.4.29}$$

where

$$\varepsilon = \max[\|g - \hat{g}\|_{s_0}, \|\Delta(g - \hat{g})\|_{s_0}, \max_{0 \le t \le T} \|f - \hat{f}\|_{s_0}(t), \max_{0 \le t \le T} \|f_t - \hat{f}_t\|_{s_0}(t)]$$

and C is a constant depending upon s_0 and R_0 only.

Remark 10.4.1. The above technique may be employed in investigating more general inverse problems and nonlinear integral equations (Romanov, 2005).

10.5 Method of the Neumann to Dirichlet maps in the half-space

Consider the problem of recovering the coefficient $q(x,y)$ of the equation

$$u_{tt} = u_{xx} + \Delta u - q(x,y)u, \quad y \in \mathbb{R}^n, \quad x \in \mathbb{R}_+, \quad t \in [0,T], \tag{10.5.1}$$

provided the behavior of a solution to the associated initial boundary value problem with the data

$$u(x, y, 0) = 0, \quad u_t(x, y, 0) = 0,$$
$$(x, y) \in \mathbb{R}^{n+1}_+ = \{(x, y) \colon y \in \mathbb{R}^n, \ x > 0\}, \tag{10.5.2}$$

$$u_x(0, y, t) = g(y, t), \quad y \in \mathbb{R}^n, \ t \in [0, T], \tag{10.5.3}$$

is observed on the plane $x = 0$:

$$u(0, y, t) = f_g(y, t), \quad y \in \mathbb{R}^n, \ t \in [0, T]. \tag{10.5.4}$$

It was shown in Section 10.4 that if $g(y, t) = -g(y)\delta'(t)$, then $q(x, y)$ is uniquely determined from (10.5.1)–(10.5.4) in the class of functions analytic in y and continuous in x. A success in demonstrating uniqueness of the solution in the class $L^\infty(\mathbb{R}^{n+1}_+)$ (see Definition A.5.6) has been achieved only for overdetermined formulations. Rakesh (1993) adapted the method proposed by J. Sylvester and G. Uhlmann (1987) for solving the inverse problem (10.5.1)–(10.5.4). We will give a brief sketch of the technique (the Sylvester–Uhlmann method will be also used in Section 11.4 for proving the uniqueness of the solution to an inverse problem for elliptic equation).

Define the Neumann–Dirichlet map

$$\Lambda_q : L_2(\mathbb{R}^n \times [0, T]) \to H^1(\mathbb{R}^n \times [0, T]),$$

which to every function $g(y, t) \in L_2(\mathbb{R}^n \times [0, T])$ associates the trace $f_g(y, t) \in H^1(\mathbb{R}^n \times [0, T])$ of the solution of the direct problem (10.5.1)–(10.5.3) on the plane $x = 0$. The function spaces $L_2(G)$ and $H^1(G) = W_2^1(G)$, $G \subset \mathbb{R}^n$, are defined in Appendix A (see Definitions A.5.4 and A.5.9).

Denote by $B(r) \subset \mathbb{R}^{n+1}$ an open ball of radius r centered at the origin:

$$B(r) = \left\{(x, y) : \sqrt{x^2 + \sum_{i=1}^{n} y_i^2} < r\right\}.$$

Theorem 10.5.1. *Suppose that functions q_1, q_2 belong to the space $L^\infty(\mathbb{R}_+ \times \mathbb{R}^n)$ and are constant (possibly different) outside the closed ball $\overline{B(r)}$ with such a radius r that $(\pi + 1)r < T$. Then the identity*

$$\Lambda_{q_1}(g) \equiv \Lambda_{q_2}(g), \quad g \in L_2(\mathbb{R}^n \times [0, T]),$$

implies $q_1 \equiv q_2$.

Proof. We start with claiming that $q_1 \equiv q_2$ if $(x, y) \notin \overline{B(r)}$, i.e., q_1 and q_2 are the same constant outside $\overline{B(r)}$. In other words, we claim that constant coefficients q can be uniquely determined from the Neumann to Dirichlet map. This assertion follows from Lemma 10.5.3 formulated below. So, from here on we assume that $q_1(x, y) = q_2(x, y) = k = \text{const}$ outside $\overline{B(r)}$.

Let $g(y, t)$ be a sufficiently smooth function which has compact support for a fixed $t \in [0, T]$. This implies that the solution to problem (10.5.1)–(10.5.3) will also have compact support for a fixed $t \in [0, T]$.

Denote $p = q_2 - q_1$, $w = u_1 - u_2$, where u_j is the solution to problem (10.5.1)–(10.5.3) with $q(x, y)$ replaced by $q_j(x, y)$, $j = 1, 2$.

From $\Lambda_{q_1}(g) = \Lambda_{q_2}(g)$ it follows that

$$w_{tt} = w_{xx} + \Delta_y w - q_1 w + p u_2, \quad x > 0,\ y \in \mathbb{R}^n,\ t \in [0, T], \tag{10.5.5}$$

$$w(x, y, 0) = 0, \quad w_t(x, y, 0) = 0, \quad x > 0,\ y \in \mathbb{R}^n, \tag{10.5.6}$$

$$w(0, y, t) = 0, \quad w_x(0, y, t) = 0, \quad y \in \mathbb{R}^n,\ t \in [0, T]. \tag{10.5.7}$$

Note that $p(x, y) = 0$ outside $\overline{B(r)}$.

Let $\varepsilon = (T - (\pi + 1)r)/6$, $T_* = T - \pi(r + \varepsilon)$ and $v(x, y, t)$ be a smooth function satisfying

$$\begin{gathered} v_{tt} = v_{xx} + \Delta_y v - q_1(x, y) v, \\ (x, y) \in H = \{(x, y) \in \overline{B(r)} : x > 0\}, \end{gathered} \tag{10.5.8}$$

$$v(x, y, T_*) = 0, \quad v_t(x, y, T_*) = 0, \quad (x, y) \in H. \tag{10.5.9}$$

Then using (10.5.5) and the divergence theorem, we get

$$\begin{aligned} \int_0^{T_*}\int_H p u_2 v \, dx\, dy\, dt &= \int_0^{T_*}\int_H (w_{tt} - w_{xx} - \Delta_y w + q_1 w) v \, dx\, dy\, dt \\ &= \int_0^{T_*}\int_H (v_{tt} - v_{xx} - \Delta_y v + q_1 v) w \, dx\, dy\, dt \\ &\quad + \int_0^{T_*}\int_{\partial H} \left(w \frac{\partial v}{\partial \nu} - v \frac{\partial w}{\partial \nu} \right) dS\, dt \\ &\quad + \int_H (w_t v - w v_t)\Big|_{t=0}^{t=T_*} dx\, dy, \end{aligned}$$

where $\frac{\partial}{\partial \nu}$ denotes the normal derivation to the surface of H.

It can be shown (see (Rakesh, 1993)) that

$$w(x, y, t) = 0, \quad \frac{\partial w}{\partial \nu}(x, y, t) = 0, \quad (x, y) \in \partial H, \quad t \in [0, T_*].$$

Hence,

$$\int_0^{T_*}\int_H p(x, y) u_2(x, y, t) v(x, y, t) \, dx\, dy\, dt = 0 \tag{10.5.10}$$

for all $g \in C^\infty(\mathbb{R}^n \times [0,T])$, for all corresponding solutions u_2 of problem (10.5.1)–(10.5.3) (with $q = q_2$), and for all smooth v satisfying (10.5.8). (10.5.9).

Equality (10.5.10) implies $p(x,y) \equiv 0$. To prove this, we will construct special solutions u_2 and v which are concentrated near a line in the space of variables x, y. The way of construction is described in the following lemma (Rakesh, 1993).

Lemma 10.5.1. *Suppose that $z \in \mathbb{R}^{n+1}$ is an arbitrary unit vector, σ is a positive real number, and $\alpha(x,y) \in C^\infty(\mathbb{R}^{n+1})$ is a finite function with*

$$\operatorname{supp}\{\alpha\} \subset N_\varepsilon = \{(x,y) \in B(2\varepsilon) : x < -\varepsilon\}. \tag{10.5.11}$$

Then for any locally bounded function $q(x,y)$ one can construct functions $v(x,y,t)$ and $u_2(x,y,t)$ of the form

$$\begin{aligned} v(x,y,t) &= \alpha((x,y)+tz)\mathrm{e}^{-i\sigma(\langle (x,y),z\rangle + t)} + \beta_1(x,y,t),\\ u_2(x,y,t) &= \alpha((x,y)+tz)\mathrm{e}^{+i\sigma(\langle (x,y),z\rangle + t)} + \beta_2(x,y,t), \end{aligned} \tag{10.5.12}$$

such that u_2 solves (10.5.1)–(10.5.3) *for a suitable chosen $g(y,t)$, v satisfies* (10.5.8), (10.5.9), *and the following estimates for β_1, β_2 hold:*

$$\|\beta_1\|_{L_2(H\times[0,T_*])} \le \frac{c}{\sigma}, \qquad \|\beta_2\|_{L_2(\mathbb{R}^{n+1}_+\times[0,T])} \le \frac{c}{\sigma},$$

where c is a constant depending only on T, α and q.

As follows from Lemma 10.5.1, we can choose v and u_2 in the form (10.5.12). Using these special solutions in (10.5.10), we obtain

$$\begin{aligned} 0 &= \int_0^{T_*}\int_H p(x,y)u_2(x,y,t)v(x,y,t)\,dx\,dy\,dt\\ &= \int_0^{T_*}\int_H p(x,y)\alpha^2((x,y)+tz)\,dx\,dy\,dt + \gamma(z,\sigma). \end{aligned}$$

The remainder $\gamma(z,\sigma)$ can be proven (Rakesh, 1993) to satisfy

$$|\gamma(z,\sigma)| \le \frac{c}{\sigma},$$

where c is a constant independent of σ. Therefore,

$$\begin{aligned} 0 &= \lim_{\sigma\to\infty}\int_0^{T_*}\int_H p(x,y)u_2(x,y,t)v(x,y,t)\,dx\,dy\,dt\\ &= \int_0^{T_*}\int_H p(x,y)\alpha^2((x,y)+tz)\,dx\,dy\,dt\\ &= \int_{\mathbb{R}^{n+1}}\Big[\int_0^{T_*} p((x,y)-tz)\,dt\Big]\alpha^2(x,y)\,dx\,dy. \end{aligned}$$

Here we have used the fact that $p(x, y) = 0$ outside H.

The class of functions $\alpha^2(x, y)$ satisfying (10.5.11) is complete in $L_2(N_\varepsilon)$. So we conclude that

$$\int_0^{T_*} p(a - tz)\, dt = 0$$

for all $a \in N_\varepsilon$, $z \in \mathbb{R}^{n+1}$, $\|z\| = 1$.

To complete the proof of the theorem, it remains to apply the following two lemmas.

Lemma 10.5.2 (Hamaker et al., 1980). *Suppose that $p(x, y)$ is a bounded function on $\mathbb{R}^{n+1}$ with compact support, and A is an infinite subset of $\mathbb{R}^{n+1}$ bounded away from the convex hull of the support of p. If, for all $a \in A$ and $z \in \mathbb{R}^{n+1}$ such that $\|z\| = 1$, the equality*

$$\int_0^{\infty} p(a + sz)\, ds = 0$$

holds, then $p(x, y) \equiv 0$.

Lemma 10.5.3 (Rakesh, 1993). *Let k_j, $j = 1, 2$, be real constants and functions $u^j(x, y, t)$, $j = 1, 2$, satisfy*

$$u^j_{tt} = u^j_{xx} + \Delta_y u^j - k_j u^j, \quad y \in \mathbb{R}^n,\ x \geq 0,\ t \in [0, T],$$

$$u^j(x, y, 0) = 0, \quad u^j_t(x, y, 0) = 0, \quad y \in \mathbb{R}^n,\ x \geq 0,$$

$$\frac{\partial u^j}{\partial x}(0, y, t) = g(y, t), \quad y \in \mathbb{R}^n,\ t \in [0, T].$$

If $u^1(0, 0, t) = u^2(0, 0, t)$ for $t \in [0, T]$, for all $g \in C^2(\mathbb{R}^n \times [0, T])$ with $\operatorname{supp} g \subset \{y : \|y\| < 2T\} \times [0, T]$, then $k_1 = k_2$.

10.6 An approach to inverse problems of acoustics using geodesic lines

In this section we present an approach to multidimensional inverse problems based on reduction to problems of integral geometry. This technique permits to prove uniqueness and conditional stability theorems for the inverse problems of recovering parameters of a medium inside a domain $D \subset \mathbb{R}^n$ when wave sources and receivers are located on the boundary $S = \partial D$ of this domain. For simplicity sake, we shall restrict our consideration to the three-dimensional acoustic equation

$$u_{tt} = c^2(x)\Delta u - c^2(x)\nabla \ln \rho \nabla u + \delta(x - x^0)\delta(t), \qquad x \in D \subset \mathbb{R}^3, \quad t \in \mathbb{R}, \tag{10.6.1}$$

with the initial data

$$u\big|_{t<0} \equiv 0. \tag{10.6.2}$$

Here $c(x) > 0$ simulates the wave velocity in the medium, $\rho(x) > 0$ denotes medium's density, $\delta(x-x^0) = \delta(x_1-x_1^0)\delta(x_2-x_2^0)\delta(x_3-x_3^0)$, $\delta(\cdot)$ is the Dirac delta-function, $D \subset \mathbb{R}^3$ is an open domain.

Inverse problems for more general equations are studied in (Romanov, 1984, 2005).

As before (Sections 5.4, 5.5), we suppose $D \subset \mathbb{R}^3$ to be an open connected bounded domain with a C^1 smooth boundary $S = \partial D$. Let $\tau(x, x^0)$ be a solution to the problem

$$|\nabla_x \tau(x, x^0)|^2 = c^{-2}(x), \quad x \in D, \tag{10.6.3}$$

provided

$$\tau(x, x^0) = O(|x - x^0|) \quad \text{as} \quad x \to x^0. \tag{10.6.4}$$

It was shown in Section 5.4 that $\tau(x, x^0)$ corresponds to the time a wave takes to run from x^0 to x. On the other hand, the function $\tau(x, x^0)$ determines Riemannian metric with the length element $d\tau = \frac{1}{c(x)}(\sum_{j=1}^{3}(dx_j)^2)^{1/2}$. The geodesic line $\Gamma(x, x^0)$ in this metric coincides with the time-minimum path along which the wave travels from x^0 to x.

10.6.1 An asymptotic expansion of the fundamental solution to the acoustic equation

The fundamental solution to a general second-order hyperbolic equation consists of a singular component and a regular one. The former is concentrated on a characteristic conoid. The support of the regular component lies inside the closed domain bounded by the conoid (Romanov, 1983, 1984). We will study the fundamental solution to the acoustic equation (10.6.1) using the technique proposed by V. G. Romanov (1984).

Definition 10.6.1. A function $U(x, t, x^0, t^0)$, $x, x^0 \in \mathbb{R}^3$, $t, t^0 \in \mathbb{R}$, is said to be a *fundamental solution to the acoustic equation* if

$$U_{tt} - c^2(x)\Delta U + c^2(x)\nabla \ln \rho(x) \nabla U = \delta(x - x^0)\delta(t - t^0), \tag{10.6.5}$$

$$U\big|_{t<0} \equiv 0. \tag{10.6.6}$$

The coefficients of equation (10.6.5) depend upon x only. Therefore,

$$U(x, t, x^0, t^0) = U(x, t - t^0, x^0, 0).$$

Without loss of generality we can put $t^0 = 0$ and denote the fundamental solution by $U(x, t, x^0)$.

Consider the Riemannian metric in $\mathbb{R}^3$

$$d\tau = \frac{1}{c(x)}\Big(\sum_{j=1}^{3}(dx_j)^2\Big)^{1/2}. \tag{10.6.7}$$

Assume that $c(x) \in C^{l+4}(\mathbb{R}^3)$, $\rho(x) \in C^{l+3}(\mathbb{R}^3)$, $l \geq 7$, and the functions $c(x)$, $\rho(x)$ are constant outside D: $c(x) = c_0 > 0$, $\rho(x) = \rho_0 > 0$, $x \in \mathbb{R}^3 \setminus D$. Suppose also that, for any $x, x^0 \in D$, there exists a unique geodesic line $\Gamma(x, x^0) \subset \mathbb{R}^3$ in metric (10.6.7) which connects the points x^0 and x. The distance between x^0 and x in metric (10.6.7) is determined by the function $\tau(x, x^0)$ satisfying (10.6.3), (10.6.4). In (Romanov, 1984) it was shown that, for a fixed x^0, the point x can be specified by Riemannian coordinates $\xi = (\xi_1, \xi_2, \xi_3)$, where $\xi = \tau(x, x^0)\tilde{\xi}$, $\tilde{\xi} = (\tilde{\xi}_1, \tilde{\xi}_2, \tilde{\xi}_3)$ is a vector tangent to $\Gamma(x, x^0)$ at the point x^0 and directed towards x. The length of $\tilde{\xi}$ is governed by the formula

$$\sum_{j=1}^{3} c^{-2}(x^0)\tilde{\xi}_j^2 = 1.$$

Denote $x = h(\xi, x^0)$. The function $h(\xi, x^0)$ is invertible in ξ: $\xi = g(x, x^0) \in C^{l+3}(\mathbb{R}^3 \times \mathbb{R}^3)$. The Jacobian for the change from Riemannian coordinates to Cartesian ones is positive: $|\frac{\partial g}{\partial x}(x, x^0)| > 0$. Moreover, $|\frac{\partial g}{\partial x}(x^0, x^0)| = 1$ and

$$\tau^2(x, x^0) = c^{-2}(x^0)\sum_{j=1}^{3} g_j^2(x, x^0). \tag{10.6.8}$$

Hence, $\tau^2 \in C^{l+3}(\mathbb{R}^3 \times \mathbb{R}^3)$.

Theorem 10.6.1. *If $l \geq 2s + 7$, $s \geq -1$, then the fundamental solution to problem* (10.6.1), (10.6.2) *has the form*

$$U(x, t, x^0) = \theta(t)\sum_{k=-1}^{s} a_k(x, x^0)\theta_k(t^2 - \tau^2(x, x^0)) + U_s(x, t, x^0). \tag{10.6.9}$$

Here $\theta(t)$ is the Heaviside function, $\theta_{-1} = \delta(t)$,

$$\theta_k(t) = \frac{t^k}{k!}\theta(t), \quad k \geq 0,$$

the functions $a_k(x, x^0)$ are calculated using the operator

$$L_x = \frac{\partial^2}{\partial t^2} - c^2(x)\Delta_x + c^2(x)\nabla_x \ln\rho\nabla_x$$

as follows:

$$a_{-1}(x,x^0) = \frac{1}{2\pi c^3(x^0)} \left| \frac{\partial}{\partial x} g(x,x^0) \right|^{1/2} \tag{10.6.10}$$
$$\times \exp\left[\frac{1}{2} \int_{\Gamma(x,x^0)} (c^2(\xi)\nabla_\xi \ln \rho(\xi) \nabla_\xi \tau(\xi,x^0) + \nabla_\xi c^2(\xi) \nabla_\xi \tau(\xi,x^0))\, d\tau\right],$$

$$a_0(x,x^0) = -\frac{a_{-1}(x,x^0)}{8\pi\tau(x,x^0)} \int_{\Gamma(x,x^0)} [a_{-1}(\xi,x^0)]^{-1} L_\xi a_{-1}(\xi,x^0)\, d\tau, \tag{10.6.11}$$

$$a_k(x,x^0) = \frac{a_{-1}(x,x^0)}{8\pi} \int_0^1 [a_{-1}(\xi,x^0)]^{-1} \eta^k L_\xi a_{k-1}(\xi,x^0)\big|_{\xi=h(\eta g(x,x^0),x^0)}\, d\eta,$$
$$k \geq 1.$$

The functions a_k belong to the spaces $C^{l-2k}(\mathbb{R}^3 \times \mathbb{R}^3)$, $U_{-1}(x,t,x^0)$ is piecewise continuous, $U_s(x,t,x^0)$ is continuous for $s \geq 0$ and has continuous partial derivatives $D_x^\alpha D_t^\beta D_{x^0}^\gamma U_s(x,t,x^0)$, $\alpha = (\alpha_1,\alpha_2,\alpha_3)$, $\gamma = (\gamma_1,\gamma_2,\gamma_3)$, for $s \geq 1$, $|\alpha|+|\gamma| < l-2s-3$, $|\alpha|+\beta+|\gamma| \leq s-1$, $|\alpha| = \alpha_1+\alpha_2+\alpha_3$, $|\gamma| = \gamma_1+\gamma_2+\gamma_3$. Moreover, the function $U_s(x,t,x^0)$ $(s \geq -1)$ is representable in the form

$$U_s(x,t,x^0) = \theta_{s+1}(t^2 - \tau^2(x,x^0))\tilde{U}_s(x,t,x^0),$$

where $\tilde{U}_s$ is a function bounded in any closed domain.

The proof of this theorem can be found in (Romanov, 1984, p. 125–134).

10.6.2 A three-dimensional inverse problem for the acoustic equation

Consider the direct problem of acoustics

$$\begin{gathered} u_{tt} = c^2(x)\Delta u - c^2(x)\nabla \ln \rho(x) \nabla u + \delta(x-x^0)\delta(t), \\ x \in D \subset \mathbb{R}^3, \quad t \in \mathbb{R}, \end{gathered} \tag{10.6.12}$$

$$u\big|_{t<0} \equiv 0 \tag{10.6.13}$$

and assume that the coefficients $c(x)$, $\rho(x)$ are unknown and an information about the solution of (10.6.12), (10.6.13) is given on the boundary $S = \partial D$:

$$u(x,t,x^0) = f(x,t,x^0), \quad x,x^0 \in S, \quad -\infty < t < \tau(x,x^0)+\varepsilon, \quad \varepsilon > 0. \tag{10.6.14}$$

We first study the problem of finding the wave velocity $c(x)$ from relations (10.6.12)–(10.6.14) with the aim of formulating conditions which will provide the uniqueness and conditional stability to its solution. It is remarkable that this problem is closely connected with the inverse kinematic problem of seismics (see Section 5.4). Indeed, let $x^0 \in S$ be fixed. If $t < \tau(x, x^0)$, then the wave has not reached the point x yet and, hence, the solution of problem (10.6.12), (10.6.13) is equal to zero. Starting with $t = \tau(x, x^0)$, the solution becomes nonzero and depends, generally speaking, upon the coefficients of equation (10.6.12). Thus, the function $\tau(x, x^0)$ is correlated with the solution of problem (10.6.12), (10.6.13). Consequently, observing behavior of the solution on the boundary S for different $x^0 \in S$, we get information about the travel times $\tau(x, x^0)$ of a signal from one point of the boundary S to another. We come thereby to the inverse kinematic problem of seismics which consists in restoring the wave velocity $c(x)$ and the associated family of time-minimum paths $\Gamma(x, x^0)$, $x, x^0 \in D$, from the equation

$$\int_{\Gamma(x,x^0)} \frac{|dy|}{c(y)} = \tau(x, x^0), \quad x, x^0 \in S. \tag{10.6.15}$$

So, let D be a simply connected bounded domain with a smooth boundary S described by the equation $F(x) = 0$, where the function $F \in C^3(\overline{D})$ is such that $F(x) < 0$, $x \in D$, and $|\nabla_x F|\big|_{x \in S} = 1$. Assume that the surface S is convex, i.e., $\Gamma(x, x^0) \subset S$ for all $x, x^0 \in S$, and that the family of paths $\Gamma(x, x^0)$ is regular in $\overline{D}$.

Denote by $\Lambda(C_1, C_2)$ a set of functions $c(x) \in C^2(D)$ which, for fixed constants C_1, C_2, satisfy the following conditions:

1) $0 < C_1 \le c^{-1}(x)$ for all $x \in D$;

2) $\|c^{-1}\|_{C^2(D)} \le C_2 < \infty$;

3) the family of paths $\Gamma(x, x^0)$ associated with the function $c(x)$ is regular in D.

Using the eikonal equation with a given function $\tau(x, x^0)$, $(x, x^0) \in S \times S$, we can calculate partial derivatives τ_{x_i}, $\tau_{x_i^0}$, $\tau_{x_i x_j^0}$, $i, j = \overline{1, 3}$, on the set $S \times S$. Indeed, the derivatives along tangent vectors can be expresses in terms of the function $\tau(x, x^0)$, while for the normal derivatives we have

$$|\nabla_x \tau(x, x^0)| = \frac{1}{c(x)}, \quad |\nabla_x \tau(x^0, x)| = \frac{1}{c(x^0)}.$$

To estimate the conditional stability of the problem of finding the velocity $c(x)$ from (10.6.15), we introduce the norm

$$\|\tau\|^2 = \int_S \int_S \Big[|\nabla_x \tau(x, x^0)|^2 + |\nabla_{x^0} \tau(x, x^0)|^2 + |x - x^0|^2 \sum_{i,j=1}^{3} \Big(\frac{\partial^2 \tau(x, x^0)}{\partial x_i \, \partial x_j^0} \Big)^2 \Big] \frac{dS_x \, dS_{x^0}}{|x - x^0|}.$$

Theorem 10.6.2 (on conditional stabilty). *For any functions $c_{(1)}, c_{(2)} \in \Lambda(C_1, C_2)$ and the corresponding functions $\tau_{(1)}(x, x^0)$, $\tau_{(2)}(x, x^0)$, the following estimate holds:*

$$\|c_{(1)} - c_{(2)}\|^2_{L_2(D)} \leq C\,\|\tau_{(1)} - \tau_{(2)}\|^2.$$

Here C is a constant depending on C_1 and C_2 only.

A proof of the theorem can be found in (Romanov, 1984).

We now turn to the discussion of the problem of recovering the density $\rho(x)$ from relations (10.6.12)–(10.6.14). We suppose that the conditions guaranteeing the conditional stability of the problem of finding $c(x)$ are fulfilled. The positive-valued functions $c(x)$, $\rho(x)$ are assumed to be smooth enough in $D \subset \mathbb{R}^3$ and constant outside D: $c(x) = c_0 > 0$, $\rho(x) = \rho_0 > 0$ for $x \in \mathbb{R}^3 \setminus D$.

In what follows we will use Theorem 10.6.1 on asymptotic expansion of the fundamental solution. We begin by simplifying formula (10.6.10) for $a_{-1}(x, x^0)$.

Lemma 10.6.1. *Let $c(x)$, $\rho(x)$ be positive-valued functions such that $c(x) \in C^{l+4}(\mathbb{R}^3)$, $\rho(x) \in C^{l+3}(\mathbb{R}^3)$, $l \geq 7$. Then, for $x, x^0 \in \mathbb{R}^3$, there holds*

$$a_{-1}(x, x^0) = A(x, x^0)\exp(R(x)/2), \tag{10.6.16}$$

where

$$\begin{aligned} R(x) &= \ln \rho(x), \\ A(x, x^0) &= \frac{1}{2\pi c^3(x^0)}\left|\frac{\partial g(x, x^0)}{\partial x}\right|^{1/2} \exp\Big(-\frac{R(x^0)}{2}\Big)\frac{c(x)}{c(x^0)}. \end{aligned} \tag{10.6.17}$$

Proof. Since $c(x) \in C^{l+4}(\mathbb{R}^3)$, $l \geq 7$, the function $|\partial g(x, x^0)/\partial x|^{1/2}$ belongs to $C^{l+3}(\mathbb{R}^3)$. Taking into account that along the geodesic $\Gamma(x, x^0)$ in metric (10.6.7) there hold

$$\begin{aligned} \frac{1}{2}c^2(\xi)\nabla_\xi \ln \rho(\xi)\nabla_\xi \tau(\xi, x^0) &= \frac{d}{d\tau}\,[\ln\sqrt{\rho(\xi)}\,], \\ c^2(\xi)\nabla_\xi \ln c^2(\xi)\nabla_\xi \tau(\xi, x^0) &= \frac{d}{d\tau}\,[\ln c^2(\xi)], \end{aligned}$$

we come to the equality

$$\begin{aligned} &\frac{1}{2}\int_{\Gamma(x,x^0)} [c^2(\xi)\nabla_\xi \ln \rho(\xi)\nabla_\xi \tau(\xi, x^0) + \nabla_\xi c^2(\xi)\nabla_\xi \tau(\xi, x^0)]\,d\tau \\ &\quad = \ln\Big[\sqrt{\frac{\rho(x)}{\rho(x^0)}\,\frac{c(x)}{c(x^0)}}\Big]. \end{aligned} \tag{10.6.18}$$

Combining (10.6.18) with (10.6.10) results in the formula

$$a_{-1}(x, x^0) = \frac{1}{2\pi c^3(x^0)} \left| \frac{\partial g(x, x^0)}{\partial x} \right|^{1/2} \sqrt{\frac{\rho(x)}{\rho(x^0)} \frac{c(x)}{c(x^0)}},$$

which, in view of (10.6.17), takes the form

$$a_{-1}(x, x^0) = A(x, x^0) \exp(R(x)/2).$$

So, the lemma is proven. □

Using Theorem 10.6.1, we can reformulate the problem of recovering $\rho(x)$ as a problem of integral geometry. For this purpose we express the integrand $\frac{L_x a_{-1}(x,x^0)}{a_{-1}(x,x^0)}$ in (10.6.11) through the functions $R(x)$ and $A(x, x^0)$. By definition of L_x,

$$L_x a_{-1}(x, x^0) = -c^2(x)\Delta_x a_{-1}(x, x^0) + c^2(x)\nabla_x R(x)\nabla_x a_{-1}(x, x^0). \quad (10.6.19)$$

From (10.6.16) it follows that

$$\begin{aligned} \Delta_x a_{-1}(x, x^0) &= [\Delta_x A(x, x^0) + \nabla_x A(x, x^0)\nabla_x R(x) \\ &\quad + (\Delta_x R(x)/2 + |\nabla_x R(x)|^2/4)A(x, x^0)] \exp(R(x)/2), \\ \nabla_x a_{-1}(x, x^0) &= \nabla_x A(x, x^0) \exp(R(x)/2) \\ &\quad + (1/2)A(x, x^0) \exp(R(x)/2)\nabla_x R(x). \end{aligned}$$

These equalities combined with (10.6.16) imply

$$\begin{aligned} \frac{\Delta_x a_{-1}(x, x^0)}{a_{-1}(x, x^0)} &= \frac{\Delta_x A(x, x^0)}{A(x, x^0)} + \frac{\Delta_x R(x)}{2} \\ &\quad + \nabla_x \ln A(x, x^0)\nabla_x R(x) + \frac{1}{4}|\nabla_x R(x)|^2, \end{aligned} \quad (10.6.20)$$

$$\frac{\nabla_x a_{-1}(x, x^0)}{a_{-1}(x, x^0)} = \nabla_x \ln A(x, x^0) + \frac{\nabla_x R(x)}{2}. \quad (10.6.21)$$

From (10.6.19)–(10.6.21) follows the desired expression for $\frac{L_x a_{-1}(x,x^0)}{a_{-1}(x,x^0)}$:

$$\frac{L_x a_{-1}(x, x^0)}{a_{-1}(x, x^0)} = \frac{c^2(x)}{2}\Big[- \Delta_x R(x) + \frac{|\nabla_x R(x)|^2}{2} - \frac{2\Delta_x A(x, x^0)}{A(x, x^0)}\Big]. \quad (10.6.22)$$

Substituting (10.6.22) into (10.6.11) results in

$$\int_{\Gamma(x,x^0)} \frac{c^2(\xi)}{2}\Big[\Delta_\xi R(\xi) - \frac{1}{2}|\nabla_\xi R(\xi)|^2\Big]\, d\tau = W(x, x^0), \quad x, x^0 \in S, \quad (10.6.23)$$

where

$$W(x,x^0) = 8\pi\tau(x,x^0)\,\frac{a_0(x,x^0)}{a_{-1}(x,x^0)} - \int_{\Gamma(x,x^0)} c^2(\xi)\,\frac{\Delta_\xi A(\xi,x^0)}{A(\xi,x^0)}\,d\tau. \quad (10.6.24)$$

Let us show that the function $W(x,x^0)$ is uniquely determined by a solution $c(x)$ of (10.6.15). Assume that the solution $c(x) \in C^{l+4}(\mathbb{R}^3)$ is known. Then we can uniquely determine $\tau(x,x^0)$ in $\mathbb{R}^3 \times S$. The functions $A(x,x^0)$ and $a_{-1}(x,x^0)$ can be found in the domains $\mathbb{R}^3 \times S$ and $S \times S$, respectively, using Lemma 10.6.1.

From (10.6.9) it follows that

$$a_0(x,x^0) = \lim_{t\to\tau(x,x^0)+0} [f(x,t,x^0) - a_{-1}(x,x^0)\delta(t^2 - \tau^2(x,x^0))],$$
$$x \in S,\ x^0 \in S.$$

This formula is correct because the term in square brackets is a function continuous for $t \geq \tau(x,x^0)$. From this formula it follows that, given $f(x,t,x^0)$ for $x, x^0 \in S$, $\tau(x,x^0) > 0$, we can specify the function $a_0(x,x^0)$ in $S \times S$. Thus, the function $W(x,x^0)$ defined by (10.6.24) is uniquely determined in $S \times S$ from $c(x)$ satisfying (10.6.15). Moreover, the family of geodesics $\Gamma(x,x^0)$, $x, x^0 \in S$, in metric (10.6.7) is also uniquely recovered from the function $c(x)$, $x \in D$.

Turning to equality (10.6.23), we now face a problem of integral geometry which consists in finding the function

$$H(x) \equiv \frac{c^2(x)}{2}\Big[\Delta_x R(x) - \frac{1}{2}\,|\nabla_x R(x)|^2\Big]$$

provided the values $W(x,x^0)$ of integrals of $H(x)$ along curves from a given family $\{\Gamma(x,x^0)\}$ are known,

The problem of integral geometry

$$\int_{\Gamma(x,x^0)} H(\xi)\,|d\xi| = W(x,x^0),$$

$$|d\xi| = \Big(\sum_{j=1}^{3}(d\xi_j)^2\Big)^{1/2}, \quad x, x^0 \in S,$$

where $W(x,x^0)$ is a known function for $x, x^0 \in S$, and $H(x)$ is to be found in $\overline{D}$, is discussed in (Romanov, 1984). The following theorem gives an estimate of stability for this problem.

Theorem 10.6.3. *Let D be a simply connected, bounded domain with a boundary S described by the equation $F(x) = 0$, where $F(x) \in C^3(\overline{D})$, $F(x) < 0$ for $x \in D$, $|\nabla_x F(x)|\big|_{x\in S} = 1$. Assume that S is convex with respect to geodesics $\Gamma(x,\alpha)$ and* *the family* $\{\Gamma(x,\alpha)\}$ *is regular in* $\overline{D}$. *Then for the above problem of integral geometry, under the assumption that*

$$c(x) \in C^2(\overline{D}), \quad 1/c(x) \geq C_1 > 0, \quad H(x) \in C^1(\overline{D}),$$

the following estimate of stability holds:

$$\int_D H^2(x)\,dx \leq \frac{1}{8\pi C_1}\int_S\int_S \Phi(W,\tau)(x,x^0)\,dS_x\,dS_{x^0},$$

where

$$\Phi(W,\tau)(x,x^0) = \begin{vmatrix} 0 & 0 & W_{x_1}(x,x^0) & W_{x_2}(x,x^0) & W_{x_3}(x,x^0) \\ 0 & 0 & F_{x_1}(x) & F_{x_2}(x) & F_{x_3}(x) \\ W_{x_1^0}(x,x^0) & F_{x_1^0}(x^0) & \tau_{x_1x_1^0}(x,x^0) & \tau_{x_2x_1^0}(x,x^0) & \tau_{x_3x_1^0}(x,x^0) \\ W_{x_2^0}(x,x^0) & F_{x_2^0}(x^0) & \tau_{x_1x_2^0}(x,x^0) & \tau_{x_2x_2^0}(x,x^0) & \tau_{x_3x_2^0}(x,x^0) \\ W_{x_3^0}(x,x^0) & F_{x_3^0}(x^0) & \tau_{x_1x_3^0}(x,x^0) & \tau_{x_2x_3^0}(x,x^0) & \tau_{x_3x_3^0}(x,x^0) \end{vmatrix}.$$

Note that the function $\Phi(W,\tau)$ involves the derivatives of $W(x,x^0)$ with respect to x, x^0 only in directions tangent to the surface S. The proof of this theorem is presented in (Romanov, 1984, p. 95–101).

So, the problem of recovering the coefficient $\rho(x)$ can be solved successively: first we find $H(x)$, $x \in \overline{D}$, satisfying (10.6.23) with a given $W(x,x^0)$, $x, x^0 \in S$; then we solve the Dirichlet problem

$$\Delta_x R(x) - |\nabla_x R(x)|^2/2 = 2H(x)/c^2(x), \tag{10.6.25}$$

where the function $R(x) \in C^{l+3}(D)$ such that $R(x) = \ln\rho_0$ for $x \in S$ is to be determined. The unique solvability of the Dirichlet problem can be demonstrated for domains of small diameters using the technique presented in (Courant and Hilbert, 1953, 1962). The questions concerning uniqueness and stability of solutions to problems for the quasilinear equation (10.6.25) can be resolved in a similar way as those for the linear elliptic equation.

Let us estimate stability of the solution to the Dirichlet problem (10.6.25).

Introduce a class of functions

$$\mathcal{R}(M,\rho_0) = \{R(x) \in C^{11}(\overline{D}) : \|R\|_{C^1(\overline{D})} \leq M,\ R(x) = \ln\rho_0 \text{ for } x \in S\},$$

where M is an arbitrary fixed number.

Let $R_1(x), R_2(x) \in \mathcal{R}(M,\rho_0)$ solve (10.6.25) for $H(x) = H_1(x)$ and $H(x) = H_2(x)$, respectively. Denote

$$\tilde{R}(x) = R_1(x) - R_2(x), \quad \tilde{H}(x) = H_1(x) - H_2(x),$$

$$\rho_j(x) = \exp\{R_j(x)\}, \quad j = 1,2, \quad a(x) = 1/\sqrt{\rho_1(x)\rho_2(x)}.$$

These functions satisfy the relations

$$\nabla_x\big(a(x)\nabla_x \tilde{R}(x)\big) = \frac{2a(x)}{c^2(x)}\,\tilde{H}(x), \quad \tilde{R}\big|_{x\in S} = 0. \tag{10.6.26}$$

For the Dirichlet problem (10.6.26) the following estimate is valid:

$$\|\tilde{R}\|^2_{L_2(D)} \le C\|\tilde{H}\|_{L_2(D)}, \tag{10.6.27}$$

where C is a constant depending on M, ρ_0, diam D, and c_0.

From Theorem 10.6.3, in view of (10.6.27), follows the stability of the inverse problem solution.

Theorem 10.6.4. *Let M, ρ_0, c_0 be fixed positive numbers such that $\rho_0 < \exp\{M\}$. Assume $c(x) \in C^{11}(\overline{D})$ to be a known positive function equal to c_0 for $x \in S$. Assume also that $\rho_1(x), \rho_2(x) \in \mathcal{M}(M, \rho_0) = \{\rho(x) : \rho(x) = \exp\{R(x)\},\ R(x) \in \mathcal{R}(M, \rho_0)\}$ solve the inverse problem* (10.6.12)–(10.6.14) *with $f(x,t,x^0)$ replaced by $f_1(x,t,x^0)$ and $f_2(x,t,x^0)$, respectively. Then there holds*

$$\|\ln\rho_1 - \ln\rho_2\|^2_{L_2(D)} \le C\int_S\int_S |\Phi(\tilde{W},\tau)|\,dS_x\,dS_{x^0},$$

where C is a constant depending on M, ρ_0, c_0, diam D*;*

$$\tilde{W}(x,x^0) = \frac{4\tau(x,x^0)c^3(x^0)}{|\partial g(x,x^0)/\partial x|^{1/2}} \lim_{t\to\tau(x,x^0)+0} [f_1(x,t,x^0) - f_2(x,t,x^0)];$$

$\Phi(W,\tau)$ is such as in Theorem 10.6.3*; the functions $g(x,x^0)$, $\tau(x,x^0)$ were defined just below formula* (10.6.7).

10.7 Two-dimensional analog of the Gelfand–Levitan–Krein equation

In this section we construct a multidimensional analog of the Gelfand–Levitan–Krein equation for an inverse problem of acoustics. We consider the problem of recovering medium's density under the assumption that on the boundary of the domain under study we can trace signals coming from an infinite number of sources. This assumption means that, as in Section 10.5, we need much more information than is usually required for demonstrating the solution uniqueness. The presented approach based on reduction of the nonlinear problem to a system of linear equations enables one to compute the desired coefficients without solving repeatedly the direct problem. The derived formulae are justified only in the class of functions analytic in y. However, as numerical experiments evidence, the sought-for function can be closely approximated using formula (10.7.13).

Consider a sequence of direct problems $\mathcal{A}_\rho u^{(k)}$, $k \in \mathbb{Z}$, such that

$$\mathcal{A}_\rho u^{(k)} = \Big[\frac{\partial^2}{\partial t^2} - \Delta_{x,y} + \nabla_{x,y} \ln \rho(x,y) \nabla_{x,y}\Big] u^{(k)} = 0, \\ x > 0, \quad y \in [-\pi, \pi], \quad t \in \mathbb{R}, \quad k \in \mathbb{Z}; \tag{10.7.1}$$

$$u^{(k)}\big|_{t<0} \equiv 0; \tag{10.7.2}$$

$$\frac{\partial u^{(k)}}{\partial x}(+0, y, t) = \mathrm{e}^{\mathrm{i}ky} \cdot \delta(t); \tag{10.7.3}$$

$$u^{(k)}\big|_{y=\pi} = u^{(k)}\big|_{y=-\pi}. \tag{10.7.4}$$

From now on we suppose that all the functions involved are smooth enough and 2π-periodic in y. Also, we assume the trace of a solution to (10.7.1)–(10.7.4) on the plane $x = 0$ exists and can be fixed:

$$u^{(k)}(0, y, t) = f^{(k)}(y, t). \tag{10.7.5}$$

The inverse problem is to reconstruct the coefficient $\rho(x, y)$ from relations (10.7.1)–(10.7.5). The values of $\rho(0, y)$ are assumed to be known.

The necessary condition for solvability of the inverse problem (10.7.1)–(10.7.5) is formulated as

$$f^{(k)}(y, +0) = -\mathrm{e}^{\mathrm{i}ky}, \quad y \in [-\pi, \pi], \quad k \in \mathbb{Z}.$$

Define an auxiliary sequence of direct problems as follows:

$$\mathcal{A}_\rho w^{(m)} = 0, \quad x > 0, \quad y \in [-\pi, \pi], \quad t \in \mathbb{R}, \quad m \in \mathbb{Z}; \tag{10.7.6}$$

$$w^{(m)}(0, y, t) = \mathrm{e}^{\mathrm{i}my} \cdot \delta(t), \quad \frac{\partial w^{(m)}}{\partial x}(0, y, t) = 0, \\ w^{(m)}\big|_{y=\pi} = w^{(m)}\big|_{y=-\pi}. \tag{10.7.7}$$

It is not difficult to verify that the solution to problem (10.7.6), (10.7.7) has the form

$$w^{(m)}(x, y, t) = S^{(m)}(x, y)[\delta(x + t) + \delta(x - t)] + \tilde{w}^{(m)}(x, y, t), \tag{10.7.8}$$

where $\tilde{w}^{(m)}(x, y, t)$ is a piecewise smooth function and

$$S^{(m)}(x, y) = \frac{1}{2}\sqrt{\frac{\rho(x, y)}{\rho(0, y)}}\,\mathrm{e}^{\mathrm{i}my}.$$

Having extended the functions $u^{(k)}(x, y, t)$ and $f^{(k)}(y, t)$ oddly in t, we come to the evident equality

$$u^{(k)}(x, y, t) = \int_{\mathbb{R}} \sum_{m=1}^{\infty} f_m^{(k)}(t - s) w^{(m)}(x, y, s)\, ds, \\ x > 0,\ y \in \mathbb{R},\ t \in \mathbb{R},\ k \in \mathbb{Z}. \tag{10.7.9}$$

Here $f_m^{(k)}(t)$ are the Fourier coefficients of $f^{(k)}(y, t)$.

Denote

$$V^m(x,y,t) = \int_0^x \frac{w^{(m)}(\xi,y,t)}{\rho(\xi,y)}\,d\xi. \tag{10.7.10}$$

The manipulation

$$\frac{\partial}{\partial t}\int_0^x \frac{(\,\cdot\,)}{\rho(\xi,y)}\,d\xi$$

converts equality (10.7.9) into

$$\begin{aligned}
\frac{\partial}{\partial t}\int_0^x \frac{u^{(k)}(\xi,y,t)}{\rho(\xi,y)}\,d\xi &= \frac{\partial}{\partial t}\int_{\mathbb{R}}\sum_{m=1}^{\infty} V^m(x,y,s) f_m^{(k)}(t-s)\,ds \\
&= 2\sum_{m=1}^{\infty} f_m^{(k)}(+0)V^m(x,y,t) + \sum_{m=1}^{\infty}\int_{-x}^{x} f_m^{(k)\prime}(t-s)V^m(x,y,s)\,ds \\
&= -2V^k(x,y,t) + \sum_{m=1}^{\infty}\int_{-x}^{x} f_m^{(k)\prime}(t-s)V^m(x,y,s)\,ds.
\end{aligned}$$

Introduce a function G^k for $x > |t|$ assuming

$$G^k(x,t) = \frac{\partial}{\partial t}\int_{-\pi}^{\pi}\int_0^x \frac{u^{(k)}(\xi,y,t)}{\rho(\xi,y)}\,d\xi\,dy.$$

Evidently, if $u^{(k)}(x,y,t)$, $\rho(x,y)$ satisfy (10.7.1)–(10.7.5), then

$$G_x^k(x,t) = 0, \quad G_t^k(x,t) = 0, \quad |t| < x.$$

Therefore, $G^k(x,t)$ is a constant for $|t| < x$. Let us find its value. Put $u^{(k)}(x,y,t) = u^{(k)}(-x,y,t)$, $\rho(x,y) = \rho(-x,y)$ for $x < 0$. Then $u^{(k)}(x,y,t)$ solves the following problem:

$$\mathcal{A}_\rho u^{(k)} = 0, \quad x \in \mathbb{R}, \quad y \in [-\pi,\pi], \quad t > 0; \quad k \in \mathbb{Z};$$

$$u^{(k)}\big|_{t=0} = 0; \quad \frac{\partial u^{(k)}}{\partial t}\bigg|_{t=0} = -2e^{iky}\delta(x), \quad u^{(k)}\big|_{y=\pi} = u^{(k)}\big|_{y=-\pi}.$$

Consequently,

$$\begin{aligned}
\lim_{t\to+0} G^k(x,t) &= \lim_{t\to+0}\int_{-\pi}^{\pi}\int_0^x \frac{u_t^{(k)}(\xi,y,t)}{\rho(\xi,y)}\,d\xi\,dy \\
&= \frac{1}{2}\lim_{t\to+0}\int_{-\pi}^{\pi}\int_{-x}^{x} \frac{u_t^{(k)}(\xi,y,t)}{\rho(\xi,y)}\,d\xi\,dy = -\int_{-\pi}^{\pi}\frac{e^{iky}}{\rho(0,y)}\,dy, \quad k \in \mathbb{Z}.
\end{aligned}$$

Let

$$\Phi^m(x,t) = \int_{-\pi}^{\pi} V^m(x,y,t)\,dy.$$

Then the functions $\Phi^m(x,t)$ satisfy the following system of integral equations

$$\Phi^k(x,t) = \frac{1}{2}\sum_{m=1}^{\infty}\int_{-x}^{x} f_m^{(k)\prime}(t-s)\Phi^m(x,s)\,ds - \frac{1}{2}\int_{-\pi}^{\pi}\frac{e^{iky}}{\rho(0,y)}\,dy,$$
$$|t|<x, \quad k\in\mathbb{Z}. \quad (10.7.11)$$

System (10.7.11) is a multidimensional analog of the Gelfand–Levitan–Krein equation. Using (10.7.7) and (10.7.10), we evaluate $V^m(x,y,t)$ for $t = x$:

$$V^m(x,y,x-0) = \frac{e^{imy}}{2\sqrt{\rho(x,y)\rho(0,y)}}. \quad (10.7.12)$$

Hence,

$$\Phi^m(x,x-0) = -\int_{-\pi}^{\pi}\frac{e^{imy}}{2\sqrt{\rho(x,y)\rho(0,y)}}\,dy.$$

The solution to the inverse problem (10.7.1)–(10.7.5) can be found by the formula

$$\frac{1}{\sqrt{\rho(x,y)}} = -\frac{\sqrt{\rho(0,y)}}{\pi}\cdot\sum_{m=1}^{\infty}\Phi^m(x,x-0)\cdot e^{-imy}. \quad (10.7.13)$$

So, for the solution $\rho(x,y)$ of the inverse problem (10.7.1)–(10.7.5) to be specified at a point $x_0 > 0$, it is required to solve system (10.7.11) with a fixed parameter $x = x_0$ and then calculate $\rho(x_0,y)$ using formula (10.7.13).

It is evident that this approach may be generalized to the case of space variables $y\in\mathbb{R}^n$, $n\in\mathbb{N}$.

Remark 10.7.1. The problem of recovering the coefficient of the equation $u_{tt} = \Delta u - q(x)u$ (see (Kabanikhin et al., 2004)) can be similarly reduced to a system of Gelfand–Levitan equations.

Chapter 11
Inverse coefficient problems for parabolic and elliptic equations

Inverse coefficient problems for parabolic and elliptic equations are the most complicated and practically important problems. As shown is Section 8.1, the problems are less stable than the corresponding inverse problems for hyperbolic equations. Therefore, optimization technique is mostly used for their numerical solutions, as well as its various modifications tailored to the properties of the corresponding direct problems (monotonicity or/and smoothness of their solutions, etc.).

An extensive literature deals with the problems of recovering coefficients of parabolic equations. We only mention three books that represent the general theory (Lavrentiev et al., 1980), numerical methods and applications to problems of heat-and-mass transfer (Alifanov et al., 1988), and applied problems (Beck et al., 1985).

We consider one-dimensional inverse coefficient problems for parabolic equations in Sections 11.1–11.3. Note that many results presented in these sections can be extended to the multidimensional case, and even to quasi-linear equations. Sections 11.4 and 11.5 contain theorems on the uniqueness of the solution of multidimensional inverse problems for elliptic equations.

Methods developed to solve the coefficient problems for heat equation are used in practice, for example, to reveal the dependence of thermophysical properties of ingot during the steel hardening. Another practical problem, which leads to the considered problems, consists in determining the characteristics of thermal protection materials collapsing in contact with high-temperature gas. Similar problems arise in geophysics in the determination of thermophysical properties of freezing (thawing) soils. The presented approach is also applied to problems of geoelectrics, geothermics and many others.

11.1 Formulation of inverse coefficient problems for parabolic equations. Association with those for hyperbolic equations

Consider an initial boundary value problem for the parabolic equation

$$u_t - (k(x)u_x)_x + q(x)u = 0, \quad t \in (0, T), \ x \in (0, 1), \tag{11.1.1}$$

$$u(x, 0) = u_0(x), \quad x \in (0, 1), \tag{11.1.2}$$

$$k(0)u_x(0,t) = \alpha_1(t), \quad t \in (0,T), \tag{11.1.3}$$

$$k(1)u_x(1,t) = \alpha_2(t), \quad t \in (0,T), \tag{11.1.4}$$

and assume that one of the coefficient $k(x)$, $q(x)$ is unknown. Given information about the solution to the direct problem (11.1.1)–(11.1.4) at $x = 0$ and $x = 1$, namely,

$$u(0,t) = f_1(t), \quad t \in [0,T], \tag{11.1.5}$$

$$u(1,t) = f_2(t), \quad t \in [0,T], \tag{11.1.6}$$

it is required to restore the unknown coefficient. In the case when both the coefficients are unknown, only a combination of them can be found under certain conditions.

One of the approaches to studying inverse problems for parabolic equations is based on relating them to inverse problems for hyperbolic ones. This method can be applied to a wide range of inverse problems with data given on a part of the boundary, including the inverse problem for Maxwell's equations in quasi-steady approximation (Romanov and Kabanikhin, 1994).

Consider the following inverse coefficient problem for the hyperbolic equation:

$$v_{tt} - (k(x)v_x)_x + q(x)v = 0, \quad x \in [0,1],\ t > 0, \tag{11.1.7}$$

$$v(x,0) = u_0(x), \quad v_t(x,0) = 0, \tag{11.1.8}$$

$$v(0,t) = g_1(t), \quad k(0)v_x(0,t) = h_1(t), \tag{11.1.9}$$

$$v(1,t) = g_2(t), \quad k(1)v_x(1,t) = h_2(t). \tag{11.1.10}$$

Introduce

$$G(t,\tau) = \frac{1}{\sqrt{\pi t}}\, \mathrm{e}^{-\tau^2/(4t)}$$

and suppose that the data of problems (11.1.7)–(11.1.10) and (11.1.1)–(11.1.6) are connected by relations

$$f_j(t) = \int_0^\infty g_j(\tau)G(t,\tau)d\tau, \quad j = 1,2, \tag{11.1.11}$$

$$\alpha_j(t) = \int_0^\infty h_j(\tau)G(t,\tau)d\tau, \quad j = 1,2. \tag{11.1.12}$$

In this case, if $v(x,t)$ satisfies (11.1.7)–(11.1.10), then the function

$$u(x,t) = \int_0^\infty v(x,t)G(t,\tau)\,d\tau \tag{11.1.13}$$

obeys (11.1.1)–(11.1.6). A proof of the fact can be found in Section 8.1.

Formula (11.1.13) establishes a one-to-one correspondence between the solutions to problems (11.1.1)–(11.1.6) and (11.1.7)–(11.1.10). Hence, the uniqueness theorems proven for the inverse coefficients problems for hyperbolic equations can be extended to the corresponding inverse problems for parabolic ones.

In particular, the problem of finding the coefficient $q(x)$ in the Cauchy problem

$$u_t = u_{xx} - q(x)u, \tag{11.1.14}$$

$$u\big|_{t=0} = \delta(x) \tag{11.1.15}$$

provided

$$u\big|_{x=0} = f_1(t), \quad u_x\big|_{x=0} = f_2(t) \tag{11.1.16}$$

is reduced to the following inverse problem:

$$\begin{gathered} \tilde{u}_{tt} = \tilde{u}_{xx} - q(x)\tilde{u}, \\ \tilde{u}\big|_{t=0} = \delta(x), \quad \tilde{u}_t\big|_{t=0} = 0, \quad \tilde{u}\big|_{x=0} = \tilde{f}_1(t), \quad \tilde{u}_x\big|_{x=0} = \tilde{f}_2(t). \end{gathered} \tag{11.1.17}$$

Here the functions $\tilde{f}_k$ are connected with f_k by the formulae

$$f_k(t) = \int_0^\infty G(t,\tau)\tilde{f}_k(\tau)\,d\tau, \quad k = 1, 2.$$

Note that problem (11.1.17) is obtained from the point-source problem (10.1.28)–(10.1.30) by differentiation with respect to t. We can therefore use the results of Section 10.1 to verify the uniqueness of the solution to the inverse problem (11.1.14)–(11.1.16).

11.2 Reducing to spectral inverse problems

Inverse coefficient problems for parabolic equations under certain conditions can be reduced to Sturm–Liouville problems (Denisov, 1994).

Consider a direct initial boundary value problem

$$u_t = u_{xx} - q(x)u, \quad x \in (0,\pi),\ t > 0, \tag{11.2.1}$$

$$u(x,0) = 0, \quad x \in (0,\pi), \tag{11.2.2}$$

$$u_x(0,t) = 0, \quad t > 0, \tag{11.2.3}$$

$$u_x(\pi,t) = g(t), \quad t > 0, \tag{11.2.4}$$

which consists in determining the function $u(x,t)$ from the given functions $q(x)$ and $g(t)$.

Suppose that the function $u(x,t)$ is smooth enough, namely, $u, u_x \in C([0,\pi] \times [0,\infty))$, $u_t, u_{xx} \in C((0,\pi) \times (0,\infty))$, the function $q(x)$ is strictly positive and continuous on the interval $[0,\pi]$. We assume also the function $g(t)$ to satisfy the following conditions: $g \in C^2[0,\infty)$; $g(+0) = g'(+0) = 0$; $g(t) = 0$ for $t \in [t_0,\infty)$, $t_0 > 0$; $g(t) \not\equiv 0$.

Introduce a function $\psi(x) = u(x,t_0)$. Note that from the facts that $g(t)$ vanishes for $t \geq t_0$ and $g(t) \not\equiv 0$ it follows that $\psi(x)$ belongs to $C^1[0,\pi]$ and satisfies the boundary conditions (11.2.3), (11.2.4). Therefore, if $q(x)$, $u(x,t)$ satisfy (11.2.1)–(11.2.4), then, for $t > t_0$, the function $u(x,t)$ is a solution of the direct problem

$$u_t = u_{xx} - q(x)u, \quad x \in (0,\pi),\ t > t_0, \tag{11.2.5}$$

$$u(x,t_0) = \psi(x), \quad x \in (0,\pi), \tag{11.2.6}$$

$$u_x(0,t) = u_x(\pi,t) = 0, \quad t \geq t_0. \tag{11.2.7}$$

Let λ_n be eigenvalues ($\lambda_i \leq \lambda_j$, $i \leq j$) and $y_n(x)$ normalized eigenfunctions of the Sturm–Liouville problem

$$\begin{gathered} y'' - q(x)y = -\lambda y, \quad x \in (0,\pi), \\ y'(0) = y'(\pi) = 0. \end{gathered} \tag{11.2.8}$$

Applying the method of separation of variables, write the solution $u(x,t)$ of problem (11.2.5)–(11.2.7) in the form

$$u(x,t) = \sum_{k=0}^{\infty} \psi_k e^{-\lambda_n (t-t_0)} y_n(x), \tag{11.2.9}$$

where $\psi_k = \int_0^\pi \psi(\xi) y_n(\xi)\, d\xi$.

Taking into account the properties of $g(t)$ and $q(x)$, one can verify that $\lambda_0 > 0$ (hence, $\lambda_n > 0$ for all $n = 1,2,\dots$). Therefore, in view of (11.2.7), the functions u, u_t and u_{xx} converge to zero as $t \to \infty$ uniformly in $x \in [0,\pi]$. Applying the Laplace transformation to $u(x,t)$, we get

$$v(x,p) = \int_0^\infty e^{-pt} u(x,t)\, dt.$$

Since (11.2.1)–(11.2.4) hold for $u(x,t)$, the function $v(x,p)$, with such $p \in \mathbb{C}$ that $\operatorname{Re} p > 0$, solves the boundary problem

$$v_{xx} - q(x)v = pv, \tag{11.2.10}$$

$$v_x(0,p) = 0, \tag{11.2.11}$$

$$v_x(\pi,p) = \nu(p), \tag{11.2.12}$$

where

$$\nu(p) = \int_0^\infty g(t)e^{-pt}\,dt.$$

Let $w(x, p)$ be a solution to the Cauchy problem

$$w_{xx} - q(x)w = pw, \tag{11.2.13}$$

$$w(0, p) = 1, \quad w_x(0, p) = 0. \tag{11.2.14}$$

From $v_x(0, p) = w_x(0, p) = 0$ it follows that $v(x, p)$ and $w(x, p)$ are linearly dependent, which means that $v(x, p) = c(p)w(x, p)$. The function $c(p)$ can be found from (11.2.12): $c(p) = \nu(p)/w_x(\pi, p)$. Hence,

$$v(x, p) = \nu(p)w(x, p)/w_x(\pi, p). \tag{11.2.15}$$

Solving problem (11.2.13), (11.2.14), we can find eigenvalues of the corresponding Sturm–Liouville problems as seen from the following lemmata.

Lemma 11.2.1. *If p_0 is a zero of the function $w(\pi, p)$, then $\lambda = -p_0$ is an eigenvalue of the Sturm–Liouville problem*

$$\begin{aligned} -y'' + q(x)y &= \lambda y, \\ y'(0) = 0, \quad y(\pi) &= 0. \end{aligned} \tag{11.2.16}$$

Lemma 11.2.2. *If p_0 is a zero of the function $w_x(\pi, p)$, then $\mu = -p_0$ is an eigenvalue of the problem*

$$\begin{aligned} -y'' + q(x)y &= \mu y, \\ y'(0) = 0, \quad y'(\pi) &= 0. \end{aligned} \tag{11.2.17}$$

One can prove the lemmata observing that if $w(x, p)$ satisfies (11.2.13), then $y(x) = w(x, p_0)$ obeys the equation $-y'' + q(x)y = -p_0 y$.

Thus, there exists a one-to-one correspondence between zeroes of the function $w(\pi, p)$ and eigenvalues of problem (11.2.16). Similarly, zeroes of $w_x(\pi, p)$ are connected with eigenvalues of problem (11.2.17).

11.3 Uniqueness theorems

The relation between problem (11.2.1)–(11.2.4) and Sturm–Liouville problems can be used in proving the uniqueness of the solution of inverse problems for parabolic equations.

Inverse problem 1. Let $u(x,t)$ be a solution of the direct problem (11.2.1)–(11.2.4) and

$$u(\pi,t) = f(t), \quad t > 0. \tag{11.3.1}$$

Given $g(t)$, $f(t)$, find a pair of functions $\{q(x), u(x,t)\}$ satisfying (11.2.1)–(11.2.4), (11.3.1).

It should be noted that the problem of determining both the functions $q(x)$ and $u(x,t)$ is easier for investigation than if we would seek only $q(x)$. Indeed, in proving uniqueness theorems the function $u(x,t)$ is required to be smooth enough. In the former case the smoothness of $u(x,t)$ can be provided by choosing an appropriate class of solutions, while otherwise we have to verify the properties of $u(x,t)$ based on those of $q(x)$.

Definition 11.3.1. A pair of functions $\{q(x), u(x,t)\}$ is said to belong to a class $\mathcal{K}_1$ if

1) $q \in C[0,\pi]$, $q(x) > 0$, $x \in [0,\pi]$;

2) $u, u_x \in C([0,\pi] \times [0,\infty))$, $u_t, u_{xx} \in C((0,\pi) \times (0,\infty))$.

Theorem 11.3.1. *Let $g(t)$ satisfy the following conditions:*

1) $g \in C^2[0,\infty)$,

2) $g(+0) = g'(+0) = 0$,

3) $g(t) = 0$ *for* $0 < t_0 \le t$,

4) $g(t) \not\equiv 0$.

Then Inverse problem 1 *has at most one solution in the class* $\mathcal{K}_1$.

Proof. Assume the contrary, that is, there exist two solutions $\{q_j(x), u_j(x,t)\}$, $j = 1,2$, of Inverse problem 1 in the class $\mathcal{K}_1$. Consider the functions $v_j(x,p)$ and $w_j(x,p)$ corresponding to the solutions. Note that the Laplace transform $\nu(p)$ of the function $g(t)$ is an analytic function over the space $\mathbb{C}$ because $g(t)$ may differ from zero only on the interval $t \in (0,t_0)$. From (11.2.15) it follows that for p such that $\operatorname{Re} p > 0$ there holds

$$\frac{w_1(\pi,p)}{w_1'(\pi,p)} = \frac{w_2(\pi,p)}{w_2'(\pi,p)}. \tag{11.3.2}$$

From here on, the prime at $w(x,p)$ denotes differentiation with respect to x. Note that the functions $w_j(\pi,p)/w_j'(\pi,p)$ are analytic and their singular points are nothing but zeroes of the functions $w_j'(\pi,p)$, respectively. Due to (11.3.2), zeroes and singular points of the function $w_1(\pi,p)/w_1'(\pi,p)$ coincide with those of $w_2(\pi,p)/w_2'(\pi,p)$. Let us show that zeroes of the function $w_j(\pi,p)$ differ from zeroes of $w_j'(\pi,p)$,

$j = 1, 2$. If it were not the case, that is, $w_j(\pi, p) = w'_j(\pi, p) = 0$ for some $p = p_0$, then the solution to the Cauchy problem

$$w''_j - (q_j(x) + p_0)w_j = 0,$$
$$w_j(\pi, p_0) = 0, \quad w'_j(\pi, p_0) = 0$$

would vanish for $x \in [0, \pi]$, which contradicts the condition $w_j(0, p_0) = 1$. Therefore, in view of (11.3.2), all the zeroes of the functions $w_1(\pi, p)$ and $w_2(\pi, p)$ coincide. The same is true for $w'_1(\pi, p)$ and $w'_2(\pi, p)$. From Lemmata 11.2.1, 11.2.2 it follows that $\lambda_n^1 = \lambda_n^2$, $\mu_n^1 = \mu_n^2$, where λ_n^j, μ_n^j, $j = 1, 2$, are eigenvalues of problems (11.2.16), (11.2.17), respectively, with $q_j(x)$ instead of $q(x)$. Applying Theorem 6.2.2, we conclude that $q_1(x) = q_2(x)$, $x \in [0, \pi]$.

Let us now verify that $u_1(x, t) = u_2(x, t)$, $x \in (0, \pi)$, $t > 0$. Because both the functions $u_j(x, t)$ solve the direct problem (11.2.1)–(11.2.4) with $q(x) = q_1(x) = q_2(x)$, their difference $\tilde{u}(x, t) = u_1(x, t) - u_2(x, t)$ satisfies

$$\tilde{u}_t = \tilde{u}_{xx} - q(x)\tilde{u}, \quad x \in (0, \pi),\ t > 0, \tag{11.3.3}$$
$$\tilde{u}(x, 0) = 0, \quad x \in (0, \pi), \tag{11.3.4}$$
$$\tilde{u}_x(0, t) = \tilde{u}_x(\pi, t) = 0, \quad t > 0. \tag{11.3.5}$$

Multiplying (11.3.3) by $\tilde{u}$ and integrating the result over the domain $(0, \pi) \times (0, t)$, we get, in view of (11.3.4), (11.3.5), the equality

$$\frac{1}{2} \int_0^\pi \tilde{u}^2(\xi, t)\, d\xi + \int_0^t \int_0^\pi \left[\tilde{u}_x^2(\xi, \tau) + q(\xi)\tilde{u}^2(\xi, \tau)\right] d\xi\, d\tau = 0.$$

Since $\{q(x), u_j(x, t)\} \in \mathcal{K}_1$, the function $q(x)$ is positive and, therefore, $\tilde{u}(x, t) \equiv 0$, which is the required result. □

Inverse problem 2. Let an additional information on the solution of the direct problem (11.2.1)–(11.2.4) be given at the left end of the interval $(0, \pi)$:

$$u(0, t) = f_0(t), \quad t > 0. \tag{11.3.6}$$

Given $g(t)$, $f_0(t)$, find a pair of functions $\{q(x), u(x, t)\}$ satisfying (11.2.1)–(11.2.4), (11.3.6).

Definition 11.3.2. A pair $\{q(x), u(x, t)\}$ is said to belong to a class $\mathcal{K}_2$ if it belongs to $\mathcal{K}_1$ and, moreover, $q(x) = q(\pi - x)$, $x \in [0, \pi]$.

Theorem 11.3.2. *If $g(t)$ satisfies the conditions of Theorem* 11.3.1, *then Inverse problem* 2 *admits no more than one solution in the class $\mathcal{K}_2$.*

Proof. This theorem is proven similarly as the previous one. Let $\{q(x), u(x,t)\}$ be a solution of Inverse problem 2 from the class $\mathcal{K}_2$. We remind that the Laplace transform $v(x,p)$ of the function $u(x,t)$ can be represented as

$$v(x,p) = \frac{\nu(p)w(x,p)}{w'(\pi,p)},$$

where $w(x,p)$ is a solution of problem (11.2.13), (11.2.14).

Assume now that in $\mathcal{K}_2$ there exist two solutions $\{q_1(x), u_1(x,t)\}, \{q_2(x), u_2(x,t)\}$ of Inverse problem 2. Then, in view of (11.3.6), for the corresponding functions $v_j(x,p)$ $w_j(x,p)$, $j = 1,2$, there holds $v_1(0,p) = v_2(0,p)$ and, hence,

$$\frac{1}{w_1'(\pi,p)} = \frac{1}{w_2'(\pi,p)}. \tag{11.3.7}$$

Using (11.3.7), one can show that the eigenvalues of problems (11.2.17) with the coefficients $q_1(x)$ and $q_2(x)$ coincide. Taking into account that the coefficients $q_j(x)$ are symmetric about the midpoint of the interval $(0,\pi)$, we can apply Theorem 6.2.3 to complete the proof. □

Assume now that the function $g(t)$ is also unknown.

Inverse problem 3. Given $f(t)$, find a triple of functions $\{q(x), g(t), u(x,t)\}$ satisfying conditions (11.2.1)–(11.2.4), (11.3.1).

It turns out that if $g(t)$ meets the conditions of Theorem 11.3.1 and, in addition, $g(t) \geq 0$, $t \geq 0$, then the solution of Inverse problem 3 is unique.

Definition 11.3.3. A triple $\{q(x), g(t), u(x,t)\}$ is said to belong to a class $\mathcal{K}_3$ if the following conditions hold:

1) $q \in C[0,\pi]$, $q(x) > 0$, $x \in [0,\pi]$;
2) $u, u_x \in C([0,\pi] \times [0,\infty))$; $u_t, u_{xx} \in C((0,\pi) \times (0,\infty))$;
3) $g \in C^2[0,\infty)$, $g(+0) = g'(+0) = 0$, $g(t) \geq 0$, $t > 0$; $g(t) = 0$ for $t \in [t_0,\infty)$, $t_0 > 0$;
4) $g(t) \not\equiv 0$.

Theorem 11.3.3. *Inverse problem 3 has at most one solution in the class $\mathcal{K}_3$.*

Proof. Suppose that two solutions $\{q_j(x), g_j(t), u_j(x,t)\}$, $j = 1,2$, of Inverse problem 3 belong to the class $\mathcal{K}_3$. Using the same line of reasoning as before, we come to

$$\frac{\nu_1(p)w_1(\pi,p)}{w_1'(\pi,p)} = \frac{\nu_2(p)w_2(\pi,p)}{w_2'(\pi,p)}. \tag{11.3.8}$$

The functions $g_1(t)$ and $g_2(t)$ vanish outside the interval $(0, t_0)$. Hence, their Laplace transforms $\nu_j(p)$, $j = 1, 2$, are analytic over the whole space $\mathbb{C}$ and have no singular points. Since $g_j(t) \geq 0$ and $g_j(t) \not\equiv 0$, we conclude that $\nu_j(p) > 0$ for all $p \in \mathbb{R}$ and $\nu_j(p)$ have neither singular points nor zeros in $\mathbb{R}$. Again, as in Theorem 11.3.1, we use (11.3.8) to deduce that the zeroes of the functions $w_1(\pi, p)$ and $w_2(\pi, p)$ coincide as well as the zeroes of $w_1'(\pi, p)$ and $w_2'(\pi, p)$. Thereby, $q_1(x) = q_2(x)$, $x \in [0, \pi]$. Then $w_1(x, p) = w_2(x, p)$ and, in view of (11.3.8), $\nu_1(p) = \nu_2(p)$. This implies $g_1(t) = g_2(t)$ and, finally, $u_1(x, t) = u_2(x, t)$, $x \in (0, \pi)$, $t > 0$. □

11.4 An overdetermined inverse coefficient problem for the elliptic equation. Uniqueness theorem

The inverse coefficient problems for elliptic equations are more difficult than the corresponding problems for parabolic ones. Indeed, if the problem is, for example, to recover only one of the coefficients of elliptic equation, we need to consider infinitely many corresponding direct problems to provide the uniqueness to its solution. Another approach to studying the problems consists in their simplification. For example, the sought-for coefficients may be independent of some variables (Section 11.5), or they are assumed to belong to a much more narrow class of solutions.

The uniqueness in the large was first shown by Sylvester and Uhlmann (1987) as to the problem of recovering the coefficient $q(x)$ of the equation

$$L_q u := \operatorname{div}(q \nabla u) = \sum_{j=1}^{n} (q(x) u_{x_j})_{x_j} = 0, \quad x \in \mathcal{D}, \tag{11.4.1}$$

where $\mathcal{D} \subset \mathbb{R}^n$ $(n \geq 3)$ is a bounded domain with smooth boundary S.

Consider the direct Dirichlet problems

$$L_q u = 0, \quad u\big|_S = g(x) \tag{11.4.2}$$

for all possible $g \in H^{1/2}(S)$ and assume that we know the functions $f_g(x)$ such that

$$q(x) \left.\frac{\partial u}{\partial \nu}\right|_S = f_g(x), \tag{11.4.3}$$

where ν denotes the unit outer normal to S. In other words, the Dirichlet to Neumann map is given by

$$\Lambda_q : g \in H^{1/2}(S) \to f_g.$$

The inverse problem is to find $q(x)$, $x \in \mathcal{D}$, knowing $\Lambda_q(g)$, $g \in H^{1/2}(S)$.

Assume that $q \in C^\infty(\overline{\mathcal{D}})$ and $q > \varepsilon > 0$. Denote

$$Q_q(g) = \int_{\mathcal{D}} q|\nabla u|^2 \, dx,$$

where u is the solution of (11.4.2). Using the divergence theorem, we have

$$Q_q(g) = \int_S \Lambda_q(g) q \, ds,$$

i.e., $Q_q(g)$ is the quadratic form associated to the linear map $\Lambda_q(g)$. Consequently, to know $\Lambda_q(g)$ or $Q_q(g)$ for all $g \in H^{1/2}(S)$ is equivalent.

Theorem 11.4.1. *Let $\mathcal{D} \subset \mathbb{R}^n$ $(n \geq 3)$ be a bounded domain with smooth boundary S. Assume that $q_1, q_2 \in C^\infty(\overline{\mathcal{D}})$ are strictly positive for $x \in \overline{\mathcal{D}}$ and*

$$Q_{q_1}(g) = Q_{q_2}(g) \tag{11.4.4}$$

for all $g \in H^{1/2}(S)$. Then $q_1(x) = q_2(x)$ for all $x \in \mathcal{D}$.

The proof of the theorem, which can be found in (Sylvester and Uhlmann, 1987), is based on using rapidly oscillatory functions g and corresponding special solutions of the Dirichlet problems (11.4.2).

11.5 An inverse problem in a semi-infinite cylinder

Let $\mathcal{D}_0$ be a semi-infinite cylinder in the space of variables x, y $(x \in \mathbb{R}^n)$ whose elements are parallel to the y-axis (Romanov, 1984):

$$\mathcal{D}_0 = \mathcal{D} \times \overline{\mathbb{R}}_+, \quad x \in \mathcal{D}, \quad \overline{\mathbb{R}}_+ = \{y : y \geq 0\}.$$

Consider in $\mathcal{D}_0$ the equation

$$u_{yy} + Lu + g(x, y) = 0, \tag{11.5.1}$$

where L is a uniformly elliptic operator in the variables $x_1, \dots, x_n$ with coefficients independent of y. Denote the boundary of $\mathcal{D}$ by S and formulate a direct problem as follows: find a solution of (11.5.1) which satisfies the conditions

$$u\Big|_{y=0} = \varphi(x), \quad \frac{\partial u}{\partial \nu}\Big|_{\Gamma} = \psi(x, y), \quad \Gamma = S \times \overline{\mathbb{R}}_+, \tag{11.5.2}$$

and tends to zero as $y \to \infty$.

Associated with (11.5.1), (11.5.2) is the following problem:

$$\tilde{u}_{tt} = L\tilde{u} + \tilde{g}(x,t),$$
$$\tilde{u}\big|_{t=0} = \varphi(x), \quad \tilde{u}_t\big|_{t=0} = 0, \quad \frac{\partial \tilde{u}}{\partial \nu}\Big|_S = \tilde{\psi}(x,t), \tag{11.5.3}$$
$$x \in \mathcal{D}, \quad t \geq 0.$$

Suppose that the functions $\tilde{g}$, $\tilde{\psi}$ are connected with g, ψ by the formulae

$$g(x,y) = \int_0^\infty \tilde{g}(x,t)H(y,t)\,dt, \quad \psi(x,t) = \int_0^\infty \tilde{\psi}(x,y)H(y,t)\,dt. \tag{11.5.4}$$

Here $H(y,t) = \frac{2}{\pi}\frac{y}{y^2+t^2}$ is a solution to the Laplace equation $H_{tt} + H_{yy} = 0$.

For the integrals in (11.5.4) to exist, we suppose $\tilde{g}$, $\tilde{\psi}$ to be bounded. The same assumption is made for the solution $\tilde{u}$ of the problem (11.5.3). Then the unique solution $u(x,t)$ of (11.5.1), (11.5.2) can be expressed through $\tilde{u}(x,t)$ as follows:

$$u(x,y) = \int_0^\infty \tilde{u}(x,t)H(y,t)\,dt. \tag{11.5.5}$$

We shall prove formula (11.5.5) assuming all the below integrals to exist.

Let a function $u(x,t)$ satisfy (11.5.5). Then

$$\begin{aligned} u_{yy} &= \int_0^\infty \tilde{u}(x,t)H_{yy}(y,t)\,dt = -\int_0^\infty \tilde{u}(x,t)H_{tt}(y,t)\,dt \\ &= (-\tilde{u}H_t + \tilde{u}_t H)\Big|_{t=0}^{t=\infty} - \int_0^\infty \tilde{u}_{tt}(x,t)H(y,t)\,dt \\ &= -\int_0^\infty \tilde{u}_{tt}(x,t)H(y,t)\,dt, \\ Lu &= \int_0^\infty H(y,t)L\tilde{u}(x,t)\,dt. \end{aligned}$$

Consequently,

$$u_{yy} + Lu + g = \int_0^\infty (-\tilde{u}_{tt} + L\tilde{u} + \tilde{g})H(y,t)\,dt = 0.$$

At the same time,

$$\begin{aligned} \frac{\partial u}{\partial \nu}\Big|_\Gamma &= \int_0^\infty \frac{\partial \tilde{u}}{\partial \nu}\Big|_S H(y,t)\,dt = \int_0^\infty \tilde{\psi}(x,t)H(y,t)\,dt = \psi(x,y), \\ u\big|_{y=0} &= \lim_{y\to 0}\int_0^\infty \tilde{u}(x,t)H(y,t)\,dt \\ &= \lim_{y\to 0}\frac{2}{\pi}\int_0^\infty \tilde{u}(x,y\tau)\frac{d\tau}{1+\tau^2} = \tilde{u}(x,0) = \varphi(x). \end{aligned}$$

Thus, the function $u(x, y)$ defined as (11.5.5) solves problem (11.5.1), (11.5.2).

Note that, for each fixed x, the function $u(x, y)$ can be regarded as an integral transform of $\tilde{u}(x,t)$ with the Cauchy kernel. Therefore, equality (11.5.5) is uniquely solvable for $\tilde{u}(x,t)$.

Consider now the problem of recovering a coefficient involved in the operator L provided the solution of problem (11.5.1), (11.5.2) satisfies

$$u\big|_{\Gamma} = f(x, y). \tag{11.5.6}$$

From the above it follows that problem (11.5.1), (11.5.2), (11.5.6) can be reduced to an equivalent inverse problem (11.5.3) with the data $\tilde{u}\big|_S = \tilde{f}(x,t)$, $t \geq 0$. The functions f and $\tilde{f}$ are connected by the invertible relation

$$f(x, y) = \int_0^\infty \tilde{f}(x,t)H(y,t)\,dt, \quad x \in S, \quad y \in \overline{\mathbb{R}}_+.$$

Thus, we have just shown how to proceed from an elliptic equation to a hyperbolic one. It should be noted, however, that in the above reduction we use essentially the following two conditions. First, the domain $\mathcal{D}$ has the form of cylinder with elements parallel to the y-axis. Second, the coefficients appearing in the operator L are independent of y. These are rigid requirements because in applied problems all the variables are usually of equal worth.

Appendix A

We all of us of education
A something somehow have obtained,
Thus, praised be God! a reputation
With us is easily attained

A. S. Pushkin.
Translated by Henry Spalding.

Appendix A contains some definitions and formulations of theorems from algebra, theory of differential and integral equations, functional analysis, and mathematical physics to study inverse and ill-posed problems.

A.1 Spaces

In modern mathematics a *space* is a set with some structure. For instance, some operations (such as summation or multiplication by a number) or functions (such as a norm or metric) may be defined on a set. Depending on the space structure, its elements are called points, vectors, events, etc. A space subset is called a *subspace* if the space structure generates on this subset a structure of the same type.

Most results of this book are presented for Hilbert and Banach spaces based on linear spaces (or, what is the same, vector ones). Therefore, first we define a linear space.

Definition A.1.1. A set $\mathcal{L}$ is called a *real linear (vector) space* if operations of summation of elements and multiplication by a scalar (that is, an element $x + y \in \mathcal{L}$ is assigned to each pair of elements x and y of the space $\mathcal{L}$, and an element $\alpha x \in \mathcal{L}$ is assigned to a real number α and every element $x \in \mathcal{L}$) are introduced with the following axioms:

1) $x + y = y + x, \alpha(\beta x) = \beta(\alpha x)$ — commutativity;
2) $(x + y) + z = x + (y + z)$, $(\alpha\beta)x = \alpha(\beta x)$ — associativity;
3) $\alpha(x + y) = \alpha x + \alpha y$, $(\alpha + \beta)x = \alpha x + \beta x$ — distributivity;
4) there exists an element $\mathbf{0} \in \mathcal{L}$, called *zero* element, such that $0 \cdot x = \mathbf{0}$ for all x;
5) $1 \cdot x = x$.

Remark A.1.1. If the scalars α and β is taken from the field of complex numbers $\mathbb{C}$, the space $\mathcal{L}$ is called a *complex linear space*.

A *linear combination* of elements $x_1, x_2, \dots, x_m \in \mathcal{L}$ is $\sum_{i=1}^{m} \alpha_i x_i \in \mathcal{L}$, $\alpha_i \in \mathbb{R}$, $1 \le i \le m$. The numbers $\alpha_1, \alpha_2, \dots, \alpha_m$ are called the *coefficients of* the *linear combination.*

A system of elements $x_1, x_2, \dots, x_m \in \mathcal{L}$ is called *linearly dependent* if there exist numbers $\alpha_1, \alpha_2, \dots, \alpha_m \in \mathbb{R}$ such that $\sum_{i=1}^{m} \alpha_i^2 \neq 0$ and $\sum_{i=1}^{m} \alpha_i x_i = \mathbf{0}$. Otherwise, the system is called *linearly independent.*

The number of elements of a maximal linearly independent system is called the dimension of the space $\mathcal{L}$ and denoted by $\dim \mathcal{L}$. If $\dim \mathcal{L} < \infty$, $\mathcal{L}$ is called a *finite-dimensional linear space*; otherwise it is called an *infinite-dimensional space*.

A.1.1 Euclidian spaces

A simple example of a *finite-dimensional real linear space* is the space $\mathbb{R}^n$ of ordered sets of n real numbers:

$$\mathbb{R}^n = \{(x_1, x_2, \dots, x_n),\ x_i \in \mathbb{R},\ 1 \le i \le n\}.$$

An element $x = (x_1, x_2, \dots, x_n) \in \mathbb{R}^n$ is called a *vector*, and the numbers x_i, its *components*. The operations of summation and multiplication by a scalar in the space $\mathbb{R}^n$ are performed componentwise:

$$\begin{aligned}(x_1, x_2, \dots, x_n) + (y_1, y_2, \dots, y_n) &= (x_1 + y_1, x_2 + y_2, \dots, x_n + y_n),\\ \alpha(x_1, x_2, \dots, x_n) &= (\alpha x_1, \alpha x_2, \dots, \alpha x_n).\end{aligned} \tag{A.1.1}$$

The dimension of $\mathbb{R}^n$ is n. To prove this, consider a system of vectors $\{e_1, e_2, \dots, e_n\}$ from $\mathbb{R}^n$ such that the ith component of vector e_i is unity and all other components are zero. It is evident that this system is linearly independent and maximal, since any vector $x = (x_1, x_2, \dots, x_n) \in \mathbb{R}^n$ can be uniquely represented as the sum $x = \sum_{i=1}^{n} x_i e_i$. The system $\{e_1, e_2, \dots, e_n\}$ is a basis in the space $\mathbb{R}^n$.

Definition A.1.2. A system of vectors $\{\varphi_1, \varphi_2, \dots, \varphi_n\}$ in the space $\mathbb{R}^n$ is called a *basis* if each vector $x \in \mathbb{R}^n$ is uniquely represented as a linear combination $x = \sum_{i=1}^{n} \alpha_i \varphi_i$.

Any linearly independent system of n vectors is a basis in $\mathbb{R}^n$. In addition to operations (A.1.1), which determine the structure of the linear space $\mathbb{R}^n$, one can define, on the set of vector pairs of this space, a function

$$\langle x, y \rangle = \sum_{i=1}^{n} x_i y_i, \tag{A.1.2}$$

where x_i and y_i are the components of vectors $x \in \mathbb{R}^n$ and $y \in \mathbb{R}^n$, respectively. It is easy to verify that the function defined by formula (A.1.2) has the following properties:

$$1)\quad \langle x, y\rangle = \langle y, x\rangle; \tag{A.1.3}$$

$$2)\quad \langle x + y, z\rangle = \langle x, z\rangle + \langle y, z\rangle; \tag{A.1.4}$$

$$3)\quad \langle \alpha x, y\rangle = \alpha\langle x, y\rangle, \quad \forall\alpha \in \mathbb{R}; \tag{A.1.5}$$

$$4)\quad \langle x, x\rangle \geq 0, \quad \text{and} \quad \langle x, x\rangle = 0 \quad \Leftrightarrow \quad x = \mathbf{0}. \tag{A.1.6}$$

Definition A.1.3. A real function $\langle x, y\rangle$ in a linear space $\mathcal{L}$ is called an *inner product* if it satisfies axioms (A.1.3)–(A.1.6).

Function (A.1.2) in the space $\mathbb{R}^n$ is an inner product.

Remark A.1.2. An inner product in the space of complex vectors $\mathbb{C}^n$ is defined by the formula

$$\langle x, y\rangle = \sum_{i=1}^{n} x_i \overline{y_i},$$

and axiom (A.1.3) is changed for $\langle x, y\rangle = \overline{\langle y, x\rangle}$.

With an inner product in a linear space $\mathcal{L}$, one can define a length $\|x\|$ of a vector x and an angle φ between vectors x and y by the formulas

$$\|x\| = \sqrt{\langle x, x\rangle}, \tag{A.1.7}$$

$$\cos\varphi = \frac{\langle x, y\rangle}{\|x\|\,\|y\|}. \tag{A.1.8}$$

The function $\|x\|$ defined by equality (A.1.7) is a norm in a linear space $\mathcal{L}$ with an inner product $\langle x, y\rangle$.

Definition A.1.4. A real function $\|x\|$ is called a *norm* in a linear space $\mathcal{L}$ if it satisfies the following axioms:

$$1)\quad \|x\| > 0, \quad \text{and} \quad \|x\| = 0 \quad \Leftrightarrow \quad x = \mathbf{0}; \tag{A.1.9}$$

$$2)\quad \|\alpha x\| = |\alpha|\,\|x\|, \quad \forall\alpha \in \mathbb{R} \quad (\text{or} \quad \alpha \in \mathbb{C}); \tag{A.1.10}$$

$$3)\quad \|x + y\| \leq \|x\| + \|y\|. \tag{A.1.11}$$

For the inner product $\langle x, y\rangle$ and the norm $\|x\|$ defined by formula (A.1.7), *the Cauchy–Schwarz–Bunyakovskii inequality* $|\langle x, y\rangle| \leq \|x\|\,\|y\|$ is valid. Notice that it follows from this inequality that formula (A.1.8) makes sense for all nonzero vectors $x \in \mathbb{R}^n$, $y \in \mathbb{R}^n$. In particular, $\varphi = \pi/2$ at $\langle x, y\rangle = 0$.

Definition A.1.5. Vectors $x \in \mathcal{L}$ and $y \in \mathcal{L}$ are called *orthogonal* if $\langle x, y \rangle = 0$.

Definition A.1.6. A basis $\{\varphi_1, \varphi_2, \dots, \varphi_n\}$ in the space $\mathbb{R}^n$ is called *orthogonal* if its vectors are pairwise orthogonal: $\langle \varphi_i, \varphi_j \rangle = 0$ for $i \neq j$. An orthogonal basis is called *orthonormal* if $\langle \varphi_i, \varphi_i \rangle = 1$ for all $1 \le i \le n$.

The basis $\{e_1, \dots, e_n\}$ of vectors e_i with all zero components except for the ith component that is equal to unity is an orthonormal basis in the space $\mathbb{R}^n$.

It should be noted that a vector norm in the space $\mathbb{R}^n$ may be defined in different ways. The norm

$$\|x\|_{\text{sph}} = \Big(\sum_{i=1}^{n} |x_i|^2\Big)^{1/2},$$

related to the inner product by formula (A.1.7) is called *Euclidean* (or *spherical*). *The cubic norm* of vector $x \in \mathbb{R}^n$ is defined by

$$\|x\|_{\text{cub}} = \max_{1 \le i \le n} |x_i|,$$

and the *octahedral norm* is calculated by the formula

$$\|x\|_{\text{oct}} = \sum_{i=1}^{n} |x_i|.$$

Remark A.1.3. In the space $\mathbb{C}^n$, the corresponding norms are defined by the same formulas.

For any $x \in \mathbb{R}^n$, the following relations are valid:

$$\|x\|_{\text{cub}} \le \|x\|_{\text{oct}} \le n\|x\|_{\text{cub}}, \tag{A.1.12}$$

$$\|x\|_{\text{cub}} \le \|x\|_{\text{sph}} \le \sqrt{n}\, \|x\|_{\text{cub}}. \tag{A.1.13}$$

Definition A.1.7. Two norms $\|\cdot\|_1$, $\|\cdot\|_2$ in a linear space $\mathcal{L}$ are called *equivalent* if there exist numbers $p, q \in \mathbb{R}$ such that for any $x \in \mathcal{L}$ the inequalities $\|x\|_1 \le p\|x\|_2$, $\|x\|_2 \le q\|x\|_1$ are valid.

It follows from relations (A.1.12) and (A.1.13) that the cubic, octahedral, and Euclidean norms are equivalent. A more general theorem is also true: all norms in a finite-dimensional space are equivalent.

Definition A.1.8. A linear space with a norm defined on it is called a *normed space*.

It should be noted that the normed spaces $(\mathbb{R}^n, \|\cdot\|_{\text{sph}})$ and $(\mathbb{R}^n, \|\cdot\|_{\text{cub}})$, which consist of the same elements (n-dimensional real vectors), differ considerably in their properties. For example, in the space $(\mathbb{R}^n, \|\cdot\|_{\text{cub}})$ it is impossible to define a scalar function $\langle x, y\rangle$ related to $\|\cdot\|_{\text{cub}}$ by (A.1.7).

In a normed space, one can define a distance between elements x and y with the function

$$\rho(x, y) = \|x - y\|. \tag{A.1.14}$$

Definition A.1.9. A real nonnegative function $\rho(x, y)$ defined on a set $\mathcal{M}$ is called a *metric* if it satisfies the following axioms for any $x, y, z \in \mathcal{M}$:

$$1)\quad \rho(x, y) = 0 \quad \Leftrightarrow \quad x = y \quad \text{(identity axiom);} \tag{A.1.15}$$

$$2)\quad \rho(x, y) = \rho(y, x) \quad \text{(symmetry axiom);} \tag{A.1.16}$$

$$3)\quad \rho(x, z) \le \rho(x, y) + \rho(y, z) \quad \text{(triangle axiom).} \tag{A.1.17}$$

Remark A.1.4. In the above axioms (A.1.15)–(A.1.17), in contrast to the norm axioms (A.1.9)–(A.1.11) and the inner product axioms (A.1.3)–(A.1.6), no operations of summation and multiplication by a scalar are used. This means that a metric may be defined not only on a linear space.

Definition A.1.10. A set $\mathcal{M}$ is called a *metric space* if for any of its elements $x, y \in \mathcal{M}$ a metric $\rho(x, y)$ is defined.

It is evident that the function $\rho(x, y) = \|x - y\|$ satisfies the metric axioms. Hence, any normed space is a metric space.

Definition A.1.11. A finite-dimensional vector space $\mathbb{E}^n$ with an inner product structure $\langle x, y\rangle$, consistent norm $\|x\| = \sqrt{\langle x, x\rangle}$ and a metric $\rho(x, y) = \|x - y\| = \sqrt{\langle x - y, x - y\rangle}$ is called a *Euclidean space.*

A simple example of a Euclidean space is the space $\mathbb{R}^n$ with the inner product function (A.1.2) and the Euclidean norm $\|x\| = (\sum_{i=1}^n x_i^2)^{1/2}$.

Now, we introduce a fundamental concept of functional analysis, convergence. Strictly speaking, a structure (inner product, norm, metric) that makes it possible to define a distance between elements of a space is introduced mainly to define and study convergence of its elements and other important properties and concepts based on convergence. It is evident that the more complicated is the structure of a space, the more diverse are the properties of this space. On the other hand, some results may have more condensed formulations and proofs in more general spaces. For instance, it is more convenient to consider spectral problems in the field of complex numbers, since, generally speaking, the spectrum of a linear continuous operator is in a complex Banach space.

Definition A.1.12. An element x of a metric space $\mathcal{M}$ is called the *limit of a sequence* $\{x_n\}$, $x_n \in \mathcal{M}$ and is denoted by $\lim_{n\to\infty} x_n = x$ if for every $\varepsilon > 0$ there exists a number $n(\varepsilon)$ such that for every $n > n(\varepsilon)$ the inequality $\rho(x_n, x) < \varepsilon$ holds. A sequence is called *convergent* (in the metric of a space $\mathcal{M}$) if it has a limit.

Definition A.1.13. A sequence $\{x_n\}$ is called a *fundamental* (or *Cauchy*) *sequence* if for every $\varepsilon > 0$ there exists a number $n(\varepsilon)$ such that for every $n > n(\varepsilon)$ and $m > 0$ the inequality $\rho(x_n, x_{n+m}) < \varepsilon$ is valid.

It is evident that any convergent sequence is fundamental. The converse is not true. A simple example is this: Consider the space of rational numbers $\mathbb{Q}$ with the metric $\rho(x, y) = |x - y|$, and let x_0 be an irrational number, that is, $x_0 \in \mathbb{R} \setminus \mathbb{Q}$. Construct a sequence of rational numbers $\{x_n\}$ that converges to x_0 in the space $\mathbb{R}$. It is evident that the sequence $\{x_n\}$ is fundamental in the space $\mathbb{Q}$ but does not converge to any rational number (let $x_n \to y$ in $\mathbb{Q}$, then $x_n \to y$ in $\mathbb{R}$, whence $y = x_0$).

Definition A.1.14. A metric space $\mathcal{M}$ is called *complete* if every fundamental sequence of points in the space $\mathcal{M}$ is convergent.

It follows from the above example that the space $\mathbb{R}$ of real numbers is complete and the space $\mathbb{Q}$ of rational numbers is not complete with respect to the metric $\rho(x, y) = |x - y|$.

A.1.2 Hilbert spaces

A special feature of Euclidean space is the presence of a finite-dimensional orthonormal basis. Removing the condition of finite-dimensionality while retaining the general structure and, in particular, the presence of an orthonormal (countable) basis in the space, we introduce the important class of separable Hilbert spaces.

Definition A.1.15. A linear space $\mathcal{H}$ is called a *Hilbert* space if a function of inner product $\langle x, y\rangle$ is defined on it (see Definition A.1.3) and if it is complete with respect to the metric $\rho(x, y) = \|x - y\| = \sqrt{\langle x - y, x - y\rangle}$ generated by the inner product.

A simple example of an infinite-dimensional Hilbert space is the space l_2 (or $\mathbb{R}^\infty$) of infinite sequences of real numbers $x = (x_1, x_2, \dots, x_i, \dots)$ with a convergent sum of squares

$$\sum_i x_i^2 < \infty.$$

In what follows, the symbol $\sum_i$ means summation over all natural i, that is, $\sum_i a_i = a_1 + a_2 + a_3 + \cdots$.

The inner product, as well as the corresponding norm and metric, in the space l_2 are calculated by the formulas

$$\langle x, y\rangle = \sum_i x_i y_i, \tag{A.1.18}$$

$$\|x\| = \Big(\sum_i x_i^2\Big)^{1/2}, \tag{A.1.19}$$

$$\rho(x, y) = \Big(\sum_i (x_i - y_i)^2\Big)^{1/2}. \tag{A.1.20}$$

The above-considered Hilbert space l_2 has the important property of separability. In the separable Hilbert spaces one can construct a countable orthonormal basis, that is, a set of vectors $e_1, e_2, \dots, e_i, \dots$ such that $\langle e_i, e_i\rangle = 1$, $\langle e_i, e_j\rangle = 0$, $i \neq j$ and every vector $x \in l_2$ can be uniquely represented as a convergent (with respect to metric (A.1.20)) series $x = \sum_{i=1}^{\infty} x_i e_i$.

To give an exact definition of separable spaces, we introduce notions of open and close sets.

Definition A.1.16. The set $B(x, r) = \{y \in \mathcal{M} : \rho(x, y) < r\}$ in a metric space $\mathcal{M}$ is called the *open sphere* of radius r with center at point x. The set $\overline{B(x, r)} = \{y \in \mathcal{M} : \rho(x, y) \leq r\}$ is called the *closed sphere*.

Definition A.1.17. The open sphere of radius ε with center at point x is called the *ε-neighborhood* of point x: $O_\varepsilon(x) = B(x, \varepsilon)$.

Definition A.1.18. A point $x \in X \subset \mathcal{M}$ is called an *inner point* of a set X if there exists $\varepsilon > 0$ such that $O_\varepsilon(x) \subset X$.

Definition A.1.19. A set $X \subset \mathcal{M}$ is called *open* if every point in X is inner.

It is evident that every open sphere is an open set.

Definition A.1.20. A point $x \in \mathcal{M}$ is called a *limit point* of a set $X \subset \mathcal{M}$ if in every neighborhood of it there are infinitely many points of the set X. A point $x \in X$ is called an *isolated point* of a set X if there exists a neighborhood of X that does not contain other points of the set X. A point x is called an *adherent point* of a set X if in every neighborhood of X there are points of the set X.

Every adherent point is either a limit point or an isolated point.

Definition A.1.21. The set $\overline{X}$ of all adherent points of a set X is called the *closure* of the set X.

Definition A.1.22. A set X is called *closed* if $\overline{X} = X$.

A closed sphere is a closed set.

Definition A.1.23. A set X is called *everywhere dense* (in a space $\mathcal{M}$) if $\overline{X} = \mathcal{M}$.

Definition A.1.24. A space that contains a countable everywhere dense subset is called *separable*.

Every separable Hilbert space has a countable orthonormal basis. The converse is also true: a Hilbert space having a countable orthonormal basis is separable.

Let $\{\varphi_i\}$ be a countable orthonormal basis of a Hilbert space $\mathcal{H}$. The expansion of an element $x \in \mathcal{H}$ in terms of the basis functions $\{\varphi_i\}$

$$x = \sum_i x_i \varphi_i$$

is called the *Fourier series*, and the numbers x_i are called Fourier coefficients of the element x in terms of the orthonormal basis functions $\{\varphi_i\}$ and calculated by the formula

$$x_i = \langle x, \varphi_i \rangle.$$

Let $\{\varphi_i\}$ be an orthonormal basis of a Hilbert space $\mathcal{H}$, then for every $x \in \mathcal{H}$ the *Parseval equality* holds:

$$\|x\|^2 = \sum_i |\langle x, \varphi_i \rangle|^2.$$

An orthonormal basis $\{\varphi_i\}$ of a Hilbert space $\mathcal{H}$ can be constructed by the following *orthogonalization process*: if $\{\psi_i\}$ is an arbitrary basis of the space $\mathcal{H}$, the elements φ_i are calculated by the recurrence formulas

$$\varphi_1 = \frac{\psi_1}{\|\psi_1\|}, \quad \varphi_i' = \psi_i - \sum_{j=1}^{i-1} \langle \psi_i, \varphi_j \rangle \varphi_j, \quad \varphi_i = \frac{\varphi_i'}{\|\varphi_i'\|}, \quad i = 2, 3, \ldots.$$

The presence of a countable orthonormal basis in a space considerably improves its structure and makes it easier to investigate it. Nevertheless, even when a Hilbert space is not separable it has some pleasant geometrical properties (see (A.1.21) below) that are not present in more general spaces (for instance, Banach spaces; see Subsection A.1.3)) without an inner product. The notion of an inner product is important, since it introduces a notion of angle between vectors and a notion of vector orthogonality.

An example of a nonseparable Hilbert space is the space $L_2(-\infty, \infty)$ (see Subsection A.1.5). It should be noted that it is due to the absence of separability in unbounded domains that the solutions to problems of mathematical physics are obtained not in the form of Fourier series related to a countable basis but in the form of Fourier integrals, which are of an uncountable nature.

Let $\mathcal{H}_0$ be a closed subspace of a Hilbert space $\mathcal{H}$. Then $\mathcal{H}_0$ is a Hilbert space with the inner product from $\mathcal{H}$. Let $\mathcal{H}_0^\perp$ denote the set of vectors that are orthogonal to all vectors from the set $\mathcal{H}_0$. The set $\mathcal{H}_0^\perp$ is called the *orthogonal complement* of the set $\mathcal{H}_0$. Since the inner product function is linear and continuous, $\mathcal{H}_0^\perp$ is a closed linear subspace of the space $\mathcal{H}$ and, hence, a Hilbert space. The intersection of the subspaces $\mathcal{H}_0$ and $\mathcal{H}_0^\perp$ has a single element, which is zero: $\mathcal{H}_0 \cap \mathcal{H}_0^\perp = \{\mathbf{0}\}$.

Lemma A.1.1. *Let $\mathcal{H}$ be a Hilbert space, $\mathcal{H}_0$ a closed subspace in $\mathcal{H}$, and $x \in \mathcal{H}$. Then there exists a unique element, $x_0 \in \mathcal{H}_0$, that is closest to x and called the (orthogonal) projection of the element x onto the subspace $\mathcal{H}_0$: $x_0 = \arg\min_{y \in \mathcal{H}_0} \|x - y\|$.*

Example A.1.1. Let $\mathcal{H} = \mathbb{R}^2$ be a plane with the Euclidean norm, $\mathcal{H}_0$ a straight line passing through the origin of coordinates, and a point $x \in \mathcal{H}$ not on this straight line. Then, the projection x_0 of the point x onto $\mathcal{H}_0$ is the foot of the perpendicular dropped from x onto $\mathcal{H}_0$.

Theorem A.1.1. *Let $\mathcal{H}$ be a Hilbert space, and $\mathcal{H}_0$ a closed subspace in $\mathcal{H}$. Then every element $x \in \mathcal{H}$ has a unique representation of the form*

$$x = x_0 + h, \quad x_0 \in \mathcal{H}_0, \quad h \in \mathcal{H}_0^\perp.$$

If $\mathcal{H}_1$ and $\mathcal{H}_2$ are Hilbert spaces that intersect only at zero point, the set $\mathcal{H} = \{x + y \colon x \in \mathcal{H}_1, y \in \mathcal{H}_2\}$ is also a Hilbert space. This set is called the *direct sum* of the spaces $\mathcal{H}_1$ and $\mathcal{H}_2$ and is denoted by $\mathcal{H} = \mathcal{H}_1 \oplus \mathcal{H}_2$.

It follows from Theorem A.1.1 that the Hilbert space $\mathcal{H}$ can be represented as the direct sum

$$\mathcal{H} = \mathcal{H}_0 \oplus \mathcal{H}_0^\perp, \tag{A.1.21}$$

where $\mathcal{H}_0$ is every closed subspace of the space $\mathcal{H}$.

In Hilbert spaces, in addition to the notion of convergence of elements with respect to the metric $\rho(x, y) = \|x - y\|$ (see Definition A.1.12), a notion of weak convergence is also introduced.

According to the definition of convergence in a metric space, a sequence $\{x_n\}$ of elements of a Hilbert space $\mathcal{H}$ converges (with respect to the norm in $\mathcal{H}$), or, what is the same, *strongly converges* to an element $x \in \mathcal{H}$ if $\lim_{n\to\infty} \|x_n - x\| = 0$. A sequence $\{x_n\}$ is called *weakly convergent* to x, and x is called the *weak limit* of the sequence $\{x_n\}$ if for every $y \in \mathcal{H}$ we have $\lim_{n\to\infty} \langle x_n - x, y\rangle = 0$.

In finite-dimensional spaces, the notions of strong and weak convergence are equivalent. In an infinite-dimensional Hilbert space, weak convergence follows from strong convergence, but the converse, generally speaking, is not true.

Theorem A.1.2. *Let a sequence of elements $\{x_n\}$ of a Hilbert space $\mathcal{H}$ weakly converge to $x \in \mathcal{H}$ and $\lim_{n\to\infty} \|x_n\| = \|x\|$. Then $\lim_{n\to\infty} x_n = x$.*

In Theorem A.1.2, the condition $\lim_{n\to\infty} \|x_n\| = \|x\|$ can be replaced by the condition $\overline{\lim}_{n\to\infty} \|x_n\| = \|x\|$.

Theorem A.1.3. *If a sequence $\{x_n\} \subset \mathcal{H}$ weakly converges to x, there exists a subsequence $\{x_{n_i}\}$ whose arithmetical means*

$$\frac{1}{k}\sum_{i=1}^{k} x_{n_i}$$

strongly converge to x.

Definition A.1.25. A set X in a metric space $\mathcal{M}$ is called (weakly) *precompact* if every sequence of elements from this set contains a (weakly) convergent (in the space $\mathcal{M}$) subsequence.

A closed set X (see Definition A.1.22) contains the limits of all its convergent subsequences.

Definition A.1.26. A set X in a metric space $\mathcal{M}$ is called (weakly) *compact* if every sequence in X contains a subsequence that is (weakly) convergent to a point from X.

Definition A.1.27. A set X in a Hilbert space is called *weakly closed* if it contains every point that is the weak limit of a subsequence $\{x_n\} \subset X$.

A (weakly) closed (weakly) precompact set is (weakly) *compact.*

A set is weakly precompact if and only if it is bounded, that is, contained in a sphere of a finite radius.

Every closed ball is weakly compact.

Theorem A.1.4. *A closed unit ball in a finite-dimensional Hilbert space is not a compact set.*

In Euclidean space every closed bounded set is compact. Weak compactness in a finite-dimensional space is equivalent to strong compactness.

A.1.3 Banach spaces

As noted above, a linear space $\mathcal{L}$ with an inner product function $\langle x, y\rangle$ is normed ($\|x\| = \sqrt{\langle x, x\rangle}$) and, hence, metric ($\rho(x, y) = \|x - y\|$). Imagine that in a linear space $\mathcal{L}$ only a norm $\|x\|$, that is, a real function satisfying relations (A.1.9)–(A.1.11), is defined. A natural question is: Can an inner product function $\langle x, y\rangle$ be reconstructed from this function so that the relation $\|x\| = \sqrt{\langle x, x\rangle}$ hold? In the general case, the answer is no. Even in a finite-dimensional space (see Subsection A.1.1) not every norm is generated by an inner product. To introduce in a normed space an inner product consistent with the norm, it is necessary and sufficient that, for every elements x, y of this space, the following parallelogram identity hold:

$$\|x + y\|^2 + \|x - y\|^2 = 2(\|x\|^2 + \|y\|^2).$$

A normed space is called *unitary* if an inner product related to the norm by $\langle x, x\rangle = \|x\|^2$ can be introduced in this space.

Definition A.1.28. A normed space that is complete with respect to the metric $\rho(x, y) = \|x - y\|$ is called a *Banach* space.

The space $\mathbb{R}^n$ with the Euclidean norm $\|x\| = (\sum_{i=1}^n x_i^2)^{1/2}$ is a Euclidean space, whereas the space $\mathbb{R}^n$ with the cubic or octahedral norm is only a Banach space.

To illustrate the differences in the geometrical properties of Hilbert and Banach spaces, consider a simple example. Let $\mathcal{B} = (\mathbb{R}^2, \|\cdot\|_{\text{oct}})$ be the space of vectors $x = (x_1, x_2) \in \mathbb{R}^2$ with the norm $\|x\|_{\text{oct}} = |x_1| + |x_2|$. In this space, the unit ball is the square with vertices at points $(1, 0)$, $(0, -1)$, $(-1, 0)$, and $(0, 1)$. As a subspace $\mathcal{B}_0 \subset \mathcal{B}$, consider the straight line passing through the origin of coordinates at an angle of 45° to the abscissa axis. Every vector $y \in \mathcal{B}_0$ has coordinates (y_1, y_1), that is, $y = (y_1, y_1)$. Let $x' = (0, 1) \in \mathcal{B}$. Find the projection of x' onto the subspace $\mathcal{B}_0$, that is, a vector $y' = (y_1', y_1')$ such that

$$\min_{y\in\mathcal{B}_0} \|x' - y\| = \|x' - y'\|. \tag{A.1.22}$$

Since $\|x' - y\| = |1 - y_1| + |y_1|$, equality (A.1.22) holds for every vector $y' \in \mathcal{B}_0$ such that $0 \le y_1' \le 1$.

Thus, the minimal distance from the vector $x' \in \mathcal{B}$ to the subspace $\mathcal{B}_0$ is achieved on an infinite set of vectors $y' \in \mathcal{B}_0$. It is clear that on the plane $\mathbb{R}^2$ with the Euclidean metric this minimum is reached on a single vector. In some normed spaces, if $\mathcal{B}_0$ is an infinite-dimensional subspace this minimum cannot be reached at all.

Definition A.1.29. Let $\mathcal{B}_1$ be a closed subspace of a space $\mathcal{B}$. If there exists a subspace $\mathcal{B}_2 \subset \mathcal{B}$ such that $\mathcal{B}_1 \cap \mathcal{B}_2 = \{\mathbf{0}\}$ and $\mathcal{B} = \mathcal{B}_1 \oplus \mathcal{B}_2$, $\mathcal{B}_1$ is said to be *complemented* to the space $\mathcal{B}$, and $\mathcal{B}_2$ is called the *direct complement* to the subspace $\mathcal{B}_1$.

The direct complement is not unique.

The direct sum is a linear subspace. The next theorem formulates conditions that are necessary and sufficient for the direct sum to be closed.

Theorem A.1.5. *Let $\mathcal{B}$ be a Banach space, and $\mathcal{B}_1$ and $\mathcal{B}_2$, subspaces of $\mathcal{B}$ intersecting only at zero. The direct sum $\mathcal{B}_1 \oplus \mathcal{B}_2 = \{x + y\colon x \in \mathcal{B}_1,\ y \in \mathcal{B}_2\}$ is a closed subspace if and only if there exists a constant $k \ge 0$ such that*

$$\|x + y\| \ge k(\|x\| + \|y\|), \quad x \in \mathcal{B}_1,\ y \in \mathcal{B}_2.$$

It follows from Theorem A.1.5 that every finite-dimensional subspace of a Banach space can be complemented to the complete space.

Recall that in a Hilbert space $\mathcal{H}$ every closed subspace $\mathcal{H}_1 \subset \mathcal{H}$ has a direct complement: $\mathcal{H} = \mathcal{H}_1 \oplus \mathcal{H}_1^{\perp}$. It turns out that this property characterizes Hilbert spaces in the class of Banach spaces.

Theorem A.1.6 (Lindenstrauss–Tzafriri). *If in a Banach space $\mathcal{B}$ every closed subspace has a direct complement, $\mathcal{B}$ is isomorphic to some Hilbert space $\mathcal{H}$.*

Definition A.1.30. A Banach space $\mathcal{B}$ is called *strictly convex* if $\|x\| \leq 1$, $\|y\| \leq 1$, and $x \neq y$ imply that $\|x + y\| < 2$. A space $\mathcal{B}$ is called *uniformly convex* if for every $\varepsilon > 0$ there exists $\delta(\varepsilon) > 0$ such that $\|x\| \leq 1$, $\|y\| \leq 1$, and $\|x - y\| \leq \varepsilon$ imply that $\|x + y\| \leq 2(1 - \delta(\varepsilon))$.

A.1.4 Metric and topological spaces

The presence of a norm function in a space means that the space is linear and metric with respect to the metric $\rho(x, y) = \|x - y\|$, which is called an *invariant metric*. Attempts to go beyond the limits of linear spaces inevitably results in the loss of the norm function, but the metric structure may be preserved (see Remark A.1.4). Recall that a space with metric structure is called a metric space. A definition of a convergent sequence of elements in a metric space was given earlier (see Definition A.1.12). Fundamental concepts of open and closed sets, ε-neighborhood, and compact sets were defined in the general case of metric spaces (see Definitions A.1.16–A.1.22, A.1.25, and A.1.27). A metric space may possess the properties of separability (see Definitions A.1.23 and A.1.24) and completeness (see Definition A.1.14).

Definition A.1.31. A complete metric space $\mathcal{M}_0$ is called the *completion* of a metric space $\mathcal{M}$ if $\mathcal{M}$ is a subspace of $\mathcal{M}_0$ and $\overline{\mathcal{M}} = \mathcal{M}_0$.

Every metric space has the completion.

Theorem A.1.7 (the principle of nested balls). *For a metric space to be complete it is necessary and sufficient that every sequence of closed nested balls in it with radii tending to zero have a nonempty intersection.*

The open and closed sets of a metric space $\mathcal{M}$ have the following properties:

1) the set $M \setminus X$, that is the complement to a closed set X, is open and the set that is the complement to an open set is closed;
2) the union of any collection and the intersection of a finite collection of open sets is open;
3) the intersection of any collection and the union of a finite collection of closed sets is closed.

Definition A.1.32. A family of open sets τ of a metric space $\mathcal{M}$ is called a *metric* (or *open*) *topology* on the space $\mathcal{M}$.

Remark A.1.5. A metric topology generated by an invariant metric is called a *normed topology*.

Now, define a more general class of spaces, topological spaces.

Definition A.1.33. To introduce a topology on a set $\mathcal{T}$ means to find τ, a collection of subsets of the set $\mathcal{T}$ such that the following axioms are satisfied:

1) the entire set $\mathcal{T}$ and the empty set $\varnothing$ are in τ;

2) the union of any collection of elements of τ is an element of τ;

3) the intersection of a finite collection of elements of τ is an element of τ.

The set $\mathcal{T}$ with the above-introduced topology τ is called a *topological space* $(\mathcal{T}, \tau)$, the elements of $\mathcal{T}$ are called *points* of the space $(\mathcal{T}, \tau)$, the elements of τ are *open sets* of the space $(\mathcal{T}, \tau)$, and the complements to the sets of τ are called *closed sets* of the space $(\mathcal{T}, \tau)$.

It is evident that a metric space $\mathcal{M}$ is a topological space with the metric topology.

Definition A.1.34. A topological space whose topology is induced by some metric is called *metrizable*.

Not every topological space is metrizable. It should be noted that the same topology may be induced by different metrics.

A *neighborhood* of a point $x \in \mathcal{T}$ of a topological space $(\mathcal{T}, \tau)$ is any open set $O(x) \in \tau$ containing the point x. A point x of a set A is called an inner point of this set if there exists a *neighborhood* of the point x that is contained in A. The closure of a set in a topological space is defined in the same way as in a metric space (see Definitions A.1.20 and A.1.21).

Theorem A.1.8. *A set X of a topological space $(\mathcal{T}, \tau)$ is closed (that is, it is the complement to an open set) if and only if $X = \overline{X}$.*

The operation of closure in a topological space possesses the following properties:

1) $\overline{X \cup Y} = \overline{X} \cup \overline{Y}$;

2) $X \subseteq \overline{X}$;

3) $\overline{\overline{X}} = \overline{X}$;

4) $\overline{\varnothing} = \varnothing$.

These properties are called *closure axioms* or *Kuratowski axioms*.

Remark A.1.6. A topological space $(\mathcal{T}, \tau)$ can be defined alternatively as a set $\mathcal{T}$ with a closure operator on its subsets satisfying axioms 1)–4). Then a closed set can be defined as a set coinciding with its closure and an open set can be defined as the complement to a closed set. As a result, we obtain the same class of topological spaces.

Definition A.1.35. A family $\mathcal{E}_x$ of neighborhoods of a point x is called a *neighborhood basis* at the point x if any neighborhood of the point x contains some neighborhood from the family $\mathcal{E}_x$. The set $\mathcal{E}$ of bases $\mathcal{E}_x$ at all possible points of a space is called the *basis of the space*.

The basis possesses the following properties:

1) any neighborhood from $\mathcal{E}_x$ contains the point x;

2) if $O_1(x), O_2(x) \in \mathcal{E}_x$, there exists $O(x) \in \mathcal{E}_x$ such that $O(x) \subset O_1(x) \cap O_2(x)$;

3) for any neighborhood $O(x) \in \mathcal{E}_x$ there exists $O'(x) \in \mathcal{E}_x$ such that $O'(x) \subset O(x)$, and for any $y \in O'(x)$ there exists $O(y) \in \mathcal{E}_y$ such that $O(y) \subset O(x)$.

It is often convenient to describe the topology of a space in terms of convergence of directions (*Moore–Smith convergence*). Introduce a notion of direction, which is a generalization of the notion of a sequence of points.

A set $\mathcal{A}$ is called *ordered* if for any pairs of its elements α_1 and α_2 a *relation* $\alpha_1 \geq \alpha_2$ is defined, and the following axioms are satisfied:

1) $\alpha \geq \alpha$ for any $\alpha \in \mathcal{A}$;

2) if $\alpha_1 \geq \alpha_2$ and $\alpha_2 \geq \alpha_3$, $\alpha_1 \geq \alpha_3$;

3) if $\alpha_1 \geq \alpha_2$ and $\alpha_2 \geq \alpha_1$, $\alpha_1 = \alpha_2$.

A map $\alpha \to x_\alpha$ from a set $\mathcal{A}$ to $\mathcal{T}$ is called *a direction* (and also *a net* or *a generalized sequence*) and denoted $\{x_\alpha\}$ if $\mathcal{A}$ is an ordered set directed in an increasing order, that is, it is a set such that for every $\alpha_1, \alpha_2 \in \mathcal{A}$ there exists $\alpha \in \mathcal{A}$ such that $\alpha \geq \alpha_1$ and $\alpha \geq \alpha_2$. An example of a direction is the set of all finite subsets in the set of natural numbers $\mathbb{N}$ ordered by inclusion. Let $\mathcal{A} = \mathbb{N}$, then $\{x_n\}$ $(n \in \mathbb{N})$ is an ordinary sequence.

A direction $\{x_\alpha\}$ $(\alpha \in \mathcal{A})$ of elements of a topological space $(\mathcal{T}, \tau)$ is called *convergent* to $x \in \mathcal{T}$ *in topology* τ if for every neighborhood $O(x)$ of the point x there exists $\alpha(O(x)) \in \mathcal{A}$ such that $x_\alpha \in O(x)$ for all $\alpha \geq \alpha(O(x))$. The point x is called the *limit* (or the *limiting point*) of the direction $\{x_\alpha\}$. The convergence of a direction is denoted as $x_\alpha \underset{\mathcal{A}}{\to} x$ or $x = \lim\limits_{\alpha} x_\alpha$.

A set $X \subset \mathcal{T}$ in a topological space $(\mathcal{T}, \tau)$ is closed if and only if the limit of any convergent direction from X belongs to X.

Remark A.1.7. If a topological space is such that its every point has a countable neighborhood basis, one can use ordinary sequences instead of directions. In particular, in a metric space the notion of convergence of a sequence with respect to the metric is equivalent to the notion of convergence of a direction in the metric topology.

Topological spaces that satisfy only the three axioms of a topological space (see Definition A.1.33) may have a very complicated structure or, vice versa, their structure may be so primitive that they cannot be studied by topological means. For instance, this is the case if we set $\tau = \{\mathcal{T}, \varnothing\}$. In this space, called antidiscrete, any point has only one neighborhood, that is, the set $\mathcal{T}$. Hence, any sequence in the antidiscrete space converges to any point. To avoid this, some additional axioms are usually introduced to specify narrower classes of topological spaces. An example is the Hausdorff separability axiom, which guarantees the important property of uniqueness for the limit of a direction.

Definition A.1.36. A topological space is called a *Hausdorff* (or *separable*) space if its any two different points have disjoint neighborhoods.

Theorem A.1.9. *In a Hausdorff space, no convergent direction may have two different limits.*

Definition A.1.37. Let $\mathcal{L}$ be a linear and at the same time topological space with such a topology that the operations of summation $(x + y)$ and multiplication by a number (λx) are continuous with respect to both arguments for any $x, y \in \mathcal{L}$, and $\lambda \in \mathbb{R}$. Then $\mathcal{L}$ is called a *linear topological space* and is denoted by $(\mathcal{L}, \tau)$.

The continuity of summation and multiplication by a scalar with respect to a topology τ means that

1) for each pair of elements, $x, y \in \mathcal{L}$ and a neighborhood $O(x + y)$ of the element $x + y$, there exist a neighborhood $O(x)$ of the element x and a neighborhood $O(y)$ of the element y such that

$$O(x) + O(y) \subset O(x + y);$$

2) for every element $x \in \mathcal{L}$, $\lambda \in \mathbb{R}$, and neighborhood $O(\lambda x)$ of the element λx, there exist a neighborhood $O(x)$ of the element x and a number $\delta > 0$ such that for any λ' such that $|\lambda' - \lambda| < \delta$ we have the inclusion

$$\lambda' O(x) \subset O(\lambda x).$$

A specific feature of a linear topological space $(\mathcal{L}, \tau)$ is the fact that each neighborhood of its any point $x \in \mathcal{L}$ has the form $x + O(\mathbf{0})$, where $O(\mathbf{0})$ is a neighborhood of the zero element $\mathbf{0}$.

Thus, the topology in a linear topological space is determined by a neighborhood basis of the zero element.

Definition A.1.38. A set X in a linear space $\mathcal{L}$ is called *convex* if it contains, for each two of its elements, $x, y \in X$, also the entire interval between them: $[x, y] = \{\lambda x + (1-\lambda)y;\ \lambda \in [0,1]\} \subset X$. A set $X \subset \mathcal{L}$ is called *balanced* if for any $x \in X$ and $\lambda \in \mathbb{R}$ such that $|\lambda| \le 1$ we have $\lambda x \in X$. A convex balanced set is called *absolutely convex*. A set $X \subset \mathcal{L}$ is called *absorbing* if for any $x \in \mathcal{L}$ there exists such a number $\lambda(x) > 0$ that $x \in \lambda X$ for all λ, $|\lambda| \ge \lambda(x)$.

In a linear topological space $(\mathcal{L}, \tau)$ there exists a neighborhood basis of zero consisting of balanced absorbing sets.

Definition A.1.39. A set X of a linear topological space $(\mathcal{L}, \tau)$ is called *bounded* if for any neighborhood of zero $O(\mathbf{0})$ in the space $(\mathcal{L}, \tau)$ there exists a number λ such that $X \subset \lambda O(\mathbf{0})$.

If X_1 and X_2 are bounded sets of a linear topological space $(\mathcal{L}, \tau)$, the following sets are also bounded in $(\mathcal{L}, \tau)$: $X_1 \cup X_2$, $X_1 + X_2$, λX_1 $(\lambda \in \mathbb{R})$, $\overline{X}_1$.

A finite set is bounded.

Theorem A.1.10. *For a set X of a linear topological space $(\mathcal{L}, \tau)$ to be bounded, it is necessary and sufficient that, for any sequence $\{x_n\} \subset X$ and any sequence $\{\lambda_n\}$ of real numbers, $\lambda_n \to 0$, the convergence $\lambda_n x_n \to \mathbf{0}$ take place.*

Theorem A.1.11. *For a linear topological space $(\mathcal{L}, \tau)$ to be normed (that is, a norm could be introduced in $(\mathcal{L}, \tau)$ so that the normed topology coincide with the topology τ), it is necessary and sufficient that there exist in $(\mathcal{L}, \tau)$ a convex bounded neighborhood of zero.*

Definition A.1.40. A Hausdorff linear topological space is called *locally convex* if it has a neighborhood basis of zero consisting of convex sets.

A normed space is locally convex, since every ball in a normed space is a convex set.

In a normed space $\mathcal{N}$ the set of open balls $B(\mathbf{0}, \varepsilon) = \{x \in \mathcal{N}\colon \|x\| < \varepsilon\}$, where ε is an arbitrary real number, forms a neighborhood basis of zero of the space $\mathcal{N}$. A neighborhood basis of zero of a locally convex space is defined by a family of seminorms.

Definition A.1.41. A real function $p(x)$ defined on a linear space $\mathcal{L}$ is called *subadditive* if for any pair of elements $x_1, x_2 \in \mathcal{L}$

$$p(x_1 + x_2) \le p(x_1) + p(x_2),$$

positive homogeneous if for $\lambda \ge 0$

$$p(\lambda x) = \lambda p(x),$$

and *homogeneous* if for any $\lambda \in \mathbb{R}$

$$p(\lambda x) = |\lambda| p(x).$$

A subadditive positive homogeneous function is called a *gauge function*. A homogeneous gauge function is called a *seminorm*.

For a nonnegative gauge function p and any $\lambda > 0$ the sets $\{x \in \mathcal{L}\colon p(x) < \lambda\}$ and $\{x \in \mathcal{L}\colon p(x) \leq \lambda\}$ are convex and absorbing. If p is a seminorm, these sets are absolutely convex.

To each convex absorbing set $V \subset \mathcal{L}$ there corresponds a nonnegative gauge function p_V, called the *Minkowski functional* (for the set V) defined by the formula

$$p_V(x) = \inf\{\lambda \in \mathbb{R}_+,\ x \in \lambda V\},$$

and

$$\{x \in \mathcal{L} : p_V(x) < 1\} \subset V \subset \{x \in \mathcal{L} : p_V(x) \leq 1\}.$$

If, in addition, V is absolutely convex, p_V is a seminorm.

Let $\mathcal{L}$ be a linear space and let $\{p_\xi\}$, $\xi \in \Theta$, be an arbitrary family of seminorms on it. Then the sets

$$\big\{x \in \mathcal{L} : \max_{1 \leq i \leq n} p_{\xi_i}(x) < \varepsilon\big\} \quad (\varepsilon > 0,\ \xi_1, \dots, \xi_n \in \Theta)$$

form a neighborhood basis of zero of some topology in $\mathcal{L}$, that is, the family of seminorms *generates* a topology on $\mathcal{L}$. If a family of seminorms $\{p_\xi\}$, ($\xi \in \Theta$) is such that for every nonzero $x \in \mathcal{L}$ there exists a number $\xi \in \Theta$ such that $p_\xi(x) \neq 0$, the topology generated by the family is the topology of a locally convex space. All seminiorms from the generating set are continuous.

The topology of any locally convex set is generated by a set of seminorms. It follows from the fact that a locally convex set has an absolutely convex neighborhood basis of zero $\mathcal{E}_\mathbf{0}$. Hence, its topology is generated by a family of seminorms $\{p_V\}$, ($V \in \mathcal{E}_\mathbf{0}$), where p_V is the Minkowski functional for the neighborhood V.

The generating set of seminorms is not unique. If a locally convex set has a countable generating set of seminorms $\{p_i\}$, it is metrizable:

$$\rho(x, y) = \sum_{k=1}^{\infty} \frac{1}{2^k} \frac{p_k(x - y)}{1 + p_k(x - y)}.$$

An important notion of general topology is a notion of compact space introduced by P. S. Alexandrov and P. S. Urysohn in 1929.

Definition A.1.42. A topological space $(\mathcal{T}, \tau)$ is called *compact* (or *bicompact*) if, for every family $\{X_k\}$ ($k \in K$) of open sets forming a *cover* of the space $(\mathcal{T}, \tau)$, that is, such that

$$\bigcup_{k \in K} X_k = \mathcal{T},$$

there exists a finite number of sets $X_{k_1}, X_{k_2}, \dots, X_{k_n}$ also covering the space $(\mathcal{T}, \tau)$. A space $(\mathcal{T}, \tau)$ is called *locally compact* if its every point has a neighborhood whose closure is compact.

A topological space $(\mathcal{T}, \tau)$ is compact if and only if every direction in $\mathcal{T}$ has a limit point. Recall that a point $x \in \mathcal{T}$ is the limit point of a direction $\{x_\alpha\}$ ($\alpha \in \mathcal{A}$) if for any neighborhood $O(x)$ of the point x and any $\alpha \in \mathcal{A}$ there exists a subscript α' such that $\alpha' \geq \alpha$ and $x_{\alpha'} \in O(x)$.

A subset $X \subset \mathcal{T}$ of a topological space $(\mathcal{T}, \tau)$ is called *compact* if it is compact when it is considered as a topological space (with the topology induced by the topology τ). A set $X \subset \mathcal{T}$ is called *precompact* if its closure is compact. It is clear that any set in a compact space is precompact.

A compact subset in a Hausdorff space (in particular, in a metric space) is closed.

It should be noted that the above-presented definitions of compact and precompact sets do not coincide with Definitions A.1.25 and A.1.26 for metric spaces. In topological spaces the property of a set to contain, in each sequence of its elements (that is, in the case $\mathcal{A} = \mathbb{N}$), a convergent subsequence is called *sequential compactness* or *sequential precompactness* depending on whether the limit points belong to this set or not. In the general case, compactness does not imply sequential compactness and vice versa. However, in metric spaces the notions of compactness and sequential compactness coincide (see Remark A.1.7).

A.1.5 Examples of Hilbert and Banach spaces

The *space* l_p consists of all infinite sequences of real numbers $x = (x_1, x_2, \dots, x_n, \dots)$ for which

$$\|x\|_{l_p} = \Big\{\sum_n |x_n|^p\Big\}^{1/p},$$

is finite, where $p \geq 1$ is a fixed number. The linear operations are defined coordinate-wise. At $p > 1$ the space l_p is a Banach space. At $p = 2$ the space l_2 having the inner product

$$\langle x, y\rangle = \sum_n x_n y_n, \quad x, y \in l_2,$$

is a Hilbert space.

The *space* $C[a, b]$ consists of real functions $x(t)$ that are continuous on the interval $[a, b]$ and have the norm

$$\|x\|_C = \max\{|x(t)| : t \in [a, b]\}.$$

The summation of functions and multiplication by a number are defined in the usual way. The space $C[a,b]$ is a Banach space.

The *space* $C^m[a,b]$ ($m \geq 1$ is a fixed number) consists of m times uniformly differentiable functions $x(t)$ with the norm

$$\|x\|_{C^m} = \|x\|_C + \sum_{k=1}^{m} \|x^{(k)}\|_C.$$

Here $x^{(k)}$ is the derivative of order k of a function $x(t)$. The spaces $C^m[a,b]$ are Banach spaces.

The *space* $L_p[a,b]$ ($1 < p < +\infty$) consists of vector-functions $x : [a,b] \to \mathbb{R}^n$ that are pth power Lebesgue integrable on the interval $[a,b]$:

$$\|x\|_{L_p} = \Big(\int_a^b |x(t)|^p \, dt\Big)^{1/p} < \infty.$$

The space $L_p[a,b]$ is a Banach space. At $p = 2$ the space $L_2[a,b]$ is a Hilbert space with the inner product

$$\langle x, y\rangle_{L_2} = \int_a^b x(t)y(t)\, dt$$

and the norm

$$\|x\|_{L_2} = \sqrt{\langle x, x\rangle_{L_2}} = \Big(\int_a^b |x(t)|^2 \, dt\Big)^{1/2}.$$

The Sobolev space $H^k(G)$ is the closure with respect to the norm $\|\cdot\|_{H^k(G)}$ of the set of functions x: $\mathbb{R}^n \to \mathbb{R}$, which have in the domain G partial derivatives up to the order k inclusive and for which the following quantity is finite:

$$\|x\|^2_{H^k(G)} = \sum_{0 \leq |\alpha| \leq k} \int_G (D^\alpha x)^2 \, dG.$$

Here $\alpha = (k_1, \dots, k_n)$ is a multiindex,

$$D^\alpha x \equiv \frac{\partial^{|\alpha|} x}{\partial x_1^{k_1} \partial x_2^{k_2} \dots \partial x_n^{k_n}}, \qquad k_i \geq 0, \qquad |\alpha| \equiv k_1 + k_2 + \dots + k_n.$$

The space $H^k(G)$ is a Hilbert space with the inner product

$$\langle x, y\rangle_{H^k(G)} = \sum_{0 \leq |\alpha| \leq k} \langle D^\alpha x, D^\alpha y\rangle_{L_2(G)} = \sum_{0 \leq |\alpha| \leq k} \int_G D^\alpha x D^\alpha y \, dG.$$

Remark A.1.8. The spaces C^m, L_p, and H^k are considered in more detail in Subsection A.5.1.

A.2 Operators

In Section A.1, a notion of space was introduced, and some important types of spaces were described. To clearly demonstrate the material, it was presented historically according to the development of mathematics: from a particular case (vector spaces $\mathbb{R}^n$) to a general one (topological spaces $(\mathcal{T}, \tau)$). This section is organized in a reverse order: definitions of an operator and its properties are given in the most general space that admits these definitions. As we go from general to specific spaces, the notions thus introduced are considered in more detail and acquire new properties.

An important notion in calculus is the following notion of mapping of one set into another:

Let X and Y be two arbitrary sets, and let A be a rule that makes each element $x \in X$ correspond to exactly one element $y \in Y$. In this case we say that we have a *mapping* A of the set X into the set Y.

The element $y = Ax \in Y$ related to the element $x \in X$ by the mapping A is called the *image of the element* x, and x is called the *inverse image* or one of the inverse images of the element y.

Let $M \subset X$. Then $A(M)$ denotes the set of elements from Y that are the images of elements $x \in M$. The set $A(M)$ is called the *image of set* M under mapping A. If $A(X) = Y$, A is said to map the set X *onto* the set Y.

A mapping A of a set X into a set Y is called *one-to-one* if $Ax_1 = Ax_2$ implies $x_1 = x_2$. A one-to-one mapping of a set X onto a set Y is called a *bijection.*

A mapping A: $X \to Y$ is called *invertible* if for any $y \in A(X)$ the equation $Ax = y$ has a unique solution. If A is invertible, the mapping A^{-1}: $Y \to X$ that relates each element $y \in A(X)$ to the unique solution of the equation $Ax = y$ is called *inverse* with respect to A. It is evident that a mapping A is invertible if and only if A bijectively maps X onto $A(X)$.

A.2.1 Operators in topological spaces

Let X and Y be two sets that belong to topological spaces $(\mathcal{T}_1, \tau_1)$ and $(\mathcal{T}_2, \tau_2)$: $X \subset \mathcal{T}_1$, $Y \subset \mathcal{T}_2$. In this case a mapping A of the set X to the set Y is often called an *operator* acting from $\mathcal{T}_1$ to $\mathcal{T}_2$ with a domain of definition X and a range of values Y. This is symbolically written in the form A: $\mathcal{T}_1 \to \mathcal{T}_2$. Generally accepted symbols for the domain of definition and the domain of values of operator A are $D_A = X$ and $R_A = A(X) = Y$, respectively.

Definition A.2.1. A mapping A: $\mathcal{T}_1 \to \mathcal{T}_2$ is called *sequentially continuous at a point* $x \in D_A$ if for every sequence $\{x_n\} \subset D_A$ convergent to x in the topology τ_1 the corresponding sequence $\{Ax_n\} \subset R_A$ converges in the topology τ_2 to the point $Ax \in R_A$.

Definition A.2.2. A mapping $A: \mathcal{T}_1 \to \mathcal{T}_2$ is called *continuous* if the inverse image of every open set is open. A mapping A is called *continuous at a point* $x \in \mathcal{T}_1$ if the inverse image of any neighborhood of the point Ax is a neighborhood of the point x.

If $A: \mathcal{T}_1 \to \mathcal{T}_2$ is a mapping from a topological space $(\mathcal{T}_1, \tau_1)$ to a topological space $(\mathcal{T}_2, \tau_2)$, the following statements are equivalent:

1. The mapping A is continuous.
2. The inverse image of every closed set is closed.
3. The mapping A is continuous at every point $x \in \mathcal{T}_1$.
4. For any point $x \in \mathcal{T}_1$ and any neighborhood $O(Ax)$ of the point Ax there exists a neighborhood $O(x)$ of the point x such that $A(O(x)) \subset O(Ax)$.

A mapping A is called (sequentially) *continuous on a set* $M \subset D_A$ if it is (sequentially) continuous at every point of this set.

A mapping $A: \mathcal{T}_1 \to \mathcal{T}_2$ is continuous if and only if for any $x \in \mathcal{T}_1$ and any direction $\{x_\alpha\}$ $(\alpha \in \mathcal{A})$ such that $x_\alpha \to_{\mathcal{A}} x$ the convergence $Ax_\alpha \to_{\mathcal{A}} Ax$ takes place.

A.2.2 Operators in metric spaces

Let $\mathcal{M}_1$ and $\mathcal{M}_2$ be metric spaces. An operator $A: \mathcal{M}_1 \to \mathcal{M}_2$ is called *continuous at a point* $x \in D_A$ if for any $\varepsilon > 0$ there exists such $\delta = \delta(\varepsilon) > 0$ that if $y \in D_A$ and $\rho(x, y) < \delta$, $\rho(Ax, Ay) < \varepsilon$.

In metric spaces, the notions of continuity and sequential continuity coincide. A continuous operator maps every convergent sequence in the space $\mathcal{M}_1$ to a convergent sequence in the space $\mathcal{M}_2$.

Definition A.2.3. A mapping $A: \mathcal{M} \to \mathcal{M}$ of a metric space $\mathcal{M}$ to itself is called *contracting* if there exists a number $0 < \alpha < 1$ such that

$$\rho(Ax, Ay) \leq \alpha\rho(x, y) \tag{A.2.1}$$

for any $x, y \in \mathcal{M}$.

S. Banach proved a theorem called the *principle of contracting mapping*. Since this theorem was often mentioned in the book, here we present a proof.

Theorem A.2.1 (the principle of contracting mapping). *Every contracting mapping of a complete metric space $\mathcal{M}$ to itself has a unique fixed point, that is, a point $x \in \mathcal{M}$ such that $Ax = x$.*

Proof. Take an arbitrary point $x_0 \in \mathcal{M}$ and construct a sequence of points $\{x_n\}$ according to the rule

$$x_1 = Ax_0, \quad x_2 = Ax_1, \quad \ldots, \quad x_n = Ax_{n-1}, \quad \ldots.$$

The sequence $\{x_n\}$ is fundamental in $\mathcal{M}$. It follows from the fact that if $m > n$, we have

$$\begin{aligned}\rho(x_n, x_m) &= \rho(Ax_{n-1}, Ax_{m-1}) \le \alpha\rho(x_{n-1}, x_{m-1}) \le \cdots \le \alpha^n \rho(x_0, x_{m-n}) \\ &\le \alpha^n\{\rho(x_0, x_1) + \rho(x_1, x_2) + \cdots + \rho(x_{m-n-1}, x_{m-n})\} \\ &\le \alpha^n \rho(x_0, x_1)\{1 + \alpha + \alpha^2 + \cdots + \alpha^{m-n-1}\} \le \frac{\alpha^n}{1-\alpha}\rho(x_0, x_1),\end{aligned}$$

where $\alpha < 1$. Hence, $\rho(x_n, x_m) \to 0$ as $n \to \infty$, $m > n$. Since the space $\mathcal{M}$ is complete, there exists a limit $x = \lim_{n\to\infty} x_n$. Since the contracting mapping A is continuous, we have $Ax = \lim_{n\to\infty} Ax_n = \lim_{n\to\infty} x_{n+1} = x$. Thus, a fixed point exists.

Now we prove that the fixed point is unique. Let $Ax = x$ and $Ay = y$. Then $\rho(x, y) \le \alpha\rho(x, y)$, that is, $\rho(x, y) = 0$, whence $x = y$. □

Exercise A.2.1. Following the proof of Theorem A.2.1, estimate the rate of convergence of the sequence $\{x_n\}$ to the fixed point.

Remark A.2.1. A frequently used operator A is such that inequality (A.2.1) holds not in the entire space $\mathcal{M}$ but only in some closed neighborhood of a point x'. Let this neighborhood be the closed ball $\overline{B(x', r)}$ of radius r with center at the point x'. The principle of contracting mapping is valid if the operator A maps the ball $\overline{B(x', r)}$ to itself. In this case, the above successive approximations $x_n \in \overline{B(x', r)}$ and reasoning are valid.

In particular, if $\rho(x', Ax') \le r(1-\alpha)$, the operator, since it satisfies condition (A.2.1), maps the ball $\overline{B(x', r)}$ to itself. It follows from the fact that for any $x \in \overline{B(x', r)}$

$$\begin{aligned}\rho(x', Ax) &\le \rho(Ax, Ax') + \rho(Ax', x') \\ &\le \alpha\rho(x, x') + r(1-\alpha) \le \alpha r + r - r\alpha = r.\end{aligned}$$

This means that the operator A performs a contracting mapping of the ball $\overline{B(x', r)}$ to itself. Therefore, by virtue of the Banach principle, the operator A has in the ball $\overline{B(x', r)}$ a unique fixed point.

Two maps, A and A_1, of a metric space $\mathcal{M}$ to itself are said to be *commutating* if for any $x \in \mathcal{M}$ the equality $A(A_1x) = A_1(Ax)$ holds.

Theorem A.2.2 (generalization of the principle of contracting mapping). *Let A and A_1 be mappings of a complete metric space $\mathcal{M}$ to itself. If A_1 is a contracting mapping and the mappings A and A_1 are commutating, the equation $Ax = x$ has a solution.*

A.2.3 Operators in linear spaces

Let $A\colon \mathcal{L}_1 \to \mathcal{L}_2$ be an operator that acts in linear topological spaces $(\mathcal{L}_1, \tau_1)$ and $(\mathcal{L}_2, \tau_2)$. Consider some peculiarities of the operations of summation and multiplication of the argument by a scalar.

Definition A.2.4. An operator $A\colon \mathcal{L}_1 \to \mathcal{L}_2$ that maps a linear space $\mathcal{L}_1$ to a linear space $\mathcal{L}_2$ is called a *linear operator* if the following axioms are satisfied:

1) A is *additive*, that is, $A(x + y) = Ax + Ay$, $\forall x, y \in \mathcal{L}_1$;
2) A is *homogeneous*, that is, $A(\alpha x) = \alpha Ax$, $\forall x \in \mathcal{L}_1$, $\forall \alpha \in \mathbb{R}$ (or $\alpha \in \mathbb{C}$).

If $\mathcal{L}_1 = \mathcal{L}_2 = \mathcal{L}$, A is called a linear operator in the space $\mathcal{L}$.

Example of a linear operator. Let

$$y(t) = \int_0^1 K(s,t)x(s)\,ds,$$

where $K(s,t)$ is a function continuous in the square $0 \le s,t \le 1$. Then, for $x(s) \in C[0,1]$ it is evident that $y(t) \in C[0,1]$. It is clear that conditions 1 and 2 of Definition A.2.4 are met. Hence, the operator

$$Ax = \int_0^1 K(s,t)x(s)\,ds$$

maps the space $C[0,1]$ to the space $C[0,1]$, and it is linear.

Introduce a notion of the *sum of linear operators* $A\colon \mathcal{L}_1 \to \mathcal{L}_2$ and $B\colon \mathcal{L}_1 \to \mathcal{L}_2$ as follows: $(A + B)x = Ax + Bx$, $x \in \mathcal{L}_1$. It is evident that $(A + B)$ is a linear operator that maps $\mathcal{L}_1$ to $\mathcal{L}_2$. The *product of a linear operator A and a scalar $\alpha \in \mathbb{R}$ (or $\alpha \in \mathbb{C}$)* is defined in a similar way, namely, $(\alpha A)x = \alpha(Ax)$ for any $x \in \mathcal{L}_1$. The *product of linear operators* $B : \mathcal{L}_2 \to \mathcal{L}_3$ and $A\colon \mathcal{L}_1 \to \mathcal{L}_2$ is defined by the rule $(BA)x = B(Ax)$, $x \in \mathcal{L}_1$. The product BA is a linear operator and maps $\mathcal{L}_1$ to $\mathcal{L}_3$.

Definition A.2.5. The set of all *linear operators* $A\colon \mathcal{L}_1 \to \mathcal{L}_2$ that map a linear space $\mathcal{L}_1$ to a linear space $\mathcal{L}_2$ forms a linear space with the above-introduced operations of summation of operators and multiplication of an operator by a scalar. This space is called the *space of linear operators*.

Definition A.2.6. A linear operator $A\colon \mathcal{L}_1 \to \mathcal{L}_2$ that maps a linear topological space $(\mathcal{L}_1, \tau_1)$ to another topological space $(\mathcal{L}_2, \tau_2)$ is called *bounded* if it transforms bounded sets to bounded sets (see Definition A.1.39).

Theorem A.2.3. *Let $(\mathcal{L}_1, \tau_1)$ and $(\mathcal{L}_2, \tau_2)$ be linear topological spaces, and let A be a linear operator that maps $\mathcal{L}_1$ to $\mathcal{L}_2$. Then boundedness of operator A follows from its continuity.*

In the general case of linear topological spaces, continuity does not follow from boundedness of an operator. However, if a linear topological space is metrizable, the following statements are equivalent:

1) operator A is continuous;
2) operator A is bounded;
3) if $x_n \to \mathbf{0}$ as $n \to \infty$, the set $\{Ax_n\}$ is bounded;
4) $Ax_n \to \mathbf{0}$ as $x_n \to \mathbf{0}$.

A.2.4 Operators in Banach spaces

Let $\mathcal{N}_1$ and $\mathcal{N}_2$ be normed spaces with norms $\|\cdot\|_{\mathcal{N}_1}$ and $\|\cdot\|_{\mathcal{N}_2}$, respectively. A linear operator $A\colon \mathcal{N}_1 \to \mathcal{N}_2$ is bounded if there exists a constant C such that for any $x \in \mathcal{N}_1$

$$\|Ax\|_{\mathcal{N}_2} \le C\|x\|_{\mathcal{N}_1}. \tag{A.2.2}$$

A linear operator $A\colon \mathcal{N}_1 \to \mathcal{N}_2$ is continuous if and only if it is bounded.

Definition A.2.7. Let A be a linear bounded operator that maps a normed space $\mathcal{N}_1$ to another normed space $\mathcal{N}_2$. The smallest of the constants C satisfying condition (A.2.2) is called the *norm* $\|A\|$ *of the operator* A.

The operator norm is calculated by the formula

$$\|A\| = \sup_{\|x\|_{\mathcal{N}_1}=1} \|Ax\|_{\mathcal{N}_2} = \sup_{x\neq \mathbf{0}} \frac{\|Ax\|_{\mathcal{N}_2}}{\|x\|_{\mathcal{N}_1}}. \tag{A.2.3}$$

The space of linear bounded operators $A\colon \mathcal{N}_1 \to \mathcal{N}_2$ that map $\mathcal{N}_1$ to $\mathcal{N}_2$ is a normed space with norm (A.2.3) and is denoted by $\mathcal{L}(\mathcal{N}_1, \mathcal{N}_2)$.

Theorem A.2.4. *The space of linear bounded operators $\mathcal{L}(\mathcal{N}, \mathcal{B})$ that map a normed space $\mathcal{N}$ to a Banach space $\mathcal{B}$ is a Banach space.*

In the space of operators one may define various types of convergence and various topologies.

Definition A.2.8. Convergence in the norm in the space of linear bounded operators $\mathcal{L}(\mathcal{N}_1, \mathcal{N}_2)$ is called *uniform convergence*. In other words, if a sequence $\{A_n\}$, $A_n \in \mathcal{L}(\mathcal{N}_1, \mathcal{N}_2)$ converges in the norm to an operator A, that is, $\|A_n - A\| \to 0$, $n \to \infty$, $\{A_n\}$ is said to *converge to A uniformly as $n \to \infty$*.

Definition A.2.9. A sequence of operators $\{A_n\}$ is said to converge pointwise as $n \to \infty$ to an operator A in the space $\mathcal{L}(\mathcal{N}_1, \mathcal{N}_2)$ if for any $x \in \mathcal{N}_1$ the relation $\lim_{n\to\infty} A_n x = Ax$ holds.

Pointwise convergence of a sequence $\{A_n\} \subset \mathcal{L}(\mathcal{N}_1, \mathcal{N}_2)$ follows from its uniform convergence. The converse is not true.

The next theorem is one of the major principles of functional analysis, the so-called *principle of uniform boundedness.*

Theorem A.2.5 (Banach). *Let a sequence $\{A_n\}$ of linear bounded operators that map a Banach space $\mathcal{B}$ to a normed space $\mathcal{N}$ converge pointwise as $n \to \infty$ to an operator A. Then the numerical sequence $\{\|A_n\|\}$ is bounded. Hence, $\lim_{x\to\mathbf{0}} A_n x = \mathbf{0}$ uniformly with respect to $n \in \mathbb{N}$ and the operator $Ax = \lim_{n\to\infty} A_n x$ is bounded.*

Theorem A.2.6. *Let $\mathcal{B}_1$ and $\mathcal{B}_2$ be Banach spaces. Then the space $\mathcal{L}(\mathcal{B}_1, \mathcal{B}_2)$ is a complete space in terms of pointwise convergence.*

Theorem A.2.7 (Banach's inverse operator theorem). *Let A be a bijective linear continuous operator that maps a Banach space $\mathcal{B}_1$ to a Banach space $\mathcal{B}_2$. Then the inverse operator A^{-1} is also continuous.*

If the inverse operator A^{-1}: $\mathcal{B}_2 \to \mathcal{B}_1$ is bounded, its norm can be found by the formula

$$\frac{1}{\|A^{-1}\|} = \inf_{x\neq\mathbf{0}} \frac{\|Ax\|_{\mathcal{B}_2}}{\|x\|_{\mathcal{B}_1}}.$$

Let $A_0 : \mathcal{L}' \to \mathcal{B}$ be a linear bounded operator defined on a linear subspace $\mathcal{L}'$ that is everywhere dense in a normed space $\mathcal{N}$ (that is, $\overline{\mathcal{L}'} = \mathcal{N}$) with values in a Banach space $\mathcal{B}$. Then the operator A_0 can be *continued to* the entire space $\mathcal{N}$ with no increase in its norm. In other words, an operator A: $\mathcal{N} \to \mathcal{B}$ such that $Ax = A_0 x$, $x \in \mathcal{L}'$, $\|A\|_{\mathcal{N}} = \|A_0\|_{\mathcal{L}'}$ can be defined.

This process of continuation is called *continuation by continuity.* If an operator is not bounded its continuation is usually called *extension.* The theory of operator extensions is a separate division of functional analysis.

Definition A.2.10. A linear continuous operator P: $\mathcal{N} \to \mathcal{N}_0$ that maps a normed space $\mathcal{N}$ onto a closed subspace $\mathcal{N}_0 \subset \mathcal{N}$ is called a *projection* (from $\mathcal{N}$ to $\mathcal{N}_0$) if P leaves the elements of $\mathcal{N}_0$ unchanged, that is, $P(x) = x$ for any $x \in \mathcal{N}_0$.

A closed subspace $\mathcal{B}_0 \subset \mathcal{B}$ of a Banach space $\mathcal{B}$ can be complemented to the space $\mathcal{B}$ (see Definition A.1.29) if there exists a projection P from $\mathcal{B}$ to $\mathcal{B}_0$.

Compact and completely continuous operators. An important class of operators approximated by finite-dimensional operators, namely, the class of completely continuous operators, was first considered by Hilbert in the space $\mathcal{L}_2$.

Definition A.2.11. An operator A: $\mathcal{N}_1 \to \mathcal{N}_2$ that maps a normed space $\mathcal{N}_1$ to a normed space $\mathcal{N}_2$ is called *compact* if it transforms any bounded set to a precompact set. If in this case A is continuous, it is called *completely continuous.*

Since a compact operator is bounded, every compact linear operator is completely continuous.

The *finite-dimensional operators*, that is, linear continuous operators with a finite-dimensional range of values, are compact.

The compact operators have the following properties:

1) the range of values, R_A, of a compact operator A is separable;
2) the linear combination $A = \alpha_1 A_1 + \alpha_2 A_2$ of compact operators $A_1 : \mathcal{N}_1 \to \mathcal{N}_2$ and A_2: $\mathcal{N}_1 \to \mathcal{N}_2$ is a compact operator;
3) let $A \in \mathcal{L}(\mathcal{N}_1, \mathcal{N}_2)$ and $A_1 \in \mathcal{L}(\mathcal{N}_2, \mathcal{N}_3)$, where $\mathcal{N}_3$ is a normed space, and let one of these operators be compact. Then the product $A_1 A$ is also compact;
4) let $\{A_n\}$ be a sequence of compact linear operators from a Banach space $\mathcal{N}_1$ to a Banach space $\mathcal{N}_2$ that converges to an operator A in the norm of the space $\mathcal{L}(\mathcal{N}_1, \mathcal{N}_2)$, that is,

$$\lim_{n \to \infty} \|A_n - A\| = 0.$$

Then A is a compact operator.

It follows from property 4 that only a completely continuous operator can be approximated by finite-dimensional operators with any accuracy.

Spectrum of an operator. One of the major problems in studying linear operators in Hilbert and Banach spaces is to find vectors that leave their directions unchanged under the action of an operator, that is, such vectors that satisfy the equation $Ax = \lambda x$, where λ is a number. It is more convenient to consider this equation in a complex space, in particular, when λ has complex values. Here we introduce some additional notions related to complex spaces in order to consider real spaces as particular cases of complex ones.

Let Z be a complex normed space. Z is said to have a *real kernel* if an operator C that maps Z to itself and is called *involution* is defined on Z with the following properties:

1) $C(\lambda_1 z_1 + \lambda_2 z_2) = \bar{\lambda}_1 C(z_1) + \bar{\lambda}_2 C(z_2)$, $\forall z_1, z_2 \in Z$, $\forall \lambda_1, \lambda_2 \in \mathbb{C}$;
2) $C^2(z) = z$, $\forall z \in Z$;
3) $\|C(z)\| = \|z\|$, $\forall z \in Z$.

The set of elements for which $C(z) = z$ is called the *real kernel* of Z and denoted by Re Z; the elements of this set are called *real*.

Every element $z \in Z$ is represented in the form

$$z = x + yi, \quad x, y \in \mathbb{R},$$

where $x = \operatorname{Re} z = [z + C(z)]/2$ is called the *real part* of the element z and $y = \operatorname{Im} z = [z - C(z)]/(2\mathrm{i})$ is called the *imaginary part* of the element z. It is evident that $C(z) = \bar{z} = x - y\mathrm{i}$.

The real kernel $X = \operatorname{Re} Z$ is a real normed space, which is complete if the initial space Z is complete.

Any real space X may be considered as the real kernel of some complex space, namely, the space Z consisting of the ordered pairs of elements from X: $z = (x, y)$ $(x, y \in X)$. The following operations are introduced in the space Z:

$$(x_1, y_1) + (x_2, y_2) = (x_1 + x_2, y_1 + y_2);$$
$$\lambda(x, y) = (\alpha x - \beta y, \beta x + \alpha y), \quad \text{where } \lambda = \alpha + \beta\mathrm{i},\ \alpha, \beta \in \mathbb{R};$$
$$\|(x, y)\| = \max_{\varphi} \|x \cos\varphi + y \sin\varphi\|.$$

To make sure that $X = \operatorname{Re} Z$, it is sufficient to introduce

$$C((x, y)) = (x, -y).$$

In this case, an element is real in the space Z if and only if it is of the form $(x, \mathbf{0})$. By identifying an element x with the element $(x, \mathbf{0})$, one can use, instead of (x, y), the conventional notation $x + y\mathrm{i}$. The space Z is called *complexification* of X.

Let Z_1 and Z_2 be complex spaces with real kernels $X_1 = \operatorname{Re} Z_1$ and $X_2 = \operatorname{Re} Z_2$. A linear continuous operator $\tilde{A} : Z_1 \to Z_2$ is called *real* if $\tilde{A}(\operatorname{Re} Z_1) \subset \operatorname{Re} Z_2$. Thus, a real operator induces a linear continuous operator from the real space X_1 to the real space X_2.

Conversely, let A be a linear continuous operator that maps the space X_1 to the space X_2, $X_1 = \operatorname{Re} Z_1$, $X_2 = \operatorname{Re} Z_2$. Defining

$$\tilde{A}(z) = \tilde{A}(x + y\mathrm{i}) = Ax + \mathrm{i}Ay, \quad z = x + y\mathrm{i},$$

we obtain a linear continuous operator $\tilde{A}$ from the complex space Z_1 to the complex space Z_2. It is evident that on X_1 the operator $\tilde{A}$ coincides with A. The operator $\tilde{A}$ is called *complex extension* of the operator A.

Thus, let $A : \mathcal{B} \to \mathcal{B}$ be a linear continuous operator that maps a complex Banach space $\mathcal{B}$ to itself. If initially the space $\mathcal{B}$ is real, we take, instead of A, its complex extension $\tilde{A}$.

Consider the equation

$$T_\lambda(x) = \lambda x - Ax = y \tag{A.2.4}$$

and investigate its behavior as a function of the complex parameter λ.

To solve equation (A.2.4), the complex plane is divided into two sets: the set $\rho(A)$ of values $\lambda \in \mathbb{C}$ at which equation (A.2.4) has a unique solution for any right-hand side $y \in \mathcal{B}$ (hence, the operator T_λ has a continuous inverse operator), and the set $\sigma(A)$ of all other values of λ. The points of set $\rho(A)$ are called *regular values* of the operator A, the set $\rho(A)$ is called the *resolvent set*, and the operator $R_\lambda = (\lambda I - A)^{-1}$ is called the *resolvent* of the operator A. The set $\sigma(A)$ is called the *spectrum* of the operator A.

If at some λ the homogeneous equation

$$T_\lambda(x) = \lambda x - Ax = \mathbf{0} \tag{A.2.5}$$

has nonzero solutions, λ is called an *eigenvalue* of the operator A. It is evident that the set $\sigma_0(A)$ of all eigenvalues of the operator A is contained in the set $\sigma(A)$. Every solution of equation (A.2.5) is called an *eigenvector* that corresponds to a given eigenvalue. The solutions to equation (A.2.5) at a fixed λ form a subspace $\mathcal{B}_\lambda$ called the *eigensubspace* (or the *root subspace*) that corresponds to the eigenvalue λ. The dimension of $\mathcal{B}_\lambda$ is called the *multiplicity of the eigenvalue* λ. If $\dim \mathcal{B}_\lambda = 1$ the eigenvalue is called *simple*.

The above definitions of an eigenvalue and an eigenvector are extensions of the well-known notions of linear algebra to the case of an infinite-dimensional space. In the Euclidean space $\mathbb{E}^n$ the spectrum of a linear operator (that is, the spectrum of a matrix) consists of all its eigenvalues. In an infinite-dimensional space, the spectrum of a linear bounded operator has a more complicated structure.

In a finite-dimensional space there are the following two possibilities:

1) equation (A.2.5) has a nonzero solution, that is, λ is an eigenvalue for the operator (matrix) A; the operator $T_\lambda^{-1} = (\lambda I - A)^{-1}$, where I is the unit matrix, does not exist in this case;
2) there exists a bounded (defined on the entire space) operator T_λ^{-1} such that $T_\lambda^{-1} = (\lambda I - A)^{-1}$, that is, λ is a regular point.

In an infinite-dimensional space there is one more possibility: the operator T_λ^{-1} exists, that is, the equation $T_\lambda x = \mathbf{0}$ has only the zero solution, but this operator is defined not on the entire space. Thus, the operator's spectrum consists of those λ-values at which the operator T_λ does not have a bounded inverse operator defined on the entire space.

The spectrum of a linear bounded operator may be divided into three non-intersecting sets:

1) the *point spectrum*, that is, the set of all eigenvalues $\sigma_0(A)$ of the operator A;
2) the *continuous spectrum*, that is, those values of λ for which the operator T_λ has an inverse operator $T_\lambda^{-1} = R_\lambda$ with a dense domain of definition that does not coincide with the entire space: $\overline{D}_{R_\lambda} = \mathcal{B}$;
3) the *residual spectrum*, that is, those values of λ at which the resolvent operator $R_\lambda = (\lambda I - A)^{-1}$ exists but its domain of definition is not dense in the entire space: $\overline{D}_{R_\lambda} \neq \mathcal{B}$.

The resolvent R_λ of a linear continuous operator A in a complex Banach space $\mathcal{B}$ is expanded into a power series in λ^{-1}:

$$R_\lambda = \frac{1}{\lambda}I + \frac{1}{\lambda^2}A + \cdots + \frac{1}{\lambda^n}A^{n-1} + \cdots \quad \Big(|\lambda| > \frac{1}{r} = \lim_{n\to\infty} \sqrt[n]{\|A^n\|}\Big). \quad \text{(A.2.6)}$$

Expansion (A.2.6) holds also at $|\lambda| > 1/r$, where $1/r$ is the radius of the least circle with a center at the origin of coordinates that contains the entire spectrum. The number $1/r$ is called the *spectral radius* of the operator A.

It should be noted that we pass from the real case to the complex one when considering the equation $Ax = \lambda x$ to guarantee that the spectrum $\sigma(A)$ of a linear continuous operator A is not empty.

A.2.5 Operators in Hilbert spaces

Consider an operator $A : \mathcal{H}_1 \to \mathcal{H}_2$ acting from a Hilbert space $\mathcal{H}_1$ to a Hilbert space $\mathcal{H}_2$.

Definition A.2.12. An operator $A : \mathcal{H}_1 \to \mathcal{H}_2$ is said to be *weakly continuous at a point* $x \in D_A$ if for every sequence $\{x_n\}$, $x_n \in D_A$, weakly convergent to x the sequence $\{Ax_n\}$ weakly converges to Ax. An operator A is said to be *demi-continuous at a point* $x \in D_A$ if for every sequence $\{x_n\} \subset D_A$ convergent to the point x the sequence $\{Ax_n\}$ weakly converges to Ax. An operator A is said to be *weakly continuous (demi-continuous)* if it is weakly continuous (demi-continuous) at every point $x \in D_A$.

For linear operators in Hilbert spaces, weak continuity follows from continuity. This is not true for nonlinear operators. For instance, the functional $J(x) = \|x\|$, $x \in \mathcal{H}_1$, is continuous but not weakly continuous.

An important class of operators in a Hilbert space is the class of orthogonal projection operators. Let $\mathcal{H}_0$ be a subspace of a Hilbert space $\mathcal{H}$. According to Lemma A.1.1 every $x \in \mathcal{H}$ is uniquely related to its projection $P(x) = x_0 \in \mathcal{H}_0$ onto the subspace $\mathcal{H}_0$. Thereby the operator $P = P_{\mathcal{H}_0}$ is defined on $\mathcal{H}$, which is called the *orthogonal projection operator* (onto the subspace $\mathcal{H}_0$).

Orthogonal projection operators in Hilbert spaces are often called projection operators, since orthogonality of projection onto a subspace is often implied. In what follows "projection operator" will also mean "orthogonal projection operator".

Projection operators are linear and bounded: $\|P\| = 1$. Hence, the operator $P_{\mathcal{H}_0}$ is a projection (see Definition A.2.10). Projection operators have the following properties:

1) for every $x \in \mathcal{H}$ the elements Px and $x - Px$ are orthogonal;

2) $x \in \mathcal{H}_0 \;\Leftrightarrow\; Px = x$;

3) $x \perp \mathcal{H}_0 \;\Leftrightarrow\; Px = \mathbf{0}$.

The class of projection operators is characterized by the following theorem.

Theorem A.2.8. *For a linear operator P defined on a Hilbert space $\mathcal{H}$ to be a projection operator, it is necessary and sufficient that it be idempotent, that is,*

$$\langle P^2 x_1, x_2\rangle = \langle P x_1, x_2\rangle \quad \forall x_1, x_2 \in \mathcal{H}, \tag{A.2.7}$$

and symmetric,

$$\langle P x_1, x_2\rangle = \langle x_1, P x_2\rangle \quad \forall x_1, x_2 \in \mathcal{H}. \tag{A.2.8}$$

Two projection operators, P_1 and P_2, in a Hilbert space $\mathcal{H}$ are called *orthogonal* if $P_1 P_2 = \mathbf{0}$. Projection operators $P_{\mathcal{H}_1}$ and $P_{\mathcal{H}_2}$ are orthogonal if and only if the subspaces $\mathcal{H}_1$ and $\mathcal{H}_2$ are orthogonal.

Theorem A.2.9. *Let $P_{\mathcal{H}_1}, P_{\mathcal{H}_2}, \dots, P_{\mathcal{H}_n}$ be projection operators in a Hilbert space $\mathcal{H}$. For the operator $P = P_{\mathcal{H}_1} + P_{\mathcal{H}_2} + \cdots + P_{\mathcal{H}_n}$ to be a projection operator, it is necessary and sufficient that the projection operators $P_{\mathcal{H}_1}, P_{\mathcal{H}_2}, \dots, P_{\mathcal{H}_n}$ be pairwise orthogonal. In this case $P = P_{\mathcal{H}_0}$, where $\mathcal{H}_0 = \mathcal{H}_1 \oplus \mathcal{H}_2 \oplus \cdots \oplus \mathcal{H}_n$.*

Theorem A.2.10. *The product $P = P_{\mathcal{H}_1} P_{\mathcal{H}_2}$ of two projection operators $P_{\mathcal{H}_1}$ and $P_{\mathcal{H}_2}$ is a projection operator if and only if they are commutative:*

$$P_{\mathcal{H}_1} P_{\mathcal{H}_2} = P_{\mathcal{H}_2} P_{\mathcal{H}_1},$$

and $P = P_{\mathcal{H}_0}$, where $\mathcal{H}_0 = \mathcal{H}_1 \cap \mathcal{H}_2$.

The following four conditions are equivalent:

a) $\|P_{\mathcal{H}_1} x\| \geq \|P_{\mathcal{H}_2} x\|$ *for every* $x \in \mathcal{H}$;

b) $\mathcal{H}_1 \supset \mathcal{H}_2$;

c) $P_{\mathcal{H}_1} P_{\mathcal{H}_2} = P_{\mathcal{H}_2}$;

d) $P_{\mathcal{H}_2} P_{\mathcal{H}_1} = P_{\mathcal{H}_2}$.

Theorem A.2.11. *The difference $\mathcal{P} = \mathcal{P}_{\mathcal{H}_1} - P_{\mathcal{H}_2}$ of two projection operators is a projection operator if and only if $\mathcal{H}_2 \subset \mathcal{H}_1$. In this case P is a projection operator on $\mathcal{H}_1 \ominus \mathcal{H}_2$.*

Theorem A.2.12. *Let $\{P_{\mathcal{H}_n}\}$ be a monotone sequence of projection operators. Then, for any $x \in \mathcal{H}$ there exists*

$$P x = \lim_{n\to\infty} P_{\mathcal{H}_n} x,$$

and the operator P is a projection operator onto a subspace $\mathcal{H}_0$, where

$$\mathcal{H}_0 = \overline{\bigcup_{n=1}^{\infty} \mathcal{H}_n},$$

if the sequence $\{P_{\mathcal{H}_n}\}$ is increasing, and

$$\mathcal{H}_0 = \bigcap_{n=1}^{\infty} \mathcal{H}_n$$

if the sequence $\{P_{\mathcal{H}_n}\}$ is decreasing.

It follows from this theorem that if $\{P_{\mathcal{H}_n}\}$ is a set of pairwise orthogonal projection operators, at any $x \in \mathcal{H}$ the series

$$Px = \sum_{n=1}^{\infty} P_{\mathcal{H}_n} x$$

converges, and the operator P is a projection operator.

It should also be noted that a projection operator $P_{\mathcal{H}_0}$ is completely continuous if and only if $\mathcal{H}_0$ is finite-dimensional.

A completely continuous operator in an infinite–dimensional Hilbert space does not have a bounded inverse operator.

An example of a linear operator in a Hilbert space that is not continuous is the *differentiation operator* that acts from the space $L_2[a,b]$ to the space $L_2[a,b]$:

$$Ux = \frac{dx}{dt}, \quad x(t) \in \mathcal{D}_L,$$

where $\mathcal{D}_L$ is the set of differentiable functions from $L_2[a,b]$. It should be noted that $\mathcal{D}_L$ is everywhere dense in $L_2[a,b]$: $\overline{\mathcal{D}_L} = L_2[a,b]$. To demonstrate that the operator U is not bounded, consider the functions

$$x_n(t) = \sin\left(n\pi \frac{t-a}{b-a}\right)$$

and calculate $\|x_n\|_{L_2}$ and $\|Ux\|_{L_2}$ as follows:

$$\|x_n\|^2_{L_2[a,b]} = \int_a^b \sin^2\left(n\pi \frac{t-a}{b-a}\right) dt = \frac{b-a}{2},$$

$$\|Ux_n\|^2_{L_2[a,b]} = \frac{n^2\pi^2}{(b-a)^2} \int_a^b \cos^2\left(n\pi \frac{t-a}{b-a}\right) dt = \frac{n^2\pi^2}{b-a} \to \infty, \quad n \to \infty.$$

One can see that the operator U is unbounded. Hence, it is not continuous. Thus, the problem of calculating the values of the operator U is unstable, that is, small deviations from the function $x(t)$ in the metric of L_2 can cause arbitrarily large deviations in the derivative $Ux = dx/dt$.

But if we consider the differentiation operator as an operator $U_1 : H^1[a,b] \to L_2[a,b]$, it is bounded, since

$$\|U_1 x\|^2_{L_2[a,b]} = \left\| \frac{dx}{dt} \right\|_{L_2[a,b]} \le \|x\|_{L_2[a,b]} + \left\| \frac{dx}{dt} \right\|_{L_2[a,b]} = \|x\|^2_{H^1[a,b]}.$$

Hence, $\|U_1\| \le 1$ and the operator U_1 is continuous.

If $\mathcal{H}$ is a separable Hilbert space, a linear bounded operator A defined on the entire space $\mathcal{H}$ admits a matrix representation.

To prove this, let $\{\varphi_i\}$ be an orthonormal basis in $\mathcal{H}$ and $\langle A\varphi_k, \varphi_i\rangle = a_{ik}$, $i,k = 1,2,\dots$. The numbers a_{ik} are Fourier coefficients in the decomposition of the vector $A\varphi_k$ in terms of the basis $\{\varphi_i\}$:

$$A\varphi_k = \sum_{i=1}^{\infty} a_{ik}\varphi_i, \quad k = 1,2,\dots;$$

$$\sum_{i=1}^{\infty} |a_{ik}|^2 = \sum_{i=1}^{\infty} |\langle A\varphi_k, \varphi_i\rangle|^2 < \infty, \quad k = 1,2,\dots.$$

Let us show that the operator A can be uniquely reconstructed from the matrix $\{a_{ik}\}_{i,k=1}^{\infty}$ and the orthonormal basis $\{\varphi_k\}$. The problem consists in finding Ax for any vector $x \in \mathcal{H}$. Let

$$x = \sum_{k=1}^{\infty} \xi_k \varphi_k \quad \text{and} \quad x_n = \sum_{k=1}^{n} \xi_k \varphi_k.$$

Then $Ax_n = \sum_{k=1}^{\infty} (\sum_{i=1}^{n} a_{ki}\xi_i)\varphi_k$. Since the operator A is continuous, we obtain

$$c_k = \langle Ax, \varphi_k\rangle = \lim_{n\to\infty} \langle Ax_n, \varphi_k\rangle = \lim_{n\to\infty} \sum_{i=1}^{n} a_{ki}\xi_i = \sum_{i=1}^{\infty} a_{ki}\xi_i.$$

Hence, the following equality holds for every vector $x \in \mathcal{H}$:

$$Ax = \sum_{k=1}^{\infty} c_k\varphi_k, \quad \text{where } c_k = \sum_{i=1}^{\infty} a_{ki}\xi_i.$$

Thus, we have obtained a matrix representation of the operator A in terms of the basis $\{\varphi_k\}$.

Now we introduce a notion of absolute operator norm for linear bounded operators defined on separable Hilbert spaces.

Let $\mathcal{H}$ be a separable Hilbert space, and let A be a linear bounded operator defined everywhere on $\mathcal{H}$. Let $\{f_k\}$ and $\{\varphi_i\}$ be two arbitrary orthonormal bases in $\mathcal{H}$. Assume that

$$\sum_{i=1}^{\infty}\sum_{k=1}^{\infty} |\langle Af_k, \varphi_i\rangle|^2 < \infty.$$

The class of operators for which this inequality holds is called the *Schmidt class*. The quantity

$$\|A\|_2 = \Big(\sum_{i=1}^{\infty}\sum_{k=1}^{\infty} |\langle Af_k, \varphi_i\rangle|^2\Big)^{1/2} \tag{A.2.9}$$

does not depend on the choice of the bases $\{f_k\}$ and $\{\varphi_i\}$ and is called the *absolute norm* of the operator A.

Some properties of the absolute norm are:

1) the norm of an operator is not larger than its absolute norm: $\|A\| \le \|A\|_2$;

2) if B is an arbitrary bounded operator, we have

$$\|BA\|_2 \le \|B\|\,\|A\|_2, \quad \|AB\|_2 \le \|B\|\,\|A\|_2;$$

3) if A and B are linear operators,

$$\|A + B\|_2 \le \|A\|_2 + \|B\|_2.$$

Setting $f_k = \varphi_k$, $k = 1, 2, \dots$, in formula (A.2.9), we obtain

$$\|A\|_2 = \Big(\sum_{i=1}^{\infty}\sum_{k=1}^{\infty} |\langle A\varphi_k, \varphi_i\rangle|^2\Big)^{1/2} = \Big(\sum_{i=1}^{\infty}\sum_{k=1}^{\infty} |a_{ik}|^2\Big)^{1/2},$$

where $a_{ik} = \langle A\varphi_k, \varphi_i\rangle$. Thus, if $\|A\|_2 < \infty$, that is, the operator A belongs to the Schmidt class, the operator A admits a matrix representation, and

$$\sum_{i=1}^{\infty}\sum_{k=1}^{\infty} |a_{ik}|^2 < \infty.$$

In addition, the Schmidt class operators are completely continuous. An example of a Schmidt class operator is a Hilbert–Schmidt integral operator.

Let $x(t) \in L_2[a, b]$ and $K(s, t) \in L_2([a, b] \times [a, b])$, that is,

$$\int_a^b\int_a^b |K^2(s, t)|\,ds\,dt < \infty. \tag{A.2.10}$$

Define an integral operator $A : L_2[a,b] \to L_2[a,b]$ as follows:

$$Ax = \int_a^b K(s,t)x(t)\,dt. \tag{A.2.11}$$

An operator $A : L_2[a,b] \to L_2[a,b]$ of form (A.2.11) with a kernel $K(s,t)$ satisfying condition (A.2.10) is called a *Hilbert–Schmidt integral operator*. The Hilbert–Schmidt integral operator belongs to the Schmidt class, and its absolute norm is calculated by the formula

$$\|A\|_2^2 = \int_a^b \int_a^b |K^2(s,t)|\,ds\,dt.$$

Hence, the operator A is completely continuous, and the following upper estimate holds for its norm:

$$\|A\| \le \Big(\int_a^b \int_a^b |K^2(s,t)|\,ds\,dt \Big)^{1/2}.$$

Such operators have no bounded inverse, that is, solving the equation $Ax = y$ is an ill-posed problem.

Finally, it should be pointed out that a linear completely continuous operator $A : \mathcal{H}_1 \to \mathcal{H}_2$ that maps a separable Hilbert space $\mathcal{H}_1$ to a separable Hilbert space $\mathcal{H}_2$ has a singular value decomposition (see Theorem 2.9.2).

A.2.6 Linear operators in finite-dimensional spaces (matrices)

A linear operator $A : \mathcal{N}_1 \to \mathcal{N}_2$ that maps a finite-dimensional normed space $\mathcal{N}_1$ ($\dim \mathcal{N}_1 = n$) to a finite-dimensional normed space $\mathcal{N}_2$ ($\dim \mathcal{N}_2 = m$) has a matrix structure: $A = [a_{ij}]_{m \times n}$.

To prove this, let $\{e_1, e_2, \dots, e_n\}$ be a basis in $\mathcal{N}_1$, and let $\{\varphi_1, \varphi_2, \dots, \varphi_m\}$ be a basis in $\mathcal{N}_2$. Then every vector $x \in \mathcal{N}_1$ can be represented in the form $x = \sum_{j=1}^n x_j e_j$ and

$$Ax = \sum_{j=1}^n x_j A e_j = y \equiv \sum_{i=1}^m y_i \varphi_i.$$

Each of the vectors $Ae_j \in \mathcal{N}_2$ can be decomposed in terms of the basis $\{\varphi_i\}_{i=1}^m$: $Ae_j = \sum_{i=1}^m a_{ij}\varphi_i$. Then the coordinates of the vectors x and $y = Ax$ are related by the matrix $A_{e,\varphi} = [a_{ij}]_{m\times n}$ as follows: $y_\varphi = A_{e,\varphi} x_e$, where x_e is the coordinate column of the vector x in terms of the basis $\{e_j\}_{j=1}^n$, and y_φ is the coordinate column of the vector y in terms of the basis $\{\varphi_i\}_{i=1}^m$.

Thus, the linear operators in finite-dimensional spaces reduce to matrices.

The norm of a matrix $A : \mathcal{N}_1 \to \mathcal{N}_2$ is interrelated with the norms of $\mathcal{N}_1$ and $\mathcal{N}_2$ as follows:

$$\|A\| = \sup_{x \neq 0} \frac{\|Ax\|_{\mathcal{N}_2}}{\|x\|_{\mathcal{N}_1}}. \tag{A.2.12}$$

If $\mathcal{N}_1$ and $\mathcal{N}_2$ are spaces $\mathbb{R}^n$ with the cubic norm $\|x\|_{\text{cub}} = \max_{1 \le i \le n} |x_i|$, $\|A\|_{\text{cub}} = \max_{1 \le i \le n} \sum_{j=1}^{n} |a_{ij}|$. If $\mathcal{N}_1 = \mathcal{N}_2 = \mathbb{R}^n$ with the octahedral norm $\|x\|_{\text{oct}} = \sum_{i=1}^{n} |x_i|$, the corresponding norm of the matrix A is calculated by the formula $\|A\|_{\text{oct}} = \max_{1 \le j \le n} \sum_{i=1}^{n} |a_{ij}|$.

Consider a linear operator $A : \mathbb{E}^n \to \mathbb{E}^m$ acting from the Euclidean space $\mathbb{E}^n$ to the Euclidean space $\mathbb{E}^m$, that is, a rectangular matrix A having m rows and n columns. Such matrices are called *$m \times n$ matrices*.

Definition A.2.13. The operation that moves the elements a_{ij} of a matrix A from the intersection of the ith row and jth column to the intersection of the jth row and ith column is called *transposition*. The matrix obtained as the result of this transposition is called the *transpose* and denoted by A^T. Let $A = [a_{ij}]_{m \times n}$, then $A^T = [a_{ij}^T]_{n \times m}$, where $a_{ij}^T = a_{ji}$.

It should be noted that if λ is an eigenvalue of an $n \times n$ square matrix A and x is the corresponding eigenvector, we have

$$|\lambda| = \frac{\|Ax\|}{\|x\|} \le \|A\|.$$

Thus, for any eigenvalue $\lambda(A)$ of a square matrix A, the following inequality holds:

$$|\lambda(A)| \le \|A\|.$$

There is another property of the matrix norm (A.2.12) interrelated with the vector norms: the norm of any rectangular submatrix Δ of the matrix A is not larger than the norm of the matrix A: $\|\Delta\| \le \|A\|$.

A square matrix A is *similar* to a matrix B if there exists a nonsingular matrix C ($\det C \neq 0$) such that $A = CBC^{-1}$.

If a matrix A is similar to a matrix B, their characteristic polynomials coincide, that is,

$$\det(A - I\lambda) = \det(B - I\lambda),$$

where $I = \operatorname{diag}(1, 1, \dots, 1)$ is the unit matrix. Hence, similar matrices have the same set of eigenvalues.

Definition A.2.14. A square matrix A is called *symmetric* if $A = A^T$.

Definition A.2.15. A square matrix A is called *orthogonal* if its transpose is the inverse matrix, that is, $AA^T = A^T A = I$.

Some properties of orthogonal matrices are:

1) the vector-columns (as well as the vector-rows) of an orthogonal $n \times n$ matrix form an orthonormal basis in the space $\mathbb{R}^n$;
2) the determinant of an orthogonal matrix is ± 1;
3) matrices are orthogonal if and only if they transform one orthogonal basis to another;
4) orthogonal transformations leave the vector norm unchanged, that is, if A is an orthogonal matrix, $\|Ax\| = \|x\|$ for any $x \in \mathbb{R}^n$;
5) the product of orthogonal matrices is an orthogonal matrix.

Definition A.2.16. Representation of a square matrix A as the product of a symmetric matrix K and an orthogonal matrix W is called *polar decomposition*: $A = KW$.

Any square matrix has a polar decomposition.

A symmetric matrix K can be reduced, by using orthogonal transformations, to diagonal form, that is, there exists an orthogonal matrix U such that $K = U\tilde{\Sigma}U^T$, where $\tilde{\Sigma} = \operatorname{diag}(\tilde{\sigma}_1, \dots, \tilde{\sigma}_n)$ is a diagonal matrix.

Introduce a diagonal orthogonal matrix $\theta = \operatorname{diag}(\theta_1, \dots, \theta_n)$ as follows:

$$\theta_i = \begin{cases} 1, & \text{if } \tilde{\sigma}_i \geq 0, \\ -1, & \text{if } \tilde{\sigma}_i < 0, \end{cases} \qquad 1 < i \leq n.$$

The equality $\tilde{\Sigma} = \Sigma\theta$, where $\Sigma = \operatorname{diag}(\sigma_1, \dots, \sigma_n)$, $\sigma_i = |\tilde{\sigma}_i| \geq 0$, $1 \leq i \leq n$, is evident.

Thus, any square matrix A can be represented in the following form:

$$A = KW = U\Sigma\theta U^T W = U\Sigma V^T,$$

where $V = W^T U\theta$. It should be noted that V is an orthogonal matrix, because it is the product of three orthogonal matrices.

Definition A.2.17. Representation of a square matrix A in the form $A = U\Sigma V^T$ is called *singular value decomposition* if U and V are orthogonal $n \times n$ matrices and Σ is a diagonal matrix with nonnegative diagonal elements, $\Sigma = \operatorname{diag}(\sigma_1, \dots, \sigma_n)$, $\sigma_i \geq 0$, $1 \leq i \leq n$.

It should be noted that, since the operation of permutation of rows and columns of a matrix is an orthogonal transformation, the orthogonal matrices U and V can always be chosen so that the diagonal elements $\sigma_1, \dots, \sigma_n$ of the matrix Σ are arranged in any required order, for instance, $\sigma_1 \geq \sigma_2 \geq \cdots \geq \sigma_n \geq 0$.

Consider the symmetric matrix $A^T A = AA^T$ and express it in terms of the matrices U, Σ, and V:

$$A^T A = V\Sigma U^T U \Sigma V^T = V\Sigma^2 V^T,$$
$$AA^T = U\Sigma V^T V \Sigma U^T = U\Sigma^2 U^T.$$

Since $U^T = U^{-1}$ and $\det U \neq 0$, the matrices AA^T and Σ^2 are similar, and the diagonal elements σ_i of the matrix Σ arranged in decreasing order are related to the eigenvalues λ_i of the matrix AA^T (it is assumed that $\lambda_1 \geq \lambda_2 \geq \cdots \geq \lambda_n \geq 0$) by means of $\sigma_i = \sqrt{\lambda_i}$, $1 \leq i \leq n$.

Thus, the elements σ_i with their multiplicities are determined uniquely from the matrix A. They are called *singular values* of the matrix A. The orthogonal matrices U and V in the singular decomposition $A = U\Sigma V^T$ are not unique.

It should be noted that, since $|\det U| = |\det V^T| = 1$, we have

$$|\det A| = \sigma_1 \sigma_2 \cdots \sigma_n.$$

Hence, the rank of a matrix A is equal to the number of its positive singular values. In particular, $\det A \neq 0$ if and only if $\min_{1 \leq i \leq n} \sigma > 0$. In this case, the singular value decomposition of the inverse matrix A^{-1} has the form

$$A^{-1} = V\Sigma^{-1} U^T,$$

where $\Sigma^{-1} = \operatorname{diag}(1/\sigma_1, 1/\sigma_2, \ldots, 1/\sigma_n)$.

It should be noted that if A is a symmetric matrix ($A = A^T$), its singular values $\sigma_1 \geq \sigma_2 \geq \cdots \geq \sigma_n \geq 0$ are the moduli of the eigenvalues that may be arranged in a different order.

Now, consider an $m \times n$ rectangular matrix A.

Definition A.2.18. Representation of an $m \times n$ rectangular matrix A in the form $A = U\Sigma V^T$, where U and V are $m \times m$ and $n \times n$ orthogonal matrices, respectively, and Σ is a diagonal $m \times n$ matrix with nonnegative elements $\sigma_{ii} = \sigma_i \geq 0$, is called *singular value decomposition* of the matrix A. The numbers $\sigma_i \geq 0$, $1 \leq i \leq \min(m, n)$ are called *singular values* of the matrix A.

The set of singular values of a rectangular matrix A is uniquely determined by the matrix A and does not depend on the choice of the orthogonal matrices U and V. Hence, the singular values of a matrix are its orthogonal invariants. In other words, multiplication of a matrix by any orthogonal matrices on the left and right does not change the set of its singular values.

Let an $m \times n$ rectangular matrix A have a set of singular values σ_i, $1 \leq i \leq p$, where $p = \min(m, n)$. In what follows, the singular values are assumed to be arranged in decreasing order: $\sigma_1 \geq \sigma_2 \geq \cdots \geq \sigma_p \geq 0$.

Now we show that for a matrix A the norm interrelated with the Euclidean norm of a vector coincides with the maximal singular value of the matrix A: $\|A\| = \sigma_1$. Recall that if A is an $m\times n$ rectangular matrix and $y = Ax$, $y \in \mathbb{R}^m$, $x \in \mathbb{R}^n$, the Euclidean norm of the vectors x and y is calculated by the formulas

$$\|x\| = \sqrt{x_1^2 + x_2^2 + \cdots + x_n^2},$$

$$\|y\| = \sqrt{y_1^2 + y_2^2 + \cdots + y_m^2},$$

and the matrix A norm interrelated with this vector norm is defined as

$$\|A\| = \max_{\|x\|=1} \|Ax\| = \sup_{x\neq 0} \frac{\|Ax\|}{\|x\|}.$$

Since orthogonal mappings leave unchanged the vector norm, they also leave unchanged the matrix norm induced by this vector norm:

$$\|UAV\| = \sup_{x\neq 0} \frac{\|UAVx\|}{\|x\|} = \sup_{x\neq 0} \frac{\|AVx\|}{\|x\|} = \sup_{x\neq 0} \frac{\|AVx\|}{\|Vx\|} = \|A\|.$$

Here A is a rectangular $m\times n$-matrix, and U and V are orthogonal $m\times m$ and $n\times n$ matrices, respectively.

Hence, if $A = U\Sigma V^T$ is the singular value decomposition of the matrix A, we have $\|\Sigma\| = \|U^T AV\| = \|A\|$. Taking into account the diagonal structure of the matrix Σ, we obtain

$$\|A\| = \|\Sigma\| = \max_{\|x\|=1} \|\Sigma x\| = \max_{\|x\|=1} \left(\sqrt{\sigma_1^2 x_1^2 + \sigma_2^2 x_2^2 + \cdots + \sigma_p^2 x_p^2}\right) = \sigma_1,$$

which was to be proved.

The relation

$$\min_{x\neq 0} \frac{\|Ax\|}{\|x\|} = \begin{cases} \sigma_p, & \text{if } m \geq n, \\ 0, & \text{if } m < n \end{cases}$$

is established in a similar way.

Let $\sigma_j(A)$ be the jth singular value of a matrix A in the ordered set $\sigma_1 \geq \sigma_2 \geq \cdots \geq \sigma_j \geq \cdots \geq \sigma_p$. Then the following statements hold (a proof may be found in [4]):

1) for any $m\times n$ matrices A and B, the following relation holds:

$$|\sigma_j(A + B) - \sigma_j(A)| \leq \|B\|, \qquad 1 \leq j \leq p;$$

2) if T, A, and R are nonsingular square $n\times n$ matrices with T and R close to the unit matrix I, that is, $\|T - I\| \leq \eta$, $\|R - I\| \leq \eta$, $\eta \geq 0$, the following inequality is valid:

$$|\sigma_j(A) - \sigma_j(TAR)| \leq \sigma_j(A)(2\eta + \eta^2), \quad 1 \leq j \leq n.$$

Definition A.2.19. Let σ be a singular value of an $m \times n$ matrix A. A vector $v \in \mathbb{R}^n$ is called a *right singular vector* and a vector $u \in \mathbb{R}^m$ is called a *left singular vector* of the matrix A corresponding to the singular value σ, if u and v are a nontrivial solution to the homogeneous system of equations

$$\begin{aligned} Av &= \sigma u, \\ A^T u &= \sigma v. \end{aligned}$$

It should be noted that if v and u are singular vectors of a matrix A corresponding to a singular number σ, we have

$$\begin{aligned} A^T A v &= A^T \sigma u = \sigma^2 v, \\ A A^T u &= A \sigma v = \sigma^2 u. \end{aligned}$$

This means that the right singular vectors of the matrix A are eigenvectors of the symmetric matrix $A^T A$ (at $\lambda = \sigma^2$) and the left singular vectors are eigenvectors of the matrix AA^T.

Let $u^{(j)}$ denote a left singular vector, and let $v^{(j)}$ be a right singular vector of the matrix A corresponding to a singular value σ_j ($1 \le j \le p$). Then $\langle u^{(j)}, u^{(i)} \rangle = 0$ and $\langle v^{(j)}, v^{(i)} \rangle = 0$ if $\sigma_i \neq \sigma_j$. In other words, the right (left) singular vectors $v^{(j)}$ and $v^{(i)}$ ($u^{(j)}$ and $u^{(i)}$) corresponding to different singular values σ_j and σ_i are orthogonal.

If A is a square nonsingular matrix of order n, there exists an orthonormal system of n right singular vectors and an orthonormal system of n left singular vectors.

For a rectangular $m \times n$ matrix A of rank r, $r \le p = \min(m, n)$, we introduce a notion of matrix kernel as the linear space of solutions of the equation $Ax = 0$. The kernel dimension of the matrix A is $n - r$, and the kernel dimension of A^T is $m - r$. We construct an orthonormal basis in the kernel of the matrix A and call the $(n - r)$ basis vectors "null" right singular vectors. Similarly, the $(m - r)$ vectors of an orthonormal basis of the kernel of A^T are called"null" left singular vectors of the matrix A. Then for every rectangular $m \times n$ matrix there exists an orthonormal system of n right singular vectors (including"null" ones) and an orthonormal system of m left singular vectors (including"null" ones).

It should be noted that if $A = U \Sigma V^T$ is the singular value decomposition of a rectangular matrix A, it follows from the equality $A^T = V \Sigma^T U^T$ that

$$AV = U\Sigma, \qquad A^T U = V \Sigma^T.$$

Taking into account the diagonal structure of the matrix $\Sigma = \operatorname{diag}\{\sigma_1, \sigma_2, \dots, \sigma_p\}$, one can see that the columns of the matrix U are left singular vectors of the matrix A, whereas the columns of the matrix V are right singular vectors.

Now, define some numerical characteristics of a matrix.

Definition A.2.20. The number

$$\mathcal{F} = \Big(\sum_{i=1}^{m}\sum_{j=1}^{n} a_{ij}^2\Big)^{1/2}$$

is called the *Frobenius norm* of an $m\times n$ matrix A.

The Frobenius norm of a matrix has the following properties:

1) $\mathcal{F}(A) = \mathcal{F}(A^T) = (\sum_{j=1}^{p} \sigma_j^2(A))^{1/2}$, where $p = \min(m, n)$;
2) The Frobenius norm is an orthogonal invariant of the matrix A, that is,

$$\mathcal{F}(UAV) = \mathcal{F}(A),$$

where $U \in \mathbb{R}^m \times \mathbb{R}^m$ and $V \in \mathbb{R}^n \times \mathbb{R}^n$ are orthogonal matrices;
3) the inequality

$$\|A\| \leq \mathcal{F}(A) = \sqrt{p}\,\|A\|,$$

where $\|A\|$ coincides with the maximal singular value of the matrix A, holds.

The norm of the matrix A can also be estimated with the numerical characteristic

$$\mathcal{M}(A) = \max\Big\{\max_i \sum_{j=1}^{n} |a_{ij}|,\ \max_j \sum_{i=1}^{m} |a_{i,j}|\Big\},$$

for which the following relation is valid:

$$\|A\| \leq \mathcal{M}(A) \leq \max(\sqrt{m}, \sqrt{n})\|A\|.$$

Definition A.2.21. The *condition number* of a square $n\times n$ matrix A is the quantity

$$\mu(A) = \sup_{x\neq 0, \xi\neq 0} \Big\{\frac{\|Ax\|\,\|\xi\|}{\|A\xi\|\,\|x\|}\Big\},$$

where $\|\cdot\|$ is the Euclidean norm.

If the matrix A is singular, we have $\mu(A) = \infty$. For a nonsingular matrix A, its condition number can be expressed in terms of singular values $\sigma_1(A) \geq \sigma_2(A) \geq \cdots \geq \sigma_n(A) > 0$ as follows:

$$\mu(A) = \frac{\sigma_1(A)}{\sigma_n(A)}.$$

The following properties of condition numbers of nonsingular matrices can be directly obtained from the above relation:

1) $\mu(A) \geq 1$, $\mu(I) = 1$, where I is the unit matrix;

2) $\mu(\text{diag}(a_1, a_2, \ldots, a_n)) = \frac{\max_{1 \leq i \leq n} |a_i|}{\min_{1 \leq i \leq n} |a_i|}$;

3) $\mu(cA) = \mu(A)$, $c = \text{const}$;

4) $\mu(A^{-1}) = \mu(A)$;

5) $\mu(A) = \|A\| \|A^{-1}\|$.

Now, consider a system of linear algebraic equations $Aq = f$ and a system with a perturbed right-hand side $A(q+\Delta q) = f+\Delta f$. Here Δq is the solution perturbation that corresponds to a perturbation Δf of the right-hand side. Then

$$\frac{\|\Delta q\|}{\|q\|} \leq \mu(A) \frac{\|\Delta f\|}{\|f\|}.$$

This inequality estimates the relative error in the solution in terms of the relative error in the right-hand side. Therefore, $\mu(A)$ is also called the condition number of the system $Aq = f$.

And now consider the system $(A + \Delta A)(q + \Delta q) = f + \Delta f$, which is obtained from the system $Aq = f$ with a perturbation Δf of the right-hand side and a perturbation ΔA of the matrix A. It will be assumed that $\det A \neq 0$, and

$$\frac{\|\Delta A\|}{\|A\|} \leq \alpha, \qquad \frac{\|\Delta f\|}{\|f\|} \leq \gamma.$$

Then, if $\alpha\mu(A) < 1$, we have

$$\frac{\|\Delta A\|}{\|A\|} \leq (\alpha + \beta) \frac{\|\mu A\|}{\|1 - \alpha\mu(A)\|}.$$

One can see from this inequality that the smaller is the condition number of a system $Aq = f$, the more stable is its solution.

The corresponding perturbation of the determinant $|\det(A + \Delta A) - \det A|$ can also be estimated in terms of the condition number $\mu(A)$:

$$\left| \frac{\det(A + \Delta A) - \det A}{\det A} \right| \leq \frac{n\mu(A)\|\Delta A\|/\|A\|}{1 - n\mu(A)\|\Delta A\|/\|A\|}.$$

A proof of the above two inequalities can be found in [4].

A.3 Dual space and adjoint operator

The notions of dual space and adjoint operator are based on the notion of a functional in a topological space.

A.3.1 Functionals

Definition A.3.1. A mapping $J : D \to \mathbb{C}$ that assigns some complex number to every $x \in D$ is called a *functional* $J(x)$ on the set D.

Since functionals are a particular case of mappings, all results of Section A.2 are valid for them.

Below we consider real-valued functionals $J : D \to \mathbb{R}$.

Definition A.3.2. A functional $J(x)$ defined on a set D is called *lower (upper) bounded* if there exists a constant C such that for every $x \in D$ the inequality $J(x) \geq C$ ($J(x) \leq C$) holds, that is, $\inf_{x \in D} J(x) > -\infty$ ($\sup_{x \in D} J(x) < \infty$). A functional $J(x)$ is called *bounded* on a set D if it is bounded on D both lower and upper.

Definition A.3.3. A functional $J(x)$ defined on a set D of a topological space $(\mathcal{T}, \tau)$ is called *τ-sequentially lower (upper) semi-continuous at a point* $x \in D$ if for any sequence $\{x_n\} \subset D$ convergent to the point x in the topology τ the inequality $\underline{\lim}_{n \to \infty} J(x_n) \geq J(x)$ ($\overline{\lim}_{n \to \infty} J(x_n) \leq J(x)$) holds.

A functional $J(x)$ is called *τ-sequentially lower (upper) semi-continuous on a set* $X \subset D$ if it is τ-sequentially lower (upper) semi-continuous at each point of this set.

A functional $J(x)$: $\mathcal{T} \to \mathbb{R}$ is *sequentially continuous* in the topology τ (or τ-sequentially continuous) at a point $x \in D$ (on a set $X \subset D$) if it is τ-sequentially lower and upper semi-continuous at the point x (on the set X).

Definition A.3.4. A sequence of elements $\{x_n\} \subset D$ is called *minimizing for a functional $J(x)$ on the set D* if $\lim_{n \to \infty} J(x_n) = J^* = \inf_{x \in D} J(x)$.

The next theorem is an analog of the well-known Weierstrass theorem.

Theorem A.3.1. *Let a functional $J(x)$ be defined on a subset D of a topological space $(\mathcal{T}, \tau)$, with D being sequentially compact in the topology τ and $J(x)$ being τ-sequentially lower semi-continuous on D. Then $J^* = \inf_{x \in D} J(x) > -\infty$ and $X_* = \{x \in D\colon J(x) = J^*\} \neq \varnothing$. The set X_* is also sequentially compact in the topology τ. Any minimizing sequence $\{x_n\} \subset D$ for the functional $J(x)$ on D is sequentially compact and converges to X_* in the topology τ.*

The convergence of a sequence $\{x_n\}$ to a set $X \subset \mathcal{T}$ of a topological space $(\mathcal{T}, \tau)$ means that for any neighborhood $O(X)$ of the set X (that is, for any set of the family τ containing X) there exists such a number $n(O(X))$ that $x_n \in O(X)$ for all $n \geq n(O(X))$.

Remark A.3.1. In topological spaces one can define lower and upper semi-continuity by passing from a sequence to a direction. If a topological space is metrizable, Definition A.3.3 and Theorem A.3.1 can be reformulated in terms of compact sets and lower semi-continuous functionals (see Remark A.1.7).

It follows from Theorem A.3.1 that a continuous functional that maps a compact metric space to $\mathbb{R}$ is bounded and reaches its upper and lower bounds.

Now, consider a functional $J\colon \mathcal{L} \to \mathbb{R}$ in a linear space $\mathcal{L}$.

Definition A.3.5. A functional $J(x)$ defined on a convex set D (see Definition A.1.38) of a linear space $\mathcal{L}$ is called *convex* if for any $x, y \in D$ and any number $\lambda \in [0, 1]$ the following inequality holds:

$$J[\lambda x + (1-\lambda)y] \le \lambda J(x) + (1-\lambda)J(y).$$

If the sign of equality in this relation is possible for $x \ne y$ only at $\lambda = 0$ and $\lambda = 1$ or at $x = y$, the functional $J(x)$ is called *strictly convex.*

Theorem A.3.2. *Let $J(x)$ be a strictly convex functional defined on a convex set D of a linear space $\mathcal{L}$. Then, if $J(x_1) = J(x_2) = J^* \equiv \inf_{x\in D} J(x)$ for $x_1, x_2 \in D$, we have $x_1 = x_2$.*

The property of convexity of a functional can be strengthened in Banach and Hilbert spaces.

Definition A.3.6. A functional $J : \mathcal{B} \to \mathbb{R}$ defined on a convex set $D \subset \mathcal{B}$ of a Banach space $\mathcal{B}$ is called *strictly uniformly convex on D* if there exists a numerical function $\delta(t)$ defined at $0 \le t \le \operatorname{diam} D \equiv \sup_{x,y\in D} \|x - y\|$ and such that $\delta(+0) = 0$, $\delta(t)$ monotonically increases on the domain of definition and the following relation holds:

$$J[\lambda x + (1-\lambda)y] \le \lambda J(x) + (1-\lambda)J(y) - \lambda(1-\lambda)\delta(\|x-y\|)$$

for all $x, y \in D$ and $\lambda \in [0, 1]$.

The function $\delta(t)$ is called the *modulus of convexity* for J on D.

Strictly uniformly convex functionals are strictly convex. A strictly uniformly convex functional preserves the property of strict uniform convexity when it is added to a convex functional.

Definition A.3.7. A functional $J : \mathcal{H} \to \mathbb{R}$ defined on a convex set D of a Hilbert space $\mathcal{H}$ is called *strongly convex* if there exists a constant $\varkappa > 0$ such that

$$J[\lambda x + (1-\lambda)y] \le \lambda J(x) + (1-\lambda)J(y) - \lambda(1-\lambda)\varkappa\|x-y\|^2$$

for any $x, y \in D$ and any $\lambda \in [0, 1]$.

It is evident that a strongly convex functional is a strictly uniformly convex functional with a modulus of convexity $\delta(t) = \varkappa t^2$ ($\varkappa > 0$). The functional $J(x) = \|x\|^2$ in a Hilbert space is strongly convex with $\varkappa = 1$.

Let J be a linear continuous functional defined on a linear subspace $\mathfrak{L}' \subset \mathfrak{L}$ which is not necessarily everywhere dense in a linear space $\mathfrak{L}$. The functional J can be extended to the entire space with the same norm. A corresponding theorem known as the *Hahn–Banach extension principle* plays an important role in mathematical analysis.

Theorem A.3.3 (Hahn–Banach extension principle). *Let a gauge function $p(x)$ be defined on a real linear space $\mathfrak{L}$ (see Definition A.1.41), and let a linear real functional $J_0(x)$ defined on a linear subspace $\mathfrak{L}' \subset \mathfrak{L}$ satisfy the condition*

$$J_0(x) \le p(x), \quad x \in \mathfrak{L}'.$$

Then there exists a linear real functional $J(x)$ defined on the entire $\mathfrak{L}$ and such that $J(x) = J_0(x)$ at $x \in \mathfrak{L}'$ and $J(x) \le p(x)$ for all $x \in \mathfrak{L}$.

If $\mathfrak{L}$ is a linear real normed space and J_0 is a bounded functional on $\mathfrak{L}' \subset \mathfrak{L}$, $p(x) = \|x\|_{\mathfrak{L}} \|J_0\|_{\mathfrak{L}'}$, $x \in \mathfrak{L}$ can be taken as the gauge function. Then the equality $\|J\|_{\mathfrak{L}} = \|J_0\|_{\mathfrak{L}'}$ holds, that is, the functional J_0 is extended to the continuous functional J on $\mathfrak{L}$ with the same norm.

The Hahn–Banach theorem is also true for complex spaces, although in a less general form.

Theorem A.3.4. *Let a seminorm $p(x)$ be defined in an arbitrary linear space $\mathfrak{L}$ (see Definition A.1.41). Let $J_0(x)$ be a linear functional defined on a linear subspace $\mathfrak{L}' \subset \mathfrak{L}$ such that*

$$|J_0(x)| \le p(x), \quad x \in \mathfrak{L}'.$$

Then there exists a linear functional $J(x)$ defined on the entire $\mathfrak{L}$ coinciding with $J_0(x)$ on $\mathfrak{L}'$ and satisfying on $\mathfrak{L}$ the following condition:

$$|J(x)| \le p(x), \quad x \in \mathfrak{L}.$$

Some corollaries follow from the Hahn–Banach theorem for the case when $\mathfrak{L}$ is a locally convex space (see Definition A.1.40).

Corollary A.3.1. *Let $J_0(x)$ be a linear continuous functional defined on a linear subspace $\mathfrak{L}' \subset \mathfrak{L}$ of a locally convex space $\mathfrak{L}$. There exists a linear continuous functional $J(x)$ on $\mathfrak{L}$ which is an extension of the functional $J_0(x)$.*

Corollary A.3.2. *For any point x_0 of a linear space $\mathfrak{L}$ and a seminorm $p(x)$ there exists a linear functional $J(x)$ on $\mathfrak{L}$ for which $|J(x)| \le p(x)$ and $J(x_0) = p(x_0)$.*

Corollary A.3.3. *Let $\mathcal{L}$ be a a locally convex space. If $J(x) = 0$ for any linear continuous functional J on $\mathcal{L}$, $x = 0$.*

Remark A.3.2. Corollary A.3.3 establishes an important property of locally convex spaces: the presence of a sufficient set of linear continuous functionals. Owing to this property, the theory of locally convex sets provides more results than the theory of linear topological spaces. In addition, almost all spaces covered by functional analysis are locally convex.

A.3.2 Dual space

As mentioned above (see Definition A.2.5), the set of linear operators mapping a linear space to another one forms a linear space. In particular, the set of all linear functionals on a general linear space forms a linear space called dual (algebraic dual space). We consider dual spaces of linear topological spaces, where the requirement of linearity of functionals is complemented by the requirement of continuity (topological dual space).

Definition A.3.8. The linear space $(\mathcal{L}, \tau)^*$ of all linear continuous functionals defined on a linear topological space $(\mathcal{L}, \tau)$ is called the *dual space of* $(\mathcal{L}, \tau)$.

By virtue of Remark A.3.2, the dual space is often considered under the assumption that $\mathcal{L}$ is a locally convex space (in particular, a normed space: in this case, according to Theorem A.1.4, $\mathcal{L}^*$ is a Banach space). If $\mathcal{L}$ is a locally convex space, $\mathcal{L}^*$ separates points on $\mathcal{L}$. For an arbitrary linear topological space $\mathcal{L}$, in some cases $\mathcal{L}^* = \{\mathbf{0}\}$ (although $\mathcal{L}^*$ may separate points on $\mathcal{L}$ also when $\mathcal{L}$ is not a locally convex space).

The dual space is a space of linear operators, and, therefore, uniform and pointwise convergences can be introduced (see Definitions A.2.8 and A.2.9).

First, consider the case when the initial space is a normed one.

Definition A.3.9. Let $\mathcal{N}$ be a normed space, and let $\mathcal{N}^*$ be its dual space. The topology arising from the norm introduced in $\mathcal{N}^*$ is called *strong topology* in the space $\mathcal{N}^*$:

$$\|l\|_{\mathcal{N}^*} = \sup_{\|x\|_{\mathcal{N}} \leq 1} |l(x)|.$$

Otherwise, as a neighborhood of zero in a space $\mathcal{N}^*$ we take the set of functionals $\{l\} \subset \mathcal{N}^*$ for which $|l(x)| < \varepsilon$ when x belongs to the closed unit ball $\{x \in \mathcal{N}: \|x\| \leq 1\}$. Choosing all possible ε, we obtain a basis of neighborhoods of zero.

Convergence in the dual space determined by the strong topology coincides with uniform convergence (see Definition A.2.8). This convergence is also called *strong convergence*.

Definition A.3.10. Let $\mathcal{N}$ be a normed space, and let $\mathcal{N}^*$ be its dual space. *Weak topology* in a space $\mathcal{N}^*$ is the topology induced by a basis of neighborhoods of zero of the space $\mathcal{N}^*$ consisting of sets of the following form:

$$\{l \in \mathcal{N}^* : |l(x_i)| < \varepsilon,\ x_i \in B_n\},$$

where ε is an arbitrary positive number, and B_n is an arbitrary finite set in $\mathcal{N}$: $B_n = \{x_i\}_{i=1}^n \subset \mathcal{N}, n < \infty$.

The weak topology determines *weak convergence* in the space $\mathcal{N}^*$: a sequence of functionals $l_m \in \mathcal{N}^*$ is called *weakly convergent* to a functional $l \in \mathcal{N}^*$ as $m \to \infty$ if for any element $x \in \mathcal{N}$ $l_m(x) \to l(x)$ as $m \to \infty$.

As in the space $\mathcal{N}^*$, in the initial normed space $\mathcal{N}$, in addition to the basic topology (*strong topology*) induced by the norm in $\mathcal{N}$, there can be another locally convex topology.

Definition A.3.11. *Weak topology* in a normed space $\mathcal{N}$ is determined by a neighborhood basis of zero of the space $\mathcal{N}$ consisting of sets of the form

$$\{x \in \mathcal{N} : |l_i(x)| < \varepsilon,\ l_i \in B_n^*\},$$

where ε is an arbitrary positive number, and B_n^* is an arbitrary finite set in $\mathcal{N}^*$: $B_n^* = \{l_i\}_{i=1}^n \subset \mathcal{N}^*, n < \infty$.

The weak topology determines weak convergence in the space $\mathcal{N}$. A sequence of elements $x_m \in \mathcal{N}$ is called *weakly convergent* to an element $x \in \mathcal{N}$ as $m \to \infty$ if for any element $l \in \mathcal{N}^*$ $l(x_m) \to l(x)$ as $m \to \infty$.

It should be noted that strong convergence of a sequence $\{x_n\} \subset \mathcal{N}$ implies weak convergence to the same limit. Strong convergence of a sequence $\{x_n\}$ to an element $x \in \mathcal{N}$ is denoted by $x_n \to x$, and weak convergence is denoted by $x_n \rightharpoonup x$.

If $\mathcal{N}$ is a uniformly convex Banach space (see Definition A.1.30), it follows from $x_n \rightharpoonup x$ in $\mathcal{N}$ and $\|x_n\| \to \|x\|$ that $x_n \to x$.

In a finite-dimensional space strong and weak convergence coincide.

Remark A.3.3. Definitions A.3.9–A.3.11 remain valid also when the initial space is a linear topological space $(\mathcal{L}, \tau)$. In this case, the unit ball in Definitions A.3.9–A.3.11 should be replaced by a bounded set in $(\mathcal{L}, \tau)$.

Compactness, boundedness, completeness, and other properties in a space $(\mathcal{L}, \tau)$ with weak topology are called weak compactness, weak boundedness, weak completeness, etc.

As mentioned above, the dual space $\mathcal{N}^*$ of a normed space $\mathcal{N}$ is a Banach space. Therefore, one can construct the dual space $\mathcal{N}^{**}$ of $\mathcal{N}^*$, and so on. The space $\mathcal{N}^{**}$ is called the *double dual space*.

Every element $x_0 \in \mathcal{N}$ defines a linear continuous functional on $\mathcal{N}^*$: $L_{x_0}(l) = l(x_0)$, where x_0 is a fixed element from $\mathcal{N}$, and l is every element from the entire space $\mathcal{N}^*$. Thus, we have a mapping of the entire space $\mathcal{N}$ on some subset of the space $\mathcal{N}^{**}$. This mapping from $\mathcal{N}$ to $\mathcal{N}^{**}$ is called *natural*. This mapping is one-to-one, isomorphic (that is, it follows from $x \leftrightarrow L_x$ and $y \leftrightarrow L_y$ that $x+y \leftrightarrow L_x + L_y$ and $\lambda x \leftrightarrow L_{\lambda x}$, where λ is a number), and isometric (that is, $\|x\| = \|L_x\|$).

Definition A.3.12. If the natural mapping of a normed space $\mathcal{N}$ maps this space onto the entire space $\mathcal{N}^{**}$, the space $\mathcal{N}$ is called *reflexive*. In this case, there is no distinction between the spaces $\mathcal{N}$ and $\mathcal{N}^{**}$: $\mathcal{N} = \mathcal{N}^{**}$.

Each uniformly convex Banach space is reflexive.

The dual space $\mathcal{B}^*$ of a Banach space $\mathcal{B}$ is complete in terms of weak convergence. In other words, if a sequence of functionals $\{l_m\} \subset \mathcal{B}^*$ is fundamental at each point $x \in \mathcal{B}$, there exists a linear functional $l(x)$ such that $l_m(x) \to l(x)$ for any $x \in \mathcal{B}$.

Some properties of weak convergence are:

A sequence of linear bounded functionals l_m mapping a Banach space $\mathcal{B}$ to $\mathbb{R}$ weakly converges to a functional l if and only if the sequence $\{\|l_m\|\}$ is bounded and $l_m(x) \to l(x)$ for any x from a set X of elements whose linear combinations lie everywhere dense in $\mathcal{B}$.

A sequence $\{x_m\} \in \mathcal{B}$ weakly converges to $x \in \mathcal{B}$ if and only if the sequence $\{\|x_m\|\}$ is bounded and $l(x_m) \to l(x)$ for any l from a set L of linear functionals whose linear combinations lie everywhere dense in $\mathcal{B}^*$.

If a linear bounded operator A maps a normed space $\mathcal{N}_1$ to a normed space $\mathcal{N}_2$ and a sequence $\{x_m\} \subset \mathcal{N}_1$ weakly converges to $x \in \mathcal{N}_1$, the sequence $\{Ax_m\} \subset \mathcal{N}_2$ weakly converges to $Ax \in \mathcal{N}_2$.

A weakly convergent sequence $\{x_m\}$ of elements of a normed space $\mathcal{N}$ is bounded, that is, the norms of elements of this sequence as a whole are bounded.

For a Banach space $\mathcal{B}$, the following analog of Theorem A.3.1 holds:

Theorem A.3.5. *Let D be a (weakly) compact set of a Banach space $\mathcal{B}$, and let a functional $J(x)$ be defined on D and lower semi-continuous on D (see Definition* A.3.3 *and Remark* A.3.1*). Then $J^* = \inf_{x\in D} J(x) > -\infty$ and the set $X_* = \{x \in D\colon J(x) = J^*\}$ is not empty and (weakly) compact. Every minimizing sequence for the functional J on D (weakly) converges to X_*.*

The following statements are useful for testing sets for weak compactness:

- every bounded weakly closed set of a reflexive Banach space is weakly compact;
- a convex closed set of a Banach space is weakly closed;
- for convex functionals there also exists a criterion of weak semi-continuity.

Theorem A.3.6. *Let D be a convex set of a Banach space. A functional $J(x)$ that is convex on D is weakly lower semi-continuous on D if and only if $J(x)$ is lower semi-continuous on D.*

It should be noted that the functional $J(x) = \|x\|^\gamma$ $(\gamma \geq 1)$ is weakly lower semi-continuous on the Banach space $\mathcal{B}$.

Every ball in the dual space $\mathcal{N}^*$ of a separable normed space $\mathcal{N}$ is weakly compact.

Theorem A.3.7. *Let D be a a convex closed set of a reflexive Banach space, and let $J(x)$ be a functional defined and strictly uniformly convex on D (see Definition* A.3.6*). Then the set $X_* = \{x \in D\colon J(x) = J^* = \inf_{x \in D} J(x)\}$ is not empty and consists of only one point $x^* \in D$, and every sequence that is minimizing for J on D converges to this point.*

Now consider linear continuous functionals defined on a Hilbert space $\mathcal{H}$.

Theorem A.3.8 (F. Riesz). *Every linear continuous functional $l(x)$ defined on a Hilbert space $\mathcal{H}$ can be represented as $l(x) = \langle l, x\rangle$, where the element $l \in \mathcal{H}$ is uniquely defined by the functional $l(x)$. In this case $\|l(x)\| = \|l\|_{\mathcal{H}}$.*

It follows from the Riesz theorem that weak convergence of a sequence $\{x_m\} \subset \mathcal{H}$ $(x_m \rightharpoonup x \in \mathcal{H})$ is equivalent to convergence of the inner products $\langle x_m, y\rangle \to \langle x, y\rangle$ as $m \to \infty$ for every $y \in \mathcal{H}$. Weak convergence in Hilbert spaces was considered in Subsection A.1.2. It also follows from Theorem A.3.8 that the dual space $\mathcal{H}^*$ can be identified with $\mathcal{H}$: $\mathcal{H} = \mathcal{H}^*$. A Hilbert space is obviously reflexive.

Examples of dual spaces

1. The dual space of a Euclidean space $\mathbb{E}$ is the space $\mathbb{E}$ itself: $\mathbb{E}^* = \mathbb{E}$.

2. Consider the space l_1 of numerical sequences $x = (x_1, x_2, \ldots)$ such that $\sum_{i=1}^{\infty} |x_i| < \infty$. The space $l_1^* = l_\infty$ of bounded numerical sequences $(\xi_1, \xi_2, \ldots)$, that is, such that $\sup_{1 \leq i < \infty} |\xi_i| < \infty$, is the dual space of this space. The space l_∞ (as well as l_1) is not reflexive.

3. The space l_q, where $1/p + 1/q = 1$, $\infty > p, q > 1$, is the dual space of the space l_p $(\infty > p > 1)$ (see Subsection A.1.5). Hence, the space l_p is reflexive: $l_p^{**} = l_p$. The space l_2 is self-dual: $l_2^* = l_2$. A similar result is true for the functional space L_p: $L_p^* = L_q$, $1/p + 1/q = 1, \infty > p, q > 1$, $L_2^* = L_2$ (see Remark A.5.1).

A.3.3 Adjoint operator

Definition A.3.13. Let A be a linear bounded operator that maps a Hilbert space $\mathcal{H}_1$ to a Hilbert space $\mathcal{H}_2$. An operator $A^* : \mathcal{H}_2 \to \mathcal{H}_1$ is called the *adjoint* of the operator A, if for all $x \in \mathcal{H}_1$, $y \in \mathcal{H}_2$ the following equality holds:

$$\langle y, Ax \rangle_{\mathcal{H}_2} = \langle A^* y, x \rangle_{\mathcal{H}_1}. \tag{A.3.1}$$

Remark A.3.4. If a linear continuous operator A is defined not on the entire space $\mathcal{H}_1$, it may have many adjoint operators. To be more exact, a bounded operator A has a unique adjoint operator if and only if its domain of definition D_A is everywhere dense in the space $\mathcal{H}_1$. Since the continuous operator A can be continuously extended to the entire space $\mathcal{H}_1$, the linear operator A is usually considered to be defined and continuous on the entire space $\mathcal{H}_1$.

If a linear operator A is continuous, A^* is also continuous, and $\|A^*\| = \|A\|$.
Some properties of adjoint operators are as follows:

1) $\|A^*A\| = \|AA^*\| = \|A\|^2$;

2) $(A^*)^* = A$;

3) $(\alpha A + \beta B)^* = \alpha A^* + \beta B^*$;

4) $(AB)^* = B^* A^*$.

Theorem A.3.9. *If A is a bounded operator defined everywhere in $\mathcal{H}$ and if A^*A is completely continuous, the operator A is completely continuous.*

It follows from Theorem A.3.9 that if A is a completely continuous operator, the operator A^* has the same property.

A sequence of operators $\{A_m\}$ in a Hilbert space $\mathcal{H}$ ($A_m : \mathcal{H} \to \mathcal{H}$) is said to be *weakly convergent* to an operator $A : \mathcal{H} \to \mathcal{H}$ if for any $x, y \in \mathcal{H}$

$$\langle A_m x, y \rangle \to \langle Ax, y \rangle.$$

It is evident that if $\{A_m\}$ weakly converges to A, $\{A_m^*\}$ weakly converges to A^* as $m \to \infty$.

Generally speaking, pointwise convergence $A_m^* \to A^*$ does not follow from pointwise convergence $A_m \to A$ as $m \to \infty$.

Below is an example of an adjoint operator.

Let an operator $A : L_2[a, b] \to L_2[a, b]$ be defined by the formula

$$Ax = \int_a^b K(t, \tau) x(\tau)\, d\tau.$$

The adjoint operator A^* is defined from the condition $\langle Ax, y\rangle_{L_2} = \langle x, A^*y\rangle_{L_2}$ for any $x, y \in L_2[a,b]$. Hence,

$$\begin{aligned}\langle Ax, y\rangle_{L_2} &= \int_a^b y(t)\int_a^b K(t,\tau)x(\tau)\,d\tau\,dt\\ &= \int_a^b x(\tau)\int_a^b K(t,\tau)y(t)\,dt\,d\tau\\ &= \int_a^b x(t)\int_a^b K^*(t,\tau)y(\tau)\,d\tau\,dt = \langle x, y^*\rangle_{L_2},\end{aligned}$$

where $y^* = A^*y = \int_a^b K^*(t,\tau)y(t)\,dt$, $K^*(t,\tau) = K(\tau,t)$.

In Euclidean spaces, where linear bounded operators have matrix form $A = [a_{ij}]_{m\times n}$, adjoint operators in an orthonormal basis correspond to *adjoint matrices* $A^* = [a^*_{ij}]_{n\times m}$, $(a^*_{ji} = \bar{a}_{ij})$. If A is a real matrix, $A^* = A^T$.

Definition A.3.14. An operator $A : \mathcal{H} \to \mathcal{H}$ that coincides with its adjoint is called *self-adjoint*.

The operator of projection P on a subspace $\mathcal{H}_0$ of a Hilbert space $\mathcal{H}$ is self-adjoint: $P = P^*$.

In a real Euclidean space, a symmetric matrix corresponds to a self-adjoint operator.

Some properties of self-adjoint operators in a Hilbert space are as follows. Let $A : \mathcal{H} \to \mathcal{H}$ be a self-adjoint operator. Then

1) at any $x \in \mathcal{H}$ the expression $\langle Ax, x\rangle$ is real;
2) the relation
$$\|A\| = \sup_{\|x\|=1} |\langle Ax, x\rangle|$$
holds;
3) the eigenvalues of the operator A are real;
4) the proper subspaces $\mathcal{H}_{\lambda_1}$ and $\mathcal{H}_{\lambda_2}$ corresponding to different eigenvalues λ_1 and λ_2 of the operator A are orthogonal;
5) a compact self-adjoint operator has at least one eigenvalue.

Theorem A.3.10. *The set of eigenvalues of a compact self-adjoint operator A is not more than countable. Specifically,*

$$A = \sum_i \lambda_i P_{\lambda_i},$$

where $\lambda_1, \lambda_2, \ldots$ are different eigenvalues of the operator A, and P_{λ_i} is the operator of projection on the proper subspace $\mathcal{H}_{\lambda_i}$ (here convergence of a series means convergence in the norm of the space of operators).

The notion of an adjoint operator is used to analyze the solvability of the equation $Ax = y$, where $A : \mathcal{H}_1 \to \mathcal{H}_2$ is a linear continuous operator from $\mathcal{H}_1$ to $\mathcal{H}_2$.

Definition A.3.15. The set $\mathrm{Ker}(A) = \{x \in \mathcal{H}_1 \colon Ax = \mathbf{0}\}$ is called the *kernel* of a linear operator $A : \mathcal{H}_1 \to \mathcal{H}_2$.

If A is bounded, $\mathrm{Ker}(A)$ is a subspace of the space $\mathcal{H}_1$.

Theorem A.3.11. *If $A : \mathcal{H}_1 \to \mathcal{H}_2$ is a linear continuous operator, $\mathcal{H}_1 = \mathrm{Ker}(A) \oplus \overline{R}_{A^*}$ and $\mathcal{H}_2 = \mathrm{Ker}(A^*) \oplus \overline{R}_A$.*

It follows from Theorem A.3.11 that a necessary condition of solvability of the equation $Ax = y$ with a linear continuous operator A is that the element y is orthogonal to the kernel $\mathrm{Ker}(A^*)$ of the operator A^*. If $R_A = \overline{R}_A$, this condition is also sufficient. If $\mathrm{Ker}(A) = \{\mathbf{0}\}$ and $R_A = \overline{R}_A$, the linear continuous operator A is a one-to-one mapping of the space $\mathcal{H}_1$ to a subspace $R_A \subset \mathcal{H}_2$ and, according to the Banach inverse operator theorem, there exists a continuous inverse operator $A^{-1} : R_A \to \mathcal{H}_1$. Thus, for a linear continuous operator not having a bounded inverse operator, in particular, for a completely continuous operator, the range of values is not closed: $R_A \neq \overline{R}_A$.

Theorem A.3.12. *Let $A : \mathcal{H}_1 \to \mathcal{H}_2$ be a linear continuous operator. There exists a continuous inverse operator A^{-1} if and only if there exists a continuous inverse operator $(A^*)^{-1}$. Specifically,*

$$(A^*)^{-1} = (A^{-1})^*.$$

Now consider, as an example, the integral equation

$$\varphi(t) = \int_a^b K(t,\tau)\varphi(\tau)\,d\tau + f(t),$$

where f and K are given and φ is the sought-for function. This equation can be written in the following form:

$$\varphi = K\varphi + f, \tag{A.3.2}$$

where K is an integral operator. To study the solvability of such equations with an arbitrary completely continuous operator K defined in a Hilbert space $\mathcal{H}$, an important theorem known as the *Fredholm alternative* is used.

Denote $A = I - K$, where I is the unit operator, and rewrite equation (A.3.2) in the following form:

$$Ax = y, \quad x, y \in \mathcal{H}.$$

Theorem A.3.13 (the Fredholm alternative). *The following statements are valid:*

a) *the nonhomogeneous equation* $Ax = y$ *is solvable only for those vectors* $y \in \mathcal{H}$ *that are orthogonal to all solutions of the homogeneous adjoint equation* $A^*y = 0$, $A^* = I - K^*$;

b) *the nonhomogeneous equation* $Ax = y$ *is uniquely solvable at any* $y \in \mathcal{H}$ *if and only if the homogeneous equation* $Ax = 0$ *has only a zero solution;*

c) *the homogeneous equations* $Ax = 0$ *and* $A^*y = 0$ *have the same finite number of linearly independent solutions.*

And, finally, it should be noted that in Definition A.3.13 the adjoint operator A^* was given for the case when the operator A maps a Hilbert space $\mathcal{H}_1$ to a Hilbert space $\mathcal{H}_2$. However, an adjoint operator can be defined for more general spaces. Let, for instance, two normed spaces, $\mathcal{N}_1$ and $\mathcal{N}_2$, and an operator $A : \mathcal{N}_1 \to \mathcal{N}_2$ be given. Let $l_2(y)$ be a linear functional on $\mathcal{N}_2$. If $y = Ax$, $x \in \mathcal{N}_1$, $y \in \mathcal{N}_2$, we have $l_2(y) = l_2(Ax) = l_1(x)$, where l_1 is a functional (evidently a linear one) on $\mathcal{N}_1$. Thus, to every $l_2 \in \mathcal{N}_2^*$ there corresponds a functional $l_1 \in \mathcal{N}_1^*$, that is, an operator $A^* : \mathcal{N}_2^* \to \mathcal{N}_1^*$ has been constructed. This operator is called the *adjoint* of the operator A, and the equality $l_2(y) = l_1(x)$ is written as $l_1 = A^*l_2$. Functionals in normed (or linear) spaces are often written in a form that is similar to that in a Hilbert space: $l_1(x) = \langle l_1, x\rangle = \langle x, l_1\rangle$, $x \in \mathcal{N}_1$, $l_1 \in \mathcal{N}_1^*$. The function $\langle \cdot, \cdot \rangle$ defined in this way on $\mathcal{N}^* \times \mathcal{N}$ is called the *inner product of* $\mathcal{N}^*$ and $\mathcal{N}$. In particular, elements x and l_1 are called *orthogonal* (in a normed space!) if $l_1(x) = \langle l_1, x\rangle = \langle x, l_1\rangle = 0$. In this form, the equality $l_2(y) = l_1(x)$ takes a form similar to (A.3.1):

$$l_2(y) = \langle y, l_2\rangle = \langle Ax, l_2\rangle = l_1(x) = \langle x, l_1\rangle = \langle x, A^*l_2\rangle.$$

For Banach spaces, all properties of adjoint operators for Hilbert spaces are valid.

A.4 Elements of differential calculus in Banach spaces

Let $A : \mathcal{B}_1 \to \mathcal{B}_2$ be an operator that acts from a Banach space $\mathcal{B}_1$ to another Banach space $\mathcal{B}_2$ having an open domain of definition D_A. Take a fixed element $x_0 \in D_A$ and assume that there exists a linear continuous operator $U \in \mathcal{L}(\mathcal{B}_1, \mathcal{B}_2)$ such that at any $x \in \mathcal{B}_1$

$$\lim_{t\to 0} \frac{A(x_0 + tx) - A(x_0)}{t} = U(x). \tag{A.4.1}$$

In this case the linear operator U is said to be the *derivative* of the operator A at the point x_0 and is denoted by

$$U = A'(x_0).$$

The derivative thus defined is called the *Gâteaux (or weak) derivative*, and the element $U(x)$ is called the *Gâteaux differential.*

Let $S(\mathbf{0}, 1)$ denote the set of all $x \in \mathcal{B}_1$ with $\|x\| = 1$. If the limiting relation (A.4.1) is uniform with respect to $x \in S(\mathbf{0}, 1)$, the operator A is said to be *differentiable* at the point x_0 and the derivative $A'(x_0)$ is called the *Fréchet (or strong) derivative.* The strong derivative is often denoted by A'_{x_0}.

In other words, the operator A is differentiable at the point x_0 if there exists a linear operator $A'_{x_0} \in \mathcal{L}(\mathcal{B}_1, \mathcal{B}_2)$ such that at any $\varepsilon > 0$ one can find such $\delta > 0$ that $\|\Delta x\| < \delta$ ($\Delta x \in \mathcal{B}_1$) implies

$$\|A(x_0 + \Delta x) - A(x_0) - A'_{x_0}(\Delta x)\| \le \varepsilon \|\Delta x\|.$$

Some properties of the derivative are:

1) If an operator A is differentiable at a point x_0, its derivative A'_{x_0} is defined uniquely.

2) If an operator A is differentiable at a point x_0, it is continuous at this point (the presence of a weak derivative is not sufficient for continuity).

3) Let $C = \alpha A + \beta B$. Then, if there exist derivatives $A'(x_0)$ and $B'(x_0)$, there also exists

$$C'(x_0) = \alpha A'(x_0) + \beta B'(x_0).$$

In this case, if A and B are differentiable (Fréchet differentiable) at a point x_0, the operator C is also differentiable at that point.

4) If $A \in \mathcal{L}(\mathcal{B}_1, \mathcal{B}_2)$, that is, A is a linear continuous operator, A is differentiable at every point $x_0 \in \mathcal{B}_1$ and $A'_{x_0} = A$.

5) Let A be an operator that maps an open set $D_A \subset \mathcal{B}_1$ to an open set $R_A \subset \mathcal{B}_2$, and let B be an operator that maps a set $D_B = R_A \subset \mathcal{B}_2$ to a Banach space $\mathcal{B}_3$. Denote $C = BA$; let B be differentiable at the point $y_0 = Ax_0$, $x_0 \in D_A$, and let A have a derivative at the point x_0. Then the operator C has a derivative at the point x_0, and

$$C'(x_0) = B'(A(x_0))A'(x_0) = B'(y_0)A'(x_0).$$

If A is differentiable at the point x_0, C is also differentiable: $C'_{x_0} = B'_{y_0} A'_{x_0}$, $y_0 = Ax_0$.

6) If under the assumptions of 5) the operator B is continuous and linear, we have

$$C'(x_0) = BA'(x_0).$$

The operator A is said to be *differentiable on a set* $X \subset D_A$ if it is differentiable at every point of this set.

Let $D_A \subset \mathcal{B}_1$ be an open set of a Banach space $\mathcal{B}_1$, and let there exist the derivative A' of an operator A mapping D_A to a Banach space $\mathcal{B}_2$. The derivative A' can be considered as an operator mapping D_A to the space $\mathcal{L}(\mathcal{B}_1, \mathcal{B}_2)$. Therefore, it makes sense to consider the derivative of this operator at a point $x_0 \in D_A$. This derivative (if it exists) is called the *second derivative* of the operator A and denoted by $A''(x_0)$. If the operator A' is differentiable, A is said to be *twice differentiable*. Similarly, we can define the third derivative, etc.

Now consider the case when an operator A acts from a Banach space $\mathcal{B}$ to the set of real numbers $\mathbb{R}$, that is, this operator is a functional.

Let $J : \mathcal{B} \to \mathbb{R}$ be a functional defined on an open set $D \subset \mathcal{B}$. Differentiability of the functional $J(x)$ at a point $x_0 \in D$ means that there exists an element $J'_{x_0} \in \mathcal{B}^*$ such that for any Δx ($\Delta x \in \mathcal{B}$, $x + \Delta x \in D$) the following equality holds:

$$J(x_0 + \Delta x) - J(x_0) = J'_{x_0}(\Delta x) + o(\Delta x), \tag{A.4.2}$$

where $o(\Delta x)/\|\Delta x\| \to 0$ as $\|\Delta x\| \to 0$.

A functional $J(x)$ is called *continuously differentiable* on a set D if it is differentiable on D and $\|J'(x+\Delta x) - J'(x)\|_{\mathcal{B}^*} \to 0$ as $\|\Delta x\|_{\mathcal{B}} \to 0$ for all $x, x+\Delta x \in D$.

An example of a continuously differentiable functional is the functional $J(x) = \|x\|^2_{\mathcal{H}}$ defined on a Hilbert space $\mathcal{H}$: $J'_x = 2x$.

The above-introduced notion of differentiability makes it possible to formulate a criterion of convexity of functionals.

Theorem A.4.1. *Let D be a convex set of a Banach space $\mathcal{B}$, and let $J(x)$ be a functional that is continuously differentiable on D. The functional is convex on D if and only if for any $x, y \in D$ one of the following inequalities holds:*

$$J(x) \geq J(y) + J'_y(x - y), \quad (J'_x - J'_y)(x - y) \geq 0.$$

Some necessary and sufficient conditions of existence of a minimum of a differentiable functional on a convex set are formulated in the theorem below.

Theorem A.4.2. *Let D be a convex set of a Banach space $\mathcal{B}$, and let $J(x)$ be a continuously differentiable functional on D. For the functional $J(x)$ to reach a minimum at a point $x^* \in D$, that is, $J(x^*) = \inf_{x \in D} J(x) \equiv J^*$, it is necessary that the inequality*

$$J'_{x^*}(x - x^*) \geq 0 \tag{A.4.3}$$

hold for all $x \in D$. *If* x^* *is an inner point of* D, *condition* (A.4.3) *is equivalent to the equality* $J'_{x^*} = 0$. *If the functional* $J(x)$ *is also convex on* D, *condition* (A.4.3) *is also sufficient.*

Let $J(x) : \mathcal{H} \to \mathbb{R}$ be a functional in a Hilbert space $\mathcal{H}$. If $J(x)$ is differentiable at a point $x \in \mathcal{H}$, its derivative J'_x is a linear continuous functional. According to the Riesz theorem, there exists only one element $y \in \mathcal{H}$ such that $J'_x(\Delta x) = \langle y, \Delta x \rangle$ for any $\Delta x \in \mathcal{H}$. This element is called the *gradient* of the functional $J(x)$ at the point x and is denoted by $J'x$.

The gradient determines the direction of steepest ascent of the functional.

Consider, as an example, the functional

$$J(x) = \frac{1}{2} \|Ax - y\|^2,$$

where $A : \mathcal{H}_1 \to \mathcal{H}_2$ is an operator differentiable at a point $x \in D_A$, $\mathcal{H}_1$ and $\mathcal{H}_2$ are Hilbert spaces, and $y \in \mathcal{H}_2$ is some fixed point. Let us find a formula for the gradient $J'x$ at the point x. Transform the difference

$$\begin{aligned} J(x + \Delta x) - J(x) &= \frac{1}{2} \|A(x + \Delta x) - y\|^2 - \frac{1}{2} \|Ax - y\|^2 \\ &= \frac{1}{2} \|Ax - y + A'_x(\Delta x) + o(\Delta x)\|^2 - \frac{1}{2} \|Ax - y\|^2 \\ &= \langle Ax - y, A'_x(\Delta x)\rangle + \frac{1}{2} \|A'_x(\Delta x) \\ &\quad + o(\Delta x)\|^2 + \langle Ax - y, o(\Delta x)\rangle \\ &= \langle A'^*_x(Ax - y), \Delta x\rangle + o(\Delta x). \end{aligned}$$

Hence, $J'x = A'^*_x(Ax - y)$. If $A : \mathcal{H}_1 \to \mathcal{H}_2$ is a linear continuous operator, $J'x = A^*(Ax - y)$.

A.5 Functional spaces

Section A.1.5 gives examples of functional spaces: the space of continuously differentiable functions $C^m[a, b]$, the space of integrable functions $L_p[a, b]$, and the Sobolev space $H^k(G) = W_2^k(G)$. In this section, the spaces C^m and L_p will be considered for vector-functions $u : \mathbb{R}^n \to \mathbb{R}$ (the spaces $C^m(G)$ and $L_p(G)$), and for functions mapping an interval $S \subset \mathbb{R}$ to an arbitrary Banach space X (the spaces $C^m(S; X)$ and $L_p(S; X)$). Subsection A.5.1 provides a definition of the Sobolev space $W_p^k(G)$ and a formulation of the Sobolev embedding theorem in the case where the domain $G \subset \mathbb{R}^n$ has a regular boundary.

A.5.1 Functional spaces for steady boundary value problems

Let $\mathbb{R}^n$ be the n-dimensional Euclidean space, and let $G \subset \mathbb{R}^n$ be an open set.

Spaces of continuously differentiable functions

Definition A.5.1. Let $C^m(\overline{G})$ denote the set of functions $u(x)$ defined on the closure $\overline{G}$ of an open set G and having the following properties:

1) the functions $u(x)$ are m times continuously differentiable on G;
2) each partial derivative $D^\alpha = \frac{\partial^{|\alpha|} u}{\partial x_1^{\alpha_1} \dots \partial x_n^{\alpha_n}}$ of order $|\alpha| = \alpha_1 + \alpha_2 + \dots + \alpha_n \leq m$, where $\alpha = (\alpha_1, \dots, \alpha_n)$ is a multi-index of nonnegative integers, can be extended to a continuous function on $\overline{G}$.

If $\overline{G}$ is compact, $C^m(\overline{G})$ is a Banach space with respect to the norm

$$\|u\|_{C^m(\overline{G})} = \sum_{|\alpha| \leq m} \sup_{x \in \overline{G}} |D^\alpha u(x)|. \tag{A.5.1}$$

Convergence of a sequence $\{u_k\} \subset C^m(\overline{G})$ with respect to the norm (A.5.1) means uniform convergence of all derivatives of order $\leq m$ on the compact $\overline{G}$.

The set $C^0(\overline{G})$ of continuous functions on $\overline{G}$ is usually denoted by $C(\overline{G})$.

Definition A.5.2. Let $C_0^m(G)$ denote the set of all functions that are m times continuously differentiable and defined on an open set $G \subset \mathbb{R}^n$ of functions with compact (in G) supports.

The set

$$\operatorname{supp} u = \overline{\{x : u(x) \neq 0\}} \cap G$$

is called the *support* of a function $u(x)$ that is continuous on G.

According to Definition A.5.2, $C_0^\infty(G)$ is the set of infinitely differentiable on an open set $G \subset \mathbb{R}^n$ functions with compact (in G) supports.

On the set $C_0^\infty(G)$, a locally convex topology can be introduced with a system of semi-norms (see Definition A.1.41):

$$p_\alpha(u) = \sum_\alpha \sup |\rho_\alpha(x) D^\alpha u(x)|, \quad u \in C_0^\infty(G). \tag{A.5.2}$$

Here $\rho = \{\rho_\alpha\}$ is an element of all families of continuous functions on G with locally finite families of supports. A family of supports $\{\operatorname{supp} \rho_\alpha\}$ of functions ρ_α that are continuous on G is called *locally finite* in G if every compact set in G has nonempty intersections only with a finite number of sets from the family.

It should be noted that the summation in (A.5.2) at every fixed $u \in C_0^\infty(G)$ holds only for a finite number of multi-indices α, since the family $\{\operatorname{supp} \rho_\alpha\}$ is locally finite.

Definition A.5.3. Let $\mathcal{D}(G)$ denote the locally convex space $C_0^\infty(G)$ with the generating system of seminorms (A.5.2).

Lemma A.5.1. *A sequence $\{u_k\} \subset \mathcal{D}(G)$ converges to zero if and only if there exists a compact set K in G such that*

1) $\operatorname{supp} u_k \subset K$ *for* $k = 1, 2, \dots;$
2) *The sequence $\{D^\alpha u_k\}$ uniformly converges to zero in K for every multiindex $\alpha = (\alpha_1, \dots, \alpha_n)$.*

Spaces of integrable functions. Before defining a space of integrable functions, recall some notions from the Lebesgue measure theory in the Euclidean space $\mathbb{R}^n$.

The number

$$\operatorname{mes}(S) = \prod_{i=1}^{n} (b_i - a_i)$$

is called the *measure* of an n-dimensional closed interval $S = [a_1, b_1] \times [a_2, b_2] \times \dots \times [a_n, b_n] \subset \mathbb{R}^n$.

Since any open set $O \subset \mathbb{R}^n$ can be represented as the union of not more than a countable number of closed intervals S_j without common internal points,

$$O = \bigcup_{j=1}^{\infty} S_j,$$

the measure of the open set O is defined as the sum of the measures of these intervals:

$$\operatorname{mes}(O) = \sum_{j=1}^{\infty} \operatorname{mes}(S_j).$$

The thus defined measure (possibly equal to $+\infty$) does not depend on the representation of the set O as the union of closed intervals without common internal points.

Let $G \subset \mathbb{R}^n$ be an arbitrary set. The *outer measure* of the set G is the number

$$\overline{\operatorname{mes}}(G) = \inf_{O \supset G} \operatorname{mes}(O),$$

where the lower bound is taken over all open sets O containing G. The set $G \subset \mathbb{R}^n$ is called *measurable* if

$$\inf_{O \supset G} \overline{\operatorname{mes}}(O \setminus G) = 0.$$

In this case $\operatorname{mes}(G) = \overline{\operatorname{mes}}(G)$.

A real function $u(x)$ defined on a measurable set $G \subset \mathbb{R}^n$ is called *measurable* if for every $a \in \mathbb{R}$ the set $\{x \in G : u(x) > a\}$ is measurable.

Any function that is continuous on an open or closed set $G \subset \mathbb{R}^n$ is measurable.

A function $u(x)$ defined on a measurable set $G \subset \mathbb{R}^n$ is called *simple* if it takes on constant values c_j on a finite number of pairwise nonintersecting measurable sets G_j, $j = 1, \dots, k$, with $\operatorname{mes}(G_j) < \infty$, and is zero on the set $G \setminus \bigcup_{j=1}^k G_j$. The *integral* of a simple function $u(x)$ on a set G is the number

$$\int_G u(x)\,dx = \sum_{j=1}^k c_j \operatorname{mes}(G_j).$$

A condition is said to be satisfied *almost everywhere* on a set $G \subset \mathbb{R}^n$ (or for *almost all* its points) if the set of points where this condition is not satisfied has a zero measure.

A function $u(x)$ defined almost everywhere on a measurable set $G \subset \mathbb{R}^n$ is called *Lebesgue integrable* over the set (or on the set) G if there exists a sequence $\{u_i(x)\}$ of simple functions that converges to $u(x)$ almost everywhere on G and for which the sequence $\{\int_G u_i(x)\,dx\}$ converges. The limit of this sequence is called the *(Lebesgue) integral* of the function $u(x)$ over G:

$$\int_G u(x)\,dx = \lim_{i\to\infty} \int_G u_i(x)\,dx.$$

Two functions, $u_1(x)$ and $u_2(x)$, defined on G are called *equivalent* if $u_1(x) = u_2(x)$ for almost all $x \in G$. If the function $u_1(x)$ is integrable on G, any function $u_2(x)$ that is equivalent to it is also integrable on G and their integrals coincide. Therefore, functions that are equivalent on G will not be distinguished. In what follows, it will also be assumed that G is a measurable set in $\mathbb{R}^n$.

Lemma A.5.2 (Fatou lemma). *Let $\{u_i\}$ be a sequence of nonnegative functions integrable on G such that*

$$\underline{\lim}_{i\to\infty} \int_G u_i(x)\,dx < \infty.$$

Then the function $u(x) = \underline{\lim}_{i\to\infty} u_i(x)$ is integrable on G, and

$$\int_G u(x)\,dx \le \underline{\lim}_{i\to\infty} \int_G u_i(x)\,dx.$$

Theorem A.5.1 (Lebesgue theorem). *Let $\{u_i\}$ be a sequence of integrable functions on G convergent almost everywhere on G to a function u. If there exists a function v* *that is integrable on G such that $|u_i(x)| \le v(x)$ $(i = 1, 2, \dots)$ almost everywhere on G, the function u is also integrable on G and*

$$\int_G u(x)\,dx = \lim_{i\to\infty} \int_G u_i(x)\,dx.$$

Definition A.5.4. Let $L_p(G)$, $1 \le p < \infty$, denote the set of all measurable functions $u : G \to \mathbb{R}$ for which

$$\int_G |u(x)|^p\,dx < \infty.$$

The set $L_p(G)$ is a Banach space with respect to the norm

$$\|u\|_{L_p(G)} = \Big(\int_G |u(x)|^p\,dx\Big)^{1/p}.$$

It should be noted that if $u(x)$ and $v(x)$ are functions that coincide almost everywhere on a set G, $\|u - v\|_{L_p(G)} = 0$. Hence, in the space $L_p(G)$ functions that coincide almost everywhere are identified.

The space $L_2(G)$ (at $p = 2$) is a Hilbert space with an inner product

$$\langle u, v\rangle = \int_G u(x)v(x)\,dx.$$

Definition A.5.5. Let $L_{p,r}(G)$, $1 \le p < \infty$, denote the set of all vector-functions $u = \{u_1, \dots, u_r\}$, $u_i \in L_p(G)$ $(i = 1, \dots, r)$.

The set $L_{p,r}(G)$ is a Banach space with respect to the norm

$$\|u\|_{L_{p,r}(G)} = \Big(\int_G \Big(\sum_{i=1}^{r} |u_i(x)|^2\Big)^{p/2} dx\Big)^{1/p}.$$

The space $L_{2,r}(G)$ is a Hilbert space with an inner product

$$\langle u, v\rangle = \int_G \sum_{i=1}^{r} u_i(x)v_i(x)\,dx.$$

Definition A.5.6. A measurable function $u : G \to \mathbb{R}$ is called *essentially bounded* on G if it is equivalent to some bounded function, that is, if there exists a constant M such that $|u(x)| \le M$ at almost all $x \in G$. The lower bound of all admissible constants M is denoted by $\operatorname{vrai\,max}_{x\in G} |u(x)|$. The set of all measurable essentially bounded functions (equivalent classes) is denoted by $L^\infty(G)$.

The set $L^{\infty}(G)$ is a Banach space with respect to the norm

$$\|u\|_{L^{\infty}(G)} = \operatorname*{vrai\,max}_{x \in G} |u(x)|.$$

Indices p and q $(1 < p < \infty)$ are called *conjugates* if $1/p + 1/q = 1$. By definition, the index that is conjugate of $p = 1$ $(p = \infty)$ is $q = \infty$ $(q = 1)$.

Lemma A.5.3 (Hölder inequality). *Let p and q be conjugate indices, $1 \leq p \leq \infty$, $u \in L_p(G)$, $v \in L_q(G)$. Then $uv \in L_1(G)$ and*

$$\left| \int_G u(x)v(x)\,dx \right| \leq \|u\|_{L_p(G)} \|v\|_{L_q(G)}.$$

At $1 < p < \infty$ the Banach spaces $L_p(G)$ and $L_{p,r}(G)$ are uniformly convex (see Definition A.1.30) and, hence, reflexive. The spaces $L_1(G)$ and $L_{\infty}(G)$ are not reflexive.

At $1 \leq p < \infty$ the Banach spaces $L_p(G)$ are separable.

To characterize weak convergence in the space $L_p(G)$, consider the dual space of $L_p(G)$. For linear functionals from the dual space $(L_p(G))^*$ of $L_p(G)$ $(1 \leq p < \infty)$, the following holds.

Lemma A.5.4. *If $\varphi \in (L_p(G))^*$, $1 \leq p < \infty$, there exists a unique element $v \in L_q(G)$ (where q is the index that is conjugate of p) such that*

$$\varphi(u) = \langle \varphi, u \rangle = \int_G u(x)v(x)\,dx \quad \forall u \in L_p(G).$$

Remark A.5.1. By virtue of the Hölder inequality, every element $v \in L_q(G)$ generates a continuous linear functional on $L_p(G)$. Therefore, the linear functionals $\varphi \in (L_p(G))^*$, $1 \leq p < \infty$, can be identified with the "generating" elements $v \in L_q(G)$ that uniquely correspond to these functionals according to Lemma A.5.4. Hence, the space $L_q(G)$ is the dual space of the Banach space $L_p(G)$. At $p = 2$ we obtain exactly the Riesz theorem on the representation of linear functionals (see Theorem A.3.8) for the Hilbert space $L_2(G)$.

By virtue of Lemma A.5.4, weak convergence of a sequence $\{u_k\} \subset L_p(G)$, $1 \leq p < \infty$, to an element $u \in L_p(G)$ implies that

$$\lim_{k \to \infty} \int_G u_k(x)v(x)\,dx = \int_G u(x)v(x)\,dx \quad \forall v \in L_q(G).$$

Since the spaces $L_p(G)$, $1 \leq p \leq \infty$, are Banach spaces, they are weakly complete, that is, they are complete with respect to weak convergence.

From every sequence that is convergent in $L_p(G)$ $(1 \leq p \leq \infty)$, one can choose a subsequence that converges almost everywhere on G.

Lemma A.5.5. *Let $1 \leq p < \infty$ and let a sequence $\{u_k\} \subset L_p(G)$ be such that $u_k \rightharpoonup u$ in $L_p(G)$ and $u_k(x) \to v(x)$ for almost all $x \in G$. Then $u = v$.*

Lemma A.5.6. *Let $G \subset \mathbb{R}^n$ be open and $1 \leq p < \infty$. Then the subset $C_0^\infty(G)$ of the space $L_p(G)$ is everywhere dense.*

Sobolev spaces. Before defining Sobolev spaces, we introduce the following notion of a generalized function (distribution):

Definition A.5.7. The elements of the dual space $\mathcal{D}^*(G)$ of $\mathcal{D}(G)$ (see Definition A.5.3) are called *distributions* (*generalized functions*) on an open set $G \subset \mathbb{R}^n$. The space $\mathcal{D}(G)$ is called the *space of test functions*.

The space $\mathcal{D}^*(G)$ becomes locally convex if a neighborhood basis of a distribution $\varphi \in \mathcal{D}^*(G)$ is introduced by means of the sets

$$\{f \in \mathcal{D}^*(G) : |f(u_i) - \varphi(u_i)| < \varepsilon_i,\ i = 1, \dots, m\}, \tag{A.5.3}$$

where $\{u_1, \dots, u_m\}$ is an arbitrary (finite) system of elements from $\mathcal{D}(G)$, and ε_i $(1 \leq i \leq m)$ are arbitrary positive numbers. The convergence of a sequence of distributions $\{\varphi_k\}$ to a distribution φ in $\mathcal{D}^*(G)$ means pointwise convergence, that is,

$$\lim_{k\to\infty} \varphi_k(u) = \varphi(u) \quad \text{for every } u \in \mathcal{D}(G).$$

The locally convex topology of the space $\mathcal{D}^*(G)$ generated by a neighborhood basis (A.5.3) will be called simple topology.

Definition A.5.8. Let D^α be the differential operator

$$D^\alpha = \frac{\partial^{|\alpha|}}{\partial x_1^{\alpha_1} \dots \partial x_n^{\alpha_n}},$$

where $\alpha = (\alpha_1, \dots, \alpha_n)$, $|\alpha| = \alpha_1 + \dots + \alpha_n$. The (*generalized*) *derivative* $D^\alpha\varphi$ of a distribution φ on G is given by the formula

$$(D^\alpha\varphi)(u) = (-1)^{|\alpha|}\varphi(D^\alpha u), \quad u \in \mathcal{D}(G).$$

Lemma A.5.7. *The mapping $\varphi \to D^\alpha\varphi$ is a continuous mapping of the space $\mathcal{D}^*(G)$ (having simple topology) to itself.*

It should be noted that if $v(x)$ is an m times continuously differentiable function on G, the derivatives of $v(x)$ of order $\leq m$ in the sense of conventional differential analysis and in the sense of distribution theory coincide, since for every multiindex α ($|\alpha| < m$) and every function $u \in \mathcal{D}(G)$ the equality

$$\int_G (D^\alpha v(x))u(x)\,dx = (-1)^{|\alpha|} \int_G v(x)(D^\alpha u(x))\,dx$$

holds.

Generally speaking, generalized functions have no values at individual points $x \in G$. Nevertheless, a generalized function can be said to be zero in a domain.

A generalized function f is said to *be zero* on an open set $Q \subset G$ if for all functions $u \in \mathcal{D}$ such that $\operatorname{supp} u \subset Q$ the equality $f(u) = 0$ holds.

This definition allows us to introduce the following notion of support of a generalized function f:

The union of all open sets $Q \subset G$ in which f is zero is called the *zero set* of a generalized function f, and its complement to G is called the *support* of the generalized function and is denoted by $\operatorname{supp} f$. A generalized function f is called *finite* if its support is bounded.

A simple example of a generalized function is the following functional generated by a function $f(x)$ that is locally integrable in $\mathbb{R}^n$:

$$f(u) = \int f(x)u(x)\,dx, \quad u \in \mathcal{D}. \tag{A.5.4}$$

A function $f(x) : \mathbb{R}^n \to \mathbb{R}$ is called *locally integrable* in $\mathbb{R}^n$ if for any bounded set $Q \subset \mathbb{R}^n$ it is absolutely integrable on the set Q. The generalized functions defined by functions that are locally integrable in $\mathbb{R}^n$ by formula (A.5.4) are called *regular generalized functions*, and the other generalized functions are called *singular generalized functions*.

An example of a singular generalized function is the Dirac δ-function (see Subsection A.6.1)

$$\delta(u) = u(\mathbf{0}), \quad u \in \mathcal{D}.$$

It is evident that $\operatorname{supp} \delta = \{\mathbf{0}\}$.

Definition A.5.9. Let G be an open connected set in $\mathbb{R}^n$. Let $W_p^k(G)$ ($1 \le p < \infty$, $k \ge 0$) denote the set of all distributions $\varphi \in \mathcal{D}^*(G)$ that are, together with their derivatives $D^\alpha\varphi$ of order $|\alpha| \le k$, functions from $L_p(G)$. These sets $W_p^k(G)$ are called *Sobolev spaces*.

At $1 \le p < \infty$ the Sobolev space $W_p^k(G)$ is a Banach space with respect to the norm

$$\|\varphi\|_{W_p^k(G)} = \Big(\int_G \Big(\sum_{|\alpha| \le k} |D^\alpha \varphi|^2\Big)^{p/2} dx\Big)^{1/p}. \tag{A.5.5}$$

At $1 \leq p < \infty$ the spaces $W_p^k(G)$ are separable. At $1 < p < \infty$ they are also uniformly convex and, hence, reflexive. The Sobolev space $W_2^k(G)$ (at $p = 2$), usually denoted by $H^k(G)$, is a Hilbert space with an inner product

$$\langle u, v\rangle_{H^k(G)} = \sum_{|\alpha|\leq k} \langle D^\alpha u, D^\alpha v\rangle = \sum_{|\alpha|\leq k} \int_G D^\alpha u\, D^\alpha v\, dx.$$

In what follows, we formulate the Sobolev embedding theorem for the case when G is a bounded domain with a sufficiently smooth boundary.

Definition A.5.10 (Calderon). A boundary Γ of a domain $G \subset \mathbb{R}^n$ is called *regular* if there exists a finite open cover $\{X_i\}$ of this boundary, a finite set of open (finite) cones K_j, and $\varepsilon > 0$ such that

1) for any point $x_0 \in \Gamma$, the ball of radius ε with center at the point x_0 lies wholly in some set X_{i_0} of the cover $\{X_i\}$: $B(x_0, \varepsilon) \subset X_{i_0}$;

2) for any point from $X_i \cap G$, the cone with vertex at this point obtained by a parallel translation of some cone K_j lies wholly in G.

A bounded domain G has a regular boundary if and only if there exist constants $R > 0$ and $L > 0$ such that for every point $x_0 \in \Gamma$ there exists a neighborhood $O(x_0)$ obtained by a parallel translation and rotation from the set

$$X_0 = \Big\{x \in \mathbb{R}^n : \sqrt{x_1^2 + \cdots + x_{n-1}^2} < R,\ |x_n| < 2LR\Big\},$$

such that the following conditions are satisfied:

1) the point $\mathbf{0} \in X_0$ is translated to x_0;

2) the intersection $O(x_0) \cap \Gamma$ corresponds to the surface

$$x_n = f(x_1, \ldots, x_{n-1}), \quad \sqrt{x_1^2 + \cdots + x_{n-1}^2} < R,$$

where the function f is Lipschitz continuous, that is,

$$|f(x_1, \ldots, x_{n-1}) - f(y_1, \ldots, y_{n-1})| \leq L\Big(\sum_{i=1}^{n-1} |x_i - y_i|^2\Big)^{1/2};$$

3) the intersection $O(x_0) \cap G$ corresponds to the set

$$\Big\{x \in X_0 : \sqrt{x_1^2 + \cdots + x_{n-1}^2} < R,\ f(x_1, \ldots, x_{n-1}) < x_n < 2LR\Big\}.$$

For a bounded domain G with a regular boundary Γ, one can define an $(n-1)$-dimensional Lebesgue surface measure σ_{n-1} by taking, as the value of this measure for a hypersurface part $O(x_0) \cap \Gamma$, the integral

$$\sigma_{n-1}(O(x_0) \cap \Gamma) = \int_{K_0} \sqrt{1 + |\text{grad } f|^2}\, dx_1 \ldots dx_{n-1},$$

where

$$K_0 = \Big\{(x_1, \ldots, x_{n-1}) : \sqrt{x_1^2 + \cdots + x_{n-1}^2} < R\Big\},$$

$$\text{grad } f = \Big\{\frac{\partial f}{\partial x_1}, \ldots, \frac{\partial f}{\partial x_{n-1}}\Big\}.$$

A domain with a regular boundary for almost all $x \in \Gamma$ (with respect to the measure σ_{n-1}) has a uniquely defined (external) *normal* vector, that is, a vector $\nu = (\nu_1, \ldots, \nu_n)$ with $|\nu| = \sqrt{\nu_1^2 + \cdots + \nu_n^2} = 1$ such that

$$\Big\langle \nu, \frac{x-y}{|x-y|} \Big\rangle \to 0 \quad \text{at } y \to x; \quad x, y \in \Gamma.$$

Here $\langle \cdot, \cdot \rangle$ denotes the inner product in the Euclidean space $\mathbb{R}^n$.

If G is a bounded domain with a regular boundary, for every vector-function $u = (u_1, \ldots, u_n)$ with $u_i \in C^1(G)$ the following *Gauss formula* (also called the *Gauss–Ostrogradskii formula* or the *Stokes formula*) is valid:

$$\int_G \sum_{i=1}^n \frac{\partial u_i}{\partial x_i}\, dx = \int_\Gamma \sum_{i=1}^n u_i \nu_i \, d\sigma_{n-1}.$$

Let $L_p(\Gamma)$, $1 \le p < \infty$, denote the set of all functions $u \colon \Gamma \to \mathbb{R}$ for which

$$\int_\Gamma |u(x)|^p \, d\sigma_{n-1} < \infty.$$

The set $L_p(\Gamma)$ is a Banach space with respect to the norm

$$\|u\|_{L_p(\Gamma)} = \Big(\int_\Gamma |u(x)|^p \, d\sigma_{n-1}\Big)^{1/p}.$$

Theorem A.5.2 (Sobolev embedding theorem). *Let G be a bounded domain with a regular boundary and $1 \le p < \infty$. Then*

$$W_p^k(\Gamma) \subset W_r^j(G)$$

for $0 \le j < k$ and every r satisfying the condition

$$\frac{1}{p} - \frac{k-j}{n} \le \frac{1}{r} < 1.$$

In addition,

$$\|\varphi\|_{W_r^j(G)} \le C\,\|\varphi\|_{W_p^k(G)} \quad \forall \varphi \in W_p^k(G),$$

where the constant C depends on G, j, k, p, and r. If $\frac{1}{p} - \frac{k-j}{n} < 0$, we have

$$W_p^k(G) \subset C^j(\overline{G})$$

and

$$\|\varphi\|_{C^j(\overline{G})} \le C\,\|\varphi\|_{W_p^k(G)} \quad \forall \varphi \in W_p^k(G).$$

Some other norms equivalent to norm (A.5.5) can be introduced in the space $W_p^1(G)$. This is formulated in the following lemma:

Lemma A.5.8. *Let G be a bounded domain with a regular boundary Γ, G_1, a subset of a positive measure in G, and Γ_1, a subset of a positive ($(n-1)$-dimensional) surface measure in Γ. Then the inequalities*

$$\int_G |\varphi|^p\,dx \le C_1\Big\{\int_G \Big(\sum_{i=1}^n \Big(\frac{\partial\varphi}{\partial x_i}\Big)^2\Big)^{p/2} dx + \Big|\int_{G_1} \varphi\,dx\Big|^p\Big\},$$

$$\int_G |\varphi|^p\,dx \le C_2\Big\{\int_G \Big(\sum_{i=1}^n \Big(\frac{\partial\varphi}{\partial x_i}\Big)^2\Big)^{p/2} dx + \Big|\int_{\Gamma_1} \varphi\,d\sigma_{n-1}\Big|^p\Big\}$$

with constants $C_1 = C_1(n,p,G_1,G)$ and $C_2 = C_2(n,p,\Gamma_1,G)$ hold for $\varphi \in W_p^1(G)$ ($p \ge 1$).

The first of these inequalities at $p = 2$ and $G_1 = G$ is known as the *Poincaré inequality*. The second estimate is often called the *Friedrichs inequality*.

A.5.2 Functional spaces for solving unsteady problems

To describe an unsteady process taking place in a spatial domain G within a time interval S, one can use time and position functions, that is, functions u that assign a real number or a vector $u(t,x)$ to a pair $\{t,x\} \in S \times G$. In this approach, time and space are equivalent. However, there is another approach to the mathematical description of unsteady processes that seems to be much more convenient. In this approach time functions are used to assign a function $u(t,\cdot)$ to a time t. For instance, a temperature or velocity distribution in a domain G is assigned to every instant of time. Thus, functions defined on a (bounded or unbounded) time interval $S \subset \mathbb{R}$, with values in some space of position functions X, are considered. Such functions are called *abstract functions*. It will be assumed that X is an arbitrary Banach space with norm $\|\cdot\|$.

To study time-dependent problems, it is reasonable to consider some spaces of functions acting from S to X, in particular, spaces of differentiable and integrable functions.

Spaces of continuously differentiable functions

Definition A.5.11. A function u: $S \to X$ is called *differentiable at a point* $t \in S$ if there exists such an element $x \in X$ for which the condition

$$\lim_{\substack{h\to 0\\ t+h\in S}} \left\| \frac{u(t+h)-u(t)}{h} - x \right\| = 0$$

is satisfied. This element x is called the *derivative of* u *at the point* t and is usually denoted by $u'(t)$.

A function $u : S \to X$ is called *differentiable* if it is differentiable at every point of the interval S. The function u': $S \to X$ that assigns to every $t \in S$ the derivative of u at the point t is called the *derivative of* u.

Definition A.5.12. A function $u : S \to X$ is *weakly differentiable at a point* $t \in S$ if there exists such an element $x \in X$ for which the condition

$$\lim_{\substack{h\to 0\\ t+h\in S}} \left\langle f, \frac{u(t+h)-u(t)}{h} - x \right\rangle = 0, \quad \forall f \in X^*,$$

where X^* is the dual space of the space X; $\langle f, y\rangle = f(y)$, $f \in X^*$, $y \in X$, is satisfied. This element x is called the *weak derivative of* u *at the point* t and is also usually denoted by $u'(t)$.

A function $u : S \to X$ is called *weakly differentiable* if it is weakly differentiable at every point of the interval S. The function $u' : S \to X$ that assigns to every $t \in S$ the weak derivative of u at the point t is called the *weak derivative of* u.

Remark A.5.2. It follows from Definitions A.5.11 and A.5.12 that the derivative of a function $u : S \to X$, if it exists, is also the weak derivative of u. Therefore, the fact that the symbol u' denotes both the derivative and the weak derivative of a function $u : S \to X$ will not lead to confusion: it is always clear from the context what derivative is meant.

From Definitions A.5.11 and A.5.12, one can, as usual, by induction introduce the notions of m-times differentiability and m-times weak differentiability, mth derivative, and mth weak derivative of a function $u : S \to X$.

Definition A.5.13. Let $C^m(S; X)$ denote the set of all functions $u : S \to X$ having continuous derivatives up to order m inclusive, and let $C^m_\omega(S; X)$ be the set of all functions $u : S \to X$ having demi-continuous weak derivatives up to order m inclusive.

Remark A.5.3. A function $u : S \to X$ is demi-continuous (see Definition A.2.12) if and only if for every $f \in X^*$ the function $\langle f, u(\cdot)\rangle : S \to \mathbb{R}$ is continuous.

The set $C^0(S;X)$ ($C^0_\infty(S;X)$) is the set of all continuous (demi-continuous) functions mapping S to X. These sets are also denoted by $C(S;X)$ ($C_\omega(S;X)$).

Note that if $u \in C^m_\omega(S;X)$

$$\langle f, u(\cdot)\rangle \in C^m(S), \quad \forall f \in X^*.$$

If X is weakly complete, the converse is also true. Here

$$\langle f, u^{(j)}(\cdot)\rangle = \frac{d^j}{dt^j}\langle f, u(\cdot)\rangle, \quad j = 0, \dots, m.$$

Lemma A.5.9. *For $u \in C^1_\omega(S;X)$ at any s and $t \in S$ such that $s < t$, we have*

$$\sup_{s \le \tau \le t} \|u'(\tau)\| < \infty$$

and

$$\|u(t) - u(s)\| \le (t-s) \sup_{s \le \tau \le t} \|u'(\tau)\|.$$

For a compact interval S, the set $C^m(S;X)$, which is a linear space with natural linear operations, becomes a Banach space by introducing the norm

$$\|u\|_{C^m(S;X)} = \sum_{j=0}^{m} \sup_{t \in S} \|u^{(j)}(t)\|.$$

Also, on the set $C^m_\omega(S;X)$ a locally convex space topology can be defined by the family of seminorms

$$p_{f,j}(u) = \sup_{t \in S} |\langle f, u^{(j)}(t)\rangle|, \quad f \in X^*, \quad j = 0, \dots, m.$$

If X is weakly complete, the locally convex space $C^m_\omega(S;X)$ is complete.

Theorem A.5.3 (Weierstrass approximation theorem). *Let S be a compact interval. Then the set of polynomials*

$$\left\{p(t) = \sum_{j=0}^{m} a_j t^j,\ t \in S,\ a_j \in X,\ j = 0, \dots, m\right\}$$

is dense in $C(S;X)$.

Space of integrable functions. First, we introduce the notions of measurability and integrability of functions from a time interval $S \subset \mathbb{R}$ to a Banach space X. These notions were introduced by Bochner.

Like the Lebesgue integral, the Bochner integral is first defined for a simple function. A function $u : S \to X$ is said to be simple if there is a finite number of disjoint

Lebesgue measurable subsets S_i in S $(i = 0, \dots, n)$ with $\operatorname{mes}(S_i) < \infty$ such that the function u on each set S_i takes a constant value x_i and $u(t) = 0$ for $t \in S \setminus \bigcup_{i=1}^{n} S_i$. The *Bochner integral* of a simple function is defined by the formula

$$\int_S u(t)\,dt = \sum_{i=1}^{n} \operatorname{mes}(S_i)x_i.$$

Definition A.5.14. If for a simple function $u : S \to X$ the set S_i can be taken as intervals, it is called a *step function.*

Definition A.5.15. A function $u : S \to X$ is said to be *Bochner measurable*, if there exists a sequence of simple functions $\{u_n\}$ such that

$$u_n(t) \to u(t) \quad \text{for almost all} \quad t \in S. \tag{A.5.6}$$

If this sequence also satisfies the condition

$$\lim_{n\to\infty} \int_S \|u(t) - u_n(t)\|\,dt = 0, \tag{A.5.7}$$

the function u is called *Bochner integrable*. The *Bochner integral* of such a function u on a Lebesgue measurable set $B \subset S$ is defined by the formula

$$\int_B u(t)\,dt = \lim_{n\to\infty} \int_S u_n(t)C_B(t)\,dt, \tag{A.5.8}$$

where C_B is the indicator function of the set B:

$$C_B(t) = \begin{cases} 1, & t \in B, \\ 0, & t \notin B. \end{cases}$$

If $B = [a, b]$, we usually write $\int_a^b u(t)\,dt$ instead of $\int_B u(t)\,dt$. At $b < a$

$$\int_a^b u(t)\,dt = -\int_b^a u(t)\,dt.$$

Note that if a function u is Bochner measurable, the real-valued function $\|u(\cdot) - u_n(\cdot)\|$ is Lebesgue measurable on S. Therefore, condition (A.5.7) makes sense. If this condition is satisfied, there exists a limit in (A.5.8), and it does not depend on the choice of a sequence of simple functions with properties (A.5.6) and (A.5.7). Hence, Definition A.5.15 is correct.

Some properties of Bochner measurable and integrable functions are as follows.

Theorem A.5.4. *Let X be a separable space. A function $u : S \to X$ is Bochner measurable if and only if for every $f \in X^*$ the function $\langle f, u(\cdot)\rangle : S \to \mathbb{R}$ is Lebesgue measurable.*

Theorem A.5.5. *Let $u : S \to X$ be a function such that there exists a sequence $\{u_n\}$ of Bochner measurable functions such that*

$$u_n(t) \rightharpoonup u(t) \quad \text{in } X \text{ for almost all } t \in S.$$

Then u is also Bochner measurable.

Theorem A.5.6. *A Bochner measurable function $u : S \to X$ is Bochner integrable if and only if the function $\|u(\cdot)\| : S \to \mathbb{R}$ is Lebesgue integrable. In this case, the following estimate holds for every (Lebesgue) measurable set $B \subset S$:*

$$\left\| \int_B u(t)\,dt \right\| \le \int_B \|u(t)\|\,dt.$$

Theorem A.5.7. *Let $u : S \to X$ be a Bochner integrable function. Then the function $v : S \to X$ defined by the rule*

$$v(t) = \int_{t_0}^{t} u(s)\,ds, \quad t_0 \in S,$$

is differentiable almost everywhere on S, and

$$v'(t) = u(t) \quad \text{for almost all} \quad t \in S.$$

Theorem A.5.8. *Let $u : S \to X$ be a Bochner integrable function, and let $B \subset S$ be a measurable set. Then for every $f \in X^*$*

$$\int_B \langle f, u(t)\rangle\,dt = \left\langle f, \int_B u(t)\,dt \right\rangle.$$

Theorem A.5.9. *Let S be a compact interval. Then every function $u \in C_\omega(S; X)$ is Bochner integrable. If $u \in C_\omega(S; X)$ and $v : S \to X$ is the function defined by the formula*

$$v(t) = \int_{t_0}^{t} u(s)\,ds, \quad t_0 \in S,$$

$v \in C^1_\omega(S; X)$ and $v' = u$. If $u \in C(S; X)$, we have $v \in C^1(S; X)$.

Theorem A.5.10. *Every function $u \in C^1_\omega(S; X)$ is differentiable almost everywhere on S.*

Definition A.5.16. $L_p(S; X)$, $1 \le p < \infty$, is the set of all Bochner measurable functions $u : S \to X$ for which

$$\int_S \|u(t)\|^p\,dt < \infty.$$

The set $L_p(S;X)$, $1 \le p < \infty$, which is a linear space with natural linear operations, transforms to a Banach space by introducing the norm

$$\|u\|_{L_p(S;X)} = \Big(\int_S \|u(t)\|^p \, dt \Big)^{1/p}.$$

The set of step functions from S to X (see Definition A.5.14) is dense in $L_p(S;X)$, $1 \le p < \infty$.

If X is a Hilbert space, the space $L_2(S;X)$ with the inner product

$$\langle u, v \rangle = \int_S \langle u(t), v(t) \rangle_X \, dt, \quad u, v \in L_2(S;X),$$

is also a Hilbert space.

Lemma A.5.10 (Hölder inequality). *Let $u \in L_p(S;X)$, $1 \le p \le \infty$, and $v \in L_q(S;X^*)$, $1/p + 1/q = 1$. Then $\langle v(\cdot), u(\cdot) \rangle \in L_1(S)$ and*

$$\int_S \langle v(t), u(t) \rangle \, dt \le \|u\|_{L_p(S;X)} \|v\|_{L_q(S;X^*)}.$$

Theorem A.5.11. *Let X be a reflexive and separable space, and $1 < p < \infty$. Then every element $f \in (L_p(S;X))^*$ admits a unique representation in the form*

$$f(u) = \int_S \langle v(t), u(t) \rangle \, dt, \quad \textit{for every} \quad u \in L_p(S;X),$$

where the function $v \in L_q(S;X^*)$ and $1/p + 1/q = 1$. The mapping $f \to v$, $f \in (L_p(S;X))^*$ is linear, and

$$\|f\|_{(L_p(S;X))^*} = \|v\|_{L_q(S;X^*)}.$$

A.6 Equations of mathematical physics

A.6.1 Dirac delta-function and its properties

This function was studied by the French mathematicians Augustin Louis Cauchy and Simeon Denis Poisson, and the German mathematician and mechanic Gustav Kirchhoff. The function was widely used by the prominent English mathematician, physicist, and engineer Oliver Heaviside in his fundamental book "Electromagnetic Theory" (vol. 2).

The name used now is delta-function, but Heaviside called it the unit impulse function. After Heaviside this function was also considered by the French mathematician Henri Lebesgue. In works by Heaviside this function was an efficient instrument for solving problems of mathematical physics. His book contains representations of the impulse function as a Fourier series, a Fourier integral, and an expansion in terms of some systems of functions. These expansions can be found in numerous textbooks on mathematical physics.

It should be noted that Heaviside considered that the delta-function is a typical function, just as all other ones. The delta-function was again introduced into mathematical physics in 1928 by the English theoretical physicist Dirac in his classical work "Principles of Quantum Mechanics". He pointed out that, strictly speaking, the delta-function cannot be considered a typical function. In the modern physical and mathematical literature, this function is named after Paul Dirac and is usually denoted by the Greek symbol $\delta(x)$.

Now, after this short introduction, we can define the Dirac delta-function and present its basic properties.

The Dirac delta-function is a representative of singular generalized functions. The notion of Dirac delta-function $\delta(x)$ as a generalized function can be obtained by considering it as the limit of functional sequences possessing certain properties.

Consider, for instance, a sequence of piecewise-constant functions $\{\delta_n(x)\}$, $x \in \mathbb{R}$, defined by the rule

$$\delta_n(x) = \begin{cases} n/2, & |x| < 1/n, \\ 0, & |x| > 1/n. \end{cases}$$

Notice that at any n the function $\delta_n(x)$ has some important properties. These are:

1) $\delta_n(x) = \delta_n(-x)$ (evenness);
2) $\int_{-\infty}^{\infty} \delta_n(x)\,dx = 1$;
3) for any continuous function $f(x)$

$$\int_{-\infty}^{\infty} f(x)\delta_n(x)\,dx = f(x^*),$$

where x^* is some point from the interval $(-1/n, 1/n)$;

4)

$$\delta_n(x) = \frac{d}{dx}\Phi_n(x), \quad x \neq \pm\frac{1}{n},$$

$$\Phi_n(x) = \int_{-\infty}^{x} dx = \begin{cases} 0, & x < -1/n, \\ (1+nx)/2, & |x| < 1/n, \\ 1, & x > 1/n. \end{cases}$$

It should be noted that as $n \to \infty$, the function $\Phi_n(x)$ tends in the limit to the Heaviside step theta function $\theta(x)$, that is,

$$\lim_{n\to\infty} \Phi_n(x) = \theta(x) = \begin{cases} 1, & x \geq 0, \\ 0, & x < 0. \end{cases}$$

The following sequences of functions $\delta_n(x)$ also have properties 1–3:

$$\delta_n(x) = \begin{cases} n - n^2|x|, & |x| < 1/n, \\ 0, & |x| > 1/n; \end{cases}$$

$$\delta_n(x) = \begin{cases} (n/2)(1 + \cos(n\pi x)), & |x| < 1/n, \\ 0, & |x| > 1/n; \end{cases}$$

$$\delta_n(x) = \frac{n}{\pi(1 + n^2x^2)}, \quad x \in \mathbb{R};$$

$$\delta_n(x) = \frac{n}{\sqrt{\pi}} e^{-n^2x^2}, \quad x \in \mathbb{R}.$$

Let n tend to infinity in the above-considered sequences of functions $\delta_n(x)$. We have

$$\lim_{n\to\infty} \delta_n(x) = \begin{cases} 0, & x \neq 0, \\ \infty, & x = 0. \end{cases}$$

This relation shows that, generally speaking, the limiting element of these functional sequences is not a function in the classical sense. Therefore, as $n \to \infty$, the limit should be understood not in terms of uniform convergence, but in terms of weak convergence. A sequence $\{g_n(x)\}$ weakly converges on an interval (a, b) if for any continuous function $f(x)$ there exists a limit,

$$\lim_{n\to\infty} \int_a^b f(x) g_n(x)\, dx = \int_a^b f(x) g(x)\, dx,$$

which determines the limiting element $g(x)$ of the weakly convergent sequence $\{g_n(x)\}$, even if this limiting element is not a function in the classical sense.

Thus, the limiting element to which the above-considered sequences $\{\delta_n(x)\}$ converge in terms of weak convergence, will be called the generalized delta-function and denoted by $\delta(x)$.

Properties 1)–3) of the functions $\delta_n(x)$ do not depend on n. Therefore, these properties can also be assigned to the limiting element, that is,

$$\delta(x) = \delta(-x);$$

$$\int_{-\infty}^{\infty} \delta(x)\, dx = 1; \tag{A.6.1}$$

and for any continuous function $f(x)$

$$\int_{-\infty}^{\infty} f(x)\delta(x)\, dx = f(0). \tag{A.6.2}$$

Formula (A.6.1) can be considered as a consequence of formula (A.6.2) at $f(x) \equiv 1$. Therefore, the integral relation (A.6.2) can be taken as the main property of the delta-function, and formula (A.6.2) can be considered a definition of the Dirac delta-function in the sense of weak convergence, that is,

$$\lim_{n\to\infty}\int_{-\infty}^{\infty} f(x)\delta_n(x)\,dx = f(0).$$

Taking into account property 4) of the above-considered sequence of piecewise-constant functions as $n \to \infty$, we obtain the formula

$$\delta(x) = \theta'(x), \quad \theta(x) = \begin{cases} 1, & x \geq 0, \\ 0, & x < 0, \end{cases}$$

which defines the generalized delta-function as the derivative of the Heaviside step theta function. This formula is naturally understood as a limiting one in the sense of weak convergence.

In addition to the delta-function $\delta(x)$, a function of the form $\delta(x - x^0)$ is also introduced to describe a concentrated action at a point $x = x^0$. The latter delta-function is defined by the following integral relation:

$$\int_{-\infty}^{\infty} f(x)\delta(x - x^0)\,dx = f(x^0),$$

where $f(x)$ is any continuous function.

A spatial delta-function $\delta(x)$, $x \in \mathbb{R}^n$, is introduced in a similar way. For instance, the delta-function in a three-dimensional space is defined by δ-sequences of products $\delta_n(x_1)\delta_n(x_2)\delta_n(x_3)$, $x = (x_1, x_2, x_3) \in \mathbb{R}^3$. Therefore, by definition,

$$\delta(x) = \delta(x_1)\delta(x_2)\delta(x_3)$$

and

$$\delta(x - x^0) = \delta(x_1 - x_1^0)\delta(x_2 - x_2^0)\delta(x_3 - x_3^0),$$

where $x^0 = (x_1^0, x_2^0, x_3^0) \in \mathbb{R}^3$.

The main property of this spatial generalized function is the integral relation

$$\iiint_\Omega f(x)\delta(x - x^0)\,dx = \begin{cases} f(x^0), & x^0 \in \Omega, \\ 0, & x^0 \notin \Omega, \end{cases}$$

where $f(x)$ is a function that is continuous in the domain Ω. The domain Ω can be the entire space $\mathbb{R}^3$.

For $f(x) \equiv 1$, we have

$$\iiint_\Omega \delta(x - x^0)\,dx = \begin{cases} 1, & x^0 \in \Omega, \\ 0, & x^0 \notin \Omega. \end{cases}$$

A.6.2 Basic equations of mathematical physics

Mathematical description of many physical processes often leads to differential, integral, or even integro-differential equations. A wide class of physical processes can be described by linear second-order differential equations.

The equation of small transversal vibrations of a string has the following form:

$$\rho \frac{\partial^2 u}{\partial t^2} = T_0 \frac{\partial^2 u}{\partial x^2} + F,$$

where $u(x, t)$ is the deviation of the string from the location of equilibrium at a point x at a time t, $\rho(x)$ is the string density, T_0 is the string tension, and the free term $F(x, t)$ is the rate of an external disturbance. If the density ρ is constant, $\rho(x) = \rho$, the equation of string vibration takes the following form:

$$\frac{\partial^2 u}{\partial t^2} = a^2 \frac{\partial^2 u}{\partial x^2} + f, \tag{A.6.3}$$

where $f = F(x, t)/\rho$ and $a^2 = T_0/\rho$ is a constant. This equation is also called the *one-dimensional wave equation.*

The equation of small transversal vibrations of a membrane,

$$\rho \frac{\partial^2 u}{\partial t^2} = T_0 \Big(\frac{\partial u^2}{\partial x_1^2} + \frac{\partial^2 u}{\partial x_2^2} \Big) + F,$$

in the case of constant density $\rho = \rho(x)$ is reduced to the form

$$\frac{\partial^2 u}{\partial t^2} = a^2 \Big(\frac{\partial^2 u}{\partial x_1^2} + \frac{\partial^2 u}{\partial x_2^2} \Big) + f, \quad a^2 = \frac{T_0}{\rho}, \quad f = \frac{F}{\rho}, \tag{A.6.4}$$

and is called the *two-dimensional wave equation.*

The *three-dimensional wave equation*

$$\frac{\partial^2 u}{\partial t^2} = a^2 \Big(\frac{\partial^2 u}{\partial x_1^2} + \frac{\partial^2 u}{\partial x_2^2} + \frac{\partial^2 u}{\partial x_3^2} \Big) + f \tag{A.6.5}$$

describes the propagation of sound waves in a homogeneous medium and the propagation of electromagnetic waves in a homogeneous nonconducting medium. The fluid density, pressure, and velocity potential, as well as the components of magnetic and electric field strength and corresponding potentials satisfy this equation.

The *wave equations* (A.6.3)–(A.6.5) can be written in a condensed form,

$$\Box_a u = f, \tag{A.6.6}$$

where $\Box_a$ is the *wave operator* (*the D'Alembert operator*)

$$\Box_a = \frac{\partial^2}{\partial t^2} - a^2 \Delta \quad (\Box = \Box_1),$$

and Δ is the *Laplace operator*

$$\Delta = \frac{\partial^2}{\partial x_1^2} + \frac{\partial^2}{\partial x_2^2} + \cdots + \frac{\partial^2}{\partial x_n^2}.$$

Heat conduction in an isotropic homogeneous medium is described by the following equation:

$$\frac{\partial u}{\partial t} = a^2 \Delta u + f, \quad a^2 = \frac{k}{c\rho}, \quad f = \frac{F}{c\rho}, \tag{A.6.7}$$

where $u(x,t)$ is the temperature of the medium at a point $x \in \mathbb{R}^3$ at a time t; the constants ρ, c, and k denote the density, specific heat, and heat conductivity coefficient of the medium; $F(x,t)$ is the rate of heat sources. Equation (A.6.7) is called the *heat conduction equation*. In this equation, there can be any number n of space variables $x_1, x_2, \dots, x_n$.

If the external perturbation $f(x,t)$ in the wave equation (A.6.6) is periodic with frequency ω and amplitude $a^2 f(x)$, that is,

$$f(x,t) = a^2 f(x) e^{i\omega t},$$

and the periodic perturbations $u(x,t)$ have the same frequency and unknown amplitude $u(x)$,

$$u(x,t) = u(x) e^{i\omega t},$$

the function $u(x)$ satisfies the steady-state equation

$$\Delta u + k^2 u = -f(x), \quad k^2 = \frac{\omega^2}{a^2}, \tag{A.6.8}$$

called the *Helmholtz equation*. At $k = 0$ equation (A.6.8) takes the form

$$\Delta u = -f \tag{A.6.9}$$

called the *Poisson equation* , and if $f = 0$, equation (A.6.9) is called the *Laplace equation*:

$$\Delta u = 0.$$

Let us now introduce some integral equations.

An equation of the form

$$q(t) = \int_a^b K(t,\tau) q(\tau)\, d\tau + f(t), \quad t \in [a,b], \tag{A.6.10}$$

is called a *Fredholm integral equation of the second kind*, and an equation of the form

$$\int_a^b K(t,\tau)q(\tau)\,d\tau = f(t), \quad t \in [a,b], \tag{A.6.11}$$

is called a *Fredholm integral equation of the first kind*. Here $q(t)$ is the sought-for function, $f(t)$ is a function given on $[a,b]$, and $K(t,\tau)$ is a function called the *kernel of the integral equation*, defined and square-integrable in the domain $[a,b] \times [a,b]$, that is, $\int_a^b \int_a^b K^2(s,\tau)\,ds\,d\tau < \infty$.

If the kernel $K(t,\tau)$ has a special form

$$K(t,\tau) = \begin{cases} k(t,\tau), & a \le \tau \le t, \\ 0, & t < \tau \le b, \end{cases}$$

equations (A.6.10) and (A.6.11) reduce to

$$q(t) = \int_a^t k(t,\tau)q(\tau)\,d\tau + f(t),$$

$$\int_a^t k(t,\tau)q(\tau)\,d\tau = f(t).$$

These are called *Volterra integral equations of the second and first kind*, respectively. The function $k(t,\tau)$, which is assumed to satisfy the condition $\int_a^b \int_a^t k^2(t,\tau)\,d\tau\,dt < \infty$, is called the *kernel of the Volterra integral equation*. In some problems, this assumption may be weakened and continuous kernels are considered.

In studying the properties of integral equations, in particular, the dependence of solutions on the initial data, the following Gronwall (Gronwall–Bellman) lemma is often useful.

Lemma (Gronwall). *Let a continuous function $y(t) \ge 0$ on an interval $J \ni t_0$ satisfy the inequality*

$$y(t) \le y_0 + \left| \int_{t_0}^t M(s)y(s)\,ds \right|,$$

where $y_0 \ge 0$ and $M : J \to \mathbb{R}$ is a continuous nonnegative function. Then

$$y(t) \le y_0 \cdot \exp\left\{ \left| \int_{t_0}^t M(s)\,ds \right| \right\}, \quad t \in J.$$

A.6.3 Classical Cauchy problem for the wave equation

The classical Cauchy problem for the wave equation has the following form:

$$\frac{\partial^2 u}{\partial t^2} - a^2 \Delta u = f(x,t), \tag{A.6.12}$$

$$u\Big|_{t=0} = u_0(x), \quad \frac{\partial u}{\partial t}\Big|_{t=0} = u_1(x), \tag{A.6.13}$$

where

$$f \in C^2(t \geq 0), \quad u_0 \in C^2(\mathbb{R}^n), \quad u_1 \in C^2(\mathbb{R}^n) \quad \text{at } n = 3, 2; \tag{A.6.14}$$

$$f \in C^1(t \geq 0), \quad u_0 \in C^2(\mathbb{R}), \quad u_1 \in C^1(\mathbb{R}) \quad \text{at } n = 1. \tag{A.6.15}$$

There exists a unique solution to the Cauchy problem (A.6.12)–(A.6.15). It is expressed by:

- the *Kirchhoff formula* at $n = 3$:

$$u(x,t) = \frac{1}{4\pi a^2} \int_{B(x,at)} \frac{f(\xi, t - |x-\xi|/a)}{|x-\xi|}\, d\xi + \frac{1}{4\pi a^2 t} \int_{S(x,at)} u_1(\xi)\, dS + \frac{1}{4\pi a^2} \frac{\partial}{\partial t}\Big[\frac{1}{t} \int_{S(x,at)} u_0(\xi)\, dS\Big];$$

- the *Poisson formula* at $n = 2$:

$$u(x,t) = \frac{1}{2\pi a} \int_0^t \int_{B(x,a(t-\tau))} \frac{f(\xi,\tau)\, d\xi\, d\tau}{\sqrt{a^2(t-\tau)^2 - |x-\xi|^2}} + \frac{1}{2\pi a} \int_{B(x,at)} \frac{u_1(\xi)\, d\xi}{\sqrt{a^2 t^2 - |x-\xi|^2}} + \frac{1}{2\pi a} \frac{\partial}{\partial t} \int_{B(x,at)} \frac{u_0(\xi)\, d\xi}{\sqrt{a^2 t^2 - |x-\xi|^2}};$$

- the *D'Alembert formula* at $n = 1$:

$$u(x,t) = \frac{1}{2a} \int_0^t \int_{x-a(t-\tau)}^{x+a(t-\tau)} f(\xi,\tau)\, d\xi\, d\tau + \frac{1}{2a} \int_{x-at}^{x+at} u_1(\xi)\, d\xi + \frac{1}{2}\,[u_0(x+at) + u_0(x-at)].$$

This solution continuously depends on the data f, u_0, and u_1 of the Cauchy problem in the following sense: if

$$|f - \tilde{f}| \leq \varepsilon, \quad |u_0 - \tilde{u}_0| \leq \varepsilon_0, \quad |u_1 - \tilde{u}_1| \leq \varepsilon_1, \quad |\mathrm{grad}(u_0 - \tilde{u}_0)| \leq \varepsilon_0'$$

(the latter inequality is not needed at $n = 1$), the corresponding solutions u and $\tilde{u}$ in any band $0 \leq t \leq T$ satisfy the following estimates:

$$|u(x,t) - \tilde{u}(x,t)| \leq \frac{T^2}{2}\varepsilon + T\varepsilon_1 + \varepsilon_0 + aT\varepsilon_0' \quad (n = 2, 3),$$

$$|u(x,t) - \tilde{u}(x,t)| \leq \frac{T^2}{2}\varepsilon + T\varepsilon_1 + \varepsilon_0 \quad (n = 1).$$

Thus, the Cauchy problem for the wave equation (A.6.12), (A.6.13) is correct under the assumptions (A.6.14), (A.6.15).

A.6.4 Fundamental solution of a differential operator

Let $\mathcal{L}(\mathbb{R}^n)$ denote all functions of the class $C^\infty(\mathbb{R}^n)$ that decrease, as $|x| \to \infty$, with all their derivatives faster than any power of $|x|^{-1}$. Convergence in the space $\mathcal{L}(\mathbb{R}^n)$ is defined as follows: a sequence of functions $\{u_k\}$ from $\mathcal{L}(\mathbb{R}^n)$ converges to a function $u \in \mathcal{L}(\mathbb{R}^n)$, $u_k \to u$, as $k \to \infty$ in $\mathcal{L}(\mathbb{R}^n)$ if for all α and β the sequence of functions $\{x^\beta D^\alpha u_k(x)\}$ uniformly converges to $x^\beta D^\alpha u(x)$ as $k \to \infty$ on $\mathbb{R}^n$. It is evident that $\mathcal{L}(\mathbb{R}^n)$ is a linear space. In addition, $\mathcal{D}(\mathbb{R}^n) \subset \mathcal{L}(\mathbb{R}^n)$, and convergence in $\mathcal{D}(\mathbb{R}^n)$ implies convergence in $\mathcal{L}(\mathbb{R}^n)$.

A *generalized function of slow growth* is any linear continuous functional on the space of basic functions $\mathcal{L}(\mathbb{R}^n)$. The generalized function of slow growth will be assumed to be complex-valued. Let $\mathcal{L}^*(\mathbb{R}^n)$ denote the set of all generalized functions of slow growth. Convergence in $\mathcal{L}^*(\mathbb{R}^n)$ is defined as weak convergence of a sequence of functionals: a sequence of generalized functions $\{f_k\}$ from $\mathcal{L}^*(\mathbb{R}^n)$ converges to a generalized function $f \in \mathcal{L}^*(\mathbb{R}^n)$ if for any $u \in \mathcal{L}(\mathbb{R}^n)$ $f_k(u) \to f(u)$, $k \to \infty$.

The linear set $\mathcal{L}^*(\mathbb{R}^n)$ with the above-introduced convergence is called the *space of generalized functions of slow growth* $\mathcal{L}^*(\mathbb{R}^n)$.

It should be noted that $\mathcal{L}^*(\mathbb{R}^n) \subset \mathcal{D}^*(\mathbb{R}^n)$ and convergence in $\mathcal{L}^*(\mathbb{R}^n)$ implies convergence in $\mathcal{D}^*(\mathbb{R}^n)$.

An important peculiarity of the class of generalized functions of slow growth is that the Fourier transform of a function from this class is a function of this class.

Since the basic functions from $\mathcal{L}(\mathbb{R}^n)$ are absolutely integrable on $\mathbb{R}^n$, one can define a function

$$F[u](\xi) = \int u(x) e^{i(\xi,x)}\, dx, \quad u \in \mathcal{L}(\mathbb{R}^n),$$

called the *Fourier transform of the function* $u(x)$. The function $F[u](\xi)$ is bounded and continuous in $\mathbb{R}^n$. The Fourier transform F maps $\mathcal{L}(\mathbb{R}^n)$ onto $\mathcal{L}(\mathbb{R}^n)$, and this is a one-to-one correspondence.

The *Fourier transform of a generalized function* $\varphi \in \mathcal{L}^*(\mathbb{R}^n)$ is defined by the formula

$$F[\varphi](u) = \varphi(F[u]), \quad \varphi \in \mathcal{L}^*(\mathbb{R}^n), \quad u \in \mathcal{L}(\mathbb{R}^n).$$

The Fourier transform is a linear and continuous operation from $\mathcal{L}^*(\mathbb{R}^n)$ to $\mathcal{L}^*(\mathbb{R}^n)$.

Consider an arbitrary linear differential equation of order m with coefficients $a_\alpha \in C^\infty(\mathbb{R}^n)$:

$$\sum_{|\alpha| \le m} a_\alpha(x) D^\alpha u = f(x), \quad u, f \in \mathcal{D}^*(\mathbb{R}^n).$$

This equation can be rewritten in operator form as follows:

$$L(x, D)u = f(x), \tag{A.6.16}$$

where

$$L(x, D) = \sum_{|\alpha|\le m} a_\alpha(x) D^\alpha.$$

A *generalized solution* of equation (A.6.16) in a domain G is any generalized function $u \in \mathcal{D}^*(\mathbb{R}^n)$ that satisfies this equation in the domain G in the following generalized sense, that is, for any $v \in \mathcal{D}(G)$:

$$(L(x, D)u)(v) = f(v).$$

If $f \in C(G)$ and a generalized solution $u(x)$ of equation (A.6.16) in a domain G belongs to the class $C^m(G)$, it is also a classical solution of this equation in the domain G.

Let $L(D)$ be a differential operator with constant coefficients:

$$L(D) = \sum_{|\alpha|\le m} a_\alpha D^\alpha.$$

A *fundamental solution* (*influence function*) of the operator $L(D)$ is a generalized function $\mathcal{E} \in \mathcal{D}^*(\mathbb{R}^n)$ that satisfies in $\mathbb{R}^n$ the following equation:

$$L(D)\mathcal{E} = \delta(x).$$

The fundamental solution $\mathcal{E}(x)$ of the operator $L(D)$ is, generally speaking, not unique. It is determined up to a term $\mathcal{E}_0(x)$, which is an arbitrary solution of the homogeneous equation $L(D)\mathcal{E}_0 = 0$.

Theorem A.6.1. *For a generalized function $\mathcal{E}$ from $\mathcal{L}^*(\mathbb{R}^n)$ to be a fundamental solution of the operator $L(D)$, it is necessary and sufficient that its Fourier transform $F[\mathcal{E}]$ satisfy the equation*

$$L(-i\xi)F[\mathcal{E}] = 1,$$

where

$$L(\xi) = \sum_{|\alpha|\le m} a_\alpha \xi^\alpha.$$

The fundamental solution $\mathcal{E}_n$ of the wave operator

$$\Box_a \mathcal{E}_n = \delta(x, t)$$

in the space $\mathcal{L}^*(\mathbb{R}^n)$ is derived by using the Fourier transform:

$$\mathcal{E}_n(x, t) = \theta(t) F_\xi^{-1}\Big[\frac{\sin(a|\xi|t)}{a|\xi|}\Big].$$

At $n = 3, 2, 1$ the fundamental solutions of the wave operator $\Box_a$ have the following form:

$$\mathcal{E}_3(x,t) = \frac{\theta(t)}{2\pi a}\,\delta(a^2t^2 - |x|^2),$$

$$\mathcal{E}_2(x,t) = \frac{\theta(at - |x|)}{2\pi a\sqrt{a^2t^2 - |x|^2}},$$

$$\mathcal{E}_1(x,t) = \frac{1}{2a}\,\theta(at - |x|).$$

The Laplace operator $\Delta = \frac{\partial^2}{\partial x_1^2} + \cdots + \frac{\partial^2}{\partial x_n^2}$ has the following fundamental solutions:

$$\mathcal{E}_2(x) = \frac{1}{2\pi}\ln|x|,$$

$$\mathcal{E}_n(x) = -\frac{1}{(n-2)\sigma_n}|x|^{-n+2}, \quad n \geq 3.$$

A derivation of these formulas and a detailed description of the theory of generalized functions as applied to equations of mathematical physics can be found in [A.1].

Appendix B

B.1 Supplementary exercises and control questions

Supplementary exercises for Chapter 1

1. Try to estimate how many different inverse problems can be formulated for the acoustic equation.

2. All the main problems of geophysics are essentially inverse problems. Try to explain why.

3. How many areas of mathematics and what other sciences were discussed in this chapter?

4. Formulate several inverse problems for the parabolic equation $u_t = u_{xx} - q(x)u$ similar to those formulated above for the acoustic equation.

5. Which of the inverse and ill-posed problems discussed in this chapter are linear and which are nonlinear?

6. Give new examples of inverse and ill-posed problems—those you might have read or thought about—and write a short account of what you understood and what was unclear to you after studying this chapter. Please feel free to send your account to the author (email: kabanikhin@mail.ru). Your feedback will be sincerely appreciated.

Control questions for Chapter 2

1. Formulate the well-posedness conditions (in the sense of Hadamard) for a problem $Aq = f$. Which approaches are used for solving the problem if one of the conditions fails?

2. Prove that the conditional stability estimate for the problem $Aq = f$ on the ball $B(\mathbf{0}, C) = \{q \in Q\colon \|q\| < C\}$ cannot be obtained if the operator A is linear and compact.

3. Make a brief diagram based on Chapter 2 that reflects the relationships and differences between the concepts of a quasi-solution, pseudo-solution, normal solution, etc.

4. Analyze the regularizing properties of gradient methods.

Control questions for Chapter 3

1. Write the general structure of Chapter 3 in the form of a diagram.
2. Compare the main concepts and methods presented in Chapter 3 with those in the previous chapter.

Control questions for Chapter 4

1. Which of the other chapters are related to the results presented in Chapter 4?
2. Give examples of integral equations of the first kind occurring in various applied problems (see Chapter 1).
3. Investigate the simplest inverse problem for the system of Maxwell's equations (see Chapter 10) after reducing it to a Volterra operator equation.

Supplementary exercises for Chapter 5

1. How are the problems considered in Chapters 4 and 5 related to each other, and what are the differences between them?
2. Find the problems in Chapter 10 that use the results of this chapter.

Supplementary exercises for Chapter 6

1. Analyze the relationship between the Gelfand–Levitan method and the method presented in Chapter 11.
2. Write the formula that relates the Jost function and the spectral function.

Supplementary exercises for Chapter 7

1. Compare the degree of ill-posedness of the problems in this section and those in Section 8.
2. Give examples of practical application of the inverse problem from Section 7.4.

Supplementary exercises for Chapter 8

1. Compare the results obtained in this chapter with analogous results of Chapter 7.
2. Try to estimate the degree of ill-posedness for the problems in this chapter and calculate the singular numbers where possible.

Supplementary exercises for Chapter 9

1. Analyze the specific characteristics of the generalized solution discussed in this chapter.
2. Verify that the operator A^* defined by formula (9.4.33) is the adjoint of the operator A.

Control questions for Chapter 10

1. Analyze the interrelations between the problems considered in Chapter 10 and those in Chapters 2, 4, 5, 6.
2. Why Chapter 10 contains more results, both theoretical and applied, as compared to the other chapters?

Control questions for Chapter 11

1. Formulate the main distinctions of inverse problems for hyperbolic, parabolic and elliptic equations.
2. How to go from parabolic or elliptic equation to hyperbolic one? Which assumptions must be met?
3. Try to construct a Hadamard model for problems of Chapter 11.

B.2 Supplementary references

Supplementary references to Chapter 1

Antyufeev, V. S. (1999). *Monte Carlo Method for Solving Inverse Problems of Radiation Transfer.* Inverse and Ill-posed Problems Series. VSP, Utrecht.

Babenko, K. I. (2002). *Foundations of Numerical Analysis.* Regular and Chaotic Dynamics, Moscow-Izhevsk (in Russian).

Bamberger, A., Chavent, G., and Lailly, P. (1979). About the stability of the inverse problem in 1-D wave equations—applications to the interpretation of seismic profiles. *Appl. Math. Optim.*, **5**(1), 1–47.

Cherednichenko, V. G. (1996). *Inverse Logarithmic Potential Problem.* Inverse and Ill-posed Problems Series. VSP, Utrecht.

Glasko, V. B. (1988). *Inverse Problems of Mathematical Physics.* American Institute of Physics Translation Series. American Institute of Physics, New York.

Godunov S. K. (1979). *Equations of Mathematical Physics.* Nauka, Moscow.

Goncharskii, A. V., Cherepashchuk, A. M., and Yagola, A. G. (1978). *Numerical Methods for the Solution of Inverse Problems of Astrophysics.* Nauka, Moscow (in Russian).

Goncharskii, A. V., Cherepashchuk, A. M., and Yagola, A. G. (1985). *Ill-posed Problems of Astrophysics.* Nauka, Moscow (in Russian).

Hasanov, A. (1995). An inverse problem for an elastoplastic medium. *SIAM J. Appl. Math.*, **55**(6), 1736–1752.

Hasanov, A. (2007). An inversion method for identification of elastoplastic properties for engineering materials from limited spherical indentation measurements. *Inverse Probl. Sci. Eng.*, **15**(6), 601–627.

Hasanov, A., Mueller, J. L., Cohn, S., and Redepenning, J. (2000). Numerical solution of a nonlocal identification problem for nonlinear ion transport. *Comput. Math. Appl.*, **39**(7–8), 225–235.

Ivanov, V. K. (1962). Integral equations of the first kind and approximate solution of the inverse problem of potential theory. *Dokl. Akad. Nauk SSSR*, **142**(5), 998–1000 (in Russian).

Kabanikhin, S. I., Martakov, S. V., Tabarovskii, L. A., and Shifon, N. Ya. (1989). The study of dielectric permittivity of the medium around a borehole at the early formation stage. In: *Studies on the Conditional Well-posedness of Problems of Mathematical Physics.* Akad. Nauk SSSR Sibirsk. Otdel., Inst. Mat., Novosibirsk, 40–59 (in Russian).

Kunetz, G. (1964). Généralisation des opérateurs d'antirésonance à un nombre quelconque de reflecteurs. *Geophys. Prospecting*, **12**, 283–289.

Novikov, P. S. (1938). On the uniqueness of the inverse problem of the potential theory. *Dokl. Akad. Nauk SSSR*, **18**(3), 165–168 (in Russian).

Prilepko, A. I. (1965). On the uniqueness of the determination of the form of a body by the values of the exterior potential. *Dokl. Akad. Nauk SSSR*, **160**(1), 40–43 (in Russian).

Romanov, V. G. (1982a). Inverse problems of seismic and electromagnetic wave propagation. In: *Methods for Solving Inverse Problems and Their Applications.* Akad. Nauk SSSR Sibirsk. Otdel., Vychisl. Tsentr, Novosibirsk, 111–118 (in Russian).

Romanov, V. G. (1982b). On the uniqueness of determination of the coefficients of the system of Maxwell's equations. In: *Nonclassical Problems for Equations of Mathe-*

matical Physics. Akad. Nauk SSSR Sibirsk. Otdel., Inst. Mat., Novosibirsk, 139–142 (in Russian).

Romanov, V. G. (1983). A linear approximation of the inverse Lamb problem. In: *Numerical Methods in Seismic Studies.* Nauka, Novosibirsk, 170–192 (in Russian).

Santosa, F. and Symes, W. W. (1989). *An Analysis of Least-Squares Velocity Inversion.* The Society of Exploration Geophysicists, Geophysical Monograph Series No. 4.

Sretenskii, L. N. (1954). On the uniqueness of determination of the form of an attracting body from the values of its external potential. *Dokl. Akad. Nauk SSSR*, **99**(1), 21–22 (in Russian).

Vasin, V. V., Prutkin I. L., and Temirkhanova L. Yu. (1996). Reconstruction of the 3D geological boundary relief from the gravitational measurements data. *Izv. Ross. Akad. Nauk*, **11**, 58–62 (in Russian).

Verzhbitskii, V. V., Kabanikhin, S. I., and Martakov, S. V. (1994). Transient nonstationary dielectric logging. *Dokl. Akad. Nauk*, **337**(3), 386–388 (in Russian).

Yagola, A. G., Kochikov, I. V., Kuramshina, G. M., and Pentin, Yu. A. (1999). *Inverse Problems of Vibrational Spectroscopy.* Inverse and Ill-posed Problems Series. VSP, Utrecht.

Supplementary references to Chapter 2

Bakushinskii, A. B. and Goncharskii, A. V. (1989). *Iterative Methods for Solving Ill-posed Problems.* Nauka, Moscow (in Russian).

Bakushinskii, A. B. and Kokurin, M. Yu. (2004). *Iterative Methods for Approximate Solution of Inverse Problems.* Math. Appl., **577**, Springer, Dordrecht.

Bakushinskii, A. B., Kokurin, M. Yu., and Smirnova, A. (2010). *Iterative Methods for Ill-Posed Problems.* De Gruyter, Berlin.

Bastay, G., Kozlov, V. A., and Turesson, B. O. (2001). Iterative methods for an inverse heat conduction problem. *J. Inverse Ill-Posed Probl.*, **9**(4), 375–388.

Beutler, F. J. and Root, W. L. (1976). The operator pseudoinverse in control and systems identification. In: *Generalized inverses and applications. Proc. Sem., Math. Res. Center, Univ. Wisconsin, Madison, Wis., (1973)*, Academic Press, New York, 397–494.

Bukhgeim, A. L. (1983). *Volterra Operator Equations.* Nauka, Novosibirsk (in Russian).

Cauchy, A. L. (1847). Méthode générale pour la résolution des systèmes d'équations simultanées. *Comptes Rendus de l'Académie des Sciences de Paris*, **25**, 536–538.

Fedotov, A. M. (1990). *Ill-posed Problems With Random Errors In The Data.* Nauka, Novosibirsk (in Russian).

Gauss, C. F. (1809). *Theoria Motus Corporum Coelestium in Sectionibus Conicis Solem Ambientium.* Perthes und Besser, Hamburg.

Hadamard, J. (1902). Sur les problèmes aux dérivées partielles et leur signification physique. *Princeton University Bulletin*, 49–52.

Hadamard, J. (1932). *Le problème de Cauchy et les équations aux dérivées partielles linéaires hyperboliques; leçons professées à l'Université Yale.* Hermann, Paris.

Hasanov, A. (1997). Inverse coefficient problems for monotone potential operators. *Inverse Problems*, **13**(5), 1265–1278.

Ivanov, V. K. (1962). On linear problems which are not well-posed. *Dokl. Akad. Nauk SSSR*, **145**(2), 270–272 (in Russian).

Ivanov, V. K. (1963). On ill-posed problems. *Mat. Sb.*, **61**(2), 211–223 (in Russian).

Ivanov, V. K. (1969). Ill-posed problems in topological spaces. *Sibirsk. Mat. Zh.*, **10**(5), 1065–1074 (in Russian).

Kabanikhin, S. I. (2002). Iterative methods of solving inverse problems for hyperbolic equations. In: *Ill-Posed and Inverse Problems*, VSP, Zeist, 201–217.

Kabanikhin, S. I. (2005). Convergence rate estimation of gradient methods via conditional stability of inverse and ill-posed problems. *J. Inverse Ill-Posed Probl.*, **13**(3–6), 259–264.

Kabanikhin, S. I. (2006). Conditional stability stopping rule for gradient methods applied to inverse and ill-posed problems. *J. Inverse Ill-Posed Probl.*, **14**(8), 805–812.

Kabanikhin, S. I. and Schieck, M. (2008). Impact of conditional stability. Convergence rates for general linear regularization methods. *J. Inverse Ill-Posed Probl.*, **16**(3), 267–282.

Kantorovich, L. V. (1952). *Functional Analysis and Applied Mathematics.* NBS Rep. 1509. U.S. Department of Commerce, National Bureau of Standards, Los Angeles, CA.

Kantorovich, L. V. and Akilov, G. P. (1982). *Functional Analysis.* Pergamon Press, Oxford-Elmsford, N.Y.

Kato, T. (1995). *Perturbation Theory for Linear Operators.* Springer-Verlag, Berlin.

Klibanov, M. V. and Timonov, A. (2004). *Carleman Estimates For Coefficient Inverse Problems And Numerical Applications.* VSP, Utrecht.

Kozlov, V. A., Maz'ya, V. G., and Fomin, A. V. (1991). An iterative method for solving the Cauchy problem for elliptic equations. *Comput. Math. Math. Phys.*, **31**(1), 45–52.

Krasnosel'skii, M. A., Vainikko, G. M., Zabreiko, P. P., Rutitskii, Ya. B., and Stetsenko, V. Ya. (1972). *Approximate Solution of Operator Equations.* Wolters-Noordhoff, Groningen.

Krein, M. G. and Krasnosel'skii, M. A. (1947). Fundamental theorems on the extension of Hermitian operators and some of their applications to the theory of orthogonal polynomials and to the problem of moments. *Uspekhi Mat. Nauk*, **2**(3), 60–106 (in Russian).

Lavrentiev, M. M. (1959). Integral equations of the first kind. *Dokl. Akad. Nauk SSSR*, **127**(1), 31–33 (in Russian).

Lawson, C. L. and Hanson, R. J. (1974). *Solving Least Squares Problems.* Prentice-Hall, Englewood Cliffs, N.J.

Legendre, A. M. (1806). *Nouvelles méthodes pour la détermination des orbites des comètes.* Courcier, Paris.

Liskovets, O. A. (1982). Theory and methods of solving ill-posed problems. *Itogi Nauki Tekh., Matem. Analiz*, **20**, 116–178 (in Russian).

Maslov, V. P. (1968). The existence of a solution of an ill-posed problem is equivalent to the convergence of a regularization process. *Uspekhi Mat. Nauk*, **23**(3), 183–184 (in Russian).

Morozov, V. A. (1987). *Regular Methods for Solving Ill-posed Problems.* Nauka, Moscow (in Russian).

Morozov, V. A. and Gordonova, V. I. (1973). Numerical algorithms for the selection of the parameter in the regularization method. *Zh. Vychisl. Mat. i Mat. Fiz.*, **13**, 539–545 (in Russian).

Nashed, M. Z. (1976). Aspects of generalized inverses in analysis and regularization. In: *Generalized inverses and applications. Proc. Sem., Math. Res. Center, Univ. Wisconsin, Madison, Wis., 1973*, Academic Press, New York, 193–244.

Nashed, M. Z. (1976). Perturbations and approximations for generalized inverses and linear operator equations. In: *Generalized inverses and applications. Proc. Sem., Math. Res. Center, Univ. Wisconsin, Madison, Wis., 1973*, Academic Press, New York, 325–396.

Nashed, M. Z. and Votruba G. F. (1976). A unified operator theory of generalized inverses. In: *Generalized inverses and applications. Proc. Sem., Math. Res. Center, Univ. Wisconsin, Madison, Wis., 1973*, Academic Press, New York, 1–109.

Nemirovskii, A. S. (1986). Regularizing properties of the conjugate gradient method in ill-posed problems. *Zh. Vychisl. Mat. Mat. Fiz.*, **26**(3), 332–347 (in Russian).

Pol'skii, N. I. (1955). On the convergence of certain approximate methods of analysis. *Ukrain. Mat. Zh.* **7**(1), 56–70 (in Russian).

Pol'skii, N. I. (1956). On a general scheme of application of approximation methods. *Dokl. Akad. Nauk SSSR*, **111**(6), 1181–1184 (in Russian).

Pol'skii, N. I. (1962). Projective methods in applied mathematics. *Dokl. Akad. Nauk SSSR*, **143**(4), 787–790 (in Russian).

Riesz, F. and Sz.-Nagy, B. (1990). *Functional Analysis.* Dover Publications, New York.

Tanana, V. P. (1997a). *Methods for Solution of Nonlinear Operator Equations.* VSP, Utrecht.

Tanana, V. P. (1997b). *Methods for Solving Operator Equations.* VSP, Utrecht.

Tikhonov, A. N. (1943). On the stability of inverse problems. *C. R. (Doklady) Acad. Sci. URSS*, **39**, 176–179.

Tikhonov, A. N. (1963a). On the solution of ill-posed problems and the method of regularization. *Dokl. Akad. Nauk SSSR*, **151**(3), 501–504 (in Russian).

Tikhonov, A. N. (1963b). On the regularization of ill-posed problems. *Dokl. Akad. Nauk SSSR*, **153**(1), 49–52 (in Russian).

Tikhonov, A. N. (1964). Stable methods for the summation of Fourier series. *Soviet Math., Dokl.*, **5**, 641–644.

Tikhonov, A. N., Goncharsky, A. V., Stepanov, V. V., and Yagola, A. G. (1995). *Numerical Methods for the Solution of Ill-posed Problems.* Math. Appl., **328**, Kluwer Acad. Publ., Dordrecht.

Tseng, Ya. Yu. (1949a). Generalized inverses of unbounded operators between two unitary spaces. *Doklady Akad. Nauk SSSR*, **67**, 431–434 (in Russian).

Tseng, Ya. Yu. (1949b). Properties and classification of generalized inverses of closed operators. *Doklady Akad. Nauk SSSR*, **67**, 607–610 (in Russian).

Tseng, Ya. Yu. (1956). Virtual solutions and general inversions. *Uspekhi Mat. Nauk*, **11**(6)(72), 213–215 (in Russian).

Vainikko, G. M. (1982). *Methods for Solving Linear Ill-posed Problems in Hilbert Spaces.* Tartu. Gos. Univ., Tartu (in Russian).

Vainikko, G. M. and Veretennikov, A. Yu. (1986). *Iteration Procedures in Ill-posed Problems.* Nauka, Moscow, (in Russian).

Vasil'ev, F. P. (1981). *Methods for Solving Extremal Problems.* Nauka, Moscow (in Russian).

Vasil'ev, F. P. (1988). *Numerical Methods for Solving Extremal Problems.* Second edition. Nauka, Moscow (in Russian).

Vasin, V. V. (1989). *Methods for Solving Unstable Problems.* Ural. Gos. Univ., Sverdlovsk (in Russian).

Vasin, V. V. and Ageev, A. L. (1995). *Ill-posed Problems with A Priori Information.* Inverse and Ill-posed Problems Series. VSP, Utrecht.

Yosida, K. (1995). *Functional Analysis.* Springer-Verlag, Berlin.

Supplementary references to Chapter 3

Cline, A. K., Moler, C. B., Stewart, G. W., and Wilkinson, J. H. (1979). An estimate for the condition number of a matrix. *SIAM J. Numer. Anal.*, **16**(2), 368–375.

Djumaev, S. (1982). On the approximation of a pseudo-solution. *Dokl. Akad. Nauk Tajik SSR*, **25**(10), 584–587 (in Russian).

Fadeev, D. K. and Fadeeva, V. N. (1963). *Computational Methods of Linear Algebra.* Fizmatgiz, Moscow (in Russian).

Fiedler, M. (1986). *Special Matrices and Their Applications in Numerical Mathematics.* Martinus Nijhoff Publishers, Dordrecht.

Forsythe, G. E., Malcolm, M. A., and Moler, C. B. (1977). *Computer Methods for Mathematical Computations.* Prentice-Hall Series in Automatic Computation. Prentice-Hall, Englewood Cliffs, N.J.

Forsythe, G. E. and Moler, C. B. (1967). *Computer Solution of Linear Algebraic Systems.* Prentice-Hall, Englewood Cliffs, N.J.

Godunov, S. K. (1980). *Solution of Systems of Linear Equations.* Nauka, Novosibirsk (in Russian).

Godunov, S. K. (1998). *Modern Aspects of Linear Algebra.* Transl. Math. Monogr., 175, Amer. Math. Soc., Providence, RI.

Godunov, S. K., Antonov, A. G., Kiriljuk, O. P., and Kostin, V. I. (1993). *Guaranteed Accuracy in Numerical Linear Algebra.* Math. Appl., 252, Kluwer Acad. Publ., Dordrecht.

Golub, G. H. (1968). Least squares, singular values and matrix approximations. *Apl. Mat.*, **13** (1), 44–51.

Greville, T. N. E. (1960). Some applications of the pseudoinverse of a matrix. *SIAM Rev.*, **2** (1), 15–22.

Ivanov, V. K., Vasin, V. V., and Tanana, V. P. (2002). *Theory of Linear Ill-posed Problems and its Applications.* Inverse and Ill-posed Problems Series, VSP, Utrecht.

Lawson, C. L. and Hanson, R. J. (1974). *Solving Least Squares Problems.* Prentice-Hall, Englewood Cliffs, N.J.

Legendre, A. M. (1806). *Nouvelles méthodes pour la détermination des orbites des comètes.* Courcier, Paris.

Mohamed, J. L. and Walsh, J. E. (Eds.) (1986). *Numerical Algorithms.* Oxford Science Publications. The Clarendon Press, Oxford University Press, New York.

Molchanov, I. N. (1983). Problems of developing linear algebra software. In: *Proc. All-Union Conference: Computational Methods of Linear Algebra. Moscow, August 23–25, 1982*, 187–202, Moscow (in Russian).

Molchanov, I. N. (1987). *Computer Methods for Solving Applied Problems. Algebra, Approximation of Functions.* Naukova Dumka, Kiev (in Russian).

Morozov, V. A. (1987). *Regular Methods for Solving Ill-posed Problems.* Nauka, Moscow (in Russian).

Morozov, V. A. and Gordonova, V. I. (1973). Numerical algorithms for the selection of the parameter in the regularization method. *Zh. Vychisl. Mat. i Mat. Fiz.*, **13** (3), 539–545 (in Russian).

Nazimov, A. B. (1986). *A Study of Regularization Using the Shift Method and its Applications.* PhD thesis, Moscow (in Russian).

Penrose, R. (1955). A generalized inverse for matrices. *Proc. Cambridge Philos. Soc.*, **51**(3), 406–413.

Reklaitis, G. V., Ravindran, A., and Ragsdell, K. M. (1983). *Engineering optimization. Methods and applications.* John Wiley & Sons, New York.

Tikhonov, A. N. (1963). On the solution of ill-posed problems and the method of regularization. *Dokl. Akad. Nauk SSSR*, **151**(3), 501–504 (in Russian).

Tikhonov, A. N. (1980). Approximate systems of linear algebraic equations. *Zh. Vychisl. Mat. i Mat. Fiz.*, **20**(6), 1373–1383 (in Russian).

Tikhonov, A. N. and Arsenin, V. Ya. (1977). *Solutions of Ill-posed Problems.* John Wiley & Sons, New York-Toronto-London.

Tikhonov, A. N., Goncharskii, A. V., Stepanov, V. V., and Yagola, A. G. (1983). *Regularizing Algorithms and A Priori Information.* Nauka, Moscow (in Russian).

Tikhonov, A. N., Goncharsky, A. V., Stepanov, V. V., and Yagola, A. G. (1995). *Numerical Methods for the Solution of Ill-posed Problems.* Math. Appl., **328**, Kluwer Acad. Publ., Dordrecht.

Vainikko, G. M. (1980). Error estimates of the successive approximation method for ill-posed problems. *Automat. Remote Control*, **3**, 84–92 (in Russian).

Vasin, V. V. (1985). *Discretization, Iterative-Approximation Algorithms for Solving Unstable Problems, and Their Applications.* PhD thesis, Sverdlovsk (in Russian).

Vasin, V. V. (1987). Iterative methods for the approximate solution of ill-posed problems with a priori information and their applications. In: *Inverse and Ill-posed Problems, Alpine-U.S. Semin. St. Wolfgang/Austria 1986, Notes Rep. Math. Sci. Engrg.*, **4**, 211–229.

Voevodin, V. V. (1966). *Numerical Methods of Algebra. Theory and Algorithms.* Nauka, Moscow (in Russian).

Voevodin, V. V. (1974). *Linear Algebra.* Nauka, Moscow (in Russian).

Voevodin, V. V. and Voevodin, Vl. V. (2006). *Encyclopaedia of Linear Algebra. The LINEAL Electronic Database.* BHV-Petersburg, St. Petersburg (in Russian).

Volovich, V. M. (1965). Solving systems of linear equations using block methods. *Computational Methods and Programming*, **3**, 106–133, Moskov. Gos. Univ. (in Russian).

Wilkinson, J. H. and Reinsch, C. (1971). *Handbook for Automatic Computation. Vol. II. Linear Algebra.* Springer-Verlag, New York-Heidelberg.

Zaikin, P. N. and Mechenov, A. S. (1971). Some aspects of the numerical solution of integral equations of the first kind using the regularization method. *Otchet Vychisl. Tsentr, Moskov. Gos. Univ.* **3**(144), Moskov. Gos. Univ. (in Russian).

Supplementary references to Chapter 4

Apartsin, A. S. (1987). Discretization methods for the regularization of some integral equations of the first kind. In: *Methods of Numerical Analysis and Optimization.* Nauka, Novosibirsk, 263–297 (in Russian).

Apartsin, A. S. (2003). *Nonclassical Linear Volterra Equations of the First Kind.* Inverse and Ill-posed Problems Series. VSP, Utrecht.

Baev, A. V. (1985). The solution of an inverse problem for the wave equation using a regularizing algorithm. *Zh. Vychisl. Mat. Mat. Fiz.*, **25**(1), 140–146 (in Russian).

Denisov, A. M. (1975). The approximate solution of a Volterra equation of the first kind. *Zh. Vychisl. Mat. Mat. Fiz.*, **15**(4), 1053–1056 (in Russian).

Gajewski, H., Gröger, K., and Zacharias, K. (1974). *Nichtlineare Operatorgleichungen und Operatordifferentialgleichungen. Mathematische Lehrbücher und Monographien, II.* Abteilung, Mathematische Monographien, **38**, Akademie-Verlag, Berlin (in German).

Kabanikhin, S. I. (1981). An approximate method for solving the inverse problem for the acoustic equation. In: *Approximate Solution Methods and Questions of the Well-posedness of Inverse Problems*, Akad. Nauk SSSR Sibirsk. Otdel., Vychisl. Tsentr, Novosibirsk, 55–62 (in Russian).

Kabanikhin, S. I. (1984). The solvability of inverse problems for differential equations. *Dokl. Akad. Nauk SSSR*, **277**(4), 788–791 (in Russian).

Kabanikhin, S. I. (1989). Regularization of Volterra operator equations of the first kind with a boundedly Lipschitz-continuous kernel. *Soviet Math. Dokl.* **39**(3), 549–552.

Lavrentiev, M. M. (1972). Inverse problems and special operator equations of the first kind. In: *Proceedings of the International Conference of Mathematicians in Nice.* Nauka, Moscow, 130–136 (in Russian).

Romanov, V. G. (1973b). A one-dimensional inverse problem for the wave equation. *Dokl. Akad. Nauk SSSR*, **211**(5), 1083–1084 (in Russian).

Romanov, V. G. (1975). Certain uniqueness classes for the solution of Volterra operator equations of the first kind. *Funktsional. Anal. i Prilozhen.*, **9**(1), 81–82 (in Russian).

Romanov, V. G., Kabanikhin, S. I., and Boboev, K. B. (1984). An inverse problem of electrodynamics for $\mathcal{P}_N$-approximation of the kinetic transport equation. *Dokl. Akad. Nauk SSSR*, **276**(2), 296–299 (in Russian).

Sergeev, V. O. (1971). Regularization of Volterra equations of the first kind. *Dokl. Akad. Nauk SSSR*, **197**(3), 531–534 (in Russian).

Vainikko, G. M. and Khyamarik, U. A. (1985). Projection methods and self-regularization in ill-posed problems. *Izv. Vyssh. Uchebn. Zaved. Mat.*, 10, 3–17 (in Russian).

Supplementary references to Chapter 5

Akhiezer, N. I. (1965). *The Classical Moment Problem And Some Related Questions In Analysis.* Hafner Publishing Co., New York.

Gel'fand, I. M., Graev, M. I., and Vilenkin, N. Ya. (1966). *Generalized Functions. Vol. 5: Integral Geometry and Representation Theory.* Academic Press, New York-London.

John, F. (1955b). *Plane Waves and Spherical Means Applied to Partial Differential Equations.* Interscience Publishers, New York-London.

Natterer, F. (2001). *The Mathematics of Computerized Tomography.* Classics Appl. Math., 32, SIAM, Philadelphia, PA.

Tikhonov, A. N., Arsenin, V. Ya., and Timonov, A. A. (1987). *Mathematical Problems in Computer Tomography.* Nauka, Moscow (in Russian).

Anikonov, D. S., Nazarov, V. G., and Prokhorov, I. V. (2002). *Poorly Visible Media in X-ray Tomography.* VSP, Utrecht.

Anikonov, D. S., Kovtanyuk, A. E., and Prokhorov, I. V. (2002). *Transport Equation And Tomography.* Inverse and Ill-posed Problems Series, VSP, Utrecht.

Radon, J. (1917). Über die Bestimmung von Funktionen durch ihre Integralwerte längs gewisser Mannigfaltigkeiten. *Ber. Verh. Sächs. Akad. Wiss. Leipzig*, **69**, 262–277.

Supplementary references to Chapter 6

Agranovich, Z. and Marchenko, V. (1963). *The Inverse Problem Of Scattering Theory.* Gordon and Breach, New York-London.

Alekseev, A. S. (1962). Some converse problems in wave propagation theory. *Bull. (Izv.) Acad. Sci. USSR Geophys. Ser.*, 944–954.

Alekseev, A. S. and Belonosov, V. S. (1998). Spectral methods in one-dimensional problems of wave propagation theory. *Mat. Model. Geofiz., Trudy Inst. Vych. Mat. Mat. Geofiz.*, **11**, 7–39 (in Russian).

Ambarzumijan, V. (1929). Über eine Frage der Eigenwerttheorie. *Zeitschr. für Physik.*, 53, 690–695.

Berezanskii, Yu. M. (1953). On the uniqueness of the determination of Schrödinger's equation from its spectral function. *Doklady Akad. Nauk SSSR*, **93**(4), 591–594 (in Russian).

Berezanskii, Yu. M. (1958). On the inverse problem of spectral analysis for Schrödinger's equation. *Tr. Mosk. Mat. Obs.*, **7**, 3–51 (in Russian).

Borg, G. (1946). Eine Umkehrung der Sturm-Liouvilleschen Eigenwertaufgabe. Bestimmung der Differentialgleichung durch die Eigenwerte. *Acta Math.*, **78**, 1–96.

Burridge, R. (1980). The Gelfand–Levitan, the Marchenko, and the Gopinath–Sondhi integral equations of inverse scattering theory, regarded in the context of inverse impulse-response problems. *Wave Motion*, **2**(4), 305–323.

Chudov, L. A. (1949). The inverse Sturm–Liouville problem. *Mat. Sbornik*, **25**(67), 451–456 (in Russian).

Colton, D. L. and Kress, R. (1983). *Integral Equation Methods In Scattering Theory. Pure Appl. Math. (N.Y.)*, John Wiley & Sons, Inc., New York.

Faddeev, L. D. (1956). Uniqueness of solution of the inverse scattering problem. *Vestnik Leningrad. Univ.*, **11**(7), 126–130 (in Russian).

Faddeev, L. D. (1959). The inverse problem in the quantum theory of scattering. *Uspekhi Mat. Nauk*, **14**(4), 57–119 (in Russian).

Gel'fand, I. M. and Levitan, B. M. (1955). On the determination of a differential equation from its spectral function. *Amer. Math. Soc. Transl. (2)*, **1**, 253–304.

Krein, M. G. (1951). Solution of the inverse Sturm–Liouville problem. *Doklady Akad. Nauk SSSR*, **76**(1), 21–24 (in Russian).

Levinson, N. (1949). The inverse Sturm–Liouville problem. *Math. Tidsskr.*, **13**, 25–30.

Levitan, B. M. (1973). *Theory Of Generalized Shift Operators.* Nauka, Moscow (in Russian).

Levitan, B. M. (1987). *Inverse Sturm–Liouville Problems.* VSP, Zeist.

Levitan, B. M. and Sargsjan, I. S. (1975). Introduction To Spectral Theory: Self-adjoint Ordinary Differential Operators. Transl. Math. Monogr., **39**, Amer. Math. Soc., Providence, RI.

Maltsev, V. P. and Semyanov, K. A. (2004). *Characterisation Of Bio-particles From Light Scattering.* Inverse and Ill-posed Problems Series, VSP, Utrecht.

Marchenko, V. A. (1952). Some questions of the theory of one-dimensional linear differential operators of the second order. *Tr. Mosk. Mat. Obs.*, **1**, 327–420 (in Russian).

Marchenko, V. A. (1986). *Sturm–Liouville Operators And Applications.* Operator Theory: Advances and Applications, **22**, Birkhäuser Verlag, Basel.

Naimark, M. A. (1967). *Linear Differential Operators.* Frederick Ungar Publishing Co., New York.

Nizhnik, L. P. (1973). *The Inverse Nonstationary Scattering Problem.* Naukova Dumka, Kiev (in Russian).

Ramm, A. G. (1992). *Multidimensional Inverse Scattering Problems.* Pitman Monogr. Surv. Pure Appl. Math., **51**, Longman Scientific & Technical, Harlow; John Wiley & Sons, Inc., New York.

Rayleigh, J. W. S. (1945). *The Theory of Sound.* Dover, New York.

Yurko, V. A. (2007). *Introduction to the Theory of Inverse Spectral Problems.* Fizmatlit, Moscow (in Russian).

Zakhariev, B. N. and Chabanov, V. M. (2007). *Submissive Quantum Mechanics: New Status Of The Theory In Inverse Problem Approach.* Nova Science Publishers, Inc., New York.

Supplementary references to Chapter 7

Kabanikhin, S. I. and Shishlenin, M. A. (2001) Linearized multidimensional inverse problem for the wave equation. *Obratnye zadachi i informatsionnye technologii*, **1**(2), 83–114 (in Russian).

Lavrentiev, M. M. and Romanov, V. G. (1966). Three linearized inverse problems for hyperbolic equations. *Dokl. Akad. Nauk SSSR*, **171**(6), 1279–1281 (in Russian).

Romanov, V. G. (1969b). A certain integral geometry problem and a linearized inverse problem for a hyperbolic equation. *Sibirsk. Mat. Zh.*, **10**(6), 1364–1374 (in Russian).

Kabanikhin S. I. Bektemesov M. A., and Nechaev D. V. (2005). Numerical solution of 2D thermoacoustic problem. *J. Inv. Ill-Posed Problems*, **13** (3), 265–276.

Supplementary references to Chapter 8

Beck, J., Blackwell, B., and St. Clair, C., Jr. (1985). *Inverse Heat Conduction: Ill-posed Problems.* John Wiley & Sons, Inc., New York.

Krein, S. G. (1957). On correctness classes for certain boundary problems. *Dokl. Akad. Nauk SSSR*, **114**(6), 1162–1165 (in Russian).

Ballani, L., Greiner-Mai, H., and Stromeyer, D. (2002). Determining the magnetic field in the core-mantle boundary zone by non-harmonic downward continuation. *Geophys. J. Intern.*, **149**, 347–389.

Danilaev, P. G. (2001). *Coefficient Inverse Problems For Parabolic Equations And Their Applications.* VSP, Utrecht.

Griedt, W. H. (1995). The determination of transient temperatures and heat transfer of a gas-metal interface applied to a 40-mm gun barrel. *Jet Propulsion*, **25**, 158–162.

Methods for inverse heat conduction problems. Univerisität-Gesamthochschule Siegen, 1995.

Kozhanov, A. I. (1999). *Composite Type Equations And Inverse Problems.* VSP, Utrecht.

Pucci, C. (1959). On the improperly posed Cauchy problems for parabolic equations. In: *Proc. Sympos. on the Numerical Treatment of Partial Differential Equations with Real Characteristics. January 1959, Rome.* Libreria Eredi Virgilio Veschi, Rome, 140–144.

Supplementary references to Chapter 9

Kabanikhin, S. I. and Karchevsky, A. L. (1995). Optimization method for solving the Cauchy problem for an elliptic equation. *J. Inverse Ill-Posed Probl.*, **3**(1), 21–46.

Lavrentiev, M. M. (1955). On the Cauchy problem for Laplace's equation. *Dokl. Akad. Nauk SSSR*, **102**(2), 205–206 (in Russian).

Lavrentiev, M. M. (1956). On the Cauchy problem for Laplace's equation. *Izv. Akad. Nauk SSSR. Ser. Mat.*, **20**(6), 819–842 (in Russian).

Ladyzhenskaya, O. A. and Uraltseva, N. N. (1973) *Linear and Quasilinear Equations of Elliptic Type.* Nauka, Moscow (in Russian).

Mergelyan, S. N. (1956). Harmonic approximation and approximate solution of Cauchy's problem for Laplace's equation. *Dokl. Akad. Nauk SSSR*, **107**(5), 644–647 (in Russian).

Chudov, L. A. (1962). Difference methods for solving the Cauchy problem for the Laplace equation. *Dokl. Akad. Nauk SSSR*, **143**(4), 798–801 (in Russian).

Aronszajn, N. (1957). A unique continuation theorem for solutions of elliptic partial differential equations or inequalities of second order. *J. Math. Pures Appl. (9)* **36**, 235–249.

Calderòn, A. P. (1958). Uniqueness in the Cauchy problem for partial differential equations. *Amer. J. Math.*, **80**, 16–36.

Carleman, T. (1926). *Les fonctions quasi-analytiques.* Gauthier-Villars, Paris.

Douglas, J. (1960). A numerical method for analytic continuation. In: *Boundary Problems in Differential Equations.* Univ. of Wisconsin Press, Madison, 179–189.

Eldén, L. (2002). Numerical solution of ill-posed Cauchy problems for PDE's. In: *Minicourse on Applied Inverse Problems. October 2002.* Pubblicazione dell'Istituto per le Applicazioni del Calcolo (Sezione di Firenze), **7**(11), 7–81.

Hào, D. N. and Lesnic, D. (2000). The Cauchy problem for Laplace's equation via the conjugate gradient method. *IMA J. Appl. Math.*, **65**(2), 199–217.

Johansson, T. (2003). An iterative procedure for solving a Cauchy problem for second order elliptic equations. *Linköping Studies in Science and Technology*, 832, 47–60.

Kabanikhin, S. I., Bektemesov, M. A., and Nechaev, D. V. (2003). Optimizational method for continuation of solution to two-dimensional elliptic equation. In: *Proceedings of the International Symposium on Inverse Problems in Engineering Mechanics (ISIP 2003)*, 447–456.

Kabanikhin, S. I. and Karchevsky, A. L. (1995). Optimization method for solving the Cauchy problem for an elliptic equation. *J. Inverse Ill-Posed Probl.*, **3**(1), 21–46.

Pucci, C. (1955). Sui problemi di Cauchy non "ben posti". *Atti Accad. Naz. Lincei. Rend. Cl. Sci. Fis. Mat. Nat. (8)*, **18**, 473–477 (in Italian).

Supplementary references to Chapter 10

Anikonov, Yu. E. (1995). *Multidimensional Inverse and Ill-Posed Problems for Differential Equations*. VSP, The Netherlands.

Ayapbergenova, A. T., Kabanikhin, S. I., and Lorenzi, A. (2003). Justification of optimization methods for inverse integro-differential hyperbolic problems. *J. Inv. Ill-Posed Problems* **12** (1).

Blagoveshchenskii, A. S. (2001). *Inverse Problems of Wave Processes* VSP, The Netherlands.

Courant, R. and Hilbert, D. (1953, 1962). *Methods of Mathematical Physics*. Vol. 1, 2. Interscience, New-York.

Hanke, M., Neubauer, A., and Scherzer, O. (1995). A convergence analysis of Landweber iteration for nonlinear ill-posed problems. *Numer. Math.* **72**, 21–37.

Hamaker, C., Smith, K. T., Solmin, D. C., and Wagner, S. C. (1980). The divergent beam x-ray transform. *Rocky Mountain Journal of Mathematics. Numer. Math.*, 253–283.

He, S. and Kabanikhin, S. I. (1995). An optimization approach to a three-dimensional acoustic inverse problem in the time domain. *J. Math. Phys.* **36** (8), 4028–4043.

Isakov, V. (1991). An inverse hyperbolic problem for the wave equation. *Communications PDE* **16**, 1183–1195.

Kabanikhin, S. I. and Ayapbergenova, A. T. (2002). Estimation of the rate of convergence of the Landweber iteration method in an inverse problem of acoustics. *Proceedings of the Steklov Institute of Mathematics* **2**, 75–97.

Kabanikhin, S. I., Iskakov, K. T., and Yamamoto, M. (2001). H_1-conditional stability with explicit Lipshitz constant for a one-dimensional inverse acoustic problem. *J. Inv. Ill-Posed Problems* **9** (3), 249–268.

Kabanikhin, S. I., Kowar, R., and Scherzer, O. (1998). On Landweber iteration for the solution of a parameter identification problem in a hyperbolic partial differential equation of second order. *J. Inv. Ill-Posed Problems* **6** (5), 403–430.

Kabanikhin, S. I., Kowar, R., Scherzer, O., and Vasin, V. V. (2001). Numerical comparison of iterative regularization methods for a parameter estimation in a hyperbolic PDE. *J. Inv. Ill-Posed Problems* **9** (6), 615–626.

Kabanikhin, S. I., Scherzer, O., and Shishlenin, M. A. (2003). Iteration methods for a solving a two-dimensional problem for hyperbolic equation by the iteration method. *J. Inv. Ill-Posed Problems* **11** (1), 1–23.

Ovsynnikov, L. V. (1965). Singular operator in the scale of Banach spaces. *Dokl. Akad. Nauk SSSR* **163** (4) (in Russian).

Rakesh. (1993). An inverse problem for the wave equation in the half plane. *Inverse Problems* **9** (3), 433–441.

Romanov, V. G. (1983). Inverse Lamb's problem in a linear approximation. In: *Numerical Methods in Seismic Investigations*. Nauka, Novosibirsk, 170–192 (in Russian).

Romanov, V. G. (1989a). On the well-posedness of the problem of determining the acoustic speed. *Sib. Mat. Zh.* **30** (4), 125–134 (in Russian).

Romanov, V. G. (1989b). On the local solvability of inverse problems for hyperbolic equations in the class of functions analytic in some variables. *Dokl. Akad. Nauk SSSR* **304** (4), 808–811 (in Russian).

Sylvester, J. and Uhlmann, G. (1987). A global uniqueness theorem for an inverse boundary value problem. *Annals of Mathematics* **125**, 153–169.

Supplementary references for Chapter 11

Danilaev, P. G. (2001). Coefficient inverse problems for parabolic type equations and their application. VSP, The Netherlands.

Hào, D. N. (1995). Methods for inverse heat conduction problems. Universität-Gesamthochschule Siegen.

Hasanov, A. (1998). Inverse coefficient problems for elliptic variational inequalities with a nonlinear monotone operator. *Inverse Problems* **14**, 1151–1169.

Hasanov, A. (2007). Simultaneous determination of source terms in a linear parabolic problem from the final overdetermination: weak solution approach. *J. Math. Anal. Appl.* **330**, 766–779.

Hasanov, A., DuChateau, P., and Pektas, B. (2006). An adjoint problem approach and coarse-fine mesh method for identification of the diffusion coefficient in a linear parabolic equation. *Inverse Problems* **14**(4), 1–29.

Ivanov, V. K. (1956). An inverse potential problem for a body close to a given one. *Izv. Akad. Nauk SSSR* **20**(6), 793–818 (in Russian).

Ivanov, V. K. (1962). Integral equations of the first kind and the approximate solution of an inverse potential problem. *Dokl. Akad. Nauk SSSR* **142**(5), 998–1000 (in Russian).

Iskenderov, A. D. (1974). On an inverse problem for quasilinear parabolic equations. *Differential Equations* **10**(5), 890–898 (in Russian).

Iskenderov, A. D. and Tagnev, R. K. (1983). Optimization problems with control in the coefficients of parabolic equations. *Differential Equations* **19**(8), 990–999 (in Russian).

Kozhanov, A. I. (1999). Composite type equations and inverse problems. VSP, The Netherlands.

Landis, E. M. (1959). Some aspects of the qualititative theory of elliptic and parabolic equations. *Uspekhi Mat. Nauk* **2**(1), 21–85 (in Russian).

Muzylev, N. V. (1980). Uniqueness theorems for some inverse heat conduction problems. *Zh. Vychisl. Mat. Mat. Fiz.* **20**(2), 388–400 (in Russian).

Supplementary references to Appendix A

Vladimirov, V. S. (1983). Equations of Mathematical Physics. Nauka, Moscow.

Gaevsky, H., Gregor, K., and Zacharias, K. (1978). Nonlinear Operator Equations and Operator Differential Equations. Mir, Moscow.

Kantorovich, L. V. and Akilov, G. P. (1977). Functional Analysis. Nauka, Moscow.

Sadovnichii, V. A. (1986). Theory of Operators. MSU, Moscow.

Epilogue

This monograph is devoted to one of the most recently developed branches of modern applied mathematics—inverse problems of natural sciences. The main difficulty in the study of inverse problems is caused by the fact that a rather extensive mathematical training is required for an in depth understanding of these problems. To make the study process easier and more effective, the author included a supplementary chapter that covers relevant topics from various areas of mathematics ranging from linear algebra to mathematical physics and functional analysis (Appendix A).

The first five chapters of the monograph are quite accessible to third-year university students. The author is absolutely convinced that the teaching of inverse and ill-posed problems should start not later than in the third year of university. The material in the next chapters gradually becomes more advanced. The problems presented become more complex and many steps have to be omitted from the proofs, the reader being referred to more specialized literature. At the same time, the presentation strategy of the second half of the monograph makes it possible to introduce the reader to a variety of modern problems being solved by numerous research groups in most developed countries.

It should be noted that a lot of important subjects are not covered in this monograph. However, books and papers on these subjects are mentioned in the supplementary references at the end of each chapter.

This monograph attempts to describe almost all the main areas of the theory of inverse and ill-posed problems under a single cover. It is therefore almost certain that the reader will notice typographical errors, a lack of clarity in some parts of the presentation, and possibly even some inaccuracies (and we hope there are none!). The author will be very grateful for any feedback, comments, and suggestions on possible corrections and/or updates. Please send them to kabanikhin@mail.ru.

Bibliography

Agranovich, Z. and Marchenko, V. (1963). *The Inverse Problem Of Scattering Theory.* Gordon and Breach, New York-London.

Aki, K. and Richards, P. G. (1980). *Quantitative Seismology. Theory and Methods.* Freeman, San Francisco.

Alekseev, A. S. (1962). Some converse problems in wave propagation theory. *Bull. (Izv.) Acad. Sci. USSR Geophys. Ser.*, 944–954.

Alekseev, A. S. (1967). Inverse dynamic problems of seismology. *Methods and Algorithms for Interpreting Geophysical Data*, Nauka, Moscow (in Russian).

Alekseev, A. S., Avdeev, A. V., Fatyanov, A. G., and Cheverda, V. A. (1991). A closed cycle of the mathematical modeling of wave processes in vertically inhomogeneous media (direct and inverse problems). *Mat. Model.*, **3**(10), 80–94 (in Russian).

Alekseev, A. S. and Belonosov, V. S. (1998). Spectral methods in one-dimensional problems of wave propagation theory. *Mat. Model. Geofiz., Trudy Inst. Vych. Mat. Mat. Geofiz.*, **11**, 7–39 (in Russian).

Alekseev, A. S. and Dobrinskii, V. I. (1975). Certain questions of the practical application of inverse dynamic problems of seismology. *Mat. Probl. Geofiz.*, **6**(2), 7–53 (in Russian).

Alekseev, A. S., Dobrinskii, V. I., Neprochnov, Ju. P., and Semenov, G. A. (1976). On the question of the practical application of the theory of inverse dynamic problems of seismology. *Dokl. Akad. Nauk SSSR*, **228**(5), 1053–1056 (in Russian).

Alekseev, A. S. and Kabanikhin, S. I. (2003). Inverse problems and new technologies in geophysics. In: *Proceedings of the International Conference "Mathematical Methods in Geophysics"*, 2003, **1**, 1–39 (in Russian).

Alifanov, O. M., Artyukhin, E. A., and Rumyantsev, S. V. (1988). *Extremal Methods for Solving Ill-posed Problems and their Applications to Inverse Problems of Heat Exchange.* Nauka, Moscow, 1988 (in Russian).

Ambarzumijan, V. (1929). Über eine Frage der Eigenwerttheorie. *Zeitschr. für Physik.*, 53, 690–695.

Androshchuk, A. A. (1986). Uniqueness of the reconstruction of the two-dimensional Schrödinger equation. *Dokl. Akad. Nauk SSSR*, **291**(5), 1033–1036 (in Russian).

Anikonov, Yu. E. (1978). *Some methods for the study of multidimensional inverse problems for differential equations*. Nauka, Novosibirsk (in Russian).

Antonenko, O. F. (1967). The inversion of a finite difference equation for solving the one-dimensional problem of seismology. *Methods and Algorithms for Interpreting Geophysical Data*. Nauka, Moscow, 92–98 (in Russian).

Antonenko, O. F. and Reznitskaya, K. G. (1978). The Newton–Kantorovich method in an inverse dynamic problem of seismology. *Inverse Problems for Differential Equations of Mathematical Physics*, Akad. Nauk SSSR Sibirsk. Otdel., Vychisl. Tsentr, Novosibirsk, 18–25, 155 (in Russian).

Aronszajn, N. (1957). A unique continuation theorem for solutions of elliptic partial differential equations or inequalities of second order. *J. Math. Pures Appl. (9)* **36**, 235–249.

Ayapbergenova, A. T., Kabanikhin, S. I., and Lorenzi, A. (2004). Justification of optimization methods for inverse integro-differential hyperbolic problems. *J. Inverse Ill-Posed Probl.*, **12**(5), 457–479.

Baev, A. V. (1985). Solution of an inverse problem for the wave equation using a regularizing algorithm. *Zh. Vychisl. Mat. Mat. Fiz.*, **25**(1), 140–146 (in Russian).

Bakushinskii, A. B. (1967). A general technique for constructing regularizing algorithms for a linear ill-posed equation in a Hilbert space. *Zh. Vychisl. Mat. Mat. Fiz.*, **7**(3), 672–677 (in Russian).

Bakushinskii A. B. (1974). Difference schemes for the solutions of incorrectly posed abstract Cauchy problems. *Differ. Equ.*, 7, 1419–1426.

Bakushinskii, A. B. (1976). A regularizing algorithm based on the Newton–Kantorovich method for solving variational inequalities. *Zh. Vychisl. Mat. Mat. Fiz.*, **16**(6), 1397–1404 (in Russian).

Bakushinskii, A. B. (1977). Methods for solving monotonic variational inequalities based on the principle of iterative regularization. *Zh. Vychisl. Mat. Mat. Fiz.*, **17**(6), 1350–1362 (in Russian).

Bakushinskii, A. B. (1979). On the principle of iterative regularization. *Zh. Vychisl. Mat. Mat. Fiz.*, **19**(4), 1040–1043 (in Russian).

Bakushinskii, A. B. (1983). An asymptotic relation for the iteratively regularized Newton–Kantorovich method. *Zh. Vychisl. Mat. Mat. Fiz.*, **23**(1), 216–218 (in Russian).

Bakushinskii, A. B. and Goncharskii, A. V. (1989a). *Iterative Methods for Solving Ill-posed Problems.* Nauka, Moscow, (in Russian).

Bakushinskii, A. B. and Goncharskii, A. V. (1989b). *Ill-posed Problems. Numerical Methods and Applications.* Moskov. Gos. Univ., Moscow, (in Russian).

Ballani, L., Greiner-Mai, H., and Stromeyer, D. (2002). Determining the magnetic field in the core-mantle boundary zone by non-harmonic downward continuation. *Geophys. J. Intern.*, **149**, 347–389.

Bamberger, A., Chavent, G., and Lailly, P. (1979). About the stability of the inverse problem in 1-D wave equations—applications to the interpretation of seismic profiles. *Appl. Math. Optim.*, **5**(1), 1–47.

Bamberger, A., Chavent, G., and Lailly, P. (1980). An optimal control solution of the inverse problem of reflection seismics. *Computing Methods in Applied Sciences and Engineering (Proc. Fourth Internat. Sympos., Versailles, 1979)*, North-Holland, Amsterdam-New York, 529–545.

Baranov, V. and Kunetz, G. (1962). Synthetic seismograms with multiple reflections. In: *Problems of Seismic Exploration*, 179–188 (in Russian).

Bastay, G., Kozlov, V. A., and Turesson, B. O. (2001). Iterative methods for an inverse heat conduction problem. *J. Inverse Ill-Posed Probl.*, **9**(4), 375–388.

Baumeister, J. (1987). *Stable Solution of Inverse Problems. Advanced Lectures in Mathematics.* Friedr. Vieweg & Sohn, Braunschweig.

Beck, J., Blackwell, B., and St. Clair, C., Jr. (1985). *Inverse Heat Conduction: Ill-posed Problems.* John Wiley & Sons, Inc., New York.

Belenkii, V. Z., Volkonskii, V. A., Ivankov, S. A., Pomanskii, A. B., and Shapiro, A. D. (1974). *Iterative Methods in Game Theory and Programming.* Nauka, Moscow (in Russian).

Belishev, M. I. (1988). An approach to multidimensional inverse problems for the wave equation. *Soviet Math. Dokl.*, **36**(3), 481–484.

Berezanskii, Yu. M. (1953). On the uniqueness of the determination of Schrödinger's equation from its spectral function. *Doklady Akad. Nauk SSSR*, **93**(4), 591–594 (in Russian).

Berezanskii, Yu. M. (1958). On the inverse problem of spectral analysis for Schrödinger's equation. *Tr. Mosk. Mat. Obs.*, **7**, 3–51 (in Russian).

Berntsson, F. and Eldén, L. (2001). Numerical solution of a Cauchy problem for the Laplace equation. *Inverse Problems*, **17**(4), 839–853.

Bimuratov, S. S. and Kabanikhin, S. I. (1992). Solution of a one-dimensional inverse problem of electrodynamics by the Newton–Kantorovich method. *Comput. Math. Math. Phys.*, **32**(12), 1729–1743.

Blagoveshchenskii, A. S. (1967). The inverse problem in the theory of seismic wave propagation. *Topics in Mathematical Physics,* **1**, New York, Consultants Bureau, 55–67.

Blagoveshchenskii, A. S. (1971). A local method for solving the nonstationary inverse problem for a nonuniform string. *Tr. Mat. Inst. Steklova*, **115**, 28–38 (in Russian).

Blagoveshchenskii, A. S. (1978). Inverse problems of the theory of elastic wave propagation. *Izv. Akad. Nauk SSSR*, **12**, 50–59 (in Russian).

Blagoveshchenskii, A. S. and Kabanikhin, S. I. (1983). Inverse problem of the theory of wave propagation in a semi-infinite nonregular waveguide. *Differ. Uravn.*, **19**(4), 603–607 (in Russian).

Bloh, A. Sh. (1953). On the determination of a differential equation by its special matrix function. *Doklady Akad. Nauk SSSR*, **92**, 209–212 (in Russian).

Borg, G. (1946). Eine Umkehrung der Sturm-Liouvilleschen Eigenwertaufgabe. Bestimmung der Differentialgleichung durch die Eigenwerte. *Acta Math.*, **78**, 1–96.

Bukhgeim, A. L. (1983). *Volterra Operator Equations.* Nauka, Novosibirsk (in Russian).

Burridge, R. (1980). The Gel'fand–Levitan, the Marchenko, and the Gopinath–Sondhi integral equations of inverse scattering theory, regarded in the context of inverse impulse-response problems. *Wave Motion*, **2**(4), 305–323.

Butkovskii, A. G. (1965). *Theory of Optimal Control of Systems with Distributed Parameters.* Nauka, Moscow (in Russian).

Calderòn, A. P. (1958). Uniqueness in the Cauchy problem for partial differential equations. *Amer. J. Math.*, **80**, 16–36.

Carasso, A. S., Sanderson, J. G., and Hyman, J. M. (1978). Digital removal of random media image degradations by solving the diffusion equation backwards in time. *SIAM J. Numer. Anal.*, **15**(2), 344–367.

Carleman, T. (1926). *Les fonctions quasi-analytiques.* Gauthier-Villars, Paris.

Cauchy, A. L. (1847). Méthode générale pour la résolution des systèmes d'équations simultanées. *Comptes Rendus de l'Académie des Sciences de Paris*, **25**, 536–538.

Chadan, K. and Sabatier, P. C. (1989). *Inverse Problems in Quantum Scattering Theory.* Springer-Verlag, New York.

Chen, Y. M. and Liu, J. Q. (1983). A numerical algorithm for solving inverse problems of two-dimensional wave equations. *J. Comput. Phys.*, **50**(2), 193–208.

Chen, Y. M. and Tsien D. S. (1977). A numerical algorithm for remote sensing of density profile of a simple ocean model by acoustic pulses. *J. Comput. Phys.*, **25**, 366–385.

Chudov, L. A. (1949). The inverse Sturm–Liouville problem. *Mat. Sbornik*, **25**(67), 451–456 (in Russian).

Chudov, L. A. (1956). A new variant of an inverse Sturm–Liouville problem on a finite interval. *Dokl. Akad. Nauk SSSR*, **109**(1), 40–43 (in Russian).

Chudov, L. A. (1962). Difference methods for solving the Cauchy problem for the Laplace equation. *Dokl. Akad. Nauk SSSR*, **143**(4), 798–801 (in Russian).

Courant, R. and Hilbert, D. (1943). *Methoden der Mathematischen Physik. Vols. I, II.* Interscience Publ., New York.

Courant, R. and Hilbert, D. (1962). *Methods of Mathematical Physics. Vol. II: Partial Differential Equations.* (Vol. II by R. Courant.) Interscience Publ., New York-London.

Denisov, A. M. (1975). The approximate solution of a Volterra equation of the first kind. *Zh. Vychisl. Mat. Mat. Fiz.*, **15**(4), 1053–1056 (in Russian).

Denisov, A. M. (1982). An inverse problem for the heat equation with a piecewise constant coefficient. *Vychisl. Metody i Programmirovanie*, **37**, 93–97 (in Russian).

Denisov, A. M. (1994). *Introduction to the Theory of Inverse and Ill-posed Problems.* Moskov. Gos. Univ., Moscow.

Denisov, A. M. (1995). Uniqueness of determination of a nonlinear coefficient of a system of partial differential equations in the small and in the large. *Russian Acad. Sci., Dokl., Math.*, **50**(2), 269–271.

Dmitriev, V. I. (1970). Inverse and direct problem of magnetotelluric sounding of a stratified medium. *Izv. Akad. Nauk SSSR, Ser. Fizika Zemli*, **1**, 64–69 (in Russian).

Douglas, J. (1960). A numerical method for analytic continuation. In: *Boundary Problems in Differential Equations.* Univ. of Wisconsin Press, Madison, 179–189.

Eldén, L. (2002). Numerical solution of ill-posed Cauchy problems for PDE's. In: *Minicourse on Applied Inverse Problems. October 2002.* Pubblicazione dell'Istituto per le Applicazioni del Calcolo (Sezione di Firenze), **7**(11), 7–81.

Engl, W., Hanke, M., and Neubauer, A. (1996). *Regularization of Inverse Problems.* Kluwer Academic Publishers, Dordrecht.

Faddeev, L. D. (1956). Uniqueness of solution of the inverse scattering problem. *Vestnik Leningrad. Univ.*, **11**(7), 126–130 (in Russian).

Faddeev, L. D. (1959). The inverse problem in the quantum theory of scattering. *Uspekhi Mat. Nauk*, **14**(4), 57–119 (in Russian).

Frank, L. S. and Chudov, L. A. (1965). Difference methods for solving the ill-posed Cauchy problem. In: Numer. Methods in Gas Dynamics. Moskov. Gos. Univ., Moscow, 3–27 (in Russian).

Fridman, V. M. (1956). Method of successive approximations for a Fredholm integral equation of the first kind. *Uspekhi Mat. Nauk*, **11**(1), 233–234 (in Russian).

Gajewski, H., Gröger, K., and Zacharias, K. (1974). *Nichtlineare Operatorgleichungen und Operatordifferentialgleichungen. Mathematische Lehrbücher und Monographien, II.* Abteilung, Mathematische Monographien, **38**, Akademie-Verlag, Berlin (in German).

Gauss K. F. (1809). *Theoria Motus Corporum Coelestium in Sectionibus Conicis Solem Ambientium.* Perthes und Besser, Hamburg.

Gel'fand, I. M., Graev, M. I., and Vilenkin, N. Ya. (1966). *Generalized Functions. Vol. 5: Integral Geometry and Representation Theory.* Academic Press, New York-London.

Gel'fand, I. M. and Levitan, B. M. (1955). On the determination of a differential equation from its spectral function. *Amer. Math. Soc. Transl. (2)*, **1**, 253–304.

Gilyazov, S. F. (1986). Regularizing algorithms based on the conjugate gradient method. *Zh. Vychisl. Mat. Mat. Fiz.* **26**(1), 15–22 (in Russian).

Glasko, V. B. (1970). The uniqueness of the solution of certain inverse problems of seismology. *Zh. Vychisl. Mat. Mat. Fiz.*, **10**(6), 1465–1480 (in Russian).

Glasko, V. B. (1972). On the question of the uniqueness of the solution of the problem of the reconstruction of the structure of the Earth's crust from the dispersion spectrum of the Rayleigh waves. *Dokl. Akad. Nauk SSSR*, **206**(6), 1345–1348 (in Russian).

Glasko, V. B. (1988). *Inverse Problems of Mathematical Physics.* American Institute of Physics Translation Series. American Institute of Physics, New York.

Godunov, S. K. (1980). *Solution of Systems of Linear Equations.* Nauka, Novosibirsk (in Russian).

Godunov, S. K. (1983). *The Matrix Exponential, the Green Matrix and Lopatinskii Conditions.* Novosibirsk. Gos. Univ., Novosibirsk (in Russian).

Godunov, S. K. (1998). *Modern Aspects of Linear Algebra.* Transl. Math. Monogr., 175, Amer. Math. Soc., Providence, RI.

Godunov, S. K., Antonov, A. G., Kiriljuk, O. P., and Kostin, V. I. (1993). *Guaranteed Accuracy in Numerical Linear Algebra.* Math. Appl., 252, Kluwer Acad. Publ., Dordrecht.

Godunov, S. K. and Ryaben'kii, V. S. (1987). *Difference Schemes. An Introduction to the Underlying Theory.* Stud. Math. Appl., 19, North-Holland, Amsterdam.

Gopinath, B. and Sondhi, M. M. (1971). Inversion of the telegraph equation and the synthesis of nonuniform lines. *Proc. IEEE*, **59**(3), 383–392.

Goupillaud, P. L. (1961). An approach to inverse filtering of near-surface layer effects from seismic records. *Geophysics*, **26**, 754–760.

Grasselli, M., Kabanikhin, S. I., and Lorenzi, A. (1989a). An inverse hyperbolic integrodifferential problem arising in geophysics. I. In: *Publ. Universita degli Studi di Milano*, **5**.

Grasselli, M., Kabanikhin, S. I., and Lorenzi, A. (1989b). An inverse hyperbolic integrodifferential problem arising in geophysics. II. *Publ. Istituto di Analisi Numerica*, **673**.

Grasselli, M., Kabanikhin, S. I., and Lorentsi, A. (1992). An inverse problem for an integro-differential equation. *Siberian Math. J.*, **33**(3), 415–426.

Griedt, W. H. (1995). The determination of transient temperatures and heat transfer of a gas-metal interface applied to a 40-mm gun barrel. *Jet Propulsion*, **25**, 158–162.

Hadamard, J. (1902). Sur les problèmes aux dérivées partielles et leur signification physique. *Princeton University Bulletin*, 49–52.

Hadamard, J. (1932). *Le problème de Cauchy et les équations aux dérivées partielles linéaires hyperboliques; leçons professées à l'Université Yale.* Hermann, Paris.

Hanke, M., Neubauer, A., and Scherzer, O. (1995). A convergence analysis of the Landweber iteration for nonlinear ill-posed problems. *Numer. Math.*, **72**(1), 21–37.

Hào, D. N. (1996). *Methods for Inverse Heat Conduction Problems.* Habilitationsschrift, University of Siegen, Siegen.

Hào, D. N. and Lesnic, D. (2000). The Cauchy problem for Laplace's equation via the conjugate gradient method. *IMA J. Appl. Math.*, **65**(2), 199–217.

Hasanov, A., Duchateau, P. and Pektaş, B. (2006). An adjoint problem approach and coarse-fine mesh method for identification of the diffusion coefficient in a linear parabolic equation. *J. Inverse Ill-Posed Probl.*, **14**(5), 435–463.

Herglotz, G. (1905). Über die Elastizität der Erde bei Berücksichtigung ihrer variablen Dichte. *Zeitschr. für Math. und Phys.*, **52**(3), 275–299.

Herglotz, G. (1911). Über Potenzreihen mit positivem, reellem Teil im Einheitskreis. *S.-B. Sächs. Akad. Wiss.*, 63, 501–511.

Iskenderov, A. D. (1984). Variational formulations of multidimensional inverse problems of mathematical physics. *Dokl. Akad. Nauk SSSR*, **274**(3), 531–533 (in Russian).

Iskenderov, A. D. and Tagiev, R. K. (1983). Problems of optimization with controls in the coefficients of a parabolic equation. *Differents. Uravn.*, **19**(8), 1324–1334 (in Russian).

Ivanov, V. K. (1956). The inverse problem of potential for a body closely approximated by a given body. *Izv. Akad. Nauk SSSR. Ser. Mat.*, **20**(6), 793–818 (in Russian).

Ivanov, V. K. (1962a). Integral equations of the first kind and approximate solution of the inverse problem of potential theory. *Dokl. Akad. Nauk SSSR*, **142**(5), 998–1000 (in Russian).

Ivanov, V. K. (1962b). On linear problems which are not well-posed. *Dokl. Akad. Nauk SSSR*, **145**(2), 270–272 (in Russian).

Ivanov, V. K. (1963). On ill-posed problems. *Mat. Sb.*, **61**(2), 211–223 (in Russian).

Ivanov, V. K. (1965). On a certain type of ill-posed linear equation in topological vector spaces. *Sibirsk. Mat. Zh.* **6**(4), 832–839 (in Russian).

Ivanov, V. K. (1966). Approximate solution of operator equations of the first kind. *Zh. Vychisl. Mat. Mat. Fiz.*, **6**(6), 1089–1094 (in Russian).

Ivanov, V. K., Vasin, V. V., and Tanana, V. P. (2002). *Theory of Linear Ill-posed Problems and its Applications.* Inverse and Ill-posed Problems Series, VSP, Utrecht.

Jobert, G. (1983). Linearized direct and inverse dynamic problem of seismic wave propagation. In: *Numerical Methods in Seismic Studies*, Nauka, Novosibirsk, 193–207 (in Russian).

Johansson, T. (2004). An iterative procedure for solving a Cauchy problem for second order elliptic equations. *Math. Nachr.*, 272, 46–54.

John, F. (1935). Abhängigkeiten zwischen den Flächenintegralen einer stetigen Funktion. *Math. Ann.*, **111**(1), 541–559.

John, F. (1955a). A note on "improper" problems in partial differential equations. *Comm. Pure Appl. Math.*, **8**(4), 591–594.

John, F. (1955b). *Plane Waves and Spherical Means Applied to Partial Differential Equations.* Interscience Publishers, New York-London.

John, F. (1959). Numerical solution of problems which are not well posed in the sense of Hadamard. *Proc. Sympos. on the Numerical Treatment of Partial Differential Equations with Real Characteristics" (Rome, January 1959)*, Libreria Eredi Virgilio Veschi, Rome, 103–116.

John, F. (1960). Continuous dependence on data for solutions of partial differential equations with a prescribed bound. *Comm. Pure Appl. Math.*, **13**, 551–585.

Kabanikhin, S. I. (1976). On a formulation of a two-dimensional inverse problem for the oscillation equation. In: *Ill-posed Mathematical Problems and Problems of Geophysics (Mathematical Problems of Geophysics)*, Akad. Nauk SSSR Sibirsk. Otdel., Vychisl. Tsentr, Novosibirsk, 64–73 (in Russian).

Kabanikhin, S. I. (1977). Finite-difference regularization of the inverse problem for an oscillation equation. In: *Problems of Well-posedness in Mathemtical Physics.*, Novosibirsk, 57–69 (in Russian).

Kabanikhin, S. I. (1979). Application of energy inequalities to an inverse problem for a hyperbolic equation. *Differents. Uravn.*, **15**(1), 61–67 (in Russian).

Kabanikhin, S. I. (1980). A finite-difference method for determining the coefficients of a hyperbolic system of the first order. In: *Uniqueness, Stability, and Methods for Solving Inverse and Ill-posed Problems*, Akad. Nauk SSSR Sibirsk. Otdel., Vychisl. Tsentr, Novosibirsk, 36–43 (in Russian).

Kabanikhin, S. I. (1981a). A problem of determining the coefficients of the acoustic equation. In: *Nonclassical Problems of Mathematical Physics*, Akad. Nauk SSSR Sibirsk. Otdel., Vychisl. Tsentr, Novosibirsk, 93–100 (in Russian).

Kabanikhin, S. I. (1981b). An approximate method for solving the inverse problem for the acoustic equation. In: *Approximate Solution Methods and Questions of the Well-posedness of Inverse Problems*, Akad. Nauk SSSR Sibirsk. Otdel., Vychisl. Tsentr, Novosibirsk, 55–62 (in Russian).

Kabanikhin, S. I. (1982). Inverse problems for the acoustic equation. In: *Boundary Value Problems for Nonlinear Equations*, Akad. Nauk SSSR Sibirsk. Otdel., Vychisl. Tsentr, Novosibirsk, 149–150 (in Russian).

Kabanikhin, S. I. (1984a). The solvability of inverse problems for differential equations. *Dokl. Akad. Nauk SSSR*, **277**(4), 788–791 (in Russian).

Kabanikhin, S. I. (1984b). A projection method for solving multidimensional inverse problems for hyperbolic equations. In: *Ill-posed Problems of Mathematical Physics and Analysis*, Nauka, Novosibirsk, 55–59 (in Russian).

Kabanikhin, S. I. (1988). *Projection-difference Methods for Determining the Coefficients of Hyperbolic Equations.* Nauka, Novosibirsk (in Russian).

Kabanikhin, S. I. (1989). Regularization of Volterra operator equations of the first kind with a boundedly Lipschitz-continuous kernel. *Soviet Math. Dokl.* **39**(3), 549–552.

Kabanikhin, S. I. (1990). Linear regularization of multidimensional inverse problems for hyperbolic equations. *Soviet Math. Dokl.*, **40**(3), 579–583.

Kabanikhin, S. I. (1990). Linear regularization of multidimensional inverse problems for hyperbolic equations. *Soviet Math. Dokl.*, **40**(3), 579–583.

Kabanikhin, S. I. (1992). Methods for solving inverse dynamic problems for hyperbolic equations. In: *Conditionally Well-posed Problems in Mathematical Physics and Analysis.* Ross. Akad. Nauk Sib. Otd., Inst. Mat., Novosibirsk, 109–123 (in Russian).

Kabanikhin, S. I. (2002). Iterative methods of solving inverse problems for hyperbolic equations. In: *Ill-Posed and Inverse Problems*, VSP, Zeist, 201–217.

Kabanikhin, S. I. (2005). Convergence rate estimation of gradient methods via conditional stability of inverse and ill-posed problems. *J. Inverse Ill-Posed Probl.*, **13**(3–6), 259–264.

Kabanikhin, S. I. and Abdiev, K. S. (1982). Modeling the initial stage of the formation of electric current and using the results in the problem of determining the conductivity tensor. In: *Questions of the Well-posedness of Inverse Problems of Mathematical*

Physics. Akad. Nauk SSSR Sibirsk. Otdel., Vychisl. Tsentr, Novosibirsk, 85–94 (in Russian).

Kabanikhin, S. I. and Akhmetov, Zh. A. (1983). Finite-difference regularization of the inverse problem for a first-order hyperbolic system. In: *Methods for Solving Inverse Problems*. Akad. Nauk SSSR Sibirsk. Otdel., Vychisl. Tsentr, Novosibirsk, 65–74 (in Russian).

Kabanikhin, S. I. and Ayapbergenova, A. T. (2002). Estimation of the rate of convergence of the Landweber iteration method in an inverse problem of acoustics. *Proc. Steklov Inst. Math.*, **2**, 75–97.

Kabanikhin, S. I., Bektemesov, M. A., and Nechaev, D. V. (2003). Optimizational method for continuation of solution to two-dimensional elliptic equation. In: *Proceedings of the International Symposium on Inverse Problems in Engineering Mechanics (ISIP 2003)*, 447–456.

Kabanikhin, S. I., Bektemesov, M. A., and Nechaev, D. V. (2005). The Cauchy problem for the wave equation with data on a time-like surface. *Proceedings of the Third International Conference "Mathematical Modeling and Information Technology in Science and Education"*, **1**, 289–294 (in Russian).

Kabanikhin, S. I., Bektemesov, M. A., and Nurseitova, A. T. (2005). *Iterative Methods for Solving Inverse and Ill-Posed Problems with Data Given on Part of the Boundary*. Kazakh Pedagogical State Univ., Almaty (in Russian).

Kabanikhin, S. I. and Boboev, K. B. (1982). A finite-difference method for determining sections in the P_1-approximation of the nonstationary kinetic transport equation. In: *Methods for Solving Ill-posed Problems and Their Applicaitons*, 213–217 (in Russian).

Kabanikhin, S. I. and Iskakov, K. T. (2000). A generalized solution of an inverse problem for an oscillation equation. *Dokl. Akad. Nauk*, **375**(1), 22–24 (in Russian).

Kabanikhin, S. I. and Iskakov, K. T. (2001). Justification of the steepest descent method in an integral formulation of an inverse problem for a hyperbolic equation. *Siberian Math. J.*, **42**(3), 478–494.

Kabanikhin, S. I., Iskakov, K. T., and Yamamoto, M. (2001). H_1-conditional stability with explicit Lipshitz constant for a one-dimensional inverse acoustic problem. *J. Inverse Ill-Posed Probl.*, **9**(3), 249–267.

Kabanikhin, S. I. and Karchevsky, A. L. (1995). Optimization method for solving the Cauchy problem for an elliptic equation. *J. Inverse Ill-Posed Probl.*, **3**(1), 21–46.

Kabanikhin, S. I., Kowar, R., and Scherzer, O. (1998). On the Landweber iteration for the solution of a parameter identification problem in a hyperbolic partial differential equation of second order. *J. Inverse Ill-Posed Probl.*, **6**(5), 403–430.

Kabanikhin, S. I., Kowar, R., Scherzer, O., and Vasin, V. V. (2001). Numerical comparison of iterative regularization methods for a parameter estimation problem in a hyperbolic PDE. *J. Inverse Ill-Posed Probl.*, **9**(6), 615–626.

Kabanikhin, S. I. and Lorenzi, A. (1999). *Identification Problems of Wave Phenomena.* VSP, Utrecht.

Kabanikhin, S. I. and Martakov, S. V. (1988). *The Study of a Projection-difference Method for Solving the Direct and Inverse Problem of Geoelectrics.* Preprint No. 13, Akad. Nauk SSSR Sibirsk. Otdel., Inst. Mat., Novosibirsk, (in Russian).

Kabanikhin, S. I., Martakov, S. V., Morgunova, T. Y. and Podgornaya, L. E. (1990). The problem of subsurface radiolocation in underground explorations at small depths. In: *Methods for Solving Inverse Problems.* Akad. Nauk SSSR Sibirsk. Otdel., Inst. Mat., Novosibirsk, 61–71 (in Russian).

Kabanikhin, S. I., Martakov, S. V., Tabarovskii, L. A., and Shifon, N. Ya. (1989). The study of dielectric permittivity of the medium around a borehole at the early formation stage. In: *Studies on the Conditional Well-posedness of Problems of Mathematical Physics.* Akad. Nauk SSSR Sibirsk. Otdel., Inst. Mat., Novosibirsk, 40–59 (in Russian).

Kabanikhin, S. I. and Satybaev, A. D. (1988). Finite-difference regularization of a linearized inverse problem for the two-dimensional wave equation. In: *Conditionally Well-posed Problems*. Akad. Nauk SSSR Sibirsk. Otdel., Inst. Mat., Novosibirsk, 39–57 (in Russian).

Kabanikhin, S. I., Satybaev, A. D., and Shishlenin, M. A. (2005). *Direct Methods of Solving Multidimensional Inverse Hyperbolic Problems.* VSP, Utrecht.

Kabanikhin, S. I., Scherzer, O., and Shishlenin, M. A. (2003). Iteration methods for solving a two dimensional inverse problem for a hyperbolic equation. *J. Inverse Ill-Posed Probl.*, **11**(1), 87–109.

Kabanikhin, S. I. and Shishlenin, M. A. (2001). Linearization of a multidimensional inverse problem. *Obratnye zadachi i informatsionnye technologii*, **1**(2), 83–114 (in Russian).

Kantorovich, L. V. (1952). *Functional Analysis and Applied Mathematics.* NBS Rep. 1509. U.S. Department of Commerce, National Bureau of Standards, Los Angeles, CA.

Karchevsky, A. L. (1997). Properties of the misfit functional for a nonlinear one-dimensional coefficient hyperbolic inverse problem. *J. Inverse Ill-Posed Probl.*, **5**(2), 139–163.

Kozlov, V. A., Maz'ya, V. G., and Fomin, A. V. (1991). An iterative method for solving the Cauchy problem for elliptic equations. *Comput. Math. Math. Phys.*, **31**(1), 45–52.

Krein, M. G. (1951). Solution of the inverse Sturm–Liouville problem. *Doklady Akad. Nauk SSSR*, **76**(1), 21–24 (in Russian).

Krein, M. G. (1954). On a method of effective solution of an inverse boundary problem. *Doklady Akad. Nauk SSSR*, **94**(6), 987–990 (in Russian).

Krein, S. G. (1957). On correctness classes for certain boundary problems. *Dokl. Akad. Nauk SSSR*, **114**(6), 1162–1165 (in Russian).

Kunetz, G. (1964). Généralisation des opérateurs d'antirésonance à un nombre quelconque de reflecteurs. *Geophys. Prospecting*, **12**, 283–289.

Ladyzhenskaja, O. A., Solonnikov, V. A., and Uraltseva, N. N. (1967). *Linear and Quasilinear Equations of Parabolic Type.* Transl. Math. Monogr., **23**, Amer. Math. Soc., Providence, R.I.

Ladyzhenskaya, O. A. and Uraltseva, N. N. (1973) *Linear and Quasilinear Equations of Elliptic Type.* Nauka, Moscow (in Russian).

Landis, E. M. (1956). On some properties of solutions of elliptic equations. *Dokl. Akad. Nauk SSSR*, **107**(5), 640–643 (in Russian).

Landis, E. M. (1962). Some questions in the qualitative theory of elliptic and parabolic equations. *Amer. Math. Soc. Transl.*, **20**, 173–238.

Lattès, R. and Lions, J.-L. (1967). *Méthode de quasi-réversibilité et applications.* Dunod, Paris.

Lavrent'ev, M. M. (1955). On the Cauchy problem for Laplace's equation. *Dokl. Akad. Nauk SSSR*, **102**(2), 205–206 (in Russian).

Lavrent'ev, M. M. (1956). On the Cauchy problem for Laplace's equation. *Izv. Akad. Nauk SSSR. Ser. Mat.*, **20**(6), 819–842 (in Russian).

Lavrent'ev, M. M. (1959). Integral equations of the first kind. *Dokl. Akad. Nauk SSSR*, **127**(1), 31–33 (in Russian).

Lavrent'ev, M. M. (1962). *Some Ill-posed Problems of Mathematical Physics.* Nauka, Novosibirsk (in Russian).

Lavrent'ev, M. M. (1964). On an inverse problem for the wave equation. *Soviet Math. Dokl.*, **5**, 970–972.

Lavrent'ev, M. M. (1972). Inverse problems and special operator equations of the first kind. In: *Proceedings of the International Conference of Mathematicians in Nice.* Nauka, Moscow, 130–136 (in Russian).

Lavrent'ev, M. M. (1973). *Conditionally Well-posed Problems for Differential Equations.* Novosibirsk. Gos. Univ., Novosibirsk.

Lavrent'ev, M. M. and Romanov, V. G. (1966). Three linearized inverse problems for hyperbolic equations. *Dokl. Akad. Nauk SSSR*, **171**(6), 1279–1281 (in Russian).

Lavrent'ev, M. M., Romanov, V. G., and Vasiliev, V. G. (1970). *Multidimensional Inverse Problems for Differential Equations.* Lecture Notes in Math., **167**, Springer-Verlag, Berlin-New York.

Lavrent'ev, M. M., Romanov, V. G., and Shishatskii, S. P. (1986). *Ill-posed Problems of Mathematical Physics and Analysis.* Transl. Math. Monogr., 64, Amer. Math. Soc., Providence, RI.

Lavrent'ev, M. M. and Savel'ev, L. Ya. (2006). *Operator Theory and Ill-posed Problems.* Inverse and Ill-posed Problems Series. VSP, Leiden.

Legendre, A. M. (1806). *Nouvelles méthodes pour la détermination des orbites des comètes.* Courcier, Paris.

Levinson, N. (1949). On the uniqueness of the potential in a Schrödinger equation for a given asymptotic phase. *Danske Vid. Selsk. Mat.-Fys. Medd.*, **25**(9), 1–25.

Levitan, B. M. (1987). *Inverse Sturm–Liouville Problems.* VSP, Zeist.

Lions, J.-L. Contrôle optimal de systèmes gouvernés par des équations aux dérivées partielles. Dunod, Paris.

Liskovets, O. A. (1967). Incorrect problems with a closed non-invertible operator. *Differents. Uravn.*, **3**(4), 636–646 (in Russian).

Liskovets, O. A. (1969). Ill-posed problems and the stability of quasisolutions. *Sibirsk. Mat. Zh.*, **10**(2), 373–385 (in Russian).

Liskovets, O. A. (1982). Theory and methods of solving ill-posed problems. *Itogi Nauki Tekh. Ser. Sovrem. Mat. Prilozh. Temat. Obz., Matem. Analiz*, **20**, 116–178 (in Russian).

Liu, P. (1996). Image reconstruction from photoacoustic pressure signals. *Proc. SPIE*, **2681**, 285–296.

Lorenzi, A. (1985). An inverse problem for a quasilinear parabolic equation. *Ann. Mat. Pura Appl. (4)*, **142**, 145–169.

Lusternik, L. A. and Sobolev, V. J. (1974). *Elements of Functional Analysis.* Halsted Press (John Wiley & Sons), New York.

Malovichko, A. K. (1956). *Methods of Analytic Continuation of Gravity Anomalies and their Application in Problems of Gravimetry.* Nedra, Moscow (in Russian).

Marchenko, V. A. (1950). Concerning the theory of a differential operator of the second order. *Doklady Akad. Nauk SSSR*, **72**, 457–460 (in Russian).

Marchenko, V. A. (1950). Transformation operators. *Doklady Akad. Nauk SSSR*, **74**, 185–188 (in Russian).

Marchenko, V. A. (1952). Some questions of the theory of one-dimensional linear differential operators of the second order. *Tr. Mosk. Mat. Obs.*, **1**, 327–420 (in Russian).

Marchenko, V. A. (1955). On reconstruction of the potential energy from phases of the scattered waves. *Dokl. Akad. Nauk SSSR*, **104**(5), 695–698 (in Russian).

Marchuk, G. I. (1964). On the formulation of certain inverse problems. *Dokl. Akad. Nauk SSSR*, **156**(3), 503–506 (in Russian).

Marchuk, G. I. (1973). *Methods of Computational Mathematics.* Nauka, Novosibirsk (in Russian).

Marquardt, D. W. (1963). An algorithm for least-squares estimation of nonlinear parameters. *J. Soc. Indust. Appl. Math.*, **11**, 431–441.

Maslov, V. P. (1968). The existence of a solution of an ill-posed problem is equivalent to the convergence of a regularization process. *Uspekhi Mat. Nauk*, **23**(3), 183–184 (in Russian).

Mendel, J. M. (1981). A two-domain approach to the normal-incidence inverse problem. *Geophys. Prospecting*, **29**, 742–757.

Mergelyan, S. N. (1956). Harmonic approximation and approximate solution of Cauchy's problem for Laplace's equation. *Dokl. Akad. Nauk SSSR*, **107**(5), 644–647 (in Russian).

Mikhlin, S. G. (1970). *Mathematical Physics, an Advanced Course.* Series in Applied Mathematics and Mechanics, **11**, North-Holland, Amsterdam-London.

Morozov, V. A. (1987). *Regular Methods for Solving Ill-posed Problems.* Nauka, Moscow (in Russian).

Murch, R. D., Tan, D. G. H., and Wall, D. J. N. (1988). Newton–Kantorovich method applied to two-dimensional inverse scattering for an exterior Helmholtz problem. *Inverse Problems*, **4**(4), 1117–1128.

Natterer, F. (2001). *The Mathematics of Computerized Tomography.* Classics Appl. Math., 32, SIAM, Philadelphia, PA.

Nemirovskii, A. S. (1986). Regularizing properties of the conjugate gradient method in ill-posed problems. *Zh. Vychisl. Mat. Mat. Fiz.*, **26**(3), 332–347 (in Russian).

Newman, D. J. (1960). Numerical method for solution of an elliptic Cauchy problem. *J. Math. and Phys.*, **39**(1), 72–75.

Nizhnik, L. P. (1973). *The Inverse Nonstationary Scattering Problem.* Naukova Dumka, Kiev (in Russian).

Novikov, P. S. (1938). On the uniqueness of the inverse problem of the potential theory. *Dokl. Akad. Nauk SSSR*, **18**(3), 165–168 (in Russian).

Novikov, S. P., Manakov, S. V., Pitaevskii, L. P., and Zakharov, V. E. (1984). *Theory of Solitons. The Inverse Scattering Method.* Contemporary Soviet Math., Consultants Bureau (Plenum), New York.

Nyambaa, Sh. and Cheverda, V. A. (1988). *An Optimization Method for Solving an Inverse Problem of Direct Current Electrical Prospecting for Vertically Inhomogeneous Media.* Preprint No. 794. Akad. Nauk SSSR Sibirsk. Otdel., Vychisl. Tsentr, Novosibirsk, (in Russian).

Pariiskii, B. S. (1968). Inverse problem for the wave equation with force at depth. In: *Certain Direct and Inverse Problems of Seismology.* Nauka, Moscow, 25–40 (in Russian).

Payne, L. E. (1975). *Improperly Posed Problems in Partial Differential Equations.* Regional Conference Series in Applied Mathematics, No. 22. Society for Industrial and Applied Mathematics, Philadelphia, PA.

Pogorelov, A. G. (1988). *Inverse Problems of Nonstationary Chemical Kinetics. A Systems Approach.* Nauka, Moscow (in Russian).

Prilepko, A. I. (1965). On the uniqueness of the determination of the form of a body by the values of the exterior potential. *Dokl. Akad. Nauk SSSR*, **160**(1), 40–43 (in Russian).

Prilepko, A. I. (1973). Inverse problems of potential theory (elliptic, parabolic, hyperbolic equation and transport equations). *Mat. Zametki*, **14**, 755–767 (in Russian).

Pucci, C. (1955). Sui problemi di Cauchy non "ben posti". *Atti Accad. Naz. Lincei. Rend. Cl. Sci. Fis. Mat. Nat. (8)*, **18**, 473–477 (in Italian).

Pucci, C. (1959). On the improperly posed Cauchy problems for parabolic equations. In: *Proc. Sympos. on the Numerical Treatment of Partial Differential Equations with Real Characteristics. January 1959, Rome.* Libreria Eredi Virgilio Veschi, Rome, 140–144.

Radon, J. (1917). Über die Bestimmung von Funktionen durch ihre Integralwerte längs gewisser Mannigfaltigkeiten. *Ber. Verh. Sächs. Akad. Wiss. Leipzig*, **69**, 262–277.

Radon, J. (1919). Über lineare Funktionaltransformationen und Funktionalgleichungen. *Sitzber. Akad. Wiss. Wien*, **128**, 1083–1121.

Rakesh (1988). A linearised inverse problem for the wave equation. *Comm. Partial Differential Equations*, **13**(5), 573–601.

Rakesh and Symes, W. W. (1988). Uniqueness for an inverse problem for the wave equation. *Comm. Partial Differential Equations*, **13**(1), 87–96.

Rayleigh, J. W. Strutt, B. (1945). *The Theory of Sound.* Dover, New York.

Reznitskaya, K. G. (1970). The existence and uniqueness of solutions to a nonlinear one-dimensional inverse dynamic problem of seismology. The Newton–Kantorovich method. In: *Proceedings of the International Symposium on Inverse Problems for Differential Equations. July, 1970. Novosibirsk.*

Reznitskaya, K. G. (1974). Relationship between the solutions to the Cauchy problem for equations of various types and ill-posed problems. *Mat. Probl. Geofiz.*, **5**, 55–62 (in Russian).

Robinson, E. A. (1973). A spectral approach to solving the inverse problem in geophysics based on the Lorentz, Fourier, and Radon transforms. *Trudy Inst. Inzh. Eelektrotech. i Radioelektr.*, **70**(9), 153–171 (in Russian).

Roger, A. (1981). Newton–Kantorovitch algorithm applied to an electromagnetic inverse problem. *IEEE Trans. Antennas and Propagation*, **29**(2), 232–238.

Romanov, V. G. (1967a). *On a Linearized Formulation of the Inverse Dynamic Problem for the Telegraph Equation.* Nauka, Moscow (in Russian).

Romanov, V. G. (1967b). Reconstruction of a function from integrals over ellipsoids of revolution with one fixed focus. *Dokl. Akad. Nauk SSSR*, **173**, 766–769 (in Russian).

Romanov, V. G. (1968a). A certain formulation of the inverse problem for a generalized wave equation. *Dokl. Akad. Nauk SSSR*, **181**(3), 554–557 (in Russian).

Romanov, V. G. (1968b). A one-dimensional inverse problem for the telegraph equation. *Differents. Uravn.*, **4**(1), 87–101 (in Russian).

Romanov, V. G. (1969a). Inverse problems of the propagation of electric oscillations in wires. *Mat. Probl. Geofiz.*, **1**, 92–102 (in Russian).

Romanov, V. G. (1969b). A certain integral geometry problem and a linearized inverse problem for a hyperbolic equation. *Sibirsk. Mat. Zh.*, **10**(6), 1364–1374 (in Russian).

Romanov, V. G. (1971). The uniqueness theorem for the one-dimensional inverse problem for the wave equation. *Mat. Probl. Geofiz.*, **2**, 100–142 (in Russian).

Romanov, V. G. (1972a). The problem of determining the one-dimensional velocity of signal propagation in a half-space from the oscillation mode of one of its points. *Mat. Probl. Geofiz.*, **3**, 164–186 (in Russian).

Romanov, V. G. (1972b). Uniqueness theorems for a certain class of inverse problems. *Soviet Physics Dokl.*, **17**(6), 525–526.

Romanov, V. G. (1972c). *Certain Inverse Problems for Equations of Hyperbolic Type.* Nauka, Novosibirsk, (in Russian).

Romanov, V. G. (1972d). An approach to solving the inverse problem of seismology with internal sources. *Mat. Probl. Geofiz.*, **3**, 296–320 (in Russian).

Romanov, V. G. (1973a). *Inverse Problems for Differential Equations.* Novosibirsk. Gos. Univ., Novosibirsk, (in Russian).

Romanov, V. G. (1973b). A one-dimensional inverse problem for the wave equation. *Dokl. Akad. Nauk SSSR*, **211**(5), 1083–1084 (in Russian).

Romanov, V. G. (1975). Certain uniqueness classes for the solution of Volterra operator equations of the first kind. *Funktsional. Anal. i Prilozhen.*, **9**(1), 81–82 (in Russian).

Romanov, V. G. (1976). An inverse problem for weakly connected hyperbolic systems of the first order. In: *Ill-posed Mathematical Problems and Problems of Geophysics (Mathematical Problems of Geophysics)*, Akad. Nauk SSSR Sibirsk. Otdel., Vychisl. Tsentr, Novosibirsk, 135–148 (in Russian).

Romanov, V. G. (1978a). The problem of determining the coefficients of a linear hyperbolic system. *Differents. Uravn.*, **14**(1), 94–103 (in Russian).

Romanov, V. G. (1978b). Inverse problems for hyperbolic systems. In: *Numerical Methods in Mathematical Physics, Geophysics and Optimal Control (Proc. Sympos., Novosibirsk, 1976)*, Nauka, Novosibirsk, 75–83 (in Russian).

Romanov, V. G. (1978c). Inverse problems for hyperbolic equations and energy inequalities. *Dokl. Akad. Nauk SSSR*, **242**(3), 541–544 (in Russian).

Romanov, V. G. (1978d). A formulation of an inverse problem for first-order symmetric hyperbolic systems. *Mat. Zametki*, **24**(2), 231–236 (in Russian).

Romanov, V. G. (1978e). The uniqueness of the solution of an inverse problem for first-order hyperbolic systems. *Mat. Zametki*, **24**(3), 359–365 (in Russian).

Romanov, V. G. (1979). Differential properties of the fundamental solution to a hyperbolic equation of the second order. In: *Ill-posed Mathematical Problems and Problems of Geophysics.* Akad. Nauk SSSR Sibirsk. Otdel., Vychisl. Tsentr, Novosibirsk, 110–121 (in Russian).

Romanov, V. G. (1982a). Inverse problems of seismic and electromagnetic wave propagation. In: *Methods for Solving Inverse Problems and Their Applications.* Akad. Nauk SSSR Sibirsk. Otdel., Vychisl. Tsentr, Novosibirsk, 111–118 (in Russian).

Romanov, V. G. (1982b). On the uniqueness of determination of the coefficients of the system of Maxwell's equations. In: *Nonclassical Problems for Equations of Mathematical Physics.* Akad. Nauk SSSR Sibirsk. Otdel., Inst. Mat., Novosibirsk, 139–142 (in Russian).

Romanov, V. G. (1983). A linear approximation of the inverse Lamb problem. In: *Numerical Methods in Seismic Studies.* Nauka, Novosibirsk, 170–192 (in Russian).

Romanov, V. G. (1984). Uniqueness of the determination of dielectric and magnetic permeabilities in an anisotropic one-dimensionally inhomogeneous medium. *Differents. Uravn.*, **20**(2), 325–332 (in Russian).

Romanov, V. G. (1985). The structure of the fundamental solution of the Cauchy problem for the system of Maxwell equations. *Dokl. Akad. Nauk SSSR*, **281**(5), 1052–1055 (in Russian).

Romanov, V. G. (1986). The structure of the fundamental solution of the Cauchy problem for a system of Maxwell equations. *Differents. Uravn.*, **22**(9), 1577–1587 (in Russian).

Romanov, V. G. (1987). *Inverse Problems of Mathematical Physics.* VNU Science Press, Utrecht.

Romanov, V. G. (1989a). Questions of the well-posedness of a problem of determining the speed of sound. *Siberian Math. J.*, **30**(4), 598–605.

Romanov, V. G. (1989b). Solvability of inverse problems for hyperbolic equations in a class of functions that are analytic with respect to some of the variables. *Soviet Math. Dokl.*, **39**(1), 160–164.

Romanov, V. G. and Bidaibekov, E. Y. (1985). On the theory of inverse problems of magnetotelluric sounding. *Dokl. Akad. Nauk SSSR*, **280**(4), 807–810 (in Russian).

Romanov, V. G. and Kabanikhin, S. I. (1987). Inverse problems of wave propagation. In: *Mathematical Modeling in Geophysics*, 151–167 (in Russian).

Romanov, V. G. and Kabanikhin, S. I. (1991). *Inverse Problems of Geoelectrics.* Nauka, Moscow (in Russian).

Romanov, V. G., Kabanikhin, S. I., and Boboev, K. B. (1984). An inverse problem for $\mathcal{P}_N$-approximation of the kinetic transport equation. *Dokl. Akad. Nauk SSSR*, **276**(2), 296–299 (in Russian).

Romanov, V. G., Kabanikhin, S. I., and Pukhnacheva, T. P. (1982). On the theory of inverse problems of electrodynamics. *Dokl. Akad. Nauk SSSR*, **266**(5), 1070–1073 (in Russian).

Romanov, V. G., Kabanikhin, S. I., and Pukhnacheva, T. P. (1984). *Inverse Problems of Electrodynamics.* Akad. Nauk SSSR Sibirsk. Otdel., Vychisl. Tsentr, Novosibirsk (in Russian).

Romanov, V. G. and Volkova, E. A. (1982). Inverse dynamic problems for an anisotropic elastic medium. *Dokl. Akad. Nauk SSSR*, **267**(4), 780–783 (in Russian).

Sacks, P. E. (1988). The inverse problem for a weakly inhomogeneous two-dimensional acoustic medium. *SIAM J. Appl. Math.*, **48**(5), 1167–1193.

Sacks, P. E. and Santosa, F. (1987). A simple computational scheme for determining the sound speed of an acoustic medium from its surface impulse response. *SIAM J. Sci. Statist. Comput.*, **8**(4), 501–520.

Sacks, P. and Symes, W. (1985). Uniqueness and continuous dependence for a multidimensional hyperbolic inverse problem. *Comm. Partial Differential Equations*, **10**(6), 635–676.

Samarskii, A. A. (1983). *Theory of Difference Schemes.* Nauka, Moscow (in Russian).

Samarskii, A. A. (1987). *An Introduction to Numerical Methods.* Nauka, Moscow (in Russian).

Samarskii, A. A. and Gulin, A. V. (1989). *Numerical Methods.* Nauka, Moscow (in Russian).

Santosa, F. (1982). Numerical scheme for the inversion of acoustical impedance profile based on the Gelfand–Levitan method. *Geophys. J. R. Astr. Soc.*, **70**, 229–244.

Santosa, F. and Schwetlick, H. (1982). The inversion of acoustical impedance profile by methods of characteristics. *Wave Motion*, **4**, 99–110.

Santosa, F. and Symes, W. W. (1989). *An Analysis of Least-Squares Velocity Inversion.* The Society of Exploration Geophysicists, Geophysical Monograph Series No. 4.

Seidman, T. and Eldén, L. (1990). An "optimal filtering" method for the sideways heat equation. *Inverse Problems*, **6**(4), 681–696.

Sergeev, V. O. (1971). Regularization of Volterra equations of the first kind. *Dokl. Akad. Nauk SSSR*, **197**(3), 531–534 (in Russian).

Sobolev, S. L. (1966). *Equations of Mathematical Physics.* Nauka, Moscow (in Russian).

Sobolev, S. L. (1991). *Some Applications of Functional Analysis in Mathematical Physics.* Transl. Math. Monogr., 90, Amer. Math. Soc., Providence, RI.

Sretenskii, L. N. (1954). On the uniqueness of determination of the form of an attracting body from the values of its external potential. *Dokl. Akad. Nauk SSSR*, **99**(1), 21–22 (in Russian).

Stoyan, G. (1978). Numerical experiments on the identification of heat conduction coefficients. In: *Theory of Nonlinear Operators (Proc. Fifth Internat. Summer School, Central Inst. Math. Mech. Acad. Sci. GDR, Berlin, 1977)*, Abh. Akad. Wiss. DDR, Abt. Math. Naturwiss. Tech., **6**, 259–268, Akademie-Verlag, Berlin.

Sun, Z. (1988). On the uniqueness of a multidimensional hyperbolic inverse problem. *Comm. Partial Differential Equations*, **13**(10), 1189–1208.

Symes, W. W. (1979). Inverse boundary value problems and a theorem of Gel'fand and Levitan. *J. Math. Anal. Appl.*, **71**(2), 379–402.

Symes, W. W. (1981). Stable solution of the inverse reflection problem for a smoothly stratified elastic medium. *SIAM J. Math. Anal.*, **12**(3), 421–453.

Symes, W. W. (1984). Some aspects of inverse problems in several-dimensional wave propagation. In: *Inverse Problems (New York, 1983), SIAM-AMS Proc.*, **14**, 97–111, Amer. Math. Soc., Providence, RI.

Symes, W. W. (1986). Linearization stability for an inverse problem in several-dimensional wave propagation. *SIAM J. Math. Anal.*, **17**(1), 132–151.

Tikhonov, A. N. (1943). On the stability of inverse problems. *C. R. (Doklady) Acad. Sci. URSS*, **39**, 176–179.

Tikhonov, A. N. (1949). On the uniqueness of the solution of the problem of electric prospecting. *Dokl. Akad. Nauk SSSR*, **69**(6), 797–800 (in Russian).

Tikhonov, A. N. (1963a). On the solution of ill-posed problems and the method of regularization. *Dokl. Akad. Nauk SSSR*, **151**(3), 501–504 (in Russian).

Tikhonov, A. N. (1963b). On the regularization of ill-posed problems. *Dokl. Akad. Nauk SSSR*, **153**(1), 49–52 (in Russian).

Tikhonov, A. N. (1965). On nonlinear equations of the first kind. *Dokl. Akad. Nauk SSSR*, **161**(5), 1023–1026 (in Russian).

Tikhonov, A. N. and Arsenin, V. Ya. (1974). *Methods for the Solution of Ill-posed Problems.* Nauka, Moscow (in Russian).

Tikhonov, A. N. and Arsenin, V. Ya. (1977). *Solutions of Ill-posed Problems.* John Wiley & Sons, New York-Toronto-London.

Tikhonov, A. N., Arsenin, V. Ya., and Timonov, A. A. (1987). *Mathematical Problems in Computer Tomography.* Nauka, Moscow (in Russian).

Tikhonov, A. N. and Glasko, V. B. (1975). Application of the regularization method in problems of geophysical interpretation. *Izv. Akad. Nauk SSSR*, 1975(1), 38–48 (in Russian).

Tikhonov, A. N., Goncharskii, A. V., Stepanov, V. V., and Yagola, A. G. (1983). *Regularizing Algorithms and A Priori Information.* Nauka, Moscow (in Russian).

Tikhonov, A. N. and Samarskii, A. A. (1990). *Equations of Mathematical Physics.* Dover Publications, New York.

Tikhonov, A. N., Vasilyeva, A. B., and Sveshnikov, A. G. (1985). *Differential Equations.* Springer-Verlag, Berlin.

Tsien, D. S. and Chen, Y. M. (1978). A pulse-spectrum technique for remote sensing of stratified media. *Radio Science*, **13**, 775–783.

Urev, M. V. (1985). An algorithm for numerical extension of generalized axially symmetric potential from the axis of symmetry. *Zh. Vychisl. Mat. Mat. Fiz.*, **25**(2), 269–282 (in Russian).

Ursin, B. and Berteussen, K. A. (1986). Comparison of some inverse methods for wave propagation in layered media. *Proc. IEEE*, **74**(3), 7–19.

Vabishchevich, I. I. (1978). On the solutions of the Cauchy problem for the Laplace equation in a doubly connected domain. *Dokl. Akad. Nauk SSSR*, **241**(6), 1257–1260 (in Russian).

Vainikko, G. M. (1982). *Methods for Solving Linear Ill-posed Problems in Hilbert Spaces*. Tartu. Gos. Univ., Tartu (in Russian).

Vainikko, G. M. and Karma, O. O. (1974). The convergence of approximate methods for the solution of linear and nonlinear operator equations. *Zh. Vychisl. Mat. Mat. Fiz.*, **14**(4), 828–837 (in Russian).

Vasil'ev, F. P. (1980). *Numerical Methods for Solving Extremal Problems*. Nauka, Moscow (in Russian).

Vasil'ev, F. P. (1981). *Methods for Solving Extremal Problems*. Nauka, Moscow (in Russian).

Vasil'ev, F. P. (1988). *Numerical Methods for Solving Extremal Problems*. Second edition. Nauka, Moscow (in Russian).

Vasin, V. V. (1989). *Methods for Solving Unstable Problems*. Ural. Gos. Univ., Sverdlovsk (in Russian).

Vasin, V. V. and Ageev, A. L. (1995). *Ill-posed Problems with A Priori Information*. Inverse and Ill-posed Problems Series. VSP, Utrecht.

Vasin, V. V. and Mokrushin, A. A. (2000). Iterative processes of Gauss–Newton type for ill-posed operator equations. *Dokl. Akad. Nauk*, **371**(1), 35–37 (in Russian).

Verzhbitskii, V. V., Kabanikhin, S. I., and Martakov, S. V. (1994). Transient nonstationary dielectric logging. *Dokl. Akad. Nauk*, **337**(3), 386–388 (in Russian).

Voevodin, V. V. (1966). *Numerical Methods of Algebra. Theory and Algorithms*. Nauka, Moscow (in Russian).

Voevodin, V. V. (1977). *Numerical Foundations of Linear Algebra*. Nauka, Moscow (in Russian).

Voronina, T. A. and Cheverda, V. A. (1994a). Full waveform inversion in processing seismic profiling data. *Dokl. Akad. Nauk*, **335**(4) (in Russian).

Voronina, T. A. and Cheverda, V. A. (1994b). Optimizational approach to data processing in vertical seismic profiling. *J. Inverse Ill-Posed Probl.*, **2**(3), 211–226.

Wiechert, E. and Zoeppritz, K. (1907). Über Erdbebenwellen. *Nachr. Königl. Gesellschaft Wiss. Göttingen*, **4**, 415–549.

Xie, G. Q. (1986). A new iterative method for solving the coefficient inverse problem of the wave equation. *Comm. Pure Appl. Math.*, **39**(3), 307–322.

Xie, G. Q. and Chen, Y. M. (1985). A modified pulse-spectrum technique for solving inverse problems of two-dimensional elastic wave equation. *Appl. Numer. Math.*, **1**(3), 217–237.

Yakhno, V. G. (1985). *One-dimensional Inverse Dynamic Problems for Anisotropic and Elastic Media.* Akad. Nauk SSSR Sibirsk. Otdel., Vychisl. Tsentr, Novosibirsk (in Russian).

Zakharov, V. E. (1980). Benney equations and quasiclassical approximation in the method of the inverse problem. *Funct. Anal. Appl.*, **14**, 89–98.

Index

Bei Fragen zur Produktsicherheit wenden Sie sich bitte an:
If you have any questions regarding product safety,
please contact:

Walter de Gruyter GmbH
Genthiner Straße 13
10785 Berlin
productsafety@degruyterbrill.com